The Adaptability of Residential Planning and Design to World-Changing Events

The Adaptability of Residential Planning and Design to World-Changing Events

Editors

Carlos C. Duarte
Nuno D. Cortiços
Anna Stefańska
Daniel Mateus
Carol Monticelli

MDPI

Basel • Beijing • Wuhan • Barcelona • Belgrade • Novi Sad • Cluj • Manchester

Editors

Carol C. Duarte
CIAUD, Research Centre for Architecture, Urbanism and Design
Universidade de Lisboa
Lisbon
Portugal

Nuno D. Cortiços
CIAUD, Research Centre for Architecture, Urbanism and Design
Universidade de Lisboa
Lisbon
Portugal

Anna Stefańska
Institute of Civil Engineering
Warsaw University of Life Science
Warsaw
Poland

Daniel Mateus
CIAUD, Research Centre for Architecture, Urbanism and Design
Universidade de Lisboa
Lisbon
Portugal

Carol Monticelli
Department of Architecture, Built Environment and Construction Engineering
Polytechnic Milan
Milan
Italy

Editorial Office
MDPI
St. Alban-Anlage 66
4052 Basel, Switzerland

This is a reprint of articles from the Special Issue published online in the open access journal *Buildings* (ISSN 2075-5309) (available at: www.mdpi.com/journal/buildings/special_issues/9M3S80BJZF).

For citation purposes, cite each article independently as indicated on the article page online and as indicated below:

Lastname, A.A.; Lastname, B.B. Article Title. *Journal Name* **Year**, *Volume Number*, Page Range.

ISBN 978-3-7258-0212-8 (Hbk)
ISBN 978-3-7258-0211-1 (PDF)
doi.org/10.3390/books978-3-7258-0211-1

Contents

About the Editors

Carlos C. Duarte

Carlos C. Duarte, Ph.D., is an architect and Adjunct Professor at Lisbon School of Architecture, Universidade de Lisboa. He graduated from the Technical University of Lisbon in 2009 and began his career in urbanism and interior design. In 2012, he joined McCullough Mulvin Architects and participated in several projects focused on heritage-building retrofit/expansions, such as Butler Gallery and St. Mary's Medieval Mile Museum in Kilkenny. This experience ignited his passion for heritage preservation, inspiring him to pursue a Postgraduate course in Conservation, Renovation, and Architectural Rehabilitation in 2013. Later, he earned his Ph.D. in Architecture from Lisbon School of Architecture, Universidade de Lisboa in 2020. Currently, he is an "Integrated Member" of the Research Center in Architecture, Urbanism, and Design (CIAUD). His research focuses on bioclimatic architecture, passive design, sustainability, NZEB, PEBs, and energy retrofit of historical buildings. As a professor, he lectures on Building Materials, Environmental Comfort, Energy Efficiency and Environment, and supervises several Master Thesis in these domains. Author and co-author of over a dozen scientific publications indexed in Scopus (MDPI and Elsevier) and an active participant in international conferences and congresses, he has been serveing, since 2013, as a Guest Editor and a member of the Topical Advisory Panel for the journal *Buildings* (MDPI), while contributing as a Reviewer for several publishers, notably MDPI, ScienceGroup, and AMPS.

Nuno D. Cortiços

Nuno D. Cortiços, PhD, is an architect and Adjunct Professor at Lisbon School of Architecture, Universidade de Lisboa. He earned his architectural license in 1999 from the Universidade Técnica de Lisboa. Initially recognized by the Portuguese Architects Council in 1999 for his academic design proposal, he later gained experience at top design studios in Lisbon, including Aires Mateus Associated Architects. Subsequently, he transitioned to construction management, overseeing projects up to 2.5 million Euros. In 2007, he entered academia, teaching construction technologies. Concurrently, he pursued his Ph.D., focusing on surveying construction anomalies in Portuguese buildings from the 1980s and 1990s, with a specific case study on higher education buildings. Successfully completing his Ph.D. in 2013, he began his professional academic journey as an assistant professor, currently teaching various subjects related to materials, anomaly patterns, construction quality, performance, energy efficiency, and environmental building updates. Specializing in professional software and collaborating with the Portuguese body for energy efficiency policy and the Energy Performance Certificates platform, he encourages student involvement. In 2015 and 2016, as vice-president of the faculty, he implemented a maintenance plan that efficiently updated building refurbishments within budget, serving as a model for other universities. Internationally, in 2022, he collaborated with the Politecnico di Milano, teaching Building Physics and workshops on tensile membrane designs applied to buildings. Since 2013, he has contributed actively as a researcher, producing articles on construction patterns, autonomous maintenance, energy efficiency, EPC reliability, COVID-19 energy pressure, and the impact of telework consumption.

Anna Stefańska

Anna Stefańska, Ph.D., is an assistant professor at the Institute of Civil Engineering, Warsaw University of Life Sciences. She holds degrees in both Architecture and Civil Engineering and has dedicated her research to the interdisciplinary optimization of design. Her extensive experience led her to obtain a Ph.D. in Architecture from Warsaw University of Technology, with her dissertation titled "Generative shaping in search for the structural form of contemporary pavilions."

Her primary research interests revolve around the interdisciplinary design process, with a particular focus on algorithmic thinking and digital fabrication within a sustainable framework. Anna is the author and co-author of numerous Scopus-indexed publications (Elsevier, Taylor and Francis, MDPI). She actively participates in international conferences and serves as a Guest Editor for a Special Issue of the journal *Buildings* (MDPI). She collaborates with research groups worldwide and, in the academic years 2021/22 and 2023/24, she visited the Lisbon School of Architecture, Universidade de Lisboa, as part of a didactic and research scholarship at the Building, Science, Technology, and Sustainability Laboratory (BSTS Lab).

Currently, Anna is a member of the Institute of Civil Engineering, Warsaw University of Life Sciences. As a professor in the Faculty of Architecture and the Faculty of Civil and Environmental Engineering, she teaches Building Design, Structural Design, Computational Methods, and supervises several diploma projects in both Architecture and Civil Engineering.

Actively engaged in architectural practice since obtaining her Masters of Architecture in 2013, Anna has been involved in numerous projects, specializing in single-family houses, infrastructure buildings, and interiors.

Daniel Mateus

Daniel Mateus is an Architect and Collaborating Researcher at CIAUD, Research Centre for Architecture, Urbanism and Design, at the Lisbon School of Architecture of the University of Lisbon, in Portugal. He graduated in Architecture from the Technical Higher Institute of the Technical University of Lisbon (IST) (2005) and obtained a doctorate in Architecture, Specialization in Design and Computation, from the Lisbon School of Architecture of the University of Lisbon (FAUL) (2020). From 2005 to 2007, he was involved in rehabilitation works in the IST Central Pavilion. Between 2006 and 2011, he worked in Portugal and Angola, collaborating in the elaboration of architecture and urbanism projects. In 2012 and 2013, he also worked in Portugal in the field of real estate. Between January and December of 2014, as well between September 2015 and May 2016, he integrated the Research Project "Tecton 3D" and developed a software for 3D modeling in an immersive virtual reality environment for the initial phase of architecture projects. During the development of his PhD (started in 2013), from December 2014 to August 2015, he was a AUSMIP+ program research fellow at Sungkyunkwan University in South Korea. Between January 2017 and June 2019, he received a research grant from CIAUD from the Lisbon School of Architecture. Between May and June 2018, he was also a research fellow of the Erasmus+ program at Penn State University, PA, USA. Daniel Mateus published and presented several international conference and journal articles. Some examples are the conference articles for eCAADe Conference 2015 and for SIGraDi Conferences 2021, 2022, and 2023, as well the journal article for the 2023 MDPI Special Issue "The Adaptability of Residential Planning and Design to World-Changing Events", where he was also guest co-editor.

Carol Monticelli

Born in 1977, with an MSc in Architecture and a PhD in Technologies for Architecture and Environment, Carol Monticelli is currently Assistant Professor at the Department of Architecture, Built Environment and Engineering Construction at Politecnico di Milano. As test Manager of the biaxial rig – Clustex, Polimi, and Member of SPACE (Experimental process for architecture and life cycle of building products), recently, her research progress has focused on deepening themes related to eco-sustainable design of buildings, environmental impact in construction, Life Cycle Assessment in architecture, and processes of technological innovation and material to analyse the application scenarios of new lightweight and textile materials in the AEC sector. Fellowships and Awards: 2008 PostDoc grant (1yr) held by Fratelli Confalonieri Foundation; 2008 Grant for PhD Thesis 2007 held by Engineers Association of Bozen Province, Italy; international committee Grant (with merit and mention about innovation in the design and construction process); 2004 Study and Improvement Grant held by Fratelli Confalonieri Foundation; 2004 Master Thesis Grant "Egisto Camerini" held by Assobeton – National Association of concrete products industries (Grant with merit and mention of the sustainability and precast of concrete). Memberships of scientific societies: TensiNet*, elected as WG LCA coordinator (2015-2016), coord. M. Mollaert (VUB Brussel); IASS, active in WG 18 Environmentally Compatible Structures (ECS) coord. L. Vegh (CTU Prague). Society of Architectural Technologists, active in the Cluster nZEB; LCA Association Network, coord. by ENEA. Lastly, she was enrolled in several research projects co-financed by Regional and International institutions (Cariplo Foundation, MIUR, Regione Lombardia, EU).

Preface

The world we inhabit is increasingly marked by volatility and uncertainty, with extreme events, both natural and human-made, posing significant threats to our communities and the built environment. As we navigate these challenging times, the role of architecture, design, and engineering in creating resilient and adaptable structures becomes paramount.

Throughout history, the residential built environment has demonstrated its ability to adapt to changing circumstances and must continue to do so. In recent times, the world has witnessed a surge in extreme events, including the COVID-19 pandemic, armed conflicts, and devastating natural disasters. These events have exposed the vulnerabilities of our built environment and the need for a paradigm shift in residential design at all levels. We must move beyond traditional approaches that focus solely on esthetics and functionality and embrace a broader understanding of concepts such as resilience, adaptability, and sustainability.

This Special Issue on "The Adaptability of Residential Planning and Design to World-Changing Events" brings together a diverse range of perspectives from researchers and practitioners to explore innovative approaches to residential design that can effectively respond to the short-, medium-, and long-term challenges we all face.

The contributions featured in this Special Issue offer a wealth of knowledge and insights, providing a roadmap for creating residential environments that are not only resilient to the immediate impact of extreme events but also adaptable to the ongoing challenges and shifts in society. By embracing this holistic approach to residential planning and design, we can pave the way for a more sustainable and resilient future for our communities.

This reprint aims to encompass a diverse range of professionals and researchers related with the architecture, urban planning, engineering, public safety, and disaster management fields. This will also be of interest to policymakers, developers, and homeowners who are committed to creating more resilient and adaptive residential environments through the development of innovative planning strategies, new buildings codes, and design guidelines.

The editors would like to express their appreciation to the authors for their insightful work, the reviewers for their rigorous evaluations, and the *Buildings* editorial team, with a special nod to Ms. Evie Ouyang for her unwavering commitment and assistance, all of which played a crucial role in the success of this Special Issue.

Carlos C. Duarte, Nuno D. Cortiços, Anna Stefańska, Daniel Mateus, and Carol Monticelli
Editors

buildings

MDPI

Editorial

The Adaptability of Residential Planning and Design to World-Changing Events

Carlos C. Duarte [1,2,*], Nuno D. Cortiços [1,2], Anna Stefańska [3], Daniel Mateus [1] and Carol Monticelli [4]

1 CIAUD, Research Centre for Architecture, Urbanism and Design, Rua Sá Nogueira, Polo Universitário do Alto da Ajuda, 1349-063 Lisbon, Portugal; ncorticos@campus.ul.pt (N.D.C.); dmateus@fa.ulisboa.pt (D.M.)
2 Lisbon School of Architecture, Universidade de Lisboa, Rua Sá Nogueira, Polo Universitário do Alto da Ajuda, 1349-063 Lisbon, Portugal
3 Institute of Civil Engineering, Warsaw University of Life Sciences SGGW, Nowoursynowska 159, 02 776 Warsaw, Poland; anna_stefanska@sggw.edu.pl
4 Department of Architecture, Built Environment and Construction Engineering, Polytechnic Milan, Via E. Bonardi 9, 20133 Milan, Italy; carol.monticelli@polimi.it
* Correspondence: carlosfcduarte@edu.ulisboa.pt

History has witnessed several events with different magnitudes that have served as catalysts for drastic changes in society, impacting people's lifestyles and reshaping the prevailing civilizational paradigm. The fields of architecture, design, and engineering have consistently played key roles in developing original and adaptive solutions to the challenges presented by extreme events, thereby leaving an indelible mark on the history of civilization. This mark is evident in the built structures in today's cities, ranging from buildings to bridges.

Therefore, it is important to frame the current moment, as the world is emerging from a pandemic (COVID-19) [1] that has left profound marks on the population; dealing with armed conflicts, occurring in Middle East [2] and at the very heart of Europe [3]; struggling with devastating seismic events, such as those witnessed in Turkey and Syria [4]; and amidst a climate crisis [5], with its far-reaching consequences, including, among others, heatwaves, forest fires, and extreme rainfall and floods.

In this way, this Special Issue presents a selection of the latest developments from the scientific community to address the issues raised by some of these events. It is focused on residential contexts, with the goal of enhancing the readiness and resilience of buildings, neighborhoods, and cities in the face of new upcoming threats. It contains fifteen original studies (fourteen articles and one revision) conducted by researchers from Italy, Poland, Portugal, Brazil, Spain, Turkey, India, China, and the Republic of Korea, showcasing its global reach. Despite the common denominator of 'World-Changing Events', the contributions vary in scope and nature, ranging from fields such as civil engineering, architecture, and urbanism to social sciences, environment, and health.

If there were any doubts regarding the disruptive effects of the COVID-19 pandemic on established concepts and paradigms, the work of Moreira and Farias [6] unequivocally clarifies it. Their study discusses the pandemic's impact on the use of residential spaces in Lisbon, drawing from an inquiry conducted during the lockdown period. The authors state that the pandemic, particularly the confinement measures, led to significant shifts in the utilization of domestic spaces, carrying enduring implications for future house designs. Furthermore, they observe that homes have begun to serve purposes for which they were not originally intended, which, in turn, underscores the need for additional spaces to accommodate these evolving requirements. Ultimately, the authors advocate for the reevaluation of post-COVID home design.

An example of the latter is the studies conducted by Kim and Kim [7] and Kim et al. [8] in South Korea, which offer valuable insights into the design of apartment spaces and minimum housing standards improvements as a direct response to the challenges posed by

Citation: Duarte, C.C.; Cortiços, N.D.; Stefańska, A.; Mateus, D.; Monticelli, C. The Adaptability of Residential Planning and Design to World-Changing Events. *Buildings* **2024**, *14*, 9. https://doi.org/10.3390/buildings14010009

Received: 4 December 2023
Accepted: 15 December 2023
Published: 20 December 2023

the last years. The first employed a mixed-method approach, combining questionnaires and photovoice, to assess the population's experiences during the pandemic period, offering valuable contributions to the prevention and management of infectious diseases within home environments. Some of the key recommendations emphasize the need for flexible layouts to effectively manage and balance work efficiency and privacy. Additionally, they propose the implementation of a storage facility at the house entrance for contaminant removal [7]. The second addresses the need to update South Korea's minimum housing standards, as the existing ones no longer reflect the evolving housing situation. The proposed standards categorize households by size and incorporate specific criteria for areas, facilities, and locations. Directly related to COVID-19 was the behavioral shift in people that resulted in the inclusion of a "dining space" and an "additional room" for one and two-person households, respectively. Furthermore, the introduction of criteria such as "hot water in bathing facilities" and "private entrance door" is also a consequence of the pandemic's impact on the population, particularly over the restrictions on access to public bathing facilities and privacy issues [8].

Likewise, the research conducted by Daglio et al. [9] underscores the need for incorporating new typological and technological strategies into post-pandemic housing design. Their study offers a fresh perspective on seamlessly integrating fixtures and furnishings into interior design, with a focus on the industrialization potential of certain components through a flexible and modular approach. The authors emphasize the importance of downsizing house areas while upholding high-quality living standards, where design is framed by sustainable and circular economy principles.

Also focusing on the exploration of residential architectural concepts is the study conducted by Racha-Pacheco et al. [10], which delves into the design of smart homes. This study focuses on the integration of cutting-edge technologies into small-scale residential units, including smart thermostats, sensors, surveillance systems, remote devices, and mobile apps, among others. The authors have developed and presented an adaptable intelligent home prototype, with the aim of raising awareness about the potential and limitations of merging these technologies while emphasizing the importance of thoughtful design in this context.

On an urban scale, Zeng et al. [11] investigate how wind patterns may contribute to the spread of the COVID-19 virus within urban areas using machine learning algorithms. The findings show that various wind patterns and building layouts can influence the spread of the virus. Areas with higher population density are more susceptible to retaining virus concentrations, thus increasing transmission rates. This study offers valuable insights for urban planners and health authorities seeking to develop strategies for preventing and controlling epidemics in urban areas.

With a different scope but still related to the pandemic, the study conducted by Ding et al. [12] analyzes the impact of COVID-19 on the utilization of self-service facilities in residential areas, particularly smart parcel lockers. Their study employs an analytical framework to evaluate over 2000 residential communities in Tianjin, China. The primary findings highlight a notable inequity and a supply–demand mismatch in accessing these facilities, leading to significant resource waste and underutilized public space. Furthermore, the authors call for the attention of stakeholders to address this issue and emphasize the need for further studies regarding the supply and demand dynamics in this context.

Events on the scale of the COVID-19 pandemic often lead to the emergence of new concepts and perspectives. In this context, the study by Quesada-Garcia et al. [13] introduces and defines a novel concept known as "healthy architecture". Furthermore, it also discusses a set of principles referred to as the "Decalogue", where buildings and environments must adhere to each other to be considered healthy. This study contributes to raising awareness that spaces in urban and building contexts should not only be guided by sustainable, functional, and aesthetic standards but also prioritize comfort, health, and safety.

The climate crisis is also a topic covered in this Special Issue, primarily through the work of Starzyk et al. [14], which addresses the efforts made to face forthcoming challenges within an urban and architectural context in Poland. In addition to referencing the consequences of climate change and the necessity for developing programs to address climate-related issues in urban development, the authors discuss the formulation of guidelines and standards aimed at mitigating adverse effects in existing buildings. Furthermore, they emphasize the incorporation of sustainable development principles into the adaptation process, highlighting the significance of integrating environmental sustainability with urban transformation and adaptation.

Climate-related issues have certainly raised awareness for the need for more studies relating outdoor thermal comfort with urban space configuration. This is the case of the work of Song et al. [15], which provides insights into the design of optimal residential block forms and layouts with the goal of enhancing thermal comfort during the hot season in Jinan, China. Their study reveals that the number of building rows affects outdoor thermal comfort, particularly on north-oriented streets. It suggests reducing the number of rows and considering a two-building column configuration or other block layouts instead.

Climate change mitigation can also occur through targeted interventions, particularly in the context of improving building energy performance. In this regard, Mateus and Henriques [16] employed computational methods to apply a parametric energy-based design model to 1940s modern buildings in Rio de Janeiro, Brazil, by upgrading and optimizing the existing external shading systems. The main goal of this improvement is to decrease the overall building energy consumption while envisioning the generation of renewable energy through solar harvesting. Iturralde et al. [17] introduce the ENSARE project, which proposes energy retrofits for residential buildings by adding automated prefabricated modules that combine RES and insulation into the building facade. The present study is centered on developing two solutions concerning automation with the potential to save time in the module's layout drafting phase, thereby accelerating the entire renovation process. Both studies contribute to a sustainable approach by integrating passive strategies with RES, aiming to reduce the overall energy consumption of residential buildings. This marks a significant step towards mitigating climate change and promoting energy efficiency.

In the wake of the devastating earthquake that occurred in Turkey and Syria, Turkish authors Yanik and Ulus [18] present a study aimed at developing a method for incorporating the combined effects of base isolation and soil–structure interaction into the mass, damping, and stiffness matrices of a structure, considering different scenarios. Their study highlights that the performance of the base isolation system may vary depending on local soil properties. Its contribution to simplifying the analytical process makes it easier for engineers to work on earthquake-resistant design and retrofitting.

In a world of rapid changes and transformations, lightweight, flexible, and easily assembled/disassembled solutions will certainly have a role in the times ahead. In this context, Cui et al. [19] developed the "Weaving Octopus", a lightweight architectural prototype that seamlessly combines textiles and hybrid structures. Highly versatile and scalable, it can function as a single unit or in aggregate configurations and is flexible enough to swiftly adapt to various scenarios, ranging from public space pavilions to facade shading mechanisms and even furniture. The authors employed a three-stage methodology, starting with conceptual design, followed by parametric simulation and optimization, and culminating in the development of physical scale models. We must underscore this study's contribution to the existing knowledge in lightweight construction design and its potential applicability range across the AEC sector.

Lastly, Vijayan et al. [20] present a comprehensive review that discusses the application of carbon fiber-reinforced polymers (CFRP) in building structural elements. The review covers various topics, including the properties and performance of CFRP, the advantages and disadvantages of their use, their potential to replace traditional construction materials, current trends, prospects, emerging materials, and their role in environmental impact within structural applications. The latter aspect holds great relevance, as these evaluations are essential for creating more resilient and sustainable building structures.

In conclusion, this Special Issue titled 'The Adaptability of Residential Planning and Design to World-Changing Events' offers a selection of studies addressing crucial challenges emerging from the most recent world-changing events. These studies, covering topics related to health, architectural design, urban planning, climate change, and structural engineering, highlight the importance of innovation, adaptation, and sustainability in our ever-evolving world. Together, they not only contribute to our collective understanding but also provide the tools to create resilient and environmentally conscious buildings, neighborhoods, and cities for the generations ahead.

The editors wish to congratulate and express their gratitude to the authors for their valuable work, to the peer reviewers for their thorough analysis, and to the assisting editors of Buildings, with a special mention of Ms. Evie Ouyang, all of whom greatly contributed to the success of this Special Issue.

Funding: This research was funded by FCT-Fundação para a Ciência e a Tecnologia, I.P., grant number UIDB/04008/2020.

Conflicts of Interest: The authors declare no conflict of interest.

References

1. The Associated Press WHO Downgrades COVID Pandemic: Says It's No Longer a Global Emergency | CBC News. Available online: https://www.cbc.ca/news/health/who-pandemic-not-emergency-1.6833321 (accessed on 5 September 2023).
2. Cohen, P. Risk of a Wider Middle East War Threatens a 'Fragile' World Economy. *The New York Times*, 1 January 2023.
3. Bigg, M.M. How Russia's War in Ukraine Has Unfolded, Month by Month. *The New York Times*, 24 February 2023.
4. Ghosh, P. Turkey Earthquake: Where Did It Hit and Why Was It so Deadly? *BBC News*, 6 February 2023.
5. Paddison, L. "The Climate Time-Bomb Is Ticking": The World Is Running out of Time to Avoid Catastrophe, New UN Report Warns. Available online: https://www.cnn.com/2023/03/20/world/ipcc-synthesis-report-climate-intl/index.html (accessed on 5 September 2023).
6. Moreira, A.; Farias, H. The Post-COVID Home. How Confinement Altered Domestic Space Use and Living Modes, in Lisbon. *Buildings* **2023**, *13*, 1195. [CrossRef]
7. Kim, M.-K.; Kim, E.-J. Apartment Space Planning Directions for Infectious Disease Prevention and Management: Insights Based on Residents' Experiences. *Buildings* **2023**, *13*, 2203. [CrossRef]
8. Kim, D.; Sim, H.; Kim, S. A Study on Recommendations for Improving Minimum Housing Standards. *Buildings* **2023**, *13*, 2708. [CrossRef]
9. Daglio, L.; Ginelli, E.; Vignati, G. Housing Design: Furniture or Fixtures? Accommodating Change through Technological and Typological Innovation. *Buildings* **2023**, *13*, 1862. [CrossRef]
10. Racha-Pacheco, P.; Ribeiro, J.T.; Afonso, J. Architecture towards Technology—A Prototype Design of a Smart Home. *Buildings* **2023**, *13*, 1859. [CrossRef]
11. Zheng, L.; Chen, Y.; Yan, L.; Zheng, J. The Impact of High-Density Urban Wind Environments on the Distribution of COVID-19 Based on Machine Learning: A Case Study of Macau. *Buildings* **2023**, *13*, 1711. [CrossRef]
12. Ding, M.; Song, Y.; Hu, Y. Supply–Demand Matching of Smart Parcel Lockers in a Residential Area: Insights from Tianjin. *Buildings* **2023**, *13*, 2088. [CrossRef]
13. Quesada-García, S.; Valero-Flores, P.; Lozano-Gómez, M. Towards a Healthy Architecture: A New Paradigm in the Design and Construction of Buildings. *Buildings* **2023**, *13*, 2001. [CrossRef]
14. Starzyk, A.; Donderewicz, M.; Rybak-Niedziółka, K.; Marchwiński, J.; Grochulska-Salak, M.; Łacek, P.; Mazur, Ł.; Voronkova, I.; Vietrova, P. The Evolution of Multi-Family Housing Development Standards in the Climate Crisis: A Comparative Analysis of Selected Issues. *Buildings* **2023**, *13*, 1985. [CrossRef]
15. Song, X.; Wang, G.; Deng, Q.; Wang, S.; Jiao, C. The Influence of Residential Block Form on Summer Thermal Comfort of Street Canyons in the Warm Temperate Zone of China. *Buildings* **2023**, *13*, 1627. [CrossRef]
16. Mateus, D.; Henriques, G.C. Energy-Based Design: Improving Modern Brazilian Buildings Performance through Their Shading Systems, the Nova Cintra Case Study. *Buildings* **2023**, *13*, 2543. [CrossRef]
17. Iturralde, K.; Das, S.; Srinivasaragavan, A.; Bock, T.; Holst, C. An Automated Prefabricated Facade Layout Definition for Residential Building Renovation. *Buildings* **2023**, *13*, 2981. [CrossRef]
18. Yanik, A.; Ulus, Y. Soil–Structure Interaction Consideration for Base Isolated Structures under Earthquake Excitation. *Buildings* **2023**, *13*, 915. [CrossRef]

19. Cui, Z.; Zhang, S.; Viscuso, S.; Zanelli, A. Weaving Octopus: An Assembly–Disassembly-Adaptable Customized Textile Hybrid Prototype. *Buildings* **2023**, *13*, 2413. [CrossRef]
20. Vijayan, D.S.; Sivasuriyan, A.; Devarajan, P.; Stefańska, A.; Wodzyński, Ł.; Koda, E. Carbon Fibre-Reinforced Polymer (CFRP) Composites in Civil Engineering Application—A Comprehensive Review. *Buildings* **2023**, *13*, 1509. [CrossRef]

buildings MDPI

Article

The Post-COVID Home. How Confinement Altered Domestic Space Use and Living Modes, in Lisbon

Ana Moreira * and Hugo Farias *

CIAUD, Research Centre for Architecture, Urbanism and Design, Lisbon School of Architecture, Universidade de Lisboa, Rua Sá Nogueira, Polo Universitário do Alto da Ajuda, 1349-063 Lisboa, Portugal
* Correspondence: ana.silva.moreira@gmail.com (A.M.); hfarias@fa.ulisboa.pt (H.F.)

Abstract: Not only is space use a result of spatial configuration, but it is also greatly determined by social patterns and society's living modes. The COVID-19 pandemic and confinement resulted, worldwide, in a 24/7 use of the domestic setting, which had to be adapted to the emergence of new needs and functions in the domestic space. The paper aims at understanding how COVID confinement altered domestic space use in Lisbon and how current society has maintained some of those changes, a reflection of new social patterns. The analysis is supported by an inquiry, carried out during the first confinement of 2020, in the city of Lisbon. Results show that domestic space use had to be adjusted to a new context, gaining new functions, and bringing into evidence the need for additional spaces. In light of the proven changes in living modes, the research concludes that the post-COVID home needs to be rethought and new housing programs should take into consideration the new social patterns and living modes.

Keywords: housing; domestic space use; living modes; Lisbon; COVID-19

Citation: Moreira, A.; Farias, H. The Post-COVID Home. How Confinement Altered Domestic Space Use and Living Modes, in Lisbon. *Buildings* **2023**, *13*, 1195. https://doi.org/10.3390/buildings13051195

Academic Editors: Carlos C. Duarte, Nuno D. Cortiços, Anna Stefańska, Daniel Mateus and Carol Monticelli

Received: 30 March 2023
Revised: 26 April 2023
Accepted: 27 April 2023
Published: 30 April 2023

1. Introduction

As if scripted for a dystopian future, the year 2020 dawned with a world-changing event—COVID-19, a global pandemic that affected peoples' lives and irreversibly transformed our modes of living and domestic space use. Worldwide, governments implemented lockdowns and confinements—schools were closed, and people were remanded home. In Portugal, the first compulsory confinement lasted two and a half months, from March 18th to May 3rd. During this time, the domestic space transcended domestic usage and functions and became an abode, school classroom, school playground, workplace, and gym, it served many alternate purposes and embodied different functions. Dwellings were thus overloaded with activities and crowded with household members, at all times, superimposing the former and the latter in the same finite space, which posed challenges to families' modes of living and domestic space use.

The research that is presented in this paper proposes an analysis of domestic space use and adaptation in Lisbon, during the first compulsory lockdown and is part of a larger investigation aimed at ascertaining the adequacy of contemporary urban dwelling programs to society's modes of living and household composition [1]. The understanding of dwelling program performance and adequacy, and the knowledge of the transformations and adaptations therein is paramount to better plan for contemporary and future society and living modes.

A worldwide subject, COVID-19's impact on housing and dwelling has been addressed by many: scholars, journalists, writers, or the *average Joe*, under different lights. Several authors [2–4] have approached the implications of confinement on domestic space use, the design features that were deemed more relevant in the lockdown period, and how lifestyles changed as a consequence of the restrictions, in different contexts. Some researchers also focused studies on specific spaces or areas, such as the use of balconies [5,6]. Addressing the issue in the Portuguese context, Moreira and Serdoura [7] discussed whether behavior and space use in a lockdown context was the cause or consequence of space properties, focusing a space syntax analysis on a particular domestic space, the living room.

The research on domestic space use during COVID-19 confinement that is here presented aims at ascertaining how households adapted their modes of living as a result of the pandemic context and also if (and how) dwellings had to be adapted (functionally or morpho-topologically) to address the demands of its inhabitants, which had changed immensely. Results show that although no significant morpho-topological alterations were accomplished (due to lack of space for some), households' living modes were impacted by COVID-19 confinement, and domestic life was disrupted by working and study activities, which normally occur outside the domestic setting, but as a consequence of the imposed lockdown was transported to the domestic sphere. Results also reveal a shift in valued aspects and qualities in a home, brought about by continuous home living.

2. Materials and Methods

The study on domestic space use alterations during the COVID-19 lockdown is founded on a single-method methodology, composed of a Post-Occupancy Evaluation (POE). This analytical approach aims at assessing product (dwelling) performance and quality by analyzing user (dweller) experience and satisfaction. For Villa and Ornstein [8] most issues affecting housing performance relate to design and layout inadequacy and the incapability to address and comply with user's real needs. POE examines environment-behavior relations, exploring environment performance, its impacts, and its effects on human behavior (and vice-versa) [9]. The selected POE method for this study was the questionnaire, one of the mostly applied in POE analyses because it allows for the assessment of uniformity and homogeneity of answers to a predetermined form [10] which can be indicative of a tendency or standard in behavior and attitudes.

The survey was carried out in the second trimester of 2020, during the first mandatory confinement, and sought to understand the impact of the round-the-clock use of the domestic setting and activities' superimposition in dweller's spatial and functional demands (in existing and future residences) and their satisfaction towards the inhabited spaces. The aim of the survey was thus to determine how dwellers adjusted modes of living (behavior) or domestic space (use or layout) to accommodate new needs and identify spatial adaptations and functional demands, but also to comprehend user satisfaction towards spaces and aspects of the lived units during confinement. The survey was available as an online inquiry (in google forms), due to public health constraints and policies, and was divulged by the Lisbon School of Architecture, Universidade de Lisboa's communication office, and by word of mouth. It comprised four sections—*Dweller and Household*, *Dwelling*, *Mode of Living*, and *Individual Questionnaire*—and was composed of closed and open questions. The closed questions addressed household and dwelling characterization and some qualitative (gradable) aspects, whilst the open questions were meant to gather opinions, giving the respondents complete liberty without influence or bias, avoiding pre-conceived ideas from the researcher.

Each section can be described briefly as follows:

- *Dweller and Household*—Questions that allow for the characterization of the household, its composition, and behavior during the pandemic context—if in lockdown, how many individuals worked from home (or were being home-schooled), if the lockdown meant moving to another dwelling (or considered) and which spaces or aspects would the respondents most value after having been confined; age and gender of the respondents were not asked, deemed irrelevant in the scope of the survey and to avoid an overly extensive questionnaire;
- *Dwelling*—Dwelling type and typology and assessment of satisfaction towards the dwelling (areas, layout, number of rooms) before and during lockdown, the identification of lacking uses/spaces, alterations (carried out or intended), and of the spaces that accumulated activities;

- *Mode of Living*—Household behavior as a unit, the spaces, and activities of family communion;
- *Individual Questionnaire*—filled out by each household member individually, focusing on the amount of time spent inside the home and spaces where the most time was spent (before and during confinement), places of work/study, and the characterization of the conditions to perform work/study activities.

The survey distribution means may have impacted and influenced some of the results, as they may express mainly the views and opinions of middle-class households, with internet access, and of people with higher levels of education (the Portuguese National Statistics Institute–INE–estimates that in 2020, 15.5% of Portuguese households, with at least one individual between the ages of 16 and 74, did not have internet access from home [11]). Nonetheless, the study defined the statistical universe of respondents to be individuals and households inhabiting Lisbon (urban setting), both in collective and single-family housing dwelling units and the collected sample was 80 surveys (188 individuals). This represents a sample of approximately 0.035% of Lisbon's population, which albeit not statistically representative of the Lisbon dwellings and households' universe, was considered a relevant subset to be investigated and thought to be indicative of urban living in a pandemic context. A similar survey, carried out in the UK by Place Alliance of UCL The Bartlett School of Planning—Home Comforts—based its results on an analogous sample of approximately 0.039% of London's population [12]. Some responses were incomplete, and, in some questions, more than one answer was permitted, hence sometimes the total percentage is other than 100%.

The collected data were combined and processed to assess congruity and recurrence and to form the image of how confinement shaped domestic space use and society's living modes in Lisbon in the first lockdown.

3. Results

The results of the survey will be presented in subsections, each corresponding to a questionnaire section, as previously described.

3.1. Dweller and Household

The majority of the respondents of the survey are part of a household composed of four people (28.9%); two and three-person households share the same representativity, in a close second, with 25%; single-person households represent 14.5% of the respondent families and lastly and very sparsely, five-person households only 6.6%. Regarding household composition, the most relevant structure is that of couples with children (60%). Couples without children (DINK (Dual Income No Kids)) represent only 17.3% of the respondent households and although present in the sample, other compositions such as single-parent households or nuclear + extended family are barely represented. Although the distribution of household compositions in the sample (number of people and kinship) is not consistent with what Moreira and Farias [13] have ascertained to be the predominant compositions in Lisbon—single-person households (34.95%), DINK (23.79%), couple with kids (23.73%) and single parent household (11.84%) (and with the 2021 Censuses [11] results—single-person household (35.24%), DINK (20.90%), couple with kids (22.63%) and single parent household (13.09%))—this discrepancy could be a consequence of the selected survey distribution method (as mentioned previously), especially regarding the representativity of the single-person households, tendentially elderly people, less technology savvy or without access to it (while more than 20% of Lisbon's population is over 65 years of age, only 39% of the Portuguese population aged between 65 and 74 have reported using the internet [11]).

Of the 80 families, 75 (or 93.75%) abided by confinement, 65% had individuals working or attending school from home (In the second trimester of 2020 (a period following the realization of the survey), INE estimates that 36% of the population of the Lisbon Metropolitan Area worked from home and that 30.8% of the Portuguese population (between the ages of 16 and 74) accessed the internet for educational purposes [11]). Only 5 families moved to a different residence during confinement due to the need for more space (40% of the

5 families), geographic move (40%) or to support family members (60%) (The total exceeds 100% because respondents were free to give more than one answer). For the other 5 families, the number of dwelling inhabitants altered (increasing in 4 families and decreasing in 1).

As a result of being remanded home, 23.75% of the inquired households considered permanently moving to another dwelling. The biggest appointed reasons for this attitude were the need for private exterior space (18.33%), the need for more interior space (8.33%), and the need for more rooms (6.67%). Some respondents also mentioned seeking better living quality outside an urban setting (5%) and layout reasons (3.33%) as motives for moving.

Having experienced confinement, almost 59% claimed to better appreciate some domestic spaces or attributes: exterior space (49%), living room (17%), home office (8.5%), bedroom (4.3%); solar intake, privacy, and multifunctionality was also mentioned as desirable traits. As for what households will most likely value in a future dwelling (Figure 1), the priority goes to a private exterior space (51.25%), followed by an indoor habitable area (48.75%), location (38.75%), and layout (25%). Albeit less relevant for the surveyed households, the number of compartments of a future dwelling was also selected and numerous compartments (17.5%) were considered to be preferred over a less compartmentalized interior (6.25%).

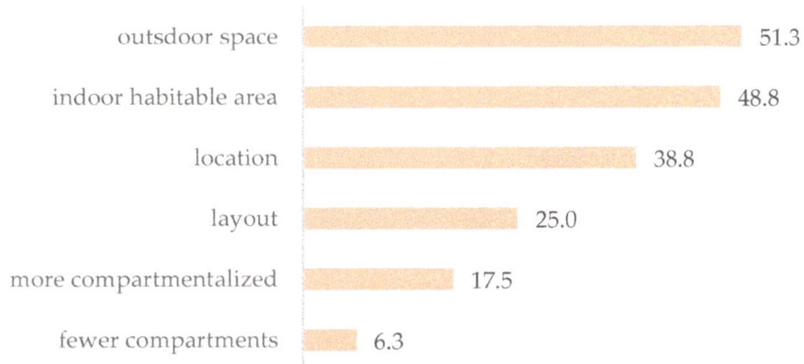

Figure 1. Mostly valued aspects or spaces in a future abode (percentage of respondents who chose the options given as first priority).

3.2. Dwelling

To establish context and better understand dwellings' performance and space use, dwelling type and typology were ascertained. Most respondents inhabit a three-bedroom unit (40%), 28.75% a two-bedroom unit, and 23.75% a four-bedroom (or more) unit; one-bedroom units are home to only 5%. As Moreira and Farias [13] have established, Lisbon's housing reality is characterized by a majority of two-bedroom and three-bedroom units, which means that the sample of the survey is in line with the municipal actuality.

When analyzing dwellers' degree of satisfaction towards numerous aspects of the home, both before and during confinement (Figure 2), it is apparent that the mean degree of satisfaction has decreased, hence demonstrating that when in constant use dwellings fall short of inhabitants' expectations, needs, and demands, and shortcomings become more noticeable [1].

A more attentive reading of Figure 2 will demonstrate that almost in all inquired aspects dweller satisfaction decreased in the period of confinement, especially of those more closely linked and impacted by constant household co-living and activities superimposition—privacy within the family, adequacy to household needs, number of rooms and area related issues (dining room, living room, kitchen and total). On the other hand, home confinement generated greater satisfaction with a few aspects and dwelling traits, that allowed for a mitigation of the pandemic situation and restrictions—balcony area, natural light, and layout. These shifts in the degree of satisfaction, namely those where the variation is highest, reveal what dwellers came to value most.

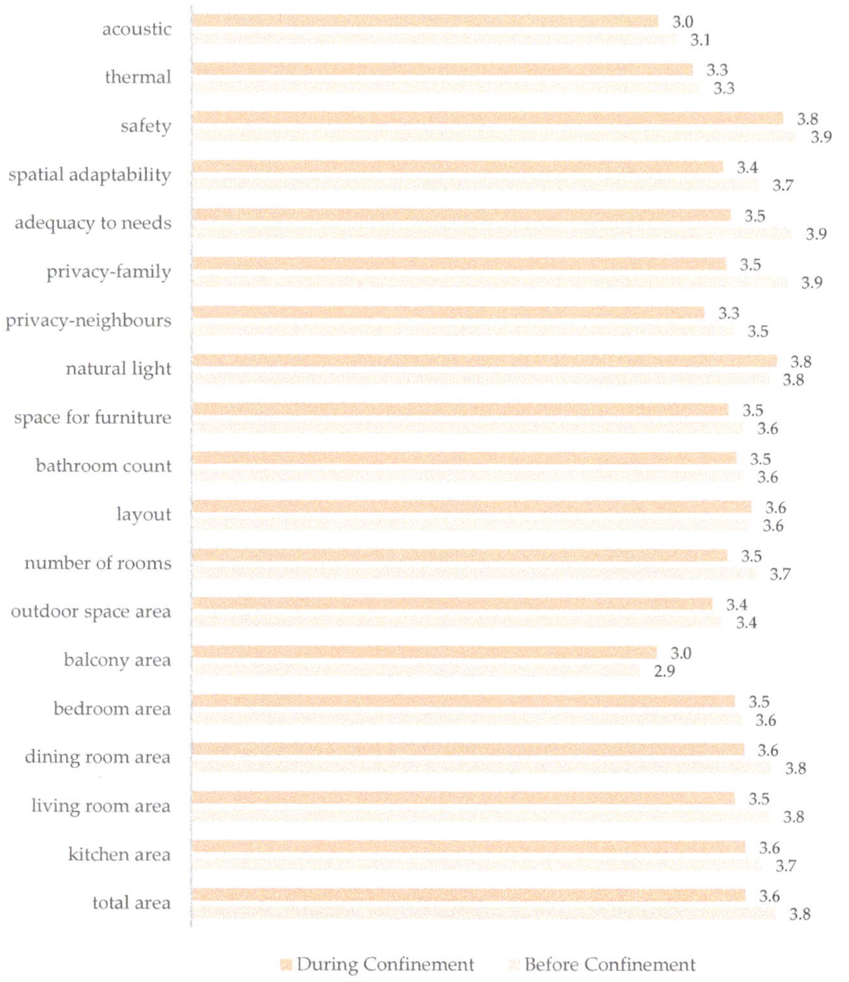

Figure 2. Dweller satisfaction (per item) before and during confinement.

Despite the alterations in satisfaction with the home, most respondents (75%) considered that their dwellings were suited for 24/7 living and that all household members' activities were feasible (70.5%). Nonetheless, 65.4% of households confirmed the need for additional space, especially a private outdoor area (for 71.9% of the former). A dedicated home office space (9%) and an extra room (7%) were also mentioned, among other less representative and significant aspects.

Addressing the stated need for additional space, 35.9% of the respondents indicated having adapted a room to fulfill a household global need or individual need, namely the integration of office space (either for adult work or children and young people study). This new domestic area and function were allocated either in the living room (51.9%) or bedroom (33.3%). Regardless of the occurrence of spatial/functional adaptation, in the majority of the inquired households (60.8%) the domestic spaces, especially the living room (59.1%) and bedroom (31.8%), accumulated functions and activities, encompassing working activities in particular. In spite of this overlapping, most respondents claimed to have been able to perform working or study tasks in privacy and in an environment with enough space which abetted concentration.

When asked if any spatial (physical) or functional alterations had been made to the dwelling during confinement, the responses relate to functional aspects alone, most of them consisting of furniture rearrangement and the creation of an office area (with adequate fittings). No profound alterations to these sample dwellings were then accomplished. As a consequence of this, more than half of the respondents claimed they would not maintain the pandemic-derived arrangements.

Another surveyed aspect regarded post-pandemic alterations, whose necessity was made apparent during confinement, and although a great disparity of answers was submitted, a few aspects were most mentioned: more space or differently laid out space (larger areas, more rooms—including a home office—more storage, and layouts that integrate kitchen and living area), outdoor area (in some cases with vegetable garden) and better sound insulation and natural light (Additionally, while the majority of answers (20%) was no change needed, some respondents indicated the urge to move to another dwelling) (Figure 3). These results are consistent with those referring to valued qualities of a future home and it is noticeable that some external space and more indoor space (more space and differently laid out or distributed space) are of great importance to a large percentage of households.

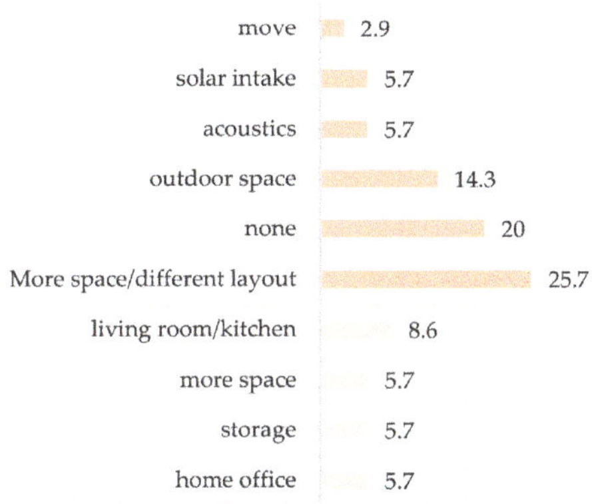

Figure 3. Aspired alterations for the post-COVID context.

3.3. Mode of Living

The mode of living section had few questions and aimed mainly at understanding family dynamics and consumption habits' shifts, to allow for a better understanding of domestic space use. Concerning consumption habits shifts, in the first lockdown, approximately 25% of the households started to outsource domestic services more often (cooking, and laundry, among others). Regarding family dynamics, the survey inquired about spaces and tasks of family gatherings, and for most the common living room is the space where the family congregates, the kitchen following suit; meal intake is the activity that mostly assembles the members of the household, as well as leisure activities. Meal preparation is a family activity in only 47.5% of the inquired households (indicative of either a segregating and compartmentalized dwelling layout where the kitchen is an isolated space from the social areas and spaces, or of social characteristics and behaviors that view cooking still a chore and mostly a feminine task [14]).

3.4. Individual Questionnaire

Before the COVID-19 lockdown, the majority of household members (49.5%) spent between 5 to 8 daily hours (excluding the sleeping period) inside their dwellings. Due to mandatory confinement, a more significant number of people (90%) were spending more than 8 daily hours in the domestic setting (Figure 4).

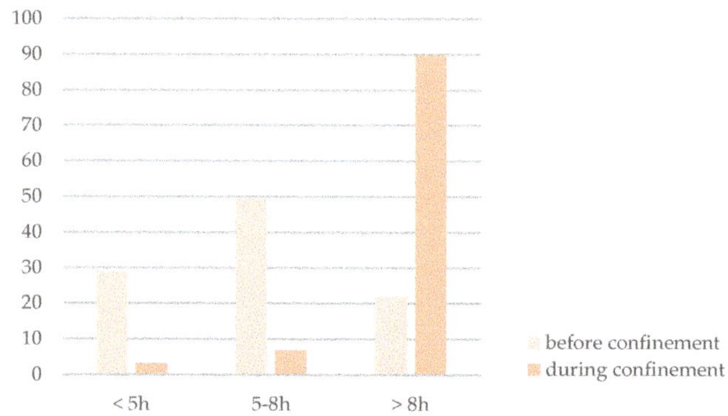

Figure 4. Time spent in the dwelling, before and during confinement.

The living room is the space mostly used by the majority of household members (both before and during confinement). Nonetheless, results show that there was a decrease in the use of this space during confinement and an increase in the use of all other domestic spaces—dining room, bedroom, kitchen, and other rooms (more than half of the answers 'other spaces' referred to an office space or study area) (Figure 5a). Albeit this reduction in usage time, the living room is the place where most respondents worked or studied during confinement. Other domestic settings that started to incorporate this new activity were the bedroom, 'other' spaces (mostly home office), dining room, and kitchen.

Figure 5. (**a**) The domestic setting where most time is spent, before and during confinement, and (**b**) the room where working or study activity is performed, during confinement.

Analyzing Figure 5a,b together, it is possible to state that the increased use of the dining room, bedroom, kitchen, and other spaces can account for the reduction in living room use (a more social and leisurely space in a normal context). This is also a consequence of confinement since a prolonged 24/7 use of the dwelling will tend to scatter inhabitants through the different compartments rather than cluster them in the same room. Moreover, it is also possible to relate the decrease in satisfaction towards the areas of these rooms

(Figure 2) with the integration of a workspace and activity in those settings—the same volume of space became a receptacle to an additional function which became exceedingly present during confinement and, at times, may have clashed with domestic life. This is noticeable in the great majority of the households that registered a significant decrease in satisfaction towards living room area, bedroom area, or both: 77% of the households where satisfaction with these aspects decreased were ménages where one person or everyone was working or studying from home, using the living room or bedroom (or both) spaces for that effect.

4. Discussion

Based on the presented results it is indisputable that public health restrictions and lockdowns have had a profound effect on the population's living modes, aspirations, and domestic space use in the city of Lisbon. Access to private outdoor space (terrace, patio, garden, balcony) that allows for effective use (adequate area and proportions) and the need for additional indoor spaces, more area or better layout have emerged as fundamental qualities that dwellers have come to value most in a dwelling (present or future) and, as such, should be considered in housing programs and models from here on in. Results furthermore demonstrate that views and aspirations of the domestic interior may have evolved, due to a shift in behavior as a consequence of prolonged confinement. Hence, new traits and aspects may have become relevant for current households. The home office dedicated space is one of those aspects. A necessity for most during lockdown, the post-covid reality maintains traces of this event and more and more people are adopting working from home, either some days per week or every day of the week, be it for practicality reasons (commuting, international networking) or employer strategy or imposition. As such, a space that can accommodate working activities in the domestic setting has become a requirement. Another relevant demand that has arisen from the confinement period is the availability of extra space or space with no pre-determined function that can accommodate dwellers' needs and be transformed as these evolve. The shift in the degree of satisfaction regarding the *adaptability of spaces* and *adequacy to household needs* as well as prioritizing *more compartmentalized interiors* in a future dwelling (against more open space plans) both reflect this viewpoint. To comply with this demand, attain *spatial resilience* [2] and greater adaptability to dwellers' needs and modes of living, the design approach should seek flexibility [1–3,15–17], either passive (without physical alteration of the space, but allowing appropriation and function allocation as needed) or active (relying on space reconfiguration—expansion and contraction—via movable elements) [15,16]. Flexibility or de-hierarchized and ambiguous [17] interiors foster adaptability and potentiate greater adequacy to household needs and demands.

The findings of the survey are consistent with other survey reports, both on a national level (Portugal) and transnational (UK). On the national level, the JLL real estate group developed a study to determine the effects of COVID-19 on housing needs [18] and concluded that 51% of the population would integrate an office space at home and 34% would transform an outdoor space (In the priority evolution scale these two aspects were also two of the three which showed bigger increase). This study also revealed a tendency of 70% of the respondents ages comprised between 25 and 40 to continue to work from home one to 3 days a week.

On a transnational level, in the UK a survey was realized by UCL The Bartlett School of Planning's Place Alliance Platform, in May 2020 [12]. The findings of this survey are very much in line with the ones presented in this paper: the importance of private outdoor space and the need for more space(s) (in rapport with occupancy levels). Moreover, although two very different societies were surveyed similar results were reported as the majority of the UK respondents also claimed to prefer greater compartmentalization, variety of spaces, and functional separation, as opposed to a more open concept, integrating a separate home office space. Storage, noise insulation, and natural light were also deemed relevant for the respondents.

5. Conclusions

In light of the presented findings, there has been a clear change in Lisbon society's modes of living as a result of the pandemic context and confinement, as well as alterations in household behaviors in the domestic setting, of which some have persisted and have seeped through and require a new approach to dwelling program and design. Home office spaces and private outdoor areas represent new household demands, which have also come to favor the possibility to adapt, appropriate, and repurpose spaces, for which extra compartments, ambiguous ones, or flexible solutions can prove to be the strategy for dwelling program design.

Further research focusing on the post-COVID home can be developed and a subsequent survey could prove useful to determine if the changes in modes of living and domestic space use were provisional or if the COVID-19 event represented an actual and permanent shift in the way we inhabit our homes.

Author Contributions: Writing—original draft preparation, A.M. and H.F.; writing—review and editing, A.M. and H.F. All authors have read and agreed to the published version of the manuscript.

Funding: This work is financed by national funds through FCT—Fundação para a Ciência e a Tecnologia, I.P., under the Strategic Project with the references UIDB/04008/2020 and UIDP/04008/2020.

Institutional Review Board Statement: Not applicable in a non-interventional study. All participants were fully informed that anonymity was assured, informed of the purpose of the research and the end of collected data. No personal data were collected.

Informed Consent Statement: Not applicable.

Data Availability Statement: Data is contained within the article.

Conflicts of Interest: The authors declare no conflict of interest. The funders had no role in the design of the study; in the collection, analyses, or interpretation of data; in the writing of the manuscript; or in the decision to publish the results.

References

1. Moreira, A. Adequação Funcional da Habitação Colectiva Urbana Contemporânea aos Novos Modos de Habitar e aos Novos Modelos Sociais. Ph.D. Thesis, Lisbon School of Architecture of the University of Lisbon, Lisbon, Portugal, April 2022.
2. Erfani, G.; Bahrami, B. COVID and the Home: The Emergence of new urban home life practiced under pandemic-imposed restrictions. *Cities Health* **2022**, 1–8. [CrossRef]
3. Alhadedy, N.; Gabr, H. Home design features post-COVID-19. *J. Eng. Appl. Sci.* **2022**, *69*, 87. [CrossRef]
4. Gür, M. Post-pandemic lifestyle changes and their interaction with resident behavior in housing and neighbourhoods: Bursa, Turkey. *J. Hous. Built Environ.* **2022**, *37*, 823–862. [CrossRef] [PubMed]
5. Grigoriadou, E. The urban balcony as the new public space for well-being in times of social distancing. *Cities Health* **2021**, *5*, S208–S211. [CrossRef]
6. Elzein, Z.; Elsemary, Y. Re-Thinking post-pandemic home design: How Covid-19 affected the perception and use of residential balconies in Egypt. *Future Cities Environ.* **2022**, *8*, 1–15. [CrossRef]
7. Moreira, A.; Serdoura, F. Confinement and House Use. Analysing the relationship between space properties and space use during pandemic isolation. In Proceedings of the 13th Space Syntax Symposium, Bergen, Norway, 20–24 June 2022.
8. Villa, S.; Ornstein, S. Avaliação do Comportamento dos Usuários no Espaço Habitacional: Métodos e Reflexões. In Proceedings of the ENTAC 2006. XI Encontro Nacional de Tecnologia do Ambiente Construído, Florianópolis, Brazil, 23–25 August 2006.
9. Romero, M.; Ornstein, S. *Avaliação Pós-Ocupação. Métodos e Técnicas Aplicados à Habitação Social*; ANTAC: Porto Alegre, Brazil, 2003.
10. Villa, S.; Ornstein, S. (Eds.) *Qualidade Ambiental na Habitação. Avaliação Pós-Ocupação*; Oficina de Textos: São Paulo, Brazil, 2013.
11. INE. Available online: www.ine.pt (accessed on 27 March 2023).
12. Place Alliance. Available online: http://placealliance.org.uk/wp-content/uploads/2020/10/Place-Alliance-Homes-and-Covid-Report_2020.pdf (accessed on 27 March 2023).
13. Moreira, A.; Farias, H. Analysing Housing and Household Evolution and Actuality in Lisbon, 2000–2020. *Curr. Urban Stud.* **2022**, *10*, 329–341. [CrossRef]
14. Moreira, A.; Farias, H. Gendered Spaces at Home. Feminine and Masculine Traits in Domestic Interiors. *Int. J. Soc. Sci. Stud.* **2022**, *10*, 91–104. [CrossRef]
15. Farias, H. A more versatile and adaptable dwelling, for a changing society. In *Architectural Research Addressing Societal Challenges, Proceedings of the EAAE ARCC 10th International Conference (EAAE ARCC 2016), Lisbon, Portugal, 15–18 June 2016*; da Costa, M.J.R.C., Roseta, F., da Costa, S.C., Lages, J.P., Eds.; CRC Press Taylor & Francis Group: Lisboa, Portugal, 2016; pp. 353–360.

16. Farias, H. Repensar A Habitação Contemporânea: Flexibilidade, Adaptabilidade, Ambiguidade Funcional E Desierarquização Do Espaço Doméstico. In Proceedings of the 3º Colóquio de Investigação em Arquitectura, Urbanismo e Design, Portugal e Brasil, Lisboa, Portugal, 22–23 September 2014.
17. Farias, H. Para um habitar mais versátil, diversificado e inclusivo. In Proceedings of the 9º Seminário Internacional Projectar—Arquitectura e Cidade: Privilégios, Conflitos e Possibilidades, Curitiba, Brazil, 22–25 October 2019.
18. Nova Vida, Nova Casa? Teve a Pandemia COVID-19 Impacto nas Necessidades Habitacionais? *JLL*, 20 July 2020.

buildings

MDPI

Article

A Study on Recommendations for Improving Minimum Housing Standards

Donghyun Kim [1], Hyunnam Sim [2] and Seiyong Kim [1,*]

[1] Department of Architecture, Korea University, Seoul 02841, Republic of Korea; ssracks@korea.ac.kr
[2] Department of Urban Regeneration, Korea University, Seoul 02841, Republic of Korea; simhn@korea.ac.kr
* Correspondence: kksy@korea.ac.kr; Tel.: +82-2-3290-3914

Abstract: Minimum housing standards aim to safeguard housing rights and enhance residential conditions. Governments develop housing policies, including provision, preservation, and support for welfare programs, guided by the status of households below these standards. Growing nations commonly utilize this approach to decrease the proportion of households not meeting these criteria. In South Korea, the ratio of households below these standards was 4.5% in 2021, down from 16.6% in 2006, as indicated by the initial Korea Housing Survey. However, this downward trend has stalled over the past decade. With the 2004 and 2011 revisions, the standards have been effective for 12 years, yet no longer mirror current housing realities due to ongoing improvements. This study aimed to propose enhancements to Korea's minimum housing standards. Through analyzing laws, prior research, present household statuses, international cases, and expert insights, recommendations emerged. Categorizing households by size, we developed precise standards covering area, facility, and location aspects. These new standards led to an 8.4% non-compliance rate in 2021. This research's findings anticipate aiding the revision of minimum housing standards, formulating pragmatic policies for enhancing residential conditions in line with present situations.

Keywords: minimum housing standard; housing rights; Korea housing survey; semi-underground house; accommodations for students studying for exams

Citation: Kim, D.; Sim, H.; Kim, S. A Study on Recommendations for Improving Minimum Housing Standards. *Buildings* **2023**, *13*, 2708. https://doi.org/10.3390/buildings13112708

Academic Editors: Carol Monticelli, Chyi Lin Lee, Anna Stefańska, Daniel Mateus, Carlos C. Duarte and Nuno D. Cortiços

Received: 29 August 2023
Revised: 23 October 2023
Accepted: 25 October 2023
Published: 27 October 2023

1. Introduction

1.1. Research Background

In the past, South Korea maintained a supply-based housing policy to address the shortage of housing inventory. With the steady increase in the housing supply rate, however, people have become more interested in the quality of housing, as well as related demands. In particular, South Korea has shown a growing interest in solving unique problems in the country, such as providing accommodations for students studying for exams and developing compact buildings with small rooms. In this sense, the Korean government has broadened the spectrum of housing policies by including housing management and welfare, rather than adhering to supply-based policies.

As the housing policy has focused more on welfare, there has been a need to create comfortable residential environments, apart from the supply of sufficient housing. The necessity for housing that satisfies minimum housing standards has been the most important factor, because if the stability and comfort of housing are not guaranteed, people would have to continuously live with anxiety [1–6]. Under these circumstances, minimum housing standards were provided for people to live in comfortable and stable residential environments, resulting in the establishment of standards regarding room configuration, area, essential facilities, and structure/performance/environment for the first time. The minimum housing standards suggested by the Ministry of Construction and Transportation (Currently, the Ministry of Land, Infrastructure, and Transport) in 2000 were enacted into law in 2004 as the "Housing Act" or "Enforcement Decree of Housing Act", and they have since been applied to housing policies along with the enhancement of area standards in 2011 [7,8].

Despite the newly strengthened standards, the ratios of households below the minimum housing standards showed a general decreasing tendency, falling to 4.5% in 2021 (Table 1) [9]. A steady supply of new housing, a decrease in the average number of household members, a drastic increase in one- to two-person households, and a decline in substandard houses have been reported as contributors to the decline. As the ratios of households below the minimum housing standards have decreased, the minimum housing standards required by people have naturally improved. Additionally, the suitability and effectiveness of the revised 2011 minimum housing standards have been in question due to the current situation of housing inventory and supply.

Table 1. Households below the minimum housing standards in South Korea [9].

	2006	2008	2010	2012	2014	2016	2017	2018	2019	2020	2021
Ratios (%)	16.6	12.7	10.6	7.2	5.4	5.4	5.9	5.7	5.3	4.6	4.5

1.2. Research Background and Purpose

This study was conducted for the following reasons:

First, investigating the areas of small houses of 60 m^2 or smaller, the area standard of the minimum housing standard in 2011 was determined while considering the area of the bottom 3% of them. However, it was assumed that the current bottom 3% standard would be higher than that in 2011. Second, as the residential area per capita generally increased from 20.2 m^2 in 2000 to 33.9 m^2 in 2021, there should be considerations regarding this tendency (Table 2) [9]. Third, as the physical characteristics of Koreans have changed, there should be new standards considering the human scale. Fourth, facility standards should be reconsidered. Finally, it is necessary to prepare standards regarding the quality of housing and housing safety, which are not included in the current minimum housing standards. In Korea, safety accidents in semi-underground houses due to flood damage have become a recent social issue, and the government has implemented a policy to remove semi-underground houses [10].

Table 2. Average residential area per capita (m^2) [9].

2000	...	2006	2008	2010	2012	2014	2016	2017	2018	2019	2020	2021
20.2	...	26.2	27.8	28.5	31.7	33.5	33.2	31.2	31.7	32.9	33.9	33.9

Accordingly, this study aims to propose new minimum housing standards (area, bedroom, facility, and location) and measures for utilizing policies.

2. Materials and Methods

2.1. Definition of Terms

2.1.1. Minimum Housing Standards

As minimum housing standards are intended to guarantee people's housing rights, the standards were categorized into area, bedroom, facility, and location criteria. While standards can be subjective depending on the era and viewpoints, we aimed to establish them in as objective a manner as possible.

2.1.2. Area Standard

This standard pertains to the minimum area required by a resident in a house. An "exclusive residential area", defined as an enclosed space usable exclusively by a resident upon opening the front door, was established as the area criterion. For multi-unit dwellings, we excluded communal areas like hallways, parking lots, elevators, and open spaces.

Regarding detached houses, non-exclusive residential areas such as yards and detached parking lots were excluded.

The essential rooms included bedrooms, kitchens or dining rooms, bathrooms, and other spaces, which were aggregated to calculate the ultimate exclusive residential area.

2.1.3. Bedroom Standard

This standard defines the quantity and types of specific rooms that must be provided for each household member. The variables of "per household member" encompass the number, age, and gender of household members.

2.1.4. Facility Standard

This standard pertains to the facilities that residents must have while living in houses. We focused on the following facilities: (1) kitchen, (2) toilet, (3) bathing facility, (4) water supply facilities, (5) sewer system (septic tank), (6) heating systems, (7) fuel for cooking, (8) entrance (front door), and (9) fire-fighting appliances.

2.1.5. Location Standard

This is the standard covered in this study for the first time, and it pertains to the location of a house. We aimed to establish a standard for the location of a house that may pose safety and health risks to residents [11]. In this study, houses were categorized into the following four types: underground, semi-underground, above-ground, and rooftop houses.

2.2. Scope and Methodology of the Study

Households ranging from one-person households to six-person households were targeted. The sub-standards to be established were categorized into area, bedroom, facility, and location standards.

This research was conducted based on five steps: analysis of the current status, overseas case studies, collection of expert opinions, derivation of improvements, and assumptions (Figure 1).

First, we investigated households falling below the minimum housing standards to analyze the current status. Since 2006, Korea has been conducting Housing Surveys, utilizing Housing Survey data from 2006 to 2021. We reviewed the situation of entire households below the minimum housing standards, as well as the status of households unable to meet sub-standards (e.g., areas, bedrooms, and facilities).

Second, in the section on overseas case studies, the focus was on the United Kingdom and Japan, which employ similar housing standards to South Korea's and their housing standards were analyzed. These countries were selected as target cases because their situations are akin to Korea's situation in terms of country size (e.g., area and population), population density, state earnings, and urbanization rates. Cases from other countries like the United States and France were also considered, but they utilized minimum housing standards (area, bedrooms, facilities, etc.) different from those of South Korea. Therefore, the United Kingdom and Japan were selected as case study destinations.

Third, expert opinions were collected by interviewing 22 experts in four fields for approximately three weeks. The questions mainly consisted of topics that cannot be identified from the current status alone, and opinions regarding the outcomes of the current analysis.

Fourth, improvements in minimum housing standards were derived through three detailed methodologies. First, a prerequisite was derived based on theoretical studies. Next, we determined the direction for specific improvements, based on the outcomes from the literature review, analysis of the current status of households unable to meet the standards, overseas case studies, and expert opinion collection. Additionally, a design simulation was conducted considering estimates of furniture sizes, universal design, and human scale. Through these three methodologies, we suggested the final improvements based on the area, facility, and location standards.

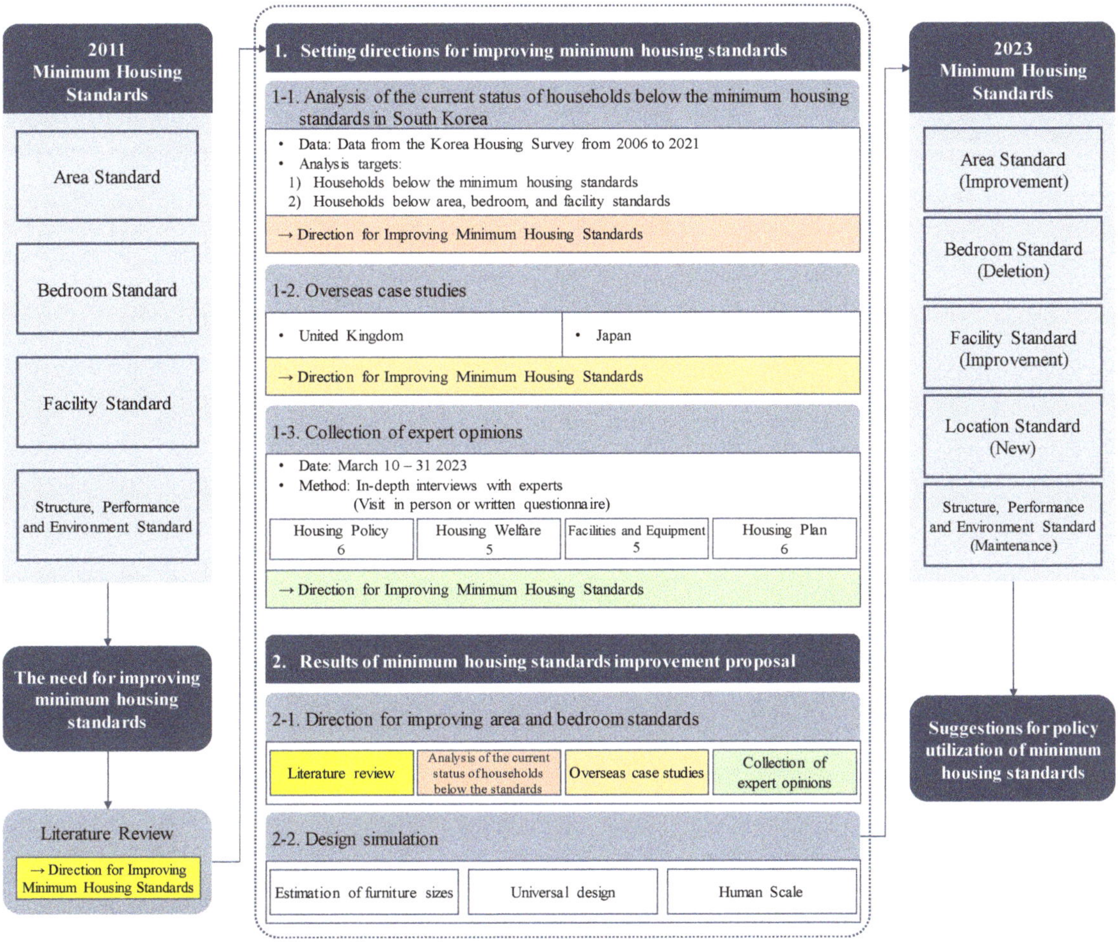

Figure 1. Research Methodology.

Fifth, we estimated households not meeting the new standard for minimum housing standards. Finally, we suggested policy utilization measures for the minimum housing standards.

3. Literature Review and Theoretical Framework

3.1. Literature Review

There are two types of studies on minimum housing standards: studies on how to utilize the standards and studies on how to improve and amend the standards.

First, there are studies on how to utilize the standards. Choi et al. [12] applied the 2011 revised standards to population and housing census data to analyze changes in households falling below the standards. They indicated how the minimum housing standards can enhance the residential environment for those with inadequate housing conditions, while also highlighting the limitations of applying the standards to vulnerable populations living in non-housing spaces [12]. Lim [13] discussed how to employ minimum housing standards in designing welfare policies, estimating households below the standards, emphasizing the need to enhance minimum housing standards, and assessing their effectiveness [13]. The above study emphasized the need not only to establish minimum housing standards but also to pursue policy measures to address households below these standards. Specific proposals, such as the establishment of housing management norms and improvement of residential environments, were suggested.

Regarding the study of improving and amending minimum housing standards, Yun [14] compared how the standards were used in South Korea with their utilization in other countries. He underscored the need to improve the current minimum housing standards in South Korea for efficient usage, citing their abstract and unclear nature [14]. Kim et al. [11] asserted that poor physical residential environments detrimentally affect residents' health, highlighting that the current minimum housing standards should consider such factors [11]. Lee [15] suggested that the existing standards are overly broad and lack specific criteria, leading to weak effectiveness. He proposed additional countermeasures, incorporating social and economic criteria [15]. Kim [16] reviewed the composition and utilization plans of the 2011 revised minimum housing standards and the Long-term Comprehensive Housing Planning of cities and provinces. The aim was to propose directions for improving minimum housing standards to utilize them as tools for housing welfare policies [16]. Lim [17] mentioned that South Korea's minimum housing standards lack effectiveness in enabling the state to take minimal protective measures. To enhance effectiveness, improving the 2011 revised minimum housing standards is necessary [17]. The aforementioned studies can be summarized as having two main opinions regarding minimum housing standards: one suggesting that they should be more "specific", and the other suggesting that they should be "strengthened". However, they only emphasized the need for these improvements without providing specific proposals.

Thus, this study examined Korean research on minimum housing standards from 2011 to 2021 to understand relevant scholars' perspectives on the standards since their first revision. While most studies highlighted issues with the current minimum housing standards and the need for improvements, they did not offer precise criteria for enhancing the standards. Therefore, efforts were made to provide specific plans for improving and utilizing the standards at a policy level through a systematic research methodology.

Furthermore, recent research has revealed a broader spectrum of keywords associated with "housing standards". In Wang et al.'s study [18], it was emphasized that "air quality" can be considered as one of the housing standards, as it was found to influence property values [18]. Wang et al. [19] indicated that both homebuyers and renters consider "air quality" a significant factor in their housing choices, asserting that clean air should be perceived as a form of environmental amenity provided to residents [19]. Boadi et al. [20] demonstrated the substantial impact of "residential satisfaction" on quality of life and resettlement [20]. Since current housing standards in Korea primarily consider physical elements, there is an opportunity to additionally incorporate the subjective aspect of residents, "residential satisfaction". Bangura et al. [21] shed light on the surge in property prices due to the COVID-19 pandemic, which subsequently contributed to housing inequality [21]. This suggests that unforeseen external factors like COVID-19 can indeed influence housing standards.

Recent research has thus explored new "factors" that were not previously given much consideration concerning "housing standards". This, in turn, provides significant direction for shaping the focus of this study.

3.2. Theoretical Framework

3.2.1. Housing Rights

According to the UN's Universal Declaration of Human Rights, housing rights pertain to economic, social, and cultural rights while safeguarding the right to live as a human being [22]. South Korea addresses housing rights in Article 35 of the "Constitution", which states: "All citizens have the right to live in a comfortable environment, and the state shall actively implement policies for this purpose" [23]. Article 2 (Housing rights) of the "Framework Act on Residence", enacted in 2015, asserts that "all citizens have the right to live in decent, comfortable, and stable residential environments free from physical and social dangers, as stipulated by related laws and ordinances" [24]. Article 3 of the same act outlines the responsibilities of the state and local governments in guaranteeing the housing rights of all citizens [24].

Several overseas studies define housing rights as follows. Pane et al. [25] provide the following definition: "The right to adequate housing is the right of all citizens without exception" [25]. Concerning housing rights, Kucharska-Stasiak et al. [26] said the following: "Adequate housing conditions are an indicator of a decent life, whereas the lack is one of the main reasons behind so-called social exclusion. The importance of housing, in ensuring the social safety of citizens, as well as supporting social equity, has been emphasized for decades" [26]. Some studies also assert that basic residential conditions or conveniences are necessary to safeguard fundamental human rights and enhance human welfare [27]. Moreover, Azarnert [28] said the following: "Minimum residential conditions requirements may also reduce population density and lead to a decline in social costs associated with population overcrowding and congestion in general" [28].

3.2.2. Minimum Housing Standards in South Korea

Article 17 of the "Framework Act on Residence" stipulates the establishment of minimum housing standards. Paragraph 1 of the Article states that the purpose of minimum housing standards is to set the necessary standards for people to maintain a pleasant and fulfilling life; the duties of the Minister of Land, Infrastructure, and Transport are to set and solidify the standards. In addition, Paragraph 3 of the Article specifies factors of minimum housing standards as follows: (1) residential area, (2) the number of rooms per use, (3) structure of a house, (4) facilities in a house, (5) performance of a house, and (6) environmental factors of a house [24].

Article 18 of the Act also stipulates the following priority supports for households below the minimum housing standards: (1) The State or a local government may give priority to supplying housing or subsidizing improvement funds for households below the minimum housing standards. (2) The State or a local government shall endeavor to reduce the number of households below the minimum housing standards. (3) The Minister of Land, Infrastructure, and Transport or the head of a local government shall take necessary measures for granting authorization and permission, such as issuing an order to supplement an application for approval for project plans in compliance with the minimum housing standard. (4) The Minister of Land, Infrastructure, and Transport or the head of a local government may take necessary measures to preferentially construct rental houses in an area densely packed with households below the minimum housing standards [24].

Although minimum housing standards play a crucial role as a policy indicator for the quality of housing, as mentioned in the Introduction, South Korea has not had a second revision since its enactment in 2004 and the first revision in 2011. Although the Ministry of Construction and Transportation (currently, the Ministry of Land, Infrastructure, and Transport) first proposed it in 2000, it was not eventually enacted into law, and it is not discussed in this study. This study aims to suggest the second revision of the minimum housing standards, which have been applied in the last 12 years since 2011.

3.2.3. Minimum Housing Standards in South Korea (in 2004 and 2011)

Minimum housing standards were initially legislated based on the "Housing Act" in 2004. The number of household members was categorized into 1–6 persons, establishing a standard household composition and specifying the minimum residential area per standard household composition and the number of rooms per use (Table 3). Concerning facilities, households are obligated to have a private walk-in kitchen, a private flush toilet, and bathing facilities, along with water supply facilities or groundwater facilities with good water quality. Furthermore, for structure, performance, and environmental standards, four criteria are proposed to ensure housing safety and comfort as follows: (1) The permanent building must possess structural strength, and principal structural parts must be heat-resistant, fire-resistant, and moisture-proof. (2) Adequate soundproofing, ventilation, lighting, and heating facilities must be provided. (3) Environmental factors like noise, vibration, odor, and air pollution must adhere to legal standards. (4) Housing should not be situated in areas at significant risk of natural disasters like tsunamis, floods, landslides, and cliff collapse [7].

Table 3. Minimum residential areas per household composition, and number of rooms per use (2004, 2011) [7,8].

Number of Household Members	Standard Household Composition [1]	Space (Room) Requirement [2]		Total Living Area (m^2)	
		2004	2011	2004	2011
1	One person household	1 K	1 K	12	14
2	Married couple	1 DK	1 DK	20	26
3	Parents + 1 Child	2 DK	2 DK	29	36
4	Parents + 2 Children	3 DK	3 DK	37	43
5	Parents + 3 Children	3 DK	3 DK	41	46
6	Grandparents + Parents + 2 Children	4 DK	4 DK	49	55

[1] Based on 1 child aged 6 years old or older in a 3-person household. Based on 2 children (1 male and 1 female) aged 8 years old or older in a 4-person household. Based on 3 children (2 males and 1 female, or 1 male and 2 females) aged 8 years old or older in a 5-person household. Based on 2 children (1 male and 1 female) aged 8 years old or older in a 6-person household. [2] K refers to the kitchen, and DK refers to a combined dining room and kitchen; the figure refers to the number of rooms that can be used as bedrooms (including areas for living rooms) or rooms that can be used as bedrooms. Note: The principle of bedroom separation for setting the number of rooms is based on the following criteria: (1) married couple shares one bedroom; (2) children aged 6 years old or older have a separate room from their parents' rooms; (3) opposite-sex children aged 8 years old or older have individual rooms; (4) grandparents use separate bedrooms.

In 2011, the first amendment was introduced. In comparison to 2004, the total residential area per household member slightly increased. The standard household composition and the number of rooms per use remain unchanged. The standard for essential facilities has been updated to include a private walk-in kitchen, a private flush toilet, and a private bathing facility, along with water supply facilities, groundwater facilities with good quality, and sewage facilities. Additionally, "safe electricity utilities and structures and facilities for safe evacuation in case of fire" were added to the previous four criteria within the structure, performance, and environmental standards [8].

4. Setting Directions for Improving Minimum Housing Standards

4.1. Analysis of the Current Status of Households below the Minimum Housing Standards in South Korea

4.1.1. Overview of the Analysis

To analyze the current situation of households below the minimum housing standards, data from the Korea Housing Survey were utilized (Table 4) [9,29]. By Article of the "Framework Act on Residence" and Article 13 of the Ordinance for Enforcement of the Act, South Korea has implemented the Korea Housing Survey since 2006. The survey was sponsored by the Ministry of Land, Infrastructure, and Transport, and the Korea Research Institute for Human Settlements under the Ministry of Land, Infrastructure, and Transport conducted the survey; considering that, these data are reliable.

The minimum housing standards in Korea, which are currently published (refer to Table 3), were utilized. Additionally, information gathered from the following statements was used to confirm the status of households below the minimum housing standards.

(1) Exclusive residential areas per number of household members:

In general, as residents do not know the exclusive residential areas of their houses, data registered in the building register was utilized.

(2) Private walk-in kitchen:

Confirmation was conducted on whether the kitchen was used exclusively and as a walk-in. If either of these two conditions were not satisfied, the households were treated as households below the standard.

(3) Private flush toilet:

The status of the toilet was confirmed, whether it was for private use or a flush toilet. If either of these two conditions was not satisfied, the households were treated as households below the standard. The presence of a Western-style toilet was not taken into consideration.

(4) Private bathrooms:

Confirmation was solely based on whether bathrooms were used exclusively. If not, the household was treated as falling below the standard. The presence and absence of hot water were not considered.

(5) Water supply and drainage facilities:

We confirmed whether the water supply and drainage facilities were installed or not. As for the water supply facility, the availability of a groundwater facility with good water quality was recognized.

(6) Number of bedrooms per household composition:

Since the minimum number of bedrooms varies depending on the household composition (e.g., a married couple), the presence or absence of children, the age of children, and the gender of children, such a factor was applied. The number of living rooms used as bedrooms was included in the number of bedrooms.

Table 4. Outline of Korea Housing Survey [9,29].

	Details		
Rationale	Article 20 of the "Framework Act on Residence" and Article 13 of the Ordinance for Enforcement of the Act		
Survey sponsor	Ministry of Land, Infrastructure, and Transport		
Survey implementer	Korea Research Institute for Human Settlements		
Survey period	2006~2016: Biennial survey in even years 2017~2021: Every year		
Survey targets and scope	General households residing nationwide		
Survey methods	Face-to-face interview		
Total number of households in South Korea	21,448,463 households (in 2021)		
Number of valid samples	Approximately 51,000 households		
Survey period	The period between July and December every year		
Weighting	The weighting applied in consideration of the sampling probability of the population		
Survey items used for this analysis	(1) Area standard	(2) Bedroom standard	(3) Facility standard
	☐ Number of household members ☐ Exclusive residential area	☐ Household composition ☐ Number of bedrooms (including areas for living rooms)	☐ Private walk-in kitchen ☐ Private flush toilet ☐ Private bathroom ☐ Water supply and drainage facility

4.1.2. Analysis Results

The current status analysis of households below the minimum housing standards from 2006 to 2021 is as follows (Figure 2). The 2022 data were excluded since the Korea Housing Survey result has not yet been released.

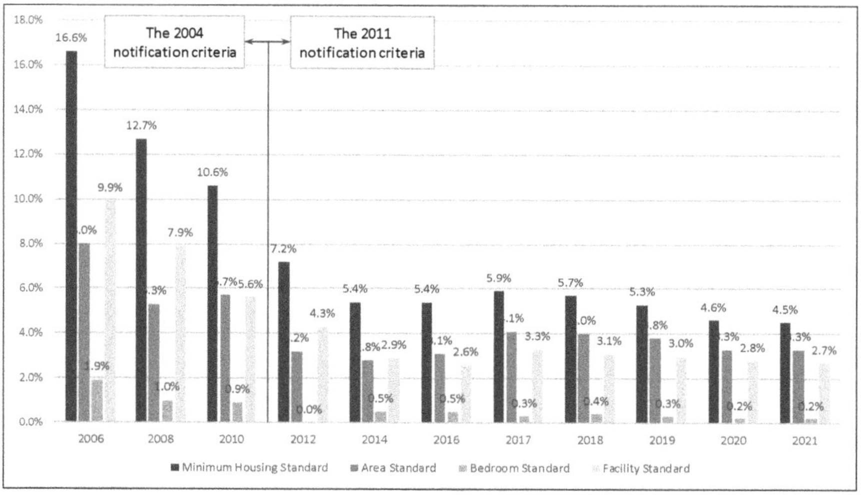

Figure 2. Status of households below the minimum housing standards.

(1) Households below the minimum housing standards:

The overall trend is downward, from 16.6% in 2006 to 4.5% in 2021. Since the trend was around 5% in 2014, it remained between 4% and 5% until 2021. Considering that the minimum housing standard was upgraded once in 2011, the rate of households below the standards since 2012 seems somewhat lower. This trend can be attributed to the following.

First, the average number of household members continuously decreased (2.7 persons in 2010 to 2.3 persons in 2021) [30]. In Korea, the main reasons for changes in family structures include nuclearization of families, children migrating for education, and spouses moving for employment opportunities. Additionally, a shift away from the traditional practice of living with elderly parents, and the trend of elderly parents residing in various facilities such as senior towns, nursing homes, and healthcare facilities, can also be considered contributing factors.

Second, only the area standard increased when the 2011 standard was upgraded (refer to Table 3). Korea possesses advanced construction technology, and the relatively high real estate values have resulted in a relatively short building lifespan that often leads to redevelopment. As a result, the quality of facilities has continuously improved due to the ongoing replacement of existing housing and redevelopment projects.

Third, the area standard of one-person households increased very limitedly, whereas the ratio of one-person households continuously increased (23.9% in 2010 to 33.4% in 2021) [31]. As mentioned earlier, Korea has witnessed a rapid and significant increase in the proportion of one-person households over a short period, driven by factors such as education, employment, the pursuit of independent livelihood by the elderly, and divorce.

Fourth, there were effects of residential environment improvements in line with the higher number of housing losses. The number of housing losses rapidly increased from 62,485 in 2010 to 99,321 in 2015 and 146,396 in 2021 [32]. The sharp increase in the number of housing losses by more than double in just 11 years is related to the real estate market. Korea has experienced a robust real estate market over the past decade, and large-scale housing supply initiatives have been pursued by private construction companies that could secure profitability.

(2) Households below the area and facility standards:

The ratio of households below the area and facility standards generally showed a decreasing trend. The households below the area standard decreased from 8.0% in 2006 to 3.3% in 2021. The households below the facility standard dropped from 9.9%

in 2006 to 2.7% in 2021. In 2021, the ratio of people residing in non-housing, including accommodations for students studying for exams, and compact buildings with compact rooms was approximately 1.7% [33]. Considering that, the ratio of households below the area and facility standards is likely to remain stuck in the range of 2–3% unless non-housing is fundamentally eliminated.

(3) Households below the bedroom standards:

The ratio of households that cannot reach bedroom standards was also downward in general. The rate decreased from 1.9% in 2006 to 0.2% in 2021, implying that most households satisfied the bedroom standard. This trend can be attributed to the following. First, since the minimum housing standards were first announced in 2004, the 2006 Korea Housing Survey, which was conducted for the first time, showed a low rate of households below the standard (1.9%); the standard was low from the beginning. Second, when the 2011 standards were upgraded, the bedroom standard was not changed. Third, living rooms that can be used as bedrooms were included as bedrooms. Compared to Western countries where bedrooms and living rooms are relatively separated, there are frequent cases where living rooms are used as bedrooms in Korea. Fourth, the average number of household members decreased, and the number of required bedrooms dropped [30].

4.1.3. Direction for Improving Minimum Housing Standards Based on the Current Status Analysis of Households below the Standards

First, considering the fact that the rate of households below the minimum housing standards dropped from 16.6% in 2006 to 4.5% in 2021, it is necessary to upgrade minimum housing standards in general. There are two contributors to the lower rate. First, the residential environment has been improved by housing losses and the supply of new housing. The average number of household members continuously decreased as well. As standards can change in each era, it is necessary to upgrade standards in line with the current situation, since South Korea can contemplate more on the residential quality such as environment and safety, rather than the quantity of housing.

Second, it is also necessary to upgrade area and facility standards. As mentioned earlier, unless non-housing types such as accommodations for students studying for exams, and compact buildings with compact rooms are removed, the rates of households below area and facility standards cannot significantly decrease, which already dropped to those rates of 3.3% and 2.7%, respectively. As for the facility standards, there should be more new standards, such as the installation of Western-style toilets, utilization of hot water in bathrooms, and security of parking lots.

Third, it is necessary to upgrade or abolish the bedroom standard. The rationale for upgrading the standard is as follows. The ratio of households below the bedroom standard was 0.2% in 2021, meaning that most households satisfied the standard. When setting the standard in 2004 and 2011, the focus was only on spaces for sleeping. Therefore, when there were a married couple and young children, sleeping together was the default. However, a bedroom is a space for sleeping, storage, and activities. One child requires some space for storage and activities, which is not smaller than a space for one adult. Therefore, it is necessary to redefine the criteria based on a bedroom per person.

The rationale for abolition is as follows: Korea had a traditional household composition in the past. Until the 1980s, Korea had a large family system with three generations living together, which could be categorized as "grandparents–parents–children". After the 1990s, as urbanization came into full swing, the households of "grandparents" who stayed in rural areas and "parents–children" who moved to cities gradually began to be divided. Here, the households of "children", who are independent of their parents due to marriage, employment, and study, began to become common. In this regard, the minimum housing standards, which were established in 2004 and 2011, respectively, included three types of standard household compositions.

However, the current Korean household composition is very different from the past. First, the ratios of one- and two-person households have increased. In particular, as for the two-person households, diverse types of households have been found, including married couples, "siblings", "lovers", and "friends". The ratio of three-generation households in which grandparents, parents, and children live together has become so low that it is hard to find such cases. The traditional type of two-generation households has been more diverse: "Married couple–children", "grandparents–married couple", and "grandparents–children". Therefore, the standard household composition cannot become "standard". If it cannot have the feature of being "standard", it may be better to exclude it from new housing standards.

4.2. Overseas Case Studies

The current status of housing standards in the United Kingdom and Japan, which have similar systems to Korea's housing standards, was analyzed. In particular, the Japan case was analyzed in more detail, as this country demonstrated a similar configuration of bedroom number and area, which are sub-standards of minimum housing standards.

4.2.1. Housing Standards in the UK

The representative housing standard in the UK is the Decent Home Standard, which was established by ex-Prime Minister Tony Blair in 2000 [34]. The most recently revised standard was announced in 2016 and is still in use today. The UK government has applied this standard to council houses, encouraging council house providers to provide houses while meeting or exceeding the Decent Home Standard. The standards can be divided into one area standard and four performance standards, and the area standard is as follows (Table 5).

Table 5. Minimum residential area of the UK (unit: m^2).

Number of Bedrooms	Persons	1-Storey Dwellings	2-Storey Dwellings	3-Storey Dwellings	Built-in Storage
1b	1p	39(37) [1,2]	-	-	1.0
	2p	50	58	-	1.5
2b	3p	61	70	-	2.0
	4p	70	79	-	
3b	4p	74	84	90	2.5
	5p	86	93	99	
	6p	95	102	108	
4b	5p	90	97	103	3.0
	6p	99	106	112	
	7p	108	115	121	
	8p	117	124	130	
5b	6p	103	110	116	3.5
	7p	112	119	125	
	8p	121	128	134	
6b	7p	116	123	129	4.0
	8p	125	132	138	

[1] Where a 1b1p has a shower room instead of a bathroom, the floor area may be reduced from 39 m^2 to 37 m^2, as shown in brackets. [2] "1b1p" means one person living in a one-bedroom house.

The Technical Housing Standards, released in May 2016, indicate the area standard [35]. This is an area standard for the interior of the house and is a standard for the supply of new housing.

Houses satisfying both area and performance standards are categorized as good housing, and if they do not meet the standards, owners or landlords must make immediate improvements. If they do not implement that, the government can take action against them, including compulsory eviction and the inability to rent houses out.

In November 2020, the UK government announced that it would completely revise the Decent Home Standard, which has been used for the last two decades [36]. A draft was completed in the fall of 2021, and stakeholder consultation was finished in the fall of 2022 [36]. It is expected that the new Decent Home Standard will be related soon.

4.2.2. Housing Standards in Japan

Japan's minimum housing standards are highly similar in format to Korea's minimum housing standards, due to the composition and area of rooms according to the number of household members. Japan first introduced the concept of a minimum housing standard in 1976, while implementing the Third Five-Year Housing Construction Plan, aiming to provide housing services above the minimum standard along with a quantitative supply of housing. There is a difference between Japan's and the UK's housing standards: in the former, the minimum housing area for having a healthy and cultural residential life was set in relation to the number of household members.

(1) 1976~2005

The areas based on the minimum housing standards, which were applied from 1976 to 2005, are as follows (Table 6) [37]. The estimation of residential areas was based on architectural planning and design, targeting one to seven households. Japan's standards had a significant difference from Korea's as it did not utilize a standard household composition; the former also had a distinction from the UK's as there was no compulsion on households below the standards, and the standards were used as a reference for the analysis of housing conditions and policies.

Table 6. Minimum housing standard of Japan (1976~2005) [37].

Number of Household Members	Space Requirement	Living Area (Room + Living)	Residential Area	Total Area
1	1 K	7.5 m^2	16 m^2	21 m^2
2	1 DK	17.5 m^2	29 m^2	36 m^2
3	2 DK	25.0 m^2	39 m^2	47 m^2
4	3 DK	32.5 m^2	50 m^2	59 m^2
5	3 DK	37.5 m^2	56 m^2	65 m^2
6	4 DK	45.0 m^2	66 m^2	76 m^2
7	5 DK	52.5 m^2	76 m^2	87 m^2

(2) 2006–Current

In 2006, Japan revised its existing minimum housing standards alongside the enactment of the Basic Law of Housing-Life (Table 7) [38]. While the standards did not undergo significant increases, the criteria for calculating the living space became more detailed. Previously, the standards considered area requirements for bedrooms, facilities, and storage spaces based on the composition of household members, aggregating the required areas for each component to derive the overall area standard. In contrast, the revised version simplifies the minimum housing standards by defining them as the area needed per individual.

Table 7. Minimum residential area standard of Japan (2006~current) [38].

Number of Household Members	Residential Area	
1	25 m^2	(1) single: 25 m^2 (2) more than 2: 10 m^2 × person + 10 m^2 (3) when calculating the number of household members ☐ Below 3 years old: 0.25 person ☐ 3 years old and above but under 6: 0.5 person ☐ 6 years old and above but under 10: 0.75 person ☐ If the calculated number of household members is less than two, it is counted as two.
2	30 m^2	
3	40 m^2	
4	50 m^2	

During this period in Japan, a housing type where multiple households share a toilet and living room spread in Japan. The concept of a "standard" household composition was also discarded, and the diversity of household compositions was acknowledged. For example, in the case of a four-person household, the previous standards defined it as "parents + 2 children", whereas the revised version eliminates the notion of a standard household composition. In the revised standards, the variables for determining the area are only "number of household members" and "age". The name of the "minimum housing standard" was changed to the "level of minimum living floor area".

The level of minimum living floor area enabled estimating necessary areas per functions such as sleeping and eating. First of all, the sleeping conditions of family members were set, and the combination of public and private spaces, sanitary spaces, and storage spaces was composed to calculate the size of the main spaces. After summing the areas up, the final residential areas were calculated through the composition of the flow of human traffic and empty room spaces. Considering the population density in Japan, as it is difficult to adopt the "one room per person" standard, the value of the number of household members is corrected in line with the age of household members.

The Japanese Ministry of Land, Infrastructure, Transport and Tourism (corresponding to Korea's Ministry of Land, Infrastructure, and Transport) analyzed households below the minimum residential standards through the "Land and Housing Survey" which is implemented every five years, and the outcomes are reflected in housing policies, and utilized as the policy basis for setting the level of minimum living floor area.

4.2.3. Direction for Improving Minimum Housing Standards Based on Analysis Results of Overseas Case Studies

The UK and Japan cases were examined as these countries have similar or better conditions in terms of country size, population density, state earnings (the per capita gross national income), and urbanization rates (Table 8). As a result of examining the housing standards of the UK and Japan, the following implications were obtained.

Table 8. Comparison of minimum housing standards (area) of UK, Japan, and South Korea.

Country		UK		Japan		South Korea	
Country size (2021)	Area	243,610 km^2		377,970 km^2		100,410 km^2	
	Population	67,326,569		125,681,593		51,744,876	
Population density (2021)		276.4/km^2		332.5/km^2		515.3/km^2	
The per capita gross national income (2021) [39]		45,380$		42,620$		34,980$	
Urbanization rates (2023) [1] [40]		84.6%		92.0%		81.5%	
Housing standard (Area standard)		**Number of Household Members**	**Residential Area (m^2)**	**Number of Household Members**	**Residential Area (m^2)**	**Number of Household Members**	**Residential Area (m^2)**
		1	40.0	1	25.0	1	14.0
		2	51.5	2	30.0	2	26.0
		3	63.0	3	40.0	3	36.0
		4	72.0	4	50.0	4	43.0
		5	88.5	5	60.0	5	46.0
		6	97.5	6	70.0	6	55.0

[1] Urbanization rates: Percentage of population dwelling in a city among a country's total population.

First, South Korea and Japan showed passive approaches, as they utilized the minimum housing standards as the design criteria for policy indicators or public housing supply. On the other hand, the UK employed the standards as an active guideline for imposing penalties on households below the standard.

Second, South Korea showed more detailed standards of housing structure, performance, and environments than the UK. The UK utilized the minimum housing standards in terms of housing management, rather than housing supply, by providing highly specific guidelines on the remodeling of houses.

Third, Japan was utilizing the most similar system to South Korea's, as it suggested residential areas per the number of household members. However, Japan's area standard is higher than South Korea's because it is assumed that the former implemented the minimum housing standard 28 years before Korea. Japan has used the 2006 revised minimum housing standard, and it seems that there would not be many changes in the area standard. Based on Japan's case, it is likely that South Korea would not change or delete the area standard after one or two revisions. Simultaneously, there would be more housing management than housing supply, that is, enhanced criteria in terms of quality.

Additionally, the area standard of South Korea is much lower than that of the UK. Although body size, gross national income, population density, and residential life culture should be considered, it is obvious that the standard is still low. In other words, South Korea's area standard should be upgraded after conducting a comparative study of overseas cases.

4.3. Collection of Expert Opinions

4.3.1. Summary of Expert Opinion Collection

The direction for specific criteria per sector (area, bedroom, facility, and location standards) was set through theoretical consideration, literature review, the analysis of the current status of households below the minimum housing standards in South Korea, and overseas case studies.

In this section, we attempted to set the direction for the parts that could not be determined earlier, by collecting experts in each field (Table 9). We targeted 22 experts in four fields, visited them in person, and used a written questionnaire to collect opinions. This process proceeded from 10 March to 31 March 2023. Experts were selected if they satisfied one of the following specific criteria: (1) participants in research projects related to housing standards; (2) authors of papers related to housing standards; (3) employees of housing-related policy organizations, research institutes, and execution organizations; (4) professors in housing and architecture-related departments.

Table 9. Outline of expert opinion collection.

	Date	10–31 March 2023			
	Method	In-depth interviews with experts (visit in person or written questionnaire)			
Expert	Field	Housing Policy	Housing Welfare	Facilities and Equipment	Housing Plan
	Number of Experts	6	5	5	6
	Participant Groups	• L Research Institute • A&U Research Institute • K University • The S Institute • K Research Institute • S Metro. Government	• N Carpenter Co., Ltd. • P University • Y University • Y Life-tech Institute • L Research Institute	• K Laboratories • M Architects • U Architecture • S Research Institute • S E&C	• U Architecture • SH Corporation • GH Corporation • Y University • Y University • S Government

4.3.2. Result of Expert Opinion Collection

Questions were posed to those experts based on the direction for minimum housing standards, area standards, bedroom standards, and facility standards, and the findings are as follows:

(1) Direction for minimum housing standards

Regarding the question "Is it necessary to improve the current minimum housing standards?" the majority of experts answered as follows: "Active improvements should be followed". Experts indicated that the current standards are outdated and cannot reflect the current situations since the rate of households below the standards is low. Furthermore, compared to the past, household composition, housing trends, and lifestyles have changed significantly, and many experts highlighted that corresponding improvements are urgently needed. On the other hand, they agreed on the idea of changing the current standards, but some experts were pessimistic about the minimum housing standards because they assumed that the standards would not be needed in the future where people's residential environments would reach a certain level in general. As for similar opinions, some experts indicated that area or bedroom standards should be removed, and related indicators of housing safety and performance should be strengthened.

Regarding the question "If standards need to be improved, to what extent should they be improved?" the majority of experts answered as follows: "The standards should be improved at a developed country's level while considering the residential life culture of South Korea". Although the US cases were reviewed, its standards were not reflected. In particular, the substantial differences in area standards made them unsuitable as reference cases. We reflected on experts' opinions that there are limitations to using the US cases since the residential life culture of the US is highly different from South Korea's. There was the minority opinion: "A standard in which the rate of household below the standard can be between 5–10%, is appropriate". Although it is inappropriate to set a standard based on the rate of households below the standard, there are several opinions that such an approach may be practical considering the role of the standard—"improvement of residential environments".

(2) Area standard

Regarding the question "A total exclusive residential area according to the number of household members is the current criteria. Do you think that it is an appropriate method?" the majority of experts said that "it is appropriate". There were two minority opinions. First, some experts expressed that considering that the rate of one- and two-person households rapidly increased, the exclusive residential area for those households should be set differently. Second, other experts indicated that as the common area for dwelling also affected the quality of life, the common area for dwelling should be considered.

Regarding the question "If you agree to increase the exclusive residential area, do you think that the UK and Japan cases, which were handled in this study, are appropriate for comparison?" the majority of experts answered "Those cases are appropriate for comparison". When setting the 2004 and 2011 standards, the UK and Japan cases were utilized as reference data, which became the rationale for this study. Japan has a similar residential life culture to South Korea; the residential life culture of the UK is different from South Korea's, but the UK case was appropriate considering it showed a target that South Korea should pursue. Regarding the upgraded level, there were many opinions that a level similar to Japan, but lower than the UK, was most appropriate. There was a minority opinion that universal design considering people with disabilities and the elderly living alone, who are relatively vulnerable, should be applied in the area estimation.

As for the question "Do you think it is appropriate that five- or six-person households should have two toilets?" most experts expressed that two toilets were appropriate. There was a minority opinion that the universal design does not have to be applied to the second toilet.

(3) Bedroom standard

Regarding the question "Do you think that the current standard for the number of bedrooms is appropriate?" most experts stated that "It is meaningless", or "The standard is low". In the past large-family era, as there was a large number of household members, parents, and children, or children slept together, but in the current era of nuclear families, it is not common anymore. Since the household composition has become more diverse than in the past, several experts indicated that the concept of "standard" is not appropriate. The current status of households below the minimum housing standards objectively indicates that the current standard is low since almost all households satisfy the bedroom standard.

There was a minority opinion that "The distinction between bedroom and living room should be removed". There were two reasons. First, as many people sleep in a living room, the living room also serves as a bedroom. Second, several one- or two-person households reside in one-room houses or two-room houses without a clear distinction between the living room and bedroom, the living room sufficiently plays the role of the bedroom.

As for the question of "What do you think about the standard household composition? (Refer to Table 3)" a great number of experts stated that "The standard household composition, which is a specific criterion, should be removed". There were two reasons. First, as mentioned earlier, as there are more nuclear families than large families today, the previous standard household composition cannot be applied anymore. Second, the situation where young children sleep with their parents was set as the standard, this implies that a space for sleeping was only considered. In the current era where nurturing and education of young children are important, a space for young children is as necessary as a space for parents. There were minority opinions that it is better for opposite-sex children to have separate bedrooms regardless of their age and that a standard separating bedrooms in line with conditions is unnecessary.

(4) Facility standard

We explained the current facility standard in detail and asked experts about the facility standards that can be added. We also indicated the names of facilities, usage patterns, and types to induce more objective responses. We utilized facility-related questions from the questions of the Korea Housing Survey, which is conducted by the government every year [9]. The names of facilities used in the questions are as follows: (1) kitchen, (2) toilet, (3) bathing facility, (4) water supply facilities, (5) drainage system (septic tank), (6) heating system, (7) fuel for cooking, (8) entrance (front door), (9) fire-fighting appliances.

The facility standards that experts would like to see added are as follows, in order of frequency of mention: (1) presence and absence and types of heating system (except for briquettes, firewood, electric heaters, etc.); (2) private bathing facilities, and absence or presence of hot water; (3) private use of an entrance (front door); (4) a private flush toilet, and absence or presence of Western-style toilets.

There were minority opinions regarding safety, evacuation, noise, and waterproofing. However, as there are practical limitations in preparing specific standards for corresponding criteria, these opinions were excluded from this study.

(5) Location standard

As for the question "As the current standards do not consider the location of housing, do you think underground or rooftop houses are appropriate?" most experts said, "They are inappropriate".

We asked more questions to experts who gave the answer: "The locations of housing can be classified into above-ground house, semi-underground house, underground house, and rooftop house. Among them, which one does not reach the minimum housing standard? (Duplicate responses possible.)" Eighteen experts thought underground houses are below the standard; 14 experts thought semi-underground houses are below the standard; 10 experts regarded rooftop houses as below the standard. Experts considered safety and health issues the most important regarding underground housing, while safety issues regarding illegal structures were most frequently mentioned in terms of rooftop houses.

The minority experts expressing an opinion of "no necessity" agreed with the need, but considering the reality that it is difficult for some people to move to over-ground houses due to housing costs, a few experts emphasized that support policies for them should be prioritized over setting standards.

5. Results of Minimum Housing Standards Improvement Proposal

5.1. Direction for Improving Area and Bedroom Standards

5.1.1. Prerequisite

Consideration was given to the prerequisites for the area and bedroom standards, and the details are as follows (refer to Section 3.2).

First, according to the "Constitution", "All citizens shall have the right to a healthy and pleasant environment. The State and all citizens shall endeavor to protect the environment". Second, according to the "Framework Act on Residence", "The people have the right to live a decent residential life in a pleasant and stable dwelling environment protected against any physical or social danger, as prescribed by relevant statutes and ordinances". Third, according to the "Framework Act on Residence", "The State has the duty of guaranteeing the people's housing rights". Fourth, according to the "Housing Act", the minimum housing standard aims to provide "a minimum standard required for the people to live a decent residential life in a pleasant dwelling environment".

5.1.2. Direction for Specific Improvements

The literature review, analysis of the current status of households below the minimum housing standards, overseas case studies, and expert opinion collection were conducted to derive the direction for improving area and bedroom standards in line with the aforementioned prerequisites. As a result, the following direction for specific improvements were derived (Table 10).

Table 10. Direction for specific improvements in minimum housing standards.

		2011	2023	
			Improvement Direction	Rationale
Area standard	1-person	14 m^2	▲	a, b, c, d
	2-person	26 m^2	△	a, b, c, d
	3-person	36 m^2	△	a, b, c, d
	4-person	43 m^2	△	a, b, c, d
	5-person	46 m^2	△	a, b, c, d
	6-person	55 m^2	△	a, b, c, d
	Universal design	Absence	▲	d
Bedroom standard [1]	1-person	1 [2] K [3]	△	b, d
	2-person	1 DK [4]	▲	b, d
	3-person	2 DK	=	b
	4-person	3 DK	=	b
	5-person	3 DK	=	b
	6-person	4 DK	=	b
	Standard household composition	Presence	×	b, d
	Standard separating bedrooms	Presence	×	b, d

[1] Bedroom standard: The direct criteria for room configuration are deleted, but it is utilized as a sub-criterion for calculating the area standard. [2] Figures in bedroom standard: Number of rooms that can be utilized as bedrooms (including spaces for living room). [3] K: Kitchen; [4] DK: Dining room and kitchen. ▲: Active improvement; △: Passive improvement; =: Maintenance; ×: Deletion. a: Literature review (refer to Section 3.1); b: Analysis of the current status of households below the standards (refer to Section 4.1.3); c: Overseas case studies (refer to Section 4.2.3); d: Expert opinion collection results (refer to Section 4.3.2).

As for the area standard, opinions supporting an upgrade of the standard were found in almost all rationales. In particular, with the increase in the rate of one-person households, there was a rationale indicating the need for active improvements. Therefore, we decided to estimate the area standard by applying universal design principles that were not reflected in the 2011 minimum housing standards.

Regarding the bedroom standard, two main directions for improvements were identified. First, the decision was made to eliminate the standard itself. As household compositions have become more diverse, the concept of a "standard" has lost its relevance.

Consequently, the "standard separating bedrooms" based on the "standard household composition" has also become less meaningful. However, the "room configuration (1 K~4 DK)" was removed but retained as an auxiliary standard for calculating the area standard. Essentially, this implies that insufficient rooms based on the number of household members will not be considered as falling below the minimum housing standard. Since the number of rooms is already included in the area standard, and configurations (K, DK) are covered in the facility standard, duplications are unnecessary.

5.1.3. Design Simulation

(1) Premise

This sector aims to calculate the size of each room that satisfies the minimum housing standards. After calculating the size of each room, it becomes possible to determine the exclusive residential area that meets the minimum housing standard per the number of household members. To achieve this, we conducted a design simulation in the following sequence. The methodology was partially based on the approach by Bae et al. [41].

(2) Estimation of Furniture Sizes

Initially, we selected the furniture that should be placed in each room to fulfill the minimum housing standards (Table 11). A bedding or bed, a wardrobe (blanket chest), and a desk are placed in the bedroom. A sink table, kitchen counter, gas table, and refrigerator are positioned in the kitchen and dining room. A toilet, washstand, and shower booth are located in the bathroom. The entrance, shoe rack, and boiler rooms were regarded as other spaces. However, for one-person households, where common heating systems are often used, boiler rooms were not taken into account. In Korea, standardized sizes known as the Korea Standard are commonly adopted in the industry [42]. Based on this standard, we computed the standard size. If checking the Korea Standard poses challenges, consideration was given to the minimum unit size of commercially available products. The dimensions from the Korea Standard and commercially available items already take into account the body sizes of Koreans.

Table 11. Standard sizes per furniture item (width × length, unit: mm).

Bedroom	Bedding	Bed (Bed frame)		Closet (Blanket chest)	Desk (Chair)
	Bedding for one person: 1000 × 2100 Bedding for two persons: 1400 × 2100	Single bed: 1000 × 2100 (1100 × 2300) Double bed: 1400 × 2100 (1500 × 2300)		900 × 700	800 × 600 (800 × 1200)
Kitchen and dining room	Sink table (free space)	Kitchen counter (free space)	Gas table	Refrigerator	Table
	840 × 550 (600)	Length: 550 (600)	2 burners: 700 × 600 4 burners: 750 × 600	1 person: 600 × 600 2 persons: 800 × 700 More than 3 persons: 1000 × 900	1~2 person: 800 × 800 More than 3 persons: 1200 × 825
Bathroom	Toilet (free space)		Washstand	Shower booth	
	550 × 700 (650 × 800)		550 × 400	800 × 800	
Others	Entrance		Shoe rack	Boiler room or Utility room	
	1 person: more than 0.5 m^2 2 persons: more than 0.8 m^2 More than 3 persons: more than 1.6 m^2		1 person: 600 × 300 2 persons: 900 × 300 More than 3 persons: 1200 × 300	1 person: more than 0.7 m^2 2 persons: more than 1.0 m^2 More than 3 persons: more than 2.0 m^2	

(3) Application of universal design

Among universal design considerations, we only took into account factors that could impact the calculation of exclusive residential areas, specifically focusing on the indoor use of wheelchairs by individuals with disabilities or the elderly living alone. In such instances, the entrance door's effective width should exceed 900 mm, and a space of over 1200 mm should be available in front of and behind the entrance door [43]. Furthermore, the minimum effective width of indoor pathways, considering wheelchair accessibility to bedrooms and kitchen, has been set at 800 mm or more. In the bathroom, handrails for individuals with disabilities have been planned to facilitate use without the need for a wheelchair.

(4) Design simulation results

Criteria for furniture arrangement and sizes, as well as universal design, were previously established. With these criteria in mind, a design simulation of a housing floor plan that meets the minimum housing standards was conducted using an AutoCAD LT 2021 program. The criteria are outlined as follows:

First, we arranged the essential furniture based on a standard for each household member. Second, we arranged openings while considering universal design principles. Third, we ensured the flow of human traffic, particularly accommodating those using wheelchairs. Fourth, we allocated free spaces to cater to the activities required for daily life, such as sleeping, cooking, washing dishes, resting, cleaning, and bathroom usage. Fifth, regarding bedrooms, the principle of one room per person was applied. However, if the number of bedrooms was fewer than the number of household members, the two people per room principle was also used. Sixth, all spaces were organized using 30 cm as the minimum unit of length, reflecting the common practice in Korean housing construction sites. Seventh, the simulation used interior dimensions, consistent with how the calculation of exclusive use area is based. Eighth, the bathroom area was assumed to be equipped with a combined shower and washbasin, adhering to minimum housing standards. Finally, the proposed residential area meeting all eight criteria was identified as the minimum housing standards' residential area (Figure 3).

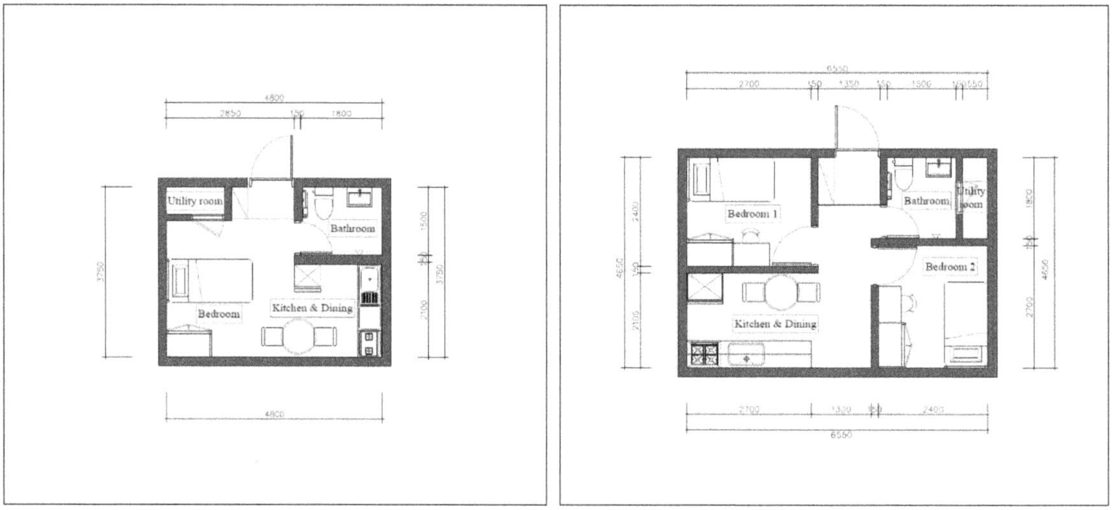

(a) (b)

Figure 3. *Cont.*

Figure 3. Design simulation of minimum housing standard per household composition (unit: mm): (**a**) 1-person (18 m^2); (**b**) 2-person (30 m^2); (**c**) 3-person (40 m^2); (**d**) 4-person (48 m^2); (**e**) 5-person (56 m^2); (**f**) 6-person (63 m^2).

5.2. Improvements in Minimum Housing Standards

5.2.1. Area Standard

The criteria in detail for calculating the area standard are as follows:

First, among the 2011 bedroom standard, the room configuration standard was utilized (1 DK~4 DK). Considering the direction of improving details in the minimum housing standards, for a one-person household, the "kitchen" was upgraded to "dining room and kitchen". This change was made since a one-person household should have a basic and minimum space for dining. Furthermore, the number of rooms for two-person households was shifted from one to two. When setting the 2011 standard, two-person households were standardized as "Married couple", and the number of rooms was set as one. However, as the composition of households became diverse, consideration was given to other types of households rather than "Married couples" and an additional room was added (refer to Table 10).

Second, the design simulation of the housing floor plan was also considered in terms of human scale. Based on these criteria, we suggest the following minimum housing standard improvements per household composition (Table 12).

Table 12. Improvements for minimum housing standards per household composition (area standard) (unit: m^2).

Number of Household Members	1	2	3	4	5	6
Spatial Composition	1 DK	2 DK	2 DK	3 DK	3 DK	4 DK
Bedroom 1	6.48 (2.4 × 2.7)	6.48 (2.4 × 2.7)	9.90 (3.0 × 3.3)	9.90 (3.0 × 3.3)	9.90 (3.0 × 3.3)	9.90 (3.0 × 3.3)
Bedroom 2	-	6.48 (2.4 × 2.7)	6.48 (2.4 × 2.7)	6.48 (2.4 × 2.7)	9.90 (3.0 × 3.3)	9.90 (3.0 × 3.3)
Bedroom 3	-	-	-	6.48 (2.4 × 2.7)	6.48 (2.4 × 2.7)	6.48 (2.4 × 2.7)
Bedroom 4	-	-	-	-	-	6.48 (2.4 × 2.7)
Kitchen & Dining	4.41 (2.1 × 2.1)	5.67 (2.1 × 2.7)	6.48 (2.4 × 2.7)	7.29 (2.7 × 2.7)	8.91 (2.7 × 3.3)	8.91 (2.7 × 3.3)
Bathroom 1	2.70 (1.5 × 1.8)	2.70 (1.5 × 1.8)	2.70 (1.5 × 1.8)	2.70 (1.5 × 1.8)	2.25 (1.5 × 1.5)	2.25 (1.5 × 1.5)
Bathroom 2	-	-	-	-	1.80 (1.2 × 1.5)	1.80 (1.2 × 1.5)
Others	4.41	9.13	14.34	15.09	16.64	16.86
Total Area	18	30	40	48	56	63

(1) As for the area of a room, the area is different by separating rooms into 1 bedroom and 2 bedrooms. (2) Total Area is rounded to zero decimal places. (3) Others: Entrance + Shoe rack + Boiler room or Utility room + Hallway inside the household + Wall inside the household.

5.2.2. Bedroom Standard

The 2011 bedroom standard is as follows:

First, the basis for the number of rooms was established: one room for one- or two-person households, two rooms for three-person households, three rooms for four- and five-person households, and four rooms for six-person households. This also included living rooms that could serve as bedrooms. However, the concept of "standard" has become less meaningful due to the increasing diversity of households compared to the past (refer to Sections 4.1.3 and 4.3.2). As a result, the criteria regarding the number of rooms were removed.

The second feature concerns room types. For one-person households, a kitchen was required, while other households needed a space that could serve as both a kitchen and a dining room. Since this criterion overlaps with the kitchen criterion in the facility standard, it was eliminated.

The third feature involves the standard for separating bedrooms. Implementing a standard for separating bedrooms presents challenges. The past standard household composition no longer fits the diverse household structures of today (refer to Sections 4.1.3 and 4.3.2). Consequently, the standard for separating bedrooms was also removed.

After a detailed analysis of bedroom standards, we decided to eliminate them altogether. Despite their removal, no issues are anticipated since they indirectly and directly affect area and facility standards. The bedroom standard was incorporated when deriving the area standard (refer to Table 12). Additionally, the previous section on households failing to meet minimum housing standards indicated that only 0.2% of households were affected, which is insignificant (refer to Section 4.1.3).

5.2.3. Facility Standard

In terms of facility standards, the United Kingdom and the United States apply facility standards termed "performance standards". However, these standards are applied as benchmarks of adequacy rather than minimum requirements.

In Korea, based on the 2011 standard, only the kitchen, toilet, bathroom, water supply, and drainage facilities were subject to the standard (refer to Section 3.2.3). Expert opinions were collected to enhance this aspect (refer to Section 4.3.2). Consequently, an overall upgrade to the standard was deemed necessary. Four facility standards for potential addition were identified while retaining existing standards, based on frequency (Table 13).

Table 13. Minimum housing standard improvements per household composition (facility standard).

Facility Standard	Usage Classification	Absence or Presence of Standard		Type Classification	Absence or Presence of Standard	
		2011	2023		2011	2023
(1) Kitchen	Private	○	○	Walk-in	○	○
(2) Toilet	Private	○	○	Flush toilet	○	○
(3) Bathroom	Private	○	○	Hot water	×	○
(4) Water supply and drainage facility	Installation	○	○	-	-	-
(5) Heating system	Installation	×	○	Fuel for heating [1]	×	○
(6) Entrance door	Private	×	○	-	-	-

[1] Heating fuels: Exclude conventional fuels such as briquettes firewood, and large electrical heaters. ○: Standard available; ×: No standard available.

First, the presence and type of the heating system were considered. While this criterion was excluded assuming that few households would not comply, an analysis of the 2011 Korea Housing Survey revealed only 0.1% of households lacked heating systems, primarily concentrated in one- to two-person households [29]. It is assumed that these households live in non-residential facilities. Nonetheless, heating systems are an indispensable requirement for housing, so they were included. Regarding heating system types, households using conventional fuels like briquettes, firewood, or large electric heaters were classified below the minimum housing standard. Large electric heaters, which might be debatable, differ from Western heating methods. Western cultures utilize electric heaters or radiators to heat spaces, whereas Korea traditionally employs floor heating. Since there are minimal space-heating facilities in Korean residences, this method is considered "temporary". Given that heating systems are essential, this criterion was included.

Second, we considered the presence or absence of hot water in the bathroom. Though this aspect is fundamental, it was omitted due to a presumed scarcity of non-compliant households. Even though the ratio of households failing to meet this standard is likely insignificant, it was incorporated, assuming these households primarily reside in non-residential facilities.

Third, the exclusive use of the entrance (front door) was added. Some housing types share a front entrance and hallway access to individual rooms, prevalent in accommodations for exam-studying students, mostly catering to one- or two-person households. This criterion pertains to family and personal privacy and safety, thus considered a fundamental right. Given its strong connection to basic human rights, it was integrated as a new standard.

Fourth, there were suggestions about Western-style toilets. Since few households lack Western-style toilets in Korea, this was omitted as a criterion, as the use of Eastern-style toilets does not pose significant inconvenience or hygiene issues.

Lastly, we deliberated on the presence or absence and type of cooking facilities. Given the reduced trend of home cooking compared to the past, this criterion was excluded.

We also considered other factors like fire-fighting appliances, structural safety, waterproofing, moisture-proofing, ventilation, natural light, noise, natural disasters, crime prevention, and sanitation. As safety and quality of life are paramount, we contemplated incorporating these criteria. However, due to the absence of objective confirmation via the current Korea Housing Survey and concerns about efficacy, we maintained the declarative essence of the 2011 minimum housing standard (refer to Section 3.2.3).

5.2.4. Location Standard

The location criterion did not exist in the past. As mentioned in the Research Background and Purpose section, there is a recent issue in Korean society about the safety of underground or semi-underground houses due to flood damage (refer to Section 1.2). In this study, we collected related opinions of experts through interviews (refer to Section 4.3.2).

(1) Underground and semi-underground houses:

First of all, underground and semi-underground houses have safety issues. In the event of a flood, if people residing in those houses are not evacuated early, it is not easy to escape on their own due to water pressure. This issue is especially found in areas with low ground levels and high land prices. The South Korean government has recognized that "underground and semi-underground houses" are inappropriate for housing, and has attempted to establish policies to provide countermeasures [10]. In addition, in the event of a fire, as there are often no emergency exits in such types of houses, people residing in such houses become more vulnerable than those living in above-ground houses.

There are also health issues. Basements are very humid and often are covered by mold. This can cause problems with respiratory diseases. Compared to above-ground houses, these types of houses are highly likely to have more cockroaches and rats, leading to a higher chance of transmitting germs to humans. As ventilation and lighting are also unfavorable, those conditions can harm the health of residents. Therefore, in this study, this criterion was set as a minimum housing standard.

(2) Rooftop houses:

Rooftop houses are also related to safety issues. "Rooftop" refers to the space at the top of a house or building. In many cases, the purpose of rooftops that were originally built for purposes other than residence are changed to residential facilities, or they are illegally expanded. This makes rooftops unsafe housing. Illegal buildings sometimes do not comply with safety-related standards, and there is no public management, which is the reason why such buildings are considered safety blind spots. Therefore, in this study, this criterion was set as a minimum housing standard.

(3) Other location standard:

As mentioned earlier, health is a critical issue directly related to housing rights. Therefore, consideration was given to whether to include location criteria related to low pollution levels in the vicinity [44]. Additionally, there was consideration regarding the incorporation of the availability of green areas (such as parks and walking trails) that can promote health in the location criteria [45]. However, as these topics fall outside the scope of this study, there is an intention to address them in more detail in future research.

5.2.5. New Minimum Housing Standards with Improvements (Synthesis)

The new minimum housing standards, which integrate the previous area standards, bedroom standards (deleted), facility standards, and location standards, are as follows (Table 14).

Table 14. Suggested minimum housing standards in 2023.

1. Area Standard		**2. Facility Standard**		**3. Location Standard**	
(1) 1-person	18 m^2	(1) Kitchen	Private and Walk-in	(1) Semi-underground house	Below the standard
(2) 2-persons	30 m^2	(2) Toilet	Private and flush toilet	(2) Underground house	Below the standard
(3) 3-persons	40 m^2	(3) Bathroom	Private and hot water	(3) Rooftop house	Below the standard
(4) 4-persons	48 m^2	(4) Water supply and drainage facility	Installation		
(5) 5-persons	56 m^2	(5) Heating system	Installation and Fuel		
(6) 6-persons	63 m^2	(6) Entrance door	Private		
4. Structure, performance, and environment standards					
(1) As a permanent building, structural strength must be secured, and materials for principal structural parts shall be heat-resistant/proof, fire-resistant, and moisture-proof.					
(2) Adequate soundproofing, ventilation, natural light, and heating facilities shall be provided.					
(3) Environmental factors such as noise, vibration, odor, and air pollution shall meet legal standards.					
(4) A house shall not be located in an area with significant risks of natural disasters such as tsunamis, floods, landslides, and cliff collapse.					
(5) It shall be equipped with safe electrical facilities, and structures and facilities for safe evacuation in case of fire.					

5.2.6. Estimation of Households below the Minimum Housing Standards Based on New Standards

We estimated new households below the minimum housing standards, based on the raw data from the 2011 Korea Housing Survey [29]. As a result, 8.4% of total households were found to fall below the minimum housing standards (Figure 4). This represents an increase of approximately 3.9% compared to the 4.5% under the previous standards. Notably, high rates were observed among six-person households (19.6%) and one-person households (15.8%). The elevated rate for six-person households could be attributed to economic challenges leading to extended family cohabitation. For one-person households, the high rate might stem from temporary situations involving work, study, or employment preparations, coupled with limited financial resources. Households below the minimum housing standards were relatively low in the case of two- to five-person households, which can be attributed to the higher proportion of independently formed families.

Regarding the area standard, 6.2% of total households were identified as falling below the standard. This marks an increase from the previous standard's 3.3%, representing a rise of approximately 2.9% under the new standard. Among the sub-standards, the area standard demonstrated the most significant impact on the minimum housing standard. While the extent of this impact might vary based on the chosen standard, it can be assumed that it reflects Korea's high population density.

Regarding the facility standard, 3.1% of total households were found to be below the standard, an increase from the 2.7% under the previous standards. This difference is attributed to the inclusion of criteria such as the presence of hot water in the bathroom, the presence/absence of heating systems and relevant fuels, as well as private entrance doors. This outcome implies that most households are already equipped with these essential facilities. Notably, one-person households displayed a notably high rate of 7.7% below the standard, primarily due to their temporary residence in places like student accommodations or compact buildings with specialized compact rooms.

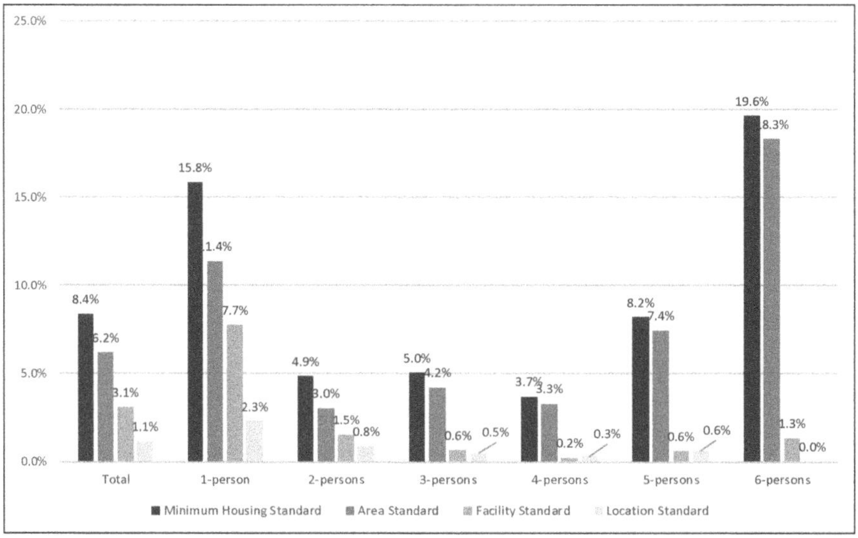

Figure 4. Households below the minimum housing standards based on new standards.

Regarding the newly proposed location standard, fewer than 1.1% of total households were found to be below the standard. This category encompasses households residing in underground, semi-underground, or rooftop houses. While constituting a small portion, such housing conditions are linked to safety and health concerns that require resolution.

Lastly, a total of 2661 households (0.013%) did not meet all three standards. These households exclusively comprised single-person households, indicative of vulnerable households in urgent need of housing improvement. Korea boasts a robust economy and a significant number of high-quality multi-unit dwellings, particularly apartments. However, the nation grapples with high population density and urbanization rates. Consequently, a relatively substantial number of households fall below the area standard, while the prevalence of households meeting facility and location standards is high.

6. Discussion

Since the enactment of the minimum housing standards in 2004, the ratio of households falling below these standards has been annually estimated through the Korea Housing Survey on a national scale. The rate of households below the minimum housing standards was 16.6% in 2006 but decreased to 4.5% in 2022, even with an update in the standards occurring in 2011.

Our study concentrated on this 4.5% rate. Even in developed countries with relatively favorable housing conditions, governments cannot fully cater to the residential needs of every individual. In this context, the 4.5% ratio could signify vulnerable segments within the population that require active government intervention to improve their housing circumstances. While this ratio may diminish assuming South Korea's continuous growth, it will eventually reach a threshold and not achieve 0%. Indeed, the percentages of households below the minimum housing standards between 2014 and 2021 have remained stagnant within the range of 4.5% and 5.9%.

Given the situation where most households, apart from those in vulnerable living conditions, meet the minimum housing standards, the standards should be elevated. In this perspective, we deliberated on the "meaning or definition of minimum housing standards" (refer to Section 5.1.1), as well as the levels of these standards. Standards are context-dependent and subject to the times and the perspectives of those defining them. Recognizing these limitations, our study aimed to create standards as objectively as possible. We derived the direction for the minimum housing standards by examining pertinent regulations, conducting research encompassing literature reviews, analysing the current state of households not meeting the minimum housing standards, examining overseas case studies, gathering expert opinions, engaging in design simulations, and estimating households not meeting the standards under the new criteria.

Furthermore, the standards were ensured to align with contemporary circumstances. First, the standard household composition was eliminated, rendered irrelevant by the diversification of household structures. For instance, the standard household composition classified a two-person household as a "married couple", whereas in this study, "siblings", "partners", "friends", and "single-parent households" can all be classified as such. Consequently, we discarded the bedroom standard predicated on the standard household composition and instead sought to supplement it with the area standard based on the number of household members.

Second, evolving societal values were incorporated. Over time, the significance of values such as human rights, safety, the environment, and quality of life has grown compared to the past. In light of this evolution, universal design principles for individuals with disabilities were integrated into the area standard. To address safety concerns, a "private entrance door" criterion was introduced within the facility standard, and "aboveground housing" was included within the location standard. As we believed the location standard played a pivotal role in safety, environmental factors, and quality of life, we introduced this new criterion in our study. To address quality of life, the area standard was elevated and living spaces were expanded, aspects intertwined with most of the other standards.

Third, we accounted for shifts in lifestyle. For one-person households, which were predominantly found in student accommodations and communal dining areas, the COVID-19 pandemic altered behavior, leading individuals to consume meals at home through delivery services. Consequently, a "dining room" criterion was introduced for one-person households. For two-person households, our previous assumption that married couples shared a bedroom was adjusted, as an increasing percentage now had separate rooms. In response, two bedrooms were assigned for two-person households. Notably, South Korea's advanced public bathing facilities like jjimjilbang (Korean dry sauna) experienced a shift during COVID-19, with more people opting to bathe at home due to infection concerns. Reflecting this change, "hot water in bathing facilities" was introduced as a new criterion. Additionally, with heightened privacy concerns, the new criterion of a "private entrance door" was introduced.

7. Conclusions

In Section 6, we examined (1) the need to introduce new minimum housing standards, (2) the assurance of objectivity, and (3) the major changes to the minimum housing standards. In this section, we will discuss the implications of this study, focusing on the potential policy applications of the newly proposed minimum housing standards.

First, these standards can serve as criteria for government housing policies. By assessing households' housing conditions, effective policies can be formulated accordingly. The minimum housing standards can be considered a highly effective tool for making precise judgments. This accurate assessment of housing conditions empowers the government to devise housing supply strategies as well as housing policies regarding supply, loss, and housing welfare programs. Housing welfare programs can use these standards as pivotal criteria for identifying eligible beneficiary households.

Second, the standards can facilitate the establishment of housing management norms. Specifically, tailored management standards can be created for households falling below the minimum housing standards. In the public realm, this can lead to the formulation of policies concerning residential mobility and housing improvements through the allocation of public resources. In the private realm, these standards can serve as voluntary management indicators. Notably, since minimum housing standards relate to fundamental rights such as safety, environment, and human rights, proactive public interventions are warranted. Immediate measures should be implemented for the 2661 households (0.013%) not meeting area, facility, and location standards, ultimately striving to improve the living conditions of 1,738,733 households (8.4%) below the minimum housing standards.

Third, the standards can be employed as criteria for the construction of public rental housing. Currently supplied public housing satisfies facility and location standards. As a result, minimum housing standards can serve as indicators for formulating area-specific design standards. Furthermore, when selecting prospective tenants for public rental housing, the number of household members intending to reside in the unit can be clearly established, with the area standard serving as a criterion.

Fourth, the standards can guide site selections for housing-redevelopment projects, housing-reconstruction projects, and urban-regeneration initiatives. These projects entail large-scale enhancements of residential environments in South Korea. Urban-regeneration projects require public funding, and housing-redevelopment and -reconstruction projects are anticipated to yield real estate gains from developmental ventures, making site selection highly competitive. Utilizing minimum housing standards for site selection can help address the issue of households below the minimum housing standards and ultimately contribute to enhancing residential environments.

Finally, the ultimate role of minimum housing standards is to "enhance living conditions". The causal link between the decreased ratio of households falling below the minimum housing standards and the formulation of these standards cannot be fully explained. The exact correlation remains indeterminate based solely on the findings of this study. However, as the government has communicated minimum housing standards and is actively establishing and implementing policies to address this concern, the standards are assumed to play a part in enhancing living conditions.

The limitations of this study and the need for further research are as follows:

First, concerning the facility standard, we omitted standards that could not be objectively verified. The Korea Housing Survey targets around 50,000 households nationwide, with most questions answered directly by householders. While subjective responses can provide insights into safety and quality of life (e.g., structural safety, waterproofing, moisture-proofing, ventilation, lighting, noise, natural disasters, crime prevention, and hygiene), survey results based on these responses may lack the validity of practical verification. Hence, this study utilized the declarative meaning of the 2011 minimum housing standard. In the future, we plan to conduct follow-up studies on relevant aspects once a survey methodology capable of objective validation is developed.

Second, we excluded the independent role of the living room. In the design simulation, each room was arranged, and the remaining space resembled a living room (refer to Figure 3). This aspect is a matter of choice rather than a limitation. The consideration of whether the living room's role is necessary in the minimum housing standards remains a topic for further research. In subsequent revisions, the living room could be categorized as an independent room.

Third, in this study, the concept of a standard household composition has been removed. This is because the standard household composition does not encompass the diversity of household compositions. However, if data allowing for a more detailed analysis of household composition distribution can be obtained, it may be possible to incorporate it into the minimum housing standards. This is a topic intended to be addressed in future research.

Lastly, a study is needed to gauge the extent to which the minimum housing standards can contribute to enhancing housing conditions. To achieve this, an analysis of government policies employing these standards is essential, uncovering correlations with improvements in housing conditions.

Author Contributions: Conceptualization, D.K.; methodology, D.K., H.S. and S.K.; software, D.K.; validation, D.K. and S.K.; analysis, D.K., H.S. and S.K.; visualization, D.K. and H.S.; supervision, S.K. All authors have read and agreed to the published version of the manuscript.

Funding: This research received no external funding.

Data Availability Statement: When citing data in the text, we have provided the address in the references section with a link.

Conflicts of Interest: The authors declare no conflict of interest.

References

1. Reyes, A.; Novoa, A.M.; Borrell, C.; Carrere, J.; Pérez, K.; Gamboa, C.; Daví, L.; Fernández, A. Living Together for a Better Life: The Impact of Cooperative Housing on Health and Quality of Life. *Buildings* **2022**, *12*, 2099. [CrossRef]
2. Rolnik, R. Late Neoliberalism: The Financialization of Homeownership and Housing Rights. *Int. J. Urban Reg. Res.* **2013**, *37*, 1058–1066. [CrossRef]
3. Wilson, S. Litigating Housing Rights in Johannesburg's Inner City: 2004–2008. *S. Afr. J. Hum. Rights* **2017**, *27*, 127–151. [CrossRef]
4. Troy, L.; Easthope, H.; Randolph, B.; Pinnegar, S. 'It depends what you mean by the term rights': Strata termination and housing rights. *Hous. Stud.* **2017**, *32*, 1–16. [CrossRef]
5. Jin, H.C.; Kang, D.W.; Lee, S.W. A Study on the Changing Factors of Minimum Housing Standard and Housing Density in Seoul Metropolitan Area, 1995–2005. *Hous. Stud. Rev.* **2010**, *18*, 41–68.
6. Xu, K.; Chai, X.; Jiang, R.; Chen, Y. Quantitative Comparison of Space Syntax in Regional Characteristics of Rural Architecture: A Study of Traditional Rural Houses in Jinhua and Quzhou, China. *Buildings* **2023**, *13*, 1507. [CrossRef]
7. Homepage of Korean Law Information Center. Available online: https://www.law.go.kr/admRulLsInfoP.do?chrClsCd=&admRulSeq=2000000055520 (accessed on 22 March 2022).
8. Homepage of Korean Law Information Center. Available online: https://www.law.go.kr/admRulLsInfoP.do?admRulSeq=2000000059613 (accessed on 22 March 2022).
9. Homepage of MOLIT Statistics System. Available online: https://stat.molit.go.kr/portal/cate/statFileView.do?hRsId=327&hFormId= (accessed on 11 July 2022).
10. Homepage of MOLIT. Available online: http://www.molit.go.kr/USR/NEWS/m_71/dtl.jsp?id=95087936 (accessed on 3 March 2023).
11. Kim, S.Y.; Kim, S.L.; Lee, J.S. The Effect on Housing on Health. *Health Soc. Sci.* **2013**, *34*, 109–133.
12. Choi, E.; Kim, Y.; Kwon, S. Spatio-Temporal Changes of Households Failing to Meet the 2011 New Minimum Housing Standard (1995–2010). *J. Korea Real Estate Anal. Assoc.* **2012**, *18*, 171–195.
13. Lim, S. The change of magnitude and characters in households below minimum housing standard (2005–2011). *Soc. Secur. Res.* **2014**, *30*, 215–244.
14. Yun, H.J. Proposal of Minimum Housing Standard for Eup-Myun Area. *J. Rural Plan.* **2011**, *17*, 13–21. [CrossRef]
15. Lee, H.S. A Study on the Minimum Housing Standards and House Rental Supplement. *Public Land Law Rev.* **2011**, *54*, 95–117.
16. Kim, J.H. Minimum Standards for Residential Accommodation (MSRA) and the Housing Welfare Policy—Focusing on the 'Housing Comprehensive Plan'. *J. Korea Plan. Assoc.* **2015**, *50*, 185–207. [CrossRef]
17. Lim, S. A Study on the Minimum Residential Standard and Guaranteeing the Housing Rights from the Perspective of Administrative Law. *Admin Law J.* **2021**, *66*, 175–204.
18. Wang, J.; Lee, C.L.; Shirowzhan, S. Macro-Impacts of Air Quality on Property Values in China—A Meta-Regression Analysis of the Literature. *Buildings* **2021**, *11*, 48. [CrossRef]
19. Wang, J.; Lee, C.L. The value of air quality in housing markets: A comparative study of housing sale and rental markets in China. *Energy Policy* **2022**, *160*, 112601. [CrossRef]
20. Boadi, E.B.; Chen, S.; Shi, G.; Li, Y.; Dodoo, R.N.A.; Mensah, F.; Dieme, I.S.M.; Appiah, R.; Mangudhla, T. The Impact of Resettlement on Quality of Life in Ghana: The Mediating Role of Residential Satisfaction. *Buildings* **2023**, *13*, 2288. [CrossRef]
21. Bangura, M.; Lee, C.L.; Schafer, B. The Unintended Consequences of COVID-19 Economic Responses on First Home Buyers? Evidence from New South Wales, Australia. *Buildings* **2023**, *13*, 1203. [CrossRef]
22. Homepage of the United Nations. Available online: https://www.un.org/en/about-us/universal-declaration-of-human-rights (accessed on 22 March 2022).
23. Homepage of Korean Law Information Center. Available online: https://www.law.go.kr/LSW/lsInfoP.do?efYd=19880225&lsiSeq=61603#0000 (accessed on 22 March 2022).
24. Homepage of Korean Law Information Center. Available online: https://www.law.go.kr/LSW/lsSc.do?section=&menuId=1&subMenuId=15&tabMenuId=81&eventGubun=060101&query=%EC%A3%BC%ED%83%9D%EB%B2%95#undefined (accessed on 22 March 2022).
25. Pane, A.; Gunawan, B.; Withaningsih, S. Development of Kampung Susun Akuarium Based on Sustainable Housing Principles. *Sustainability* **2023**, *15*, 8673. [CrossRef]
26. Kucharska-Stasiak, E.; Źróbek, S.; Żelazowski, K. European Union Housing Policy. An Attempt to Synthesize the Actions Taken. *Sustainability* **2022**, *14*, 39. [CrossRef]
27. Bacter, C.; Săveanu, S.; Buhaş, R.; Marc, C. Housing for Sustainable Societies. Children's Perception and Satisfaction with Their House in Countries around the World. *Sustainability* **2021**, *13*, 8779. [CrossRef]
28. Azarnert, L.V. Migration, congestion and growth. *Macroecon. Dyn.* **2019**, *23*, 3035–3064. [CrossRef]

29. Homepage of Micro Data Integrated Service. Available online: https://mdis.kostat.go.kr/dwnlSvc/ofrSurvSearch.do?curMenuNo=UI_POR_P9240 (accessed on 9 January 2023).
30. Homepage of Korea Indicator. Available online: https://www.index.go.kr/unify/idx-info.do?idxCd=4229 (accessed on 2 February 2023).
31. Homepage of Korea Indicator. Available online: https://www.index.go.kr/unify/idx-info.do?idxCd=5065 (accessed on 2 February 2023).
32. Homepage of Korean Statistical Information Service. Available online: https://kosis.kr/statHtml/statHtml.do?orgId=116&tblId=DT_MLTM_5416&vw_cd=MT_ZTITLE&list_id=I1_11&scrId=&seqNo=&lang_mode=ko&obj_var_id=&itm_id=&conn_path=MT_ZTITLE&path=%252FstatisticsList%252FstatisticsListIndex.do (accessed on 2 February 2023).
33. Homepage of Korean Statistical Information Service. Available online: https://kosis.kr/search/search.do?query=%EA%B0%80%EA%B5%AC%EC%88%98 (accessed on 2 February 2023).
34. Homepage of Government of U.K. Available online: https://www.gov.uk/guidance/decent-homes-standard-review (accessed on 15 April 2022).
35. Homepage of Department for Communities and Local Government. Available online: https://www.gov.uk/government/publications/technical-housing-standards-nationally-described-space-standard (accessed on 15 April 2022).
36. Homepage of Department for Communities and Local Government. Available online: https://www.gov.uk/guidance/decent-homes-standard-review#contents (accessed on 8 May 2023).
37. Homepage of Ministry of Land, Infrastructure, Transport and Tourism. Available online: https://www.mlit.go.jp/en/index.html (accessed on 2 May 2022).
38. Homepage of Ministry of Land, Infrastructure, Transport and Tourism. Available online: https://www.mlit.go.jp/common/001098415.pdf (accessed on 2 May 2022).
39. Homepage of Korean Statistical Information Service. Available online: https://kosis.kr/statHtml/statHtml.do?orgId=101&tblId=DT_2AS017&conn_path=I2 (accessed on 16 February 2023).
40. Homepage of Korean Statistical Information Service. Available online: https://kosis.kr/statHtml/statHtml.do?orgId=101&tblId=DT_2KAA204 (accessed on 16 February 2023).
41. Bae, S.S.; Chun, H.S.; Kim, M.C.; Seo, J.A. *Study on Improving Minimum Housing Standards*; Korea Research Institute for Human Settlements: Sejong, Republic of Korea, 2010.
42. Homepage of Korean Standards & Certification. Available online: https://standard.go.kr/KSCI/portalindex.do (accessed on 11 April 2023).
43. Homepage of Ministry of Government Legislation. Available online: https://www.law.go.kr/LSW/lsSc.do?dt=20201211&subMenuId=15&menuId=1&query=%EC%9E%A5%EC%95%A0%EC%9D%B8#undefined (accessed on 11 April 2023).
44. Liu, Y.; Lan, B.; Shirai, J.; Austin, E.; Yang, C.; Seto, E. Exposure to air pollution and noise from multi-modal commuting in a Chinese city. *Int. J. Environ. Res. Public Health* **2019**, *16*, 2539. [CrossRef] [PubMed]
45. Dahmann, N.; Wolch, J.; Joassart-Marcelli, P.; Reynolds, K.; Jerrett, M. The active city? Disparities in the provision of urban recreation resources. *Health Place* **2010**, *16*, 431–445. [CrossRef] [PubMed]

buildings

MDPI

Article

Housing Design: Furniture or Fixtures? Accommodating Change through Technological and Typological Innovation

Laura Daglio *, Elisabetta Ginelli and Giulia Vignati

Department ABC, Politecnico di Milano, 20133 Milano, Italy; elisabetta.ginelli@polimi.it (E.G.); giulia.vignati@polimi.it (G.V.)
* Correspondence: laura.daglio@polimi.it

Abstract: The recent global pandemic has sped up architectural research in residential design aimed at rethinking housing layouts, services, and construction methods to accommodate the changing needs of the rapidly evolving contemporary society. New typological and technological design approaches are required to address, on the one hand, the adaptability of the plan as a result of higher flexibility and temporariness in familiar and working patterns, together with a downsizing of the layouts to ensure affordability and quality of life. On the other hand, the issues of sustainability and circular economy require specific attention to interpret the resilience of the building and the reuse/recycle of the fit-out systems. The paper aims at interpreting the notion of integration between fixtures and furnishing in housing design, based on a comprehensive literature review enriched with a case study analysis that shows design concepts and approaches rooted in theories and experiences of 20th-century architecture. Principles, potentials, and barriers to the development of integrated systems are highlighted and the possible implementation of industrialised production components, the potential for modularity, flexibility, and assembly are discussed.

Keywords: micro-living; built-in furniture; prefabricated fit-out; dry assembly construction systems

Citation: Daglio, L.; Ginelli, E.; Vignati, G. Housing Design: Furniture or Fixtures? Accommodating Change through Technological and Typological Innovation. *Buildings* **2023**, *13*, 1862. https://doi.org/10.3390/buildings13071862

Academic Editors: Carlos C. Duarte, Nuno D. Cortiços, Anna Stefańska, Daniel Mateus and Carol Monticelli

Received: 17 June 2023
Revised: 15 July 2023
Accepted: 18 July 2023
Published: 22 July 2023

1. Introduction

The recent COVID-19 pandemic has added new momentum to research and experimentation in housing design already committed to responding to a dramatically increased and evolving residential demand and to new needs emerging from the cultural, societal, and economic changes of the new millennium.

The quantitative increase in the demand is the result of urban population growth, especially in metropolitan areas, and in the rise in the number of households due to societal and demographic shifts towards single-person families [1].

The recession and economic uncertainty caused by the spread of COVID-19 has accelerated the growth of the rental market leading to high fees and unaffordable living spaces, because the target has enlarged from low-mid income and young tenants to progressively include a larger percentage of the population affected by the economic crisis and forced to abandon the idea of purchasing a home [2]. The greater difficulty of access to the housing market is due to financial restrictions, increasing prices and changes in life habits, especially among the new generations. In fact, the so-called Millennials were born in the age of the sharing economy and thus conceive the house rather as a service than a social good as a consequence of the emerging duality between ownership and usage.

Following this transformation of the house tenet, higher performances are required at low management costs, within a single "package" that includes property, furniture/equipment, energy supply contracts, maintenance, and management [3]. These housing requests find a response in the private rental real estate market and also through the development of online platforms, which enhance the level of services provided and offer a "ready-to-use product", where temporariness and flexibility are integrated into the lease [4,5].

Such a transformation, together with the recent forced indoor living during the pandemic, is questioning the size, the programme [6], and the indoor/outdoor, private/public relationship of residential spaces.

In addition, fluctuating working patterns stemming from the recent developments in corporate culture paved the way, on the one hand, for a more flexible business environment leading to higher occupational mobility (on average, 3% of European workers change jobs every year [7]), and, on the other hand, for the need of homeworking facilities and spaces also sped up by the remote activities determined by the lockdown periods.

In this sense, the residential market—the Living sector reached USD 83.4 billion in investments in 2020 [8]—diversifies investment strategies to address segments of the housing demand focused on short-term rental agreements. These transformations involve the sharing of spaces and services within homes and new users such as digital nomads, city users, young professionals, off-site students, singles, and young people in general [9,10].

Accordingly, issues of housing and mainly rental affordability, as well as of appropriate size and a diverse programme, orient housing design trends towards downsizing and adaptability as the main spatial features to simultaneously respond to global cost control and quality of life [11]. In large metropolitan areas, the convergence of these factors requires a cost-effective housing response that integrates, in a small space, a high technological level based on functionality [12]. A demand is emerging linked to flexible spaces and more fluid market conditions, gauged on a temporary work or study horizon [13]. The concept of micro-living [14–20] in combination with real-time adaptation [21], transformability [22], and multifunctionality [23] may represent a possible solution requiring an investigation on both the layout organisation as well as the fit-out and the equipment of the space, to be thus developed from a typological and technological point of view.

The research on the design of space-saving housing units equipped with multipurpose built-in furniture can be traced back in the history of 20th-century architecture to provide for "internal variability of the dwelling [. . .] driven by the desire to make minimum sized apartments as tolerable and as cheap as possible" [24] or by the exploration of flexibility as the expression of the relationship between human experience and space [25]. It is a field of investigation which periodically emerges in times of urgent housing demands, especially after the World Wars [26], also according to a pragmatic [27], common-sense approach [28] and in combination with studies on prefabricated dwelling units [29].

Hence, this paper sets out to investigate this typological and technological design concept, to understand its current viability [30] and to update it in order to address contemporary issues of space and resources' consumption. In particular, the work drives from the research developed in the late 1970s in Italy [31,32] focused on the building sector industrialisation and on the efforts to merge standardisation and prefabrication of the construction system with the furniture equipment of the house to pursue quality and cost efficiency in order to respond to a high housing demand. This paper shares the belief that off-site manufacturing and industrialisation in the construction sector, boosted by recent innovations in digital technologies and new materials, represent a possible means of improved quality and cost reduction through labour and time cuts [33–37]. In addition, following the principles of circular economy to simultaneously achieve environmental and economic benefits, the research focuses on mechanical fastening and dry assembly solutions [38] to allow for the possible reuse and recycling of the building components through disassembly and reversibility [39]. The goal is to understand how such an integration between the construction system and furniture/fixtures can deliver the equipment of all-inclusive micro-housing units by a rationalised cost-effective supply chain based on principles of transformability and disassembly.

Therefore, the paper explores the following Research Questions: How can this integration be interpreted and conceived (RQ1)? Which technological elements of the construction system can provide an integration with the fit-out and furnishing of living spaces (RQ2)? What are the design problem variables, encompassing limits and opportunities to explore further developments for standardised mass-produced integrated systems for equipped ready-to-use micro-living solutions (RQ3)?

Accordingly, the proposed methodology encompasses a comprehensive literature review addressing the concept of built-in furniture, prefabricated furniture, and technological and typological strategies for housing downsizing with the aim of clarifying, deepening, and updating the notion of integration between fit-out and construction systems.

In order to answer RQ1, different design approaches emerging in different historical social and economic contexts are thus disclosed, showing how the integration tenet has been interpreted and applied throughout the recent architectural history. A new approach is then introduced arguing how it responds to current social, environmental, and industrial demands.

Secondly, addressing RQ2, a selection and analysis of case studies was carried out mainly focusing on contemporary architecture but rooted in significant examples from the past to investigate the many categories of integration based on the building classification system.

Finally, in order to answer RQ3, these categories are evaluated and discussed according to design feasibility criteria in order to highlight barriers and potentials of the possible implementation as standardised solutions for the construction industry. A problem setting methodology is applied to list the design variables involved, to provide initial guidelines for the future development of the solutions.

2. The Integration between Furniture and Architecture

In the following section, categories of interpretation of the notion of integration are listed to highlight how the cultural, societal, and economic backgrounds have influenced and determined its insurgence, especially since the early 20th century. Finally, the paper attempts to develop and propose an updated definition, resulting from the issues of the contemporary context.

2.1. A Tectonic Concept

A first interpretation of the integration can be considered as embedded in the construction principle and stemming from a strong functional and practical purpose which also includes the enhancement of indoor comfort performances. In ancient massive architecture, the thickness of loadbearing walls still allows for the inclusion of cavities to equip the interiors with storage space as well as with fireplaces and chimneys. In historic masonry buildings, the window opening allows for the inclusion of shelving and cabinets as well as alcoves to sit, giving the benefit of more direct daylighting. The medieval towers [40], which are also known to have inspired Louis Kahn for the marked organisation of his layout design, are a clear example and demonstrate how the boundaries between fitting and servant space can be blurred, especially when including building services and appliances in the equipment category.

The fireplace nooks of the Arts and Crafts houses and manors, the built-in benches, and shelves to fill in the niches around the mantelpieces expand the heating device enclosure to exploit the indoor thermal comfort and provide a family gathering place. These buildings, intended as a celebration of design and craftmanship, represent a display of integrated woodwork to include the building as well as the fitting and furniture [41].

Moreover, North American examples of the Arts and Crafts movement show how in some cases cabinets were a hinged section of the board and panelling [42], thus encompassed in the framed timber construction system. Since American millwork companies started to launch a standardised prefabricated production listing all the wooden components from doors, to colonnades, to sideboards and cabinets [42] (ibidem) together, such an integration as belonging to a general construction system clearly emerges.

Although, in the work of Frank Lloyd Wright, the functional decorative, design, and craft elements are still present to realise a "whole as an integral unit", taking inspiration from traditional residential Japanese architecture making use of integrated furniture [43], the architectural design pursues a dynamic spatial concept that introduces a different idea of integration [44].

2.2. A Spatial Concept

As the application of the innovative reinforced concrete or steel frame construction at the turn of the 20th century allowed for the realisation of the continuity of the interior layout and the open plan, the functional and residential prevailed in the structural requirements, allowing for free separation of the rooms. A second possible design concept at the basis of the integration of the equipment into architecture focuses on efficiency as well as on perception of the space in a seamless continuity between indoor and outdoor areas. In sharp contrast with the preceding excesses of fitting, stuff, and decoration, the Modernist ideal conceives "furniture as a necessity but also, perhaps, as something that gets in the way of the flow of space" [45]. The process of rationalisation of function did not eliminate furniture but simply triggered a complete redesign of the elements [46] to be placed freely and provide a humane connotation to the abstract interiors without "hindering neither the movement of the body nor the eye" [47]. "Le Corbusier, for example . . . came to view furniture as 'equipment' [responding to utilitarian needs] which was both conceived in architectural terms and integrated with the architecture" [47] (ibidem, p. 7). The notion was the result of the collaboration with Charlotte Perriand, who continued experimenting with "incorporated equipment" and "utilitarian walls", drawing inspiration from her experience in Japan [48]. Although Eileen Gray designed extensive built-in furnishings in the E. 1027 house that was never realised, it is in the tiny Badovici apartment in Paris that the concept was applied to a micro-living scenario anticipating some approaches of contemporary design to use every bit of available surfaces [49]. Richard Neutra exported and reinterpreted the Modernist approach in the United States, also including this concept of built-in equipment [50].

In addition, the concept of flexibility was explored to respond to the changing needs of the residents, both for a long-term reconfiguration and for real-time adaptation [24] (ibidem). Transformable and movable elements, from the well-known sliding partitions of Gerrit Rietveld's Schröder House (Utrecht, 1924) to Le Corbusier's Cabanon with its rich set of mobile devices equipping his micro-living design, were proposed not only to address the *existenzminimum* issue for affordable homes but also to rethink the ample residences of higher classes following canons of modernity [51].

Finally, within the idea of the *raumplan* beyond the economy of space, the built-in storage and sofa elements of Adolf Loos villas contribute to enhance the perception of the separated alcoves they create as "theatre boxes" within the fluid interior, as spatial psychological devices to provide comfort through a perception of intimacy and control [52] (pp. 5–15). In his projects for fitted furniture, conceived for efficiency and durability, practical reasons prevail in the consistency with the rationale of the whole, in evident opposition with many of his contemporaries committed to fully designing interiors.

2.3. A Design Approach Concept

Inspired by Richard Wagner's mid-19th century theory of *Gesamkunstwerk*, the notion of total design refers to a design approach extended to every detail of the architectural shell and of the fitting, aimed at providing a singular intense interior experience, an overall seamless multiscale creation. From the Arts and Crafts movement to Secession and the Deutscher Werkbund, even, in a way, to Bauhaus, the "absence of a firm distinction between the frame and the artifacts being framed is, of course, the whole point of the total work of art" [53]. In fact, as Wigley points out [53] (ibidem), such a romantic attitude encompasses both an implosive and explosive concept: the former relying on the collaboration between architects and artists to resist to the homogenising consequences of standardisation and industrialisation, the latter to embrace progressive machine-age reproduction as a means to spread the salvific powers of design.

A related idea in the continuous concept of architecture and furniture can be rooted in the application of the grid as a common design ratio to establish relationships, define principles, and generate form [54]. "The grid is not only an organizational principle and a tectonic structural device, but also entails a frame that absorbs variation, where multiplicity, heterogeneity, and diversity are resources capable of yielding singularity." [54] (ibidem, p. 2). This integration in terms of a modular, geometrical matrix to synthetise architecture and furniture into an organic whole can perhaps be recognised in Mies van der Rohe's work, from the early residential projects to the Farnsworth House [55], and it is undoubtedly linked to prefabrication and serialised modular systems in construction and equipment production.

2.4. A (Mass) Production Concept

Although the relationship between industrialised construction and prefabricated interiors has yet to be thoroughly investigated, the possibility to exploit the properties of new, especially lightweight, materials and innovative industrial fabrication technologies to mass produce off-site components and systems to respond to the requirements of the increasing housing demand in quantitative and qualitative terms informs both interiors as well as architectural design history. In fact, according to Schneiderman, "prefabrication in the built environment owes much of its advancement to concepts first investigated for use in the interior" [56]. Anticipated by the early explorations of the Modern Movement, the post-Second World War research achieved significant results in both fields, especially focusing on two main objects: the equipped wall and the service core. The first line of investigation aims at integrating room dividers and equipment in a single industrialised system exploring the potentials of modularity, flexibility, assembly, and performances of the solutions and of their joints [57]. The second stems from the innovations entailed by the introduction of an increasing set of building services (plumbing, sewing, electrical systems) and appliances in the home, which fostered dramatic changes in domestic culture [58] and from the parallel efforts into the spatial and comfort improvements of bathroom and kitchen design following hygienic and spatial efficiency issues. A third line focusing on the house as a prefabricated infrastructure implemented as a lightweight, temporary, and disposable capsule can be traced back to the first proposals for the Dymaxion House (1927) by Buckminster Fuller [59]. It was then continued and encouraged in the 1950s by the moulding potentials of plastics as a disrupting new material to revolutionise the world of interiors and architecture. Similarly, in The House of the Future (1956) and the Appliances House (1957–1958), the Smithsons explore on the one hand the "influence of furniture and household appliances on the architecture of the house" [60] and, on the other hand, the plasticity of the new synthetic materials to create a continuous curvaceous surface. However, the investigations on a single building solution to construct equipped architecture as a whole "habitat" [61] are mainly theoretical and generate provocative hypotheses such as the visions by Archigram and Radical Design.

However, following these abundant research studies and experimentations, the social context, and the economic push from the industry—including, on the one hand, component manufacturers and construction companies and, on the other hand, furniture, equipment, and appliance companies interested in expanding their markets—in the late 1970s, the perspective of a possible convergence of the two sectors towards industrialisation seemed on the horizon.

The goal of offering an all-inclusive, built, equipped, and working home releasing the future inhabitants from the burden of providing for the fitting out appeared as an interesting opportunity, especially for low-cost housing. The question of industrialisation had already led to the breaking down of the construction system to deeply analyse the performances of the built environment to address the issue of dwellings as a whole. Therefore, equipment cannot be considered simply the addition of furniture in an enclosed space. A similar path had been previously undertaken by school and health facilities designed to reduce costs and the time of production of finished ready-to-use spaces. Moreover, experimental projects culminating in the Italian "New domestic landscapes" (MOMA, 1972) could pave the way to explore the application of these ideas from something for the cultural elite to mass production [62].

2.5. The Proposed Integration: A Circular Economy Concept

The concept of integration between equipment, furniture, and construction that this paper addresses is based on the diverse spatial and technological notions previously drafted to understand whether the different systems of the building can simultaneously provide for the organisation, equipment, and comfort of residential interiors. It is on the one hand grounded in the breakdown of the building system in different levels to adapt to change. The initial Infill and Support separation [63] developed and implemented afterwards as Open Building theory [64] experimented over the years with the involvement of inhabitants in the fit-out and reorganisation stages through technological and design innovation. The Shearing layers [65] have different levels of longevity of its systems to enable the increase in the circularity potential from the start and for an easy conversion and adaptable reuse over time, thus boosting the resilience of the building. In both cases, a special design approach to allow for the physical and technological separation of the layers/levels is necessary focusing on the interface (the joints between two different building systems) and detailing.

On the other hand, it draws on the assumption that industrialised lightweight dry assembly construction can respond to the same principles of circular economy in terms of construction waste reduction, recycle of building components and materials, disassembly, and life cycle control, in addition to the time, economic, and environmental costs decreasing and having the quality enhancement typical of prefabrication [66]. This technological integration in the different subsystems of construction in combination with the spatial and typological concepts of transformability, multifunctionality, and downsizing can additionally respond to the needs of the contemporary housing demand while simultaneously pursuing sustainability goals [67].

Moreover, as a possible implementation, the easy disassembly and reuse of such systems, in line with the aforementioned concept of house as a service, can allow for the delivery of the integrated fitting-out solutions within a product service system delivery, to be temporarily rented by the real estate investor/owner of the asset and afterwards returned to the manufacturer to be further let.

3. A Phenomenology of Integration

In this section, the paper, mainly focusing on the building solutions, aims at understanding which technical elements of the construction system can be revisited to allow for the integration with furniture/equipment and thus achieve the technological and spatial multifunctionality pursued by the research study. The study drafts a possible taxonomy of the integration between the equipment and the technical elements of the construction system through the selection of contemporary case studies and projects. The collection includes prototypes, exhibition installations, micro-living examples (smaller than 40 sqm), and larger flats searched through a literature review and websites. A total of 50 contemporary case studies were catalogued, built between 1980 and 2023 (Table 1), which adds to a selection of more than 40 historic examples. Although the case studies are mainly customised designed solutions, the understanding of their design has contributed to highlight principles and suggestions for a possible future implementation into components of industrialised production to pursue the reduction in environmental and economic costs.

The selection criteria considered the following aspects:

- Multifunctionality to allow for the downsizing of the home.
- Space enhancement, exploiting the three dimensions to identify space-saving solutions, and integrating furniture in the horizontal/vertical shell of the room.
- Transformability, thanks to mobile devices, which can instantly change the character and perception of the interior environment depending on the position.

3.1. The Integrated Structure

The structural system might host cavities and spaces to integrate storage/equipment and to increase the liveability of the domestic space in terms of use as well as to improve the thermal performances of the enclosure. In "The Box House", a 20 sqm cabin designed by Ralph Erskine in the 1940s (Figure 1), the north loadbearing façade is expanded to accommodate stacked firewood on the external surface and storage on the internal one [68]. This contributes to improve the thermal insulation of the wall.

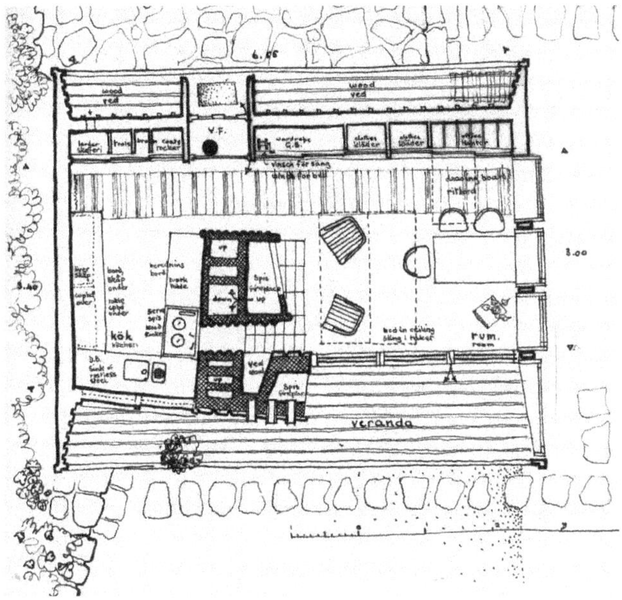

Figure 1. The integrated structure. The Box House, by Ralph Erskine, Lissima, Sweden 1941–1942 (https://fraserfolio.ca/the-box) (accessed on 10 July 2023).

In his holiday house built in 1945, Jean Prouvé [69,70] experiments with lightweight standardised components prefabricated in his workshop to build a framed wall structure where the aluminium battens are used to support cabinets and shelves. New materials and technological transfer from the nautical industry are at the base of the dry assembly easily dismountable building.

In the Shigeru Ban's Nine Grid Square House (1997), the open plan, divided by sliding partitions, is furnished by storage spaces enclosed between the steel studs of the loadbearing wall [71].

The concept can be related to a tectonic principle developing the spatial potentials of filigree construction which can be easily conceived with industrialised solutions taking direct inspiration from the Prouvé home; however, it is mainly employed for low-rise construction.

3.2. The Integrated Envelope/Window

The idea to replace the infill wall with a system of ancillary spaces/equipment finds an outstanding example in the "Domus Demain" project (1984) by Yves Lyon and François Leclerq [72], where a transparent and translucent strip replaces the façade and encompasses the kitchen, bathroom, and furniture components (Figure 2). Like the external loadbearing wall, the equipped envelope is expanded and can also be used to improve the insulation of the external wall.

Figure 2. The integrated envelope/window. The Domus Demain, by Yves Lyon and François Leclerq, 1984 (http://hiddenarchitecture.net/domus-demain/) (accessed on 10 July 2023).

In the "Tempe à pailla" villa (1932), Eileen Gray [73] anticipated the "Finestra attrezzata" (equipped window) by Giò Ponti, where the opening becomes an opportunity to play with light and transparency to enrich shelving and store space with a visual backlit effect. In fact, in Gray's retreat, the equipment is simply juxtaposed to the window, whereas in the "Alloggio Uniambientale", developed for the X Triennale di Milano (Milan, 1954) by Gio Ponti [74], the shrinkage of the liveable surface is balanced by the amount of natural light and continuity with the outdoor environment thanks to the glazed façade integrated with furniture and paintings fixed on and in-between windows, creating the impression of a transparent wall.

An analogous concept can be recognised In the "Weekend House Straume" (Knut Hjeltnes, Alesund, 2016) [75], where the kitchen cabinets are integrated in the ribbon window enclosed between two layers of fixed transparent and internal translucent sliding glass.

A different approach towards the integration mainly focused on transformability relies on the presence/addition of volumes colonising the façade to host sliding furniture/storage space that disappears from the interiors, thus in turn changing its function.

This is the case of the "Penthouse T.O" of 18 sqm by Pool Architektur ZT GmbH [76] (p. 62), where the bed, desk, and wardrobe can vanish in the thickness of the wall, thanks to a system of niches added to the envelope [77]. Similarly, in the installation project "Azioni a scomparsa" Giovanni Lauda created for the exhibition "Italia e Giappone. Design come stile di vita" (Japan, 2001) [76] (ibidem, p. 152), a central neutral space is enclosed between two 2 m thick walls from which mobile furniture can be extracted and the kitchen and bathroom are located.

This augmented notion of the building envelope can offer interesting applications, especially for the renovation of the existing buildings with a stress on energy upgrades. The possibility to replace the windows with equipped solutions as well as the infill façade or dress it with a prefabricated system that can integrate furniture, insulation, and a new vertical distribution of the building services can open up possible new production sectors to work on the built environment.

3.3. The Integrated Raised Floor

Inspired by the higher proximity to the floor areas in domestic traditional culture, interpreting the floor not as a continuous plain surface but as a system of articulated varied levels to accommodate furniture can become an opportunity to rethink the standardised technologies of the raised floor for an advanced use. In fact, such a tradition emerges in the "Newton House", by Yukawa design lab (Tokyo, 2016) [78], or in the "Tatsumi apartment" house by Hiroyuki Ito architects (Tokyo 2016) [79], organised on different levels employed as sitting areas, desks, and storage space [80]. The multifunctional (home/office) space,

"Womb" (an acronym for work, office, home, base), is designed by Johnson Chou [76] (ibidem, p. 20) to be four rooms in one, offering different automated disappearing/pivoting devices including the bed that can be sunk into a raised platform. In the "Apartment in Tessaloniki" (Figure 3) by 27 Architects (Greece, 2009) through the creation of recesses and extrusions in the floor different activities can take place (sitting, circulation, library) [81].

Figure 3. The integrated raised floor. Apartment in Thessaloniki, by 27 Architects, Thessaloniki, Greece, 2009 (https://www.archdaily.com/52530/apartment-in-thessaloniki-27-architects) (accessed on 10 July 2023).

In the "Suitcase house" (2001) by Gary Chang [77] (ibidem), the floor conceals inhabitable spaces that are accessible thanks to pneumatically assisted floor panels that completely liberate the interiors of furniture and activate the open plan room.

The project "Occultamento" by Ugo La Pietra (1974) [82] questions the codified fit-out middle-class culture to propose a new integrated system intended for social housing programmes and thus targeting the downsizing issue, including the raised usable floor in the rethinking strategy. In the "Optibo" proposal, a 25 sqm accommodation, White Design explore dining area equipment that vanish underneath the wooden floor, thanks to a mechanical system, to cater for a real-time change of function [76,83] (ibidem, p. 186).

The development of this concept for a standardised production raises questions of fruition within a design-for-all concept and of the related compliance with building regulations, including minimum ceiling height, especially in building renovations. However, the combined integration of the horizontal distribution of the building services' networks can offer interesting suggestions both for new construction as well as for refurbishment projects.

3.4. *The Integrated False Ceiling*

Since the 1909 patent of the disappearing bed by James Rountree, the integration of furniture and equipment in the ceiling is already offered in the market as a space-saving solution with different types of manual or automated mechanisms for retractable beds and tables/desks more or less concealed in the false ceiling and facing the issue of the presence of the suspension ties in the room space. The American manufacturer Bumblebee [84] has already targeted the market of temporary micro-living apartments delivering sophisticated high-end products, competing with other minor local companies.

However, some recent projects can suggest a further development of this idea of integration focusing on the coincidence between the false ceiling grid and a system of rails to suspend and slide equipment/furniture.

The "Circuit Box" by Studio X Design Group [76] (ibidem, p. 36) is composed of a series of rings equipped with accessories compacted in box in the wall and hanging to a rail system along which they slide passing one through the other. Similarly, in the

"AMRA7 apartment", by Piratininga Arquitetos Associados + Bruno Rossi Arquitetos (São Paulo, 2019) [84], a shelf system of modular steel panels, sometimes filled with wood cabinets, moves along rails to reconfigure and dynamise the spaces, creating different living environments.

In the perspective of the implementation of an industrialised system, an equipped modular grid possibly providing an electrical connection can offer a diverse set of applications from the sliding (equipped) or rolling partitions to the inclusion of suspended plug-in appliances, albeit to be thoroughly investigated and integrated with the structure of the flooring system above.

3.5. The Integrated Partition

Since the initial proposals by Modern architects, the integrated partition has represented a field for innumerable explorations, although it is still neglected in contemporary culture [85]. It includes examples of storage/equipped interior walls that, when separating different rooms, are faced with the issue of providing acoustic insulation. The case studies also show different sliding and pivoting solutions to allow for instantaneous transformability, to realise the multifunctionality of interiors, extracting or hiding support surfaces or containers, and modifying the spatial layout by opening or closing compartments between the rooms.

The market offers a set of diverse solutions mainly intended for office or health facilities in spite of the huge amount of research and experimentations not only for the application of affordable and public housing but also to definitely replace traditional partitions in housing design for a new modern style of living. The "Nelson's StorageWall" [86] is an example of this concept together with the sophisticated research by Angelo Mangiarotti, culminating in the Cub8 equipped wall [87]. The design challenge is represented by the development of modular systems providing for acoustical performances, diversity of equipment integrations, and ranges of transformability. The building services integration as for the office products should be included, perhaps encompassing in addition the mechanical and plumbing/sewage systems. However, issues of customisation concerning decoration and materials emerge as well as rooted domestic cultures.

3.6. The Service Cores

In contrast to the previous categories, service cores are not embedded in the shells that enclose the rooms but rather entail a different relationship with the spaces they are located in. These units are "elements that are created in their entirety as single all-inclusive pieces" [38] (ibidem, p.21). Mainly intended as prefabricated bathroom or kitchen units, these products are already available on the market and are delivered in different materials and finishes.

Recent projects show efforts in terms of transformability such as the "Wildbrook" intervention, which proposes a kitchen block with a variable (rotating) position, bound in a point of the floor to the building systems [76] (ibidem, p. 145).

However, an interesting approach towards the development of industrialised integrated systems also aimed at micro-living solutions can be that of expanding the concept of all-inclusive pieces to include a diverse set of fixtures and fittings to equip a neutral space. This unit can also be moved around the house to provide a specific quality and function to the areas of the room. The concept drives from the "Spazio abitabile" (Figure 4) proposal by Bruno Munari that [88], addressing the economy of space, gathers all the furniture of the house in one single container or, also, although more aimed at investigating the modern living culture for higher classes, the Ettore Sottsass or Joe Colombo projects portrayed for the Emilio Ambasz curated Moma exhibition "Italy. The new domestic landscape" (1972) [62] (ibidem).

Figure 4. The service core. Spazio abitabile, by Bruno Munari. (https://www.lombardiabeniculturali.it/fotografie/schede/IMM-3u040-0002975/) (accessed on 10 July 2023).

Still, anticipated by Bruno Munari's "Abitacolo" for kids and by other projects, the idea of also using the roof of such cores to increase the available space can offer some inspiration. An example, which exploits the position of the bathroom and the height of the rooms to create a mezzanine space, is the project "Abitazione per una giovane", a house of 17 sqm by Cini Boeri in 1978 [89]. Also, in the "Naked House" by Shigeru Ban [71] (ibidem), the surface otherwise occupied by three-dimensional mobile elements is restored on their roof.

Accordingly, the possibility to explore a modular prefabricated system that can either integrate the ancillary servant spaces or just the furniture with transformable devices appears as another path for addressing downsizing and multifunctionality in a micro-living accommodation.

Table 1. The list of case studies analysed classified according to category of integration.

N	Project Denomination	Architect	Location, Year	Surface Sqm.	Reference	Category of Integration
1.	Domus Demain	Yves Lyon and François Leclerq	1984	70 sqm	[72] (ibidem)	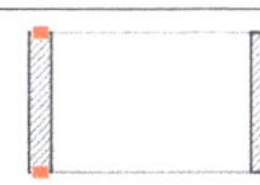
2.	Casa Insinga	Umberto Riva	Milan, Italy, 1989	70 sqm	[90]	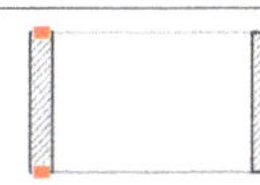
3.	Built-in furniture Rothman Apartment	John Pawson	London, UK, 1990	-	[91] (ibidem, p. 26)	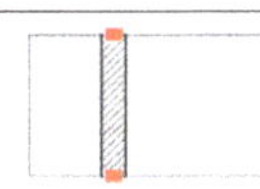
4.	Void Space/Hinged Space Housing	Steven Holl Architects, Hideaki Ariizumi, Pier Luigi Copat	Fukuoka, Japan, 1991	45 sqm	[92] (ibidem)	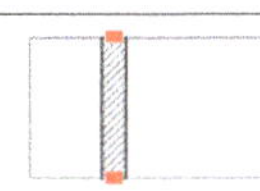
5.	Vinyl Milford	Allan Wexler	Katonah, New York, USA, 1994	12 sqm	[76] (ibidem, p. 158)	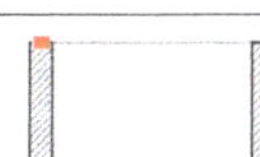

Table 1. *Cont.*

N	Project Denomination	Architect	Location, Year	Surface Sqm.	Reference	Category of Integration
6.	Apartment in London	Guard Tillman Pollock	London, UK, 1996	70 sqm	[93]	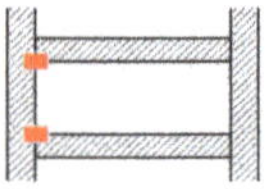
7.	Furniture House	Shigeru Ban Associates	Lake Yamanaka, Japan, 1996	60 sqm	[71] (ibidem)	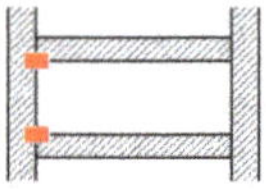
8.	Nine Grid Square House	Shigeru Ban Associates	Hadano City, Japan, 1997	60 sqm	[71] (ibidem)	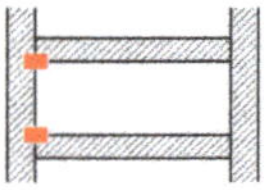
9.	The Montecarlo Apartment	Lazzarini & Pickering	Montecarlo, France, 1997	38 sqm	[91] (ibidem, p. 292)	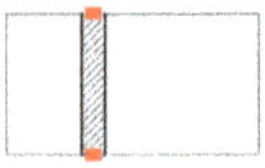
10.	Petit maison du weekend	Patkau architects	Columbus, Ohio, USA,1998	8 sqm	[94] (ibidem)	
11.	Portable construction training centre	Office for mobile design	Los Angeles, California, USA, 1999	30 sqm	[95] (ibidem)	
12.	Penthouse T.O.	pool Architektur ZT GmbH	Wien, Austria, 1999	18 sqm	[76] (ibidem, p. 62)	
13.	Sarphatistrat Offices	Steven Holl Architects	Amsterdam, The Netherlands, 2000	-	[92] (ibidem)	
14.	Villa les Roses	Fabienne Couvert & Guillaume Terver	Aix-en-Provence, France,2000	60 sqm	[76] (ibidem)	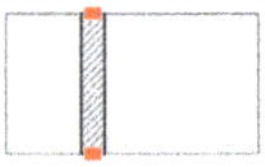
15.	Azioni a scomparsa	Dante Donegani, Jae Kyu Lee, Elena Mattei	Exhibition "Italia e Giappone. Design come stile di vita", Tokyo, Japan, 2001	20 sqm	[76] (ibidem, p. 152)	
16.	Suitcase house	Edge design institute	Beijing, China, 2002	-	[95]	
17.	Wildbrook	Urs Hartmann, Markus Wetzel	Zurich, Switzerland, 2002	4 sqm	[76] (ibidem, p. 145)	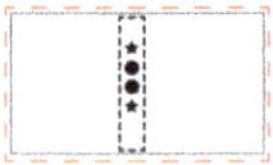
18.	Drawer-House	Nendo	Mejiro, Tokyo, Japan, 2003	50 sqm	[76] (ibidem, p. 196)	

Table 1. *Cont.*

N	Project Denomination	Architect	Location, Year	Surface Sqm.	Reference	Category of Integration
19.	CircuitBox	STUDIO X DESIGN GROUP Lara Rettondini, Oscar Brito	"Tokyo Design Week 2004", Tokyo, Japan, 2004	16 sqm	[76] (ibidem, p. 136)	
20.	Social housing	Aranguren + Gallegos Arquitectos	Madrid, Spain, 2005	80 sqm	[96]	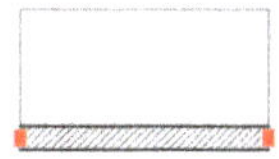
21.	Optibo	White Design	Goteborg, Sweden, 2006	25 sqm	[76] (ibidem, p. 186)	
22.	Home/office for a graphic designer	Roger Hirsch Architect + Myriam Corti	New York, USA, 2006	56 sqm	[76] (ibidem, p. 70)	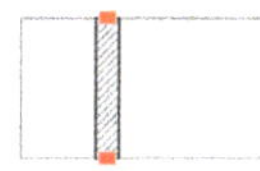
23.	Womb: work, office, meditation, base	Johnson Chou	Toronto, Canada, 2006	30 sqm	[76] (ibidem, p. 20)	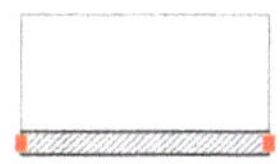
24.	Black Treefrog	SPLTTERWERK	Bad Waltersdorf, Austria, 2006	200 sqm	[76] (ibidem, p. 105)	
25.	Salon Blauraum	Blauraum Architekten	Hamburg, Germany, 2006	-	[76] (ibidem, p. 137)	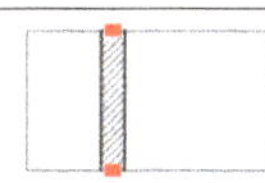
26.	Casa Engadina	Elisabetta Terragni	Samedan, Switzerland, 2006	95 sqm	[97]	
27.	Life Style	L.M. Rojo, R. Montero, M. Dominguez, E. Ontiveros, C. Vélez, G. Morale	Barcelona, Spain, 2007	-	[98]	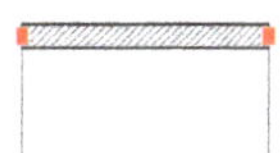
28.	Apartment in Thessaloniki	27 Architects	Thessaloniki, Greece, 2009	100 sqm	[81] (ibidem)	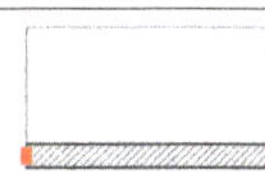
29.	BIQ	Schenk + Waiblinger Architects	Hamburg, Germany, 2011	50 sqm	[99]	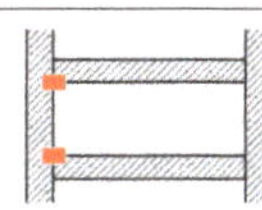
30.	Darlinghurst Apartment	Brad Swartz Architect	Darlinghurst, Australia, 2014	27 sqm	[100]	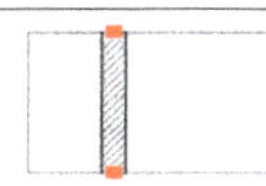
31.	Susaloon	Elii architects	Madrid, Spain, 2014	24 sqm	[101]	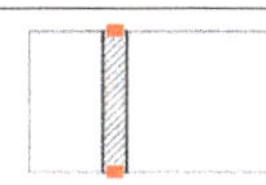

Table 1. *Cont.*

N	Project Denomination	Architect	Location, Year	Surface Sqm.	Reference	Category of Integration
32.	Estradenhaus 1 Berlin	Wolfram Popp	Berlin, Germany, 2015	80 sqm	[102]	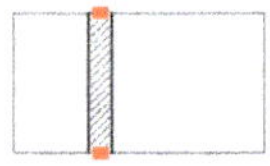
33.	Kramergasse 13	OLK Ruf	Wien, Austria, 2015	-	[94] (ibidem)	
34.	Cell bricks	Atelier Tekuto	Tokyo, Japan, 2015	-	[94] (ibidem)	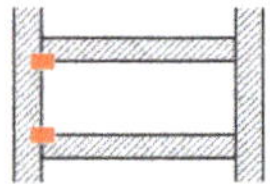
35.	Appartamento Viadutos	Vão	São Paulo, Brazil, 2016	35 sqm	[103]	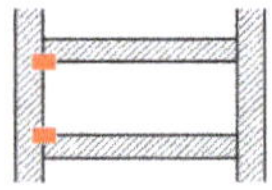
36.	Studio flat	CIAO	London, UK, 2016	35 sqm	[104]	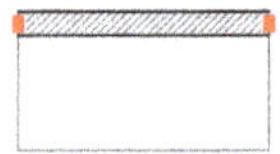
37.	Weekend House Straume	Knut Hjeltnes	Ålesund, Norway, 2016	140 sqm	[75] (ibidem)	
38.	Newton House	Yukawa design lab	Tokyo, Japan 2016	99 sqm	[78] (ibidem)	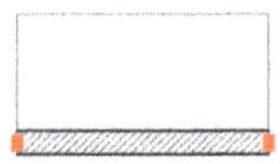
39.	Tatsumi apartment	Hiroyuki Ito architects	Tokyo, Japan 2016	34 sqm	[79] (ibidem)	
40.	Micro dwellings	Kasita	Austin, Texas USA, 2017	33 sqm	[105]	
41.	Micro home	Ana White	Alaska, USA, 2017	16 sqm	[106]	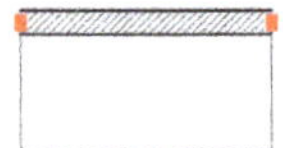
42.	ScopeHome	MINI LIVING Urban Cabin, Penda	"Exhibition House Vision", Beijing, China, 2018	30 sqm	[107]	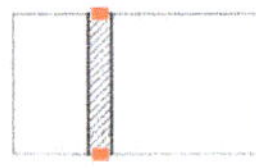
43.	Apartamento Consolação	Canoa Arquitetura	São Paulo, Brazil, 2018	33 sqm	[108]	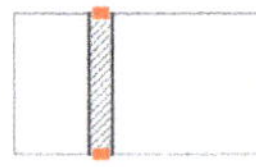
44.	Appartamento AMRA7	Piratininga Arquitetos Associados + Bruno Rossi Arquitetos	São Paulo, Brazil, 2019	300 sqm	[84] (ibidem)	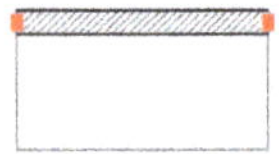

Table 1. *Cont.*

N	Project Denomination	Architect	Location, Year	Surface Sqm.	Reference	Category of Integration
45.	Studio Brasilia 27	Fabio Cherman	Brasilia, Brazil, 2019	27 sqm	[109]	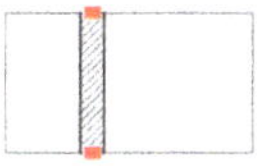
46.	Andradas Apartment	OCRE arquitetura	Porto Alegre, Brazil, 2020	29 sqm	[110]	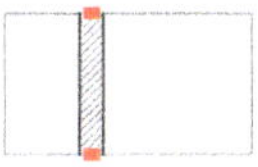
47.	Micro-apartment	Proctor and Shaw	London, UK, 2021	29 sqm	[111]	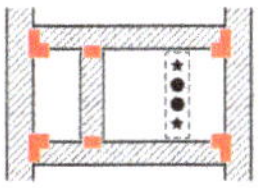
48.	Smart Zendo apartment	Sim-Plex	Hong Kong, China, 2021	45 sqm	[112]	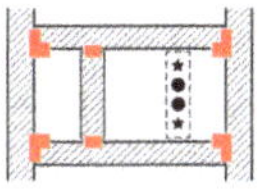
49.	Urban Loft	NeuronaLab	Poblenou, Barcelona, Spain, 2022	55 sqm	[113]	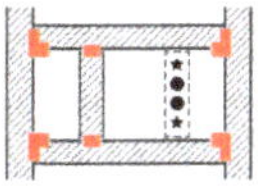
50.	Doméstico	Juan Alberto Andrade, María José Váscones	Quito, Ecuador, 2022	27.5 sqm	[114]	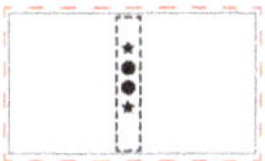

4. Setting the Boundaries for the Development of Integrated Systems

The analysis of the historic and contemporary case studies collected illustrates a rich phenomenology (cf. Table 1) encompassing both furniture and equipment as well as kitchen and bathroom fixtures, which has been classified according to defined integration categories (cf. Table 2). The technological multifunctionality of these systems allows for spatial multifunctionality when implemented with a transformability feature. Hence, furniture and fit-out disappear behind shutters, cavities, or pivoting devices enabling the shrinkage of the living space.

The integration can characterise either the shell (vertical and horizontal surfaces) enclosing the interior space or ancillary space-making cores, which can additionally be inhabitable, that can articulate and organise the room layout determining its function; movable elements allow for real-time flexibility. Accordingly, the design focus shifts from construction to spatial layout and from architecture to interior design, exploring an intermediate range of competences, sometimes referred to as "furnitecture" [95] (ibidem).

This work mainly addresses the problem setting for the design of advanced products which can replace ordinary building subsystems and which complement the primary purpose and performances with the added value of place making, delivering an augmented fruition of the dwelling spaces. The focus will be on the shell to cater for the basic equipment of the interiors as it offers a more original field of investigation and wider ranges of integrability in comparison with the cores. In fact, prefabricated bathrooms and kitchen units are already available on the market and furniture cores seldom involve the integration with construction subsystems.

Following the case studies' classification, in this section, the paper sets out to deepen the understanding of the integrated systems between construction subsystems and furniture/fixtures to offer a basic problem setting and analysis for the development of spatial and technological design.

First, the integrated system is examined per se, considering its place making potential, the enhanced performances, and the time variable. In fact, as combined components, these structures entail, on the one hand, the packing of the sequence of building activities to reduce construction time; on the other hand, as adaptive and responsive elements, they allow real-time flexibility of the space.

Accordingly, the multifaceted categories of integrability, transformability, and multi-functionality are proposed and defined (cf. Table 3) as criteria for the breakdown structure of the design problem and to allow for its analysis.

Table 2. Categories of integration.

N	Category	Place of Integration	Interface between the Technical Elements	
1.	Integrated structure	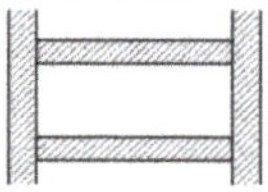	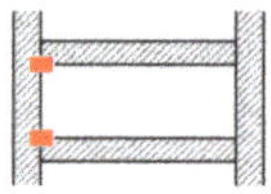	
2.	Integrated envelope/window			Lower flooring system Upper flooring system
3.a	Integrated raised floor	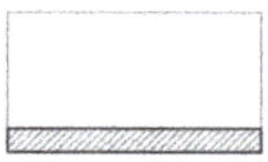		Vertical walls (enclosure) Interior partition walls
3.b	Integrated false ceiling		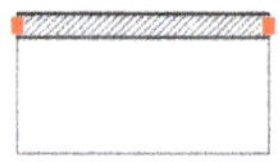	Vertical walls (enclosure) Internal partition walls
4.	Integrated partition wall	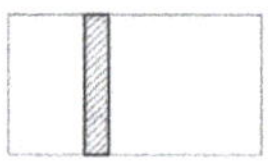	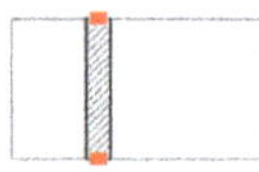	Lower flooring system Upper flooring system
5.	Service cores	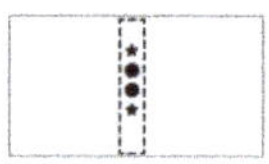	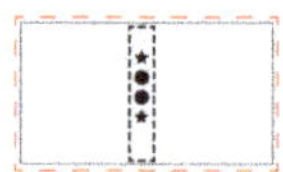	Whole building
6.	Total integration	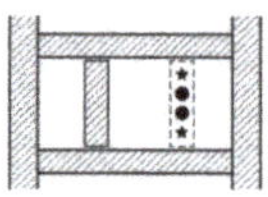	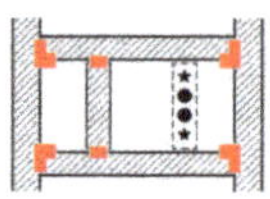	Whole building

Table 3. Integrated systems. Parameters and feasibility conditions. Definitions and performance specifications.

Parameters		**Definitions and Performance Specifications**	
Integrated systems		Industrialised dry assembly prefabricated systems integrating furniture/equipment and construction subsystems, in compliance with circular economy principles	
Typologies of integrable furniture/equipment	Multipurpose 3D elements	Functions: storage, support, stowage, ...	
Integrable components	Cabinets, wardrobe, shelving, appliances, installations, bed, ...		
Degree of integrability	Total	Relationship/integration with all the technical elements of the construction subsystem	
	Partial	Relationship/integration with some of the technical elements of the construction subsystem	e.g., partial (with a limited extension) replacement of the false ceiling of the room

Table 3. *Cont.*

Parameters		Definitions and Performance Specifications	
Degree of transformability of the integrated furniture/equipment	Partial	Appearance variation through movement of components (turning, tilting, sliding, retractable, . . .)	Disassembly and reassembly of components with the same purpose
	Total	Arrangement variation as components change place and position	Changing position and replacing components with different purpose
Multifunctionality of the solution	Real-time multipurpose characteristic	Technological purposes Functional purposes Spatial purposes	Multifunctionality varies according to position (cf. Table 2. Categories of integration)
Feasibility conditions of the solutions	Regulatory	According to regulation requirements	Planning/building
	Design	According to new construction/renovation According to products/materials employed for the construction of the adjacent subsystems	Factors related to building technologies
		According to the degree of integration with the adjacent subsystems	Structural system Partition walls Upper and lower flooring system Fixtures and installations
		According to the location, installation, usage, and accessibility of the integrated system	e.g.,: integration with the enclosure, fixed to floor or ceiling, with or without floor and ceiling finishes, recessed, . . .
	Construction	Building technologies employed for the installation of the integrated system layering/components	Dry assembly technologies Mechanical fastening Chemical bonding technologies
		Installation and usage of the integrated system according to the layering of the adjacent subsystems	Shape, dimensions, weight, components and joints
	Spatial	Real-time flexibility	Quick changeability of the spatial arrangement Reversibility Easy modification of the components Diversity of the usage conditions Customisation, ...
	Technological	Compliance with the mandatory or design performances	Acoustic Hygrothermal Insulation Sound absorption Fireproofing Impact strength Compressive strength Stability Anti-seismic Durability Morphological dimensional Compatibility Modularity Easy disassembly/reassembly, . . .

Table 3. *Cont.*

Parameters		**Definitions and Performance Specifications**
Added systemic value of the solution	Enhanced value through the integration	Degree of integrability Degree of transformability Degree of multifunctionality Feasibility conditions and compliance

Secondly, as the object of the research encompasses subsystems, albeit augmented, belonging to an architectural whole, the complexity of the problem is addressed considering both the elements and the interfaces with the adjacent technological subsystems of the building, as outlined in Table 2. In fact, the joints between the different subsystems represent a significant aspect of the problem to hinder the overall quality of the home if neglected. It is a design challenge to implement these detailed design issues anticipating faults during the assembly process and envisaging the aggregability/modularity control of components in relation to the other construction systems.

Finally, design environment variables are mapped to cater for the multiscale dimension of the problem including the regulatory, industrial, and economic context of architectural design. Hence, Table 4 also includes a breakdown of feasibility conditions as an attempt to map the complexity of the problem.

Table 4. Integrated false ceiling, main design requirements.

Place of Integration Localisation and Movement of Integrable Equipment/Built-in Furniture	**Integrated False Ceiling** Integrable Equipment/Built-in Furniture, with Vertical up/down Movement	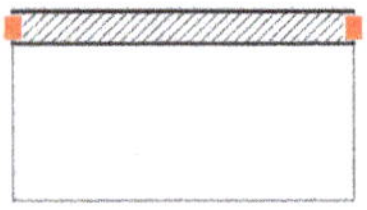
Feasibility conditions	Regulatory	Determined by the relationship with interior heights set by local planning/building regulations
	Design	Determined by the integration with the flooring system above and with the building systems; by the relationship (existing or non-existent) with the enclosure; by the relationship with other technical elements (internal partition, etc.)
	Construction	Determined by technical building solutions; design choices and decisions in relation to reduce/reuse/recycle principle; degree of eco-compatibility of materials; mechanised and/or digital movability
	Spatial	Determined by ensuring a flexible range of products; ensuring spatial flexibility; ensuring quick mechanised movement
	Technological	Determined by compliance with regulatory, production, installation, and circularity requirements both of the component and of the integrated system (product/technical elements + interfaces)

Table 4. *Cont.*

Place of Integration Localisation and Movement of Integrable Equipment/Built-in Furniture	**Integrated False Ceiling** Integrable Equipment/Built-in Furniture, with Vertical up/down Movement	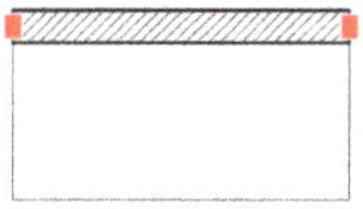
Multifunctionality combined performance	Technological performance	Acoustic Sound absorbing Anti-seismic Structural Cleanability/Maintainability, ...
	Functional performance	Accessibility Ease of use Usability Operational ease, ...
	Spatial performance	Usage space of the integrated system Functional overlap Dimensional checks Real-time transformation, . . .

In Table 4, the breakdown structure is applied to one of the categories of integration to exemplify and illustrate the complex interrelated issues that the design phase has to address, encompassing the many opportunities as well as highlighting possible issues and warnings concerning the integrated system and its interfaces.

5. Discussion

The breakdown analysis of the different categories of integration between the furniture/equipment of the house and the construction subsystems displays design issues and paths for future development. In particular, it sets a range of spatial and technological interdependent variables which represent an abstract setting of the problem. However, according to specific design goals and strategies, to the specific programme of the building (e.g., according to end users), or to the context, a hierarchy can be set to determine changeable factors and invariants and thus better orientate design. It is the case of the possible application of the integrated system to the renovation of the built heritage where geometric and technological features of the building as a whole are already predetermined. It is a growing market especially in Europe [115], where these multifunctional solutions can efficiently respond to the issues of requalification and refurbishment, providing enhanced performances to the existing building [116] and equipping the interiors with easily accessible vertical and horizontal distribution networks of the building services.

Following the application of the breakdown structure to each category of integration, some considerations emerge from the understanding of the feasibility conditions related to building regulations and the geometry of the integrated transformable system. In fact, on the one hand, these solutions allow for space-saving strategies through the superimposition of the functions and movement spaces; on the other hand, the minimum height and area of inhabitable rooms, as defined by hygienic regulations, may represent a limit to the downsizing of the space. In addition, the integrated system, embedding equipment, and fixtures constitute an increased version of the original simple construction system, resulting in a higher occupied volume. This leads to issues of possible redundancy of the solutions to allow for flexibility and adaptability, which requires to think "more specifically about the type(s) of change that might occur and about how they can be accommodated" [117]. Such a deeper simultaneous understanding of the spatial, technological, and building system variables as set by the problem analysis presented demonstrates how a new comprehensive design approach is required merging specialised competences. Such a new design culture should also encompass the skills required for the development as well as the application of the modular, industrialised systems [118].

6. Conclusions

The paper reports the partial results of our ongoing research on the topic moved, on the one hand, by the awareness of a paradigm change in the building sector where, finally, the goals of circular economy, rationalisation of the industry, and technological innovation converge to push forward a long-standing demand for off-site manufacturing.

On the other hand, signals can be detected in the market of a slow change towards ready-made dry assembly mechanical systems emerging from the building industry manufacturers and of a mild entrance of the fixtures and fit-out companies, for example, into the more traditional sector of room dividers and finally from the growing market of high-end albeit still high-cost automated transformable furniture manufacturers [119]. In fact, a possible impeding barrier to be further investigated is represented by the two supply chains traditionally involved in the fitting out and furniture delivery and assembly, characterised by different organisational structures.

The rising demand for transformability and adaptability significantly contributes to this change. Accordingly, a systematic investigation of the ongoing industry trends to evaluate the market feasibility for the introduction of the integrated systems investigated should also be carried out to thoroughly understand the economic feasibility of the concept.

Finally, domestic culture should be addressed, not only questioning consolidated habits or expectations but also modular design [120] and production [121] and the customisation vs. standardisation trade-off, investigating the decoration potentials of these solutions to enrich a catalogue of possible materials, finishes, and combinations. In this direction, significant potentials arise from processes of digital manufacturing and mass customisation [122,123] to understand whether the target of temporary housing and micro-living can be expanded and how to encompass a larger demand.

Accordingly, the cultural shift from extraordinary to ordinary solutions, from handicraft to reversible technology [124], entailed by the embedding of the basic equipment/fixtures in the construction of buildings, can found a possible parallel in the recent history of the introduction of installations and building services in the home, expanding the path of mechanisation [125].

Author Contributions: Conceptualization, L.D. and E.G.; Methodology, L.D. and E.G.; Investigation, G.V.; Writing—review & editing, G.V. All authors have read and agreed to the published version of the manuscript.

Funding: This research received no external funding.

Data Availability Statement: No new data were created or analysed in this study. Data sharing does not apply to this article.

Conflicts of Interest: The authors declare no conflict of interest.

References

1. European Economy Discussion Papers. Housing Market Developments in the Euro Area: Focus on Housing Affordability. Discussion p. 171, September 2022. Available online: https://economy-finance.ec.europa.eu/publications/housing-market-developments-euro-area-focus-housing-affordability_en (accessed on 17 April 2023).
2. Deloitte. Real Estate Predictions 2021. Understanding the Private Rented Sector Property 2021. Available online: https://www2.deloitte.com/content/dam/Deloitte/nl/Documents/real-estate/deloitte-nl-real-estate-prs-real-estate-predictions.pdf (accessed on 8 May 2023).
3. Glumac, B.; Islam, N. Housing preferences for adaptive re-use of office and industrial buildings. *Sustain. Cities Soc.* **2020**, *62*, 102379. [CrossRef]
4. Rogers, D. Uploading Real Estate: Home as a digital, global commodity. In *Housing and Home Unbound: Intersections in Economics, Environment and Politics in Australia*, 1st ed.; Cook, N., Davison, A., Crabtree, L., Eds.; Routledge: London, UK, 2016; p. 23, ISBN 9781315669342.
5. Fields, D. Unwilling subjects of financialization. *IJURR* **2017**, *41*, 588–603. [CrossRef]
6. Cuerdo-Vilches, T.; Navas-Martín, M.Á.; Oteiza, I. Working from Home: Is Our Housing Ready? *Int. J. Environ. Res. Public Health* **2021**, *18*, 7329. [CrossRef]

7. IZA institute of Labor Economy. Occupational Mobility in Europe: Extent, Determinants and Consequences. 2019. Available online: https://www.iza.org/publications/dp/12679/occupational-mobility-in-europe-extent-determinants-and-consequences (accessed on 8 May 2023).
8. Aberdeen Standard Investments & JLL. European Living Investor Survey 2020. Available online: https://www.jll.de/content/dam/jll-com/documents/pdf/research/jll-european-living-investor-survey-2020-jll-and-asi.pdf (accessed on 17 April 2023).
9. Oorschot, L.; Asselbergs, T. New Housing Concepts: Modular, Circular, Biobased, Reproducible, and Affordable. *Sustainability* **2021**, *13*, 13772. [CrossRef]
10. Elgani, E.; Scullica, F. Living, working and travelling. In *New Processes of Hybridization for the Spaces of Hospitality and Work*, 1st ed.; Franco Angeli Editore: Milano, Italy, 2019; ISBN 9788891777201.
11. Engerstam, S.; Warsame, A.; Wilhelmsson, M. Exploring the Effects of Municipal Land and Building Policies on Apartment Size in New Residential Construction in Sweden. *J. Risk Financ. Manag.* **2023**, *16*, 220. [CrossRef]
12. Heath, S.; Cleaver, E. Young, free and single? Young adults and independent living arrangements. In *Handbook of Youth and Young Adulthood*, 2nd ed.; Furlong, A., Ed.; Routledge: London, UK, 2016; ISBN 9780203881965.
13. Bergan, T.L.; Gorman-Murray, A.; Power, E.R. Coliving housing: Home cultures of precarity for the new creative class. *Soc. Cult. Geogr.* **2020**, *22*, 1204–1222. [CrossRef]
14. Karatseyeva, T.; Akhmedova, A. Modern tendency of development of architectural typology on the example of micro-apartment for Almaty city. *Eng. Constr. Archit. Manag.* **2022**. [CrossRef]
15. Gronostajska, B.E.; Szczegielniak, A. Inside a Microapartment: Design Solutions to Support Future Sustainable Lifestyles. *Buildings* **2021**, *11*, 654. [CrossRef]
16. Tervo, A.; Hirvonen, J. Solo dwellers and domestic spatial needs in the Helsinki Metropolitan Area. *Finl. Hous. Stud.* **2020**, *35*, 1194–1213. [CrossRef]
17. Clinton, E. Micro-living: Why occupants choose to live in very small dwellings. *Aust. Plan.* **2018**, *55*, 189–197. [CrossRef]
18. Hoang, M.; Vandal, A. Micro-MACRO living in the global high-rise. *CTBUH J.* **2017**, *3*, 32–37. Available online: https://www.jstor.org/stable/90020855 (accessed on 17 April 2023).
19. Rack, F. *Micro-Apartments as an Emerging Real Estate Market*, 1st ed.; GRIN Verlag: Munich, Germany, 2017; ISBN-10: 3668404437.
20. Waite, R. Micro-homes: Part of the solution or part of the problem? *Archit. J.* **2015**. Available online: https://www.architectsjournal.co.uk/news/micro-homes-part-ofthe-solution-or-partof-the-problem/8675105.article (accessed on 17 April 2023).
21. Skar, R.; Bragança, L.; Gervásio, H. Adaptability of Buildings: A Critical Review on the Concept Evolution. *Appl. Sci.* **2021**, *11*, 4483. [CrossRef]
22. Tarpio, J.; Huuhka, S.; Vestergaard, I. Barriers to implementing adaptable housing: Architects' perceptions in Finland and Denmark. *J. Hous. Built Environ.* **2022**, *37*, 1859–1881. [CrossRef]
23. European Commission. MORE-CONNECT Development and Advanced Prefabrication of Innovative, Multifunctional Building Envelope Elements for Modular Retrofitting and Smart Connections. 2019. Available online: https://ec.europa.eu/research/participants/documents/downloadPublic?documentIds=080166e5c5ffc66e&appId=PPGMS (accessed on 3 July 2023).
24. Schneider, T.; Till, J. *Flexible Housing*, 1st ed.; Routledge: London, UK, 2007; ISBN-10: 0750682027.
25. Kronenburg, R. *Flexible: Architecture that Responds to Change*, 1st ed.; Laurence King Pub: London, UK, 2007; ISBN-10: 1856694615.
26. Gibberd, F. *Built-in Furniture in Great Britain*; Alec Tiranti Ltd.: London, UK, 1948; ASIN: B0007J185K.
27. Brown, A. *Practical Designs For Built-In Furniture*, 1st ed.; Crosby Lockwood & Son Ltd.: London, UK, 1960; ASIN: B003U2IXZY.
28. Braddell, D. Common sense in furniture design. *J. R. Soc. Arts* **1942**, *90*, 4610. Available online: https://www.jstor.org/stable/41359878 (accessed on 10 July 2023).
29. Albert Farwell Bemis Foundation; Kelly, B. *The Prefabrication of Houses*, 1st ed.; Cambridge, Technology Press of the Massachusetts Institute of Technology and Wiley: New York, NY, USA, 1951; ISBN 9780262110013.
30. Caille, P.J.-F. Details. Cloisons, In *AMC*, Actualités, April 2006. Available online: https://www.amc-archi.com/article/cloisons.17203 (accessed on 10 July 2023).
31. Mandolesi, E. *Edilizia. Le Finalità. Il Processo Edilizio*; Utet: Torino, Italy, 1978; Volume 1, ISBN-10: 8802029881.
32. Petrignani, A.; Blachère, G.; Chiaia, V.; Ciribini, G.; De Vita, R.; Gorio, F.; Grisotti, M.; Guiducci, R.; Petrignani, M.; Salvati, M.; et al. *L'industrializzazione Dell'edilizia*, 1st ed.; Dedalo Libri: Bari, Italy, 1965.
33. Haas, C.T.; O'Connor, J.T.; Tucker, R.L.; Eickmann, J.A.; Fagerlund, W.R. *Prefabrication and Pre-Assembly Trends and Effects on the Construction Workforce*; Center for Construction Industry Studies, University of Texas: Austin, TX, USA, 2000.
34. Housing Communities and Local Government Committee. *Modern Methods of Construction. 15th Report of Session 2017–2019*; House of Commons: London, UK, 2019; Available online: https://committees.parliament.uk/work/1939/modern-methods-of-construction-inquiry/publications/ (accessed on 8 May 2023).
35. Jaillon, L.; Poon, C.S. Sustainable construction aspects of using prefabrication in dense urban environment: A Hong Kong case study. *Constr. Manag. Econ.* **2008**, *26*, 953–966. [CrossRef]
36. Jiang, R.; Mao, C.; Hou, L.; Wu, C.; Tan, J. A SWOT analysis for promoting off-site construction under the backdrop of China's new urbanisation. *J. Clean. Prod.* **2018**, *173*, 225–234. [CrossRef]
37. Nazir, F.; Edwards, D.; Shelbourn, M.; Martek, I.; Thwala, W.; El-Gohary, H. Comparison of modular and traditional UK housing construction: A bibliometric analysis. *J. Eng. Des. Technol.* **2020**, *19*, 164–186. [CrossRef]

38. Tichelmann, K.; Pfau, J. *Detail Practice. Dry Construction: Principles, Details, Example*, 1st ed.; Birkhäuser: Basel, Switzerland, 2008; ISBN 9783764388089.
39. Durmisevic, E. Circular Economy in Construction Design. In *Strategies for Reversible Buildings*, 1st ed.; TU-Delft: Delft, The Netherlands, 2019; ISBN 9789082169843. Available online: https://www.bamb2020.eu/wp-content/uploads/2019/05/Reversible-Building-Design-Strateges.pdf (accessed on 10 July 2023).
40. Biedermann, N.; Deplazes, A. Sculpted architecture. The Scottish tower house. In *Construction Architecture. Materials, Processes, Structures*; Deplazes, A., Ed.; Birkhauser: Basel, Switzerland, 2005; ISBN-10: 3038214523.
41. Todd, P. *William Morris and the Arts and Crafts Home*, 1st ed.; Thames & Hudson: London, UK, 2005; ISBN-10: 0500512523.
42. Powell, J. The Legacy of Arts & Crafts Built-Ins. *Oldhouse J.* 2009. Available online: https://www.oldhouseonline.com/interiors-and-decor/the-legacy-of-arts-crafts-built-ins/ (accessed on 17 April 2023).
43. Drexler, A. *The Architecture of Japan*, 1st ed.; The Museum of Modern Art: New York, NY, USA, 1955.
44. Cromley, E.C. Frank Lloyd Wright in the Kitchen. In *Buildings & Landscapes: Journal of the Vernacular Architecture Forum*; University of Minnesota Press: Minneapolis, MN, USA, 2012; Volume 19, pp. 18–42. [CrossRef]
45. Heathcote, E. Is furniture architecture? *Archit. Rev.* **2018**, *1453*, 68–73. Available online: https://www.jstor.org/stable/887689 (accessed on 17 April 2023).
46. Sendai, S. Realization of the standard cabinet as "equipment" by Le Corbusier: The transformation of the "wall". *Jpn. Archit. Rev.* **2019**, *2*, 494–506. [CrossRef]
47. Benton, C. 'L'Aventure du Mobilier': Le Corbusier's Furniture Designs of the 1920s. *J. Decor. Arts Soc. 1890–1940* **1982**, *6*, 7–22. Available online: https://www.jstor.org/stable/41806690 (accessed on 17 April 2023).
48. Sendai, S. The conception of "equipment" by Charlotte Perriand: Cross-over between Le Corbusier and Japan. *J. Asian Archit. Build. Eng.* **2019**, *18*, 1–9. [CrossRef]
49. Rault, J. *Eileen Gray and the Design of Sapphic Modernity*, 1st ed.; Taylor & Francis: Milton Park, UK, 2017; ISBN 9781351568579.
50. Berdasco, V.A. Mobiliario built-in como articulador del espacio en la arquitectura residencial de Richard Neutra. *Res. Mobilis* **2021**, *10*, 94–108. [CrossRef]
51. Alison, F. (Ed.) *Le Corbusier-Interior of the Cabanon: Le Corbusier 1952*, 1st ed.; Triennale Electa: Milano, Italy, 2006; EAN: 9788837043940.
52. Colomina, B. Intimacy and spectacle: The interiors of Adolf Loos. *AA Files* **1990**, *20*, 5–15. Available online: https://www.jstor.org/stable/29543700 (accessed on 17 April 2023).
53. Wigley, M. Whatever happened to total design. *Harv. Des. Mag.* **2001**, *47*, 21–24. Available online: https://www.harvarddesignmagazine.org/issues/47/whatever-happened-to-total-design-issue-47 (accessed on 17 April 2023).
54. Capomaggi, J. The houses that don't exist: Openness in domestic modular systems. *Interiors* **2023**, 1–23. [CrossRef]
55. Fu, X.L. Analysis on the integrated design of architecture and furniture: Taking the Farnsworth House as an example. *J. Phys. Conf. Ser.* **2019**, *1168*, 032102. [CrossRef]
56. Schneiderman, D. *Inside Prefab: The Ready-Made Interior*; Princeton Architectural Press: New York, NY, USA, 2012; ISBN-10:1568989873.
57. Stolper, H. Einbauten. In *Planung form Fonstruktion. Built-in Forniture*, 1st ed.; Julius Hoffmann: Stuttgart, Germany, 1960; ISBN VEA0051688.
58. Lupton, E.; Miller, J.A. *The Bathroom the Kitchen and the Aesthetic of Waste A Process of Elimination*; Princeton Architectural Press: New York, NY, USA, 1992; ISBN-10: 0938437429.
59. Steiner, H.A. The architecture of the well-serviced Environment. *Archit. Res. Q.* **2005**, *9*, 133–143. [CrossRef]
60. Zeinstra, J. Houses of the Future Ideal Homes. *OASE 32*. Available online: https://www.oasejournal.nl/en/Downloads/528362bfc79eb3b5f5000753/OASE%2075%20-%20203%20Houses%20of%20the%20Future.pdf (accessed on 17 April 2023).
61. Chenut, D. *Ipotesi Per un Habitat Contemporaneo*; Il Saggiatore: Milano, Italy, 1968.
62. Ambasz, E. *Italy: The New Domestic Landscape. Achievements and Problems of Italian Design*, 1st ed.; The Museum of Modern Art: New York, NY, USA, 1972; ISBN-10: 0870703935.
63. Habraken, N.J. *Supports: An Alternative to Mass Housing*; The Architectural Press: London, UK, 1999. [CrossRef]
64. Kendall, S.H. (Ed.) *Residential Architecture as Infrastructure: Open Building in Practice*, 1st ed.; Routledge: London, UK, 2021. [CrossRef]
65. Brand, S. *How Buildings Learn*; Viking: New York, NY, USA, 1994; ISBN-10: 0140139966.
66. Durmisevic, E. *Transformable Building Structures: Design for Dissassembly as a Way to Introduce Sustainable Engineering to Building Design & Construction*; TU-Delft: Delft, The Netherlands, 2006; ISBN: 90-902-0341-9; ISBN 978-90-9020341-6.
67. Geraedts, R. FLEX 4.0, A Practical Instrument to Assess the Adaptive Capacity of Buildings. *Energy Procedia* **2016**, *96*, 568–579. [CrossRef]
68. Egelius, M. *Ralph Erskine, Architect*, 1st ed.; Coronet Books Inc.: Philadelphia, PA, USA, 1990; ISBN-10: 9179880460.
69. Hubert, B.; Steinegger, J.C. (Eds.) *Jean Prouvé. Une Architecture par L'industrie*; Les Editions d'Architecture Artemis: Zurich, Swi-tzerland, 1971.
70. Lees-Maffei, G. *Jean Prouvé. The Poetics of the Technical Object*, 1st ed.; Vitra Design Museum, Skira: Losanna, Switzerland, 2006; ISBN 887624967.
71. McQuaid, M. *Shigeru Ban*, 1st ed.; Phaidon Press: New York, NY, USA, 2006; ISBN-10: 0714846295.

72. Hidden Architecture. Domus Demain. *Hidden Architecture Atlas*, December 2017. Available online: http://hiddenarchitecture.net/domus-demain/ (accessed on 10 July 2023).
73. Constant, C. *Eileen Gray*, 1st ed.; Phaidon: New York, NY, USA, 2000; EAN: 9780714839059.
74. *Catalogo-X Triennale 1954*; S.A.M.E.: Milano, Italy, 1954.
75. González, M.F. (Ed.) Weekend House Straume/Knut Hjeltnes. *Archdaily*, September 2019. Available online: https://www.archdaily.com/924181/weekend-house-straume-knut-hjeltnes (accessed on 10 July 2023).
76. Mostaedi, A. *Great Spaces: Flexible Homes*; Links Internacional: Barcelona, Spain, 2006; ISBN-10: 9788496263543.
77. Chang, G. In the Age of Indeterminacy: Towards a Non-visual Pragmaticism. *Archit. Des.* **2003**, *73*, 58–67.
78. Newtown House/Yukawa Design Lab. *Archdaily*, July 2017. Available online: https://www.archdaily.com/874950/newtown-house-kohei-yukawa-plus-hiroto-kawaguchi (accessed on 10 July 2023).
79. Hiroyuki Ito Architects. Tatsumi Apartment House. *Divisare*, March 2017. Available online: https://divisare.com/projects/339022-hiroyuki-ito-architects-masao-nishikawa-tatsumi-apartement-house (accessed on 10 July 2023).
80. O'Keefe, D. Floor polish: Floor as furniture in Japan. *Archit. Rev.* **2018**. Available online: https://www.architectural-review.com/essays/floor-polish-floor-as-furniture-in-japan (accessed on 10 July 2023).
81. Apartment in Thessaloniki. *Archdaily*, March 2010. Available online: https://www.archdaily.com/52530/apartment-in-thessaloniki-27-architects (accessed on 10 July 2023).
82. Rui, A. *Ugo La Pietra. Disequilibrating Design*; Corraini: Milano, Italy, 2014; EAN: 9788875704759.
83. Reuter, R. *Space-Saving Techniques by the Use of Transformable Architecture*; Delft University: Delft, The Netherlands, 2006; Available online: https://efaidnbmnnnibpcajpcglclefindmkaj/https://repository.tudelft.nl/islandora/object/uuid:4f324dc7-f510-4ea3-8949-e35ba526c8ed/datastream/OBJ/downloadusq (accessed on 10 July 2023).
84. Pereira, M. (Ed.) Apartment AMRA7/Piratininga Arquitetos Associados+Bruno Rossi Arquitetos. *Archdaily*, July 2020. Available online: https://www.archdaily.com/944321/apartment-amra7-piratininga-arquitetos-associados-plus-bruno-rossi-arquitetos (accessed on 10 July 2023).
85. Frampton, K. *The Poetics of Space in the Late Modern Dwelling*; Ecole Polytechnique Fèdèrale de Lausanne: Lausanne, Switzerland, 1996.
86. Nelson, G.; Wright, H.N. *Tomorrow's House. How to Plan Your Post-War Home Now*; Simon and Schuster: New York, NY, USA, 1945.
87. Burkhardt, F. (Ed.) *Angelo Mangiarotti. Complete Works*, 1st ed.; Motta Architettura: Milano, Italy, 2010; ISBN 978886116115.
88. Munari, B. *Spazio Abitabile*; Stampa Alternativa: Roma, Italy, 1996; EAN: 9788872264751.
89. Boeri, C. *Le Dimensioni Umane Dell'abitazione*; Franco Angeli: Milano, Italy, 1980; ISBN 9788820422479.
90. Umberto Riva. Casa Insinga, Milan 1989; *Atlas of Interiors*, November 2015. Available online: https://www.domusweb.it/en/news/2014/11/04/rooms_you_may_have_missed.html (accessed on 10 July 2023).
91. Ottolini, G.; Di Prizio, V. *La Casa Attrezzata. Qualità Dell'abitare e Rapporti di Integrazione fra Arredamento e Architettura*, 2nd ed.; Liguori Editore: Napoli, Italy, 2005; EAN: 9788820720261.
92. Holl, S. *PARALLAX*, 1st ed.; Priceton Architectural Press: New York, NY, USA, 2000; ISBN 9781568982618.
93. Guard Tillman Pollock. Apartment in London. Soho, London, UK. 1996. Available online: http://www.guardtillmanpollock.com/soho-1/ (accessed on 10 July 2023).
94. Kronenburg, R. *Flexible. Une Architecture Pour Répondre au Changement*, 1st ed.; Editions Norm: Paris, France, 2007; ISBN-10: 2915542082.
95. Yudina, A. *Furnitecture: Funiture That Transforms Space*, 1st ed.; Thames & Hudson: London, UK, 2015; ASIN: B01JXTTY2S.
96. Aranguren, M.J.; González Gallegos, J. *Acciones Aranguren y Gallegos, 1995–2004*, 1st ed.; Editorial Rueda: Madrid, Spain, 2004; ISBN 9788472071452.
97. *Abitare, Architettura Degli Interni: Progettare a Colori* n. 461, May 2006; pp. 102-110. 2006.
98. Lapuerta, J.M. *Collective Housing: A Manual*, 1st ed.; Actar Publishers: Barcelona, Spain, 2007; ISBN 9788496954151.
99. IBA Hamburg, Smart Material Houses. BIQ. Available online: https://www.internationale-bauausstellung-hamburg.de/en/projects/the-building-exhibition-within-the-building-exhibition/smart-material-houses/biq/projekt/biq.html (accessed on 10 July 2023).
100. Darlinghurst Apartment. *Archdaily*, August 2015. Available online: https://www.archdaily.com/772121/darlinghurst-apartment-brad-swartz-architect?ad_medium=gallery (accessed on 3 July 2023).
101. Susaloon. *Archdaily*, June 2015. Available online: https://www.archdaily.com/644104/076-susaloon-elii (accessed on 3 July 2023).
102. Estradenhaus 1 Berlin. Available online: http://www.planpopp.com/choriner_str_56_55/choriner_1.html (accessed on 3 July 2023).
103. Pereira, M. (Ed.) Appartamento Viadutos. *Archdaily*, February 2018. Available online: https://www.archdaily.com/877123/viadutos-apartment-vao (accessed on 3 July 2023).
104. Mull, O. CIAO Adds Space-Saving Custom Furniture to London Micro Apartment. *Dezeen*, December 2016. Available online: https://www.dezeen.com/2016/12/29/ciao-micro-apartment-islington-london-space-saving-custom-furniture-interior-design/ (accessed on 3 July 2023).
105. McKnight, J. Kasita Launches Sales of High-Tech Micro Dwellings at SXSW. *Dezeen*, March 2017. Available online: https://www.dezeen.com/2017/03/14/kasita-aims-solve-us-housing-crisis-high-tech-micro-dwellings/ (accessed on 3 July 2023).
106. White, A. Micro Home. January 2017. Available online: https://www.ana-white.com/blog/2017/01/open-concept-rustic-modern-tiny-house-photo-tour-and-sources (accessed on 3 July 2023).

107. Shuang, H. MINI LIVING Urban Cabin/Penda. *Archdaily*, October 2018. Available online: https://www.archdaily.com/903077/mini-living-urban-cabin-penda (accessed on 3 July 2023).
108. Pereira, M. (Ed.) Apartamento Consolação. *Archdaily*, July 2023. Available online: https://www.archdaily.com.br/br/949301/apartamento-consolacao-canoa-arquitetura/5f80d17063c017485c000092-apartamento-consolacao-canoa-arquitetura-corte?next_project=no (accessed on 3 July 2023).
109. Studio Brasilia 27. *Archdaily*, November 2014. Available online: https://www.archdaily.com.br/br/756524/studio-brasilia-27-fabio-cherman?ad_medium=gallery (accessed on 3 July 2023).
110. Pereira, M. (Ed.) Andradas Apartment. *Archdaily*, August 2020. Available online: https://www.archdaily.com/946001/andradas-apartment-ocre-arquitetura (accessed on 3 July 2023).
111. Astbury, J. Micro-Apartment. *Dezeen*, September 2021. Available online: https://www.dezeen.com/2021/09/25/shoji-apartment-proctor-and-shaw-london/ (accessed on 3 July 2023).
112. Borgobello, B. Hong Kong Micro Apartment Packs Smart Tech into Transformable Spaces. *New Atlas*, November 2019. Available online: https://newatlas.com/architecture/smart-zendo-china-micro-apartment-sim-plex/ (accessed on 6 July 2023).
113. Frearson, A. NeuronaLab Reorganises Barcelona Loft with Blue Stair Storage Unit. *Dezeen*, June 2022. Available online: https://www.dezeen.com/2022/06/10/loft-in-poblenou-neuronalab-blue-storage-unit-mezzanine/ (accessed on 6 July 2023).
114. McKnight, J. Doméstico Uses "Habitable Artifact" to Organize Micro Apartment in Quito. *Dezeen*, May 2022. Available online: https://www.dezeen.com/2022/05/26/bed-desk-domestico-micro-apartment-quito/ (accessed on 6 July 2023).
115. Business Wire. Europe Building Renovation Market Report 2022. Available online: https://www.businesswire.com/news/home/20220523005805/en/Europe-Building-Renovation-Market-Report-2022-Market-Predicted-to-Grow-from-956.88-Billion-in-2021-to-1012.47-Billion-by-2026{-}{-}-ResearchAndMarkets.com (accessed on 17 April 2023).
116. Kendall, S. Four Decades of Open Building Implementation: Realising Individual Agency in Architectural Infrastructures Designed to Last. In *Loose-Fit Architecture: Designing Buildings for Change*, 1st ed.; Lifschutz, A., Ed.; Wiley: New York, NY, USA, 2017; pp. 54–63. [CrossRef]
117. Schmidt, R., III; Austin, S. *Adaptable Architecture Theory and Practice*, 1st ed.; Routledge: London, UK, 2016; p. 68, ISBN 9780415522588.
118. Kieran, S.; Timberlake, J. *Refabricating Architecture. How Manufacturing Methodologies Are Poised to Transform Building Construction*; McGraw-Hill: New York, NY, USA, 2004.
119. Goessler, T.; Kaluarachchi, Y. Smart Adaptive Homes and Their Potential to Improve Space Efficiency and Personalisation. *Buildings* **2023**, *13*, 1132. [CrossRef]
120. Staib, G.; Drrhfer, A. *Elemente und Systeme: Modulares Bauen–Entwurf, Konstruktion, Neue Technologien*, 1st ed.; Birkhäuser: Basel, Switzerland, 2013; ISBN-10: 376438655X.
121. Starr, M.K. Modular production-a 45-year-old concept. *Int. J. Oper. Prod. Manag.* **2010**, *30*, 7–19. [CrossRef]
122. Piroozfar, P.A.E.; Piller, F.T. *Mass Customisation and Personalisation in Architecture and Construction*, 1st ed.; Routledge: London, UK, 2013; ISBN 9780415622844.
123. Larsen, M.S.S.; Lindhard, S.M.; Brunoe, T.D.; Nielsen, K.; Larsen, J.K. Mass Customization in the House Building Industry: Literature Review and Research Directions. *Front. Built Environ.* **2019**, *5*, 115. [CrossRef]
124. Canal Architecture; Ruben, P. *Costruire Réversible*; Canal Architecture: Paris, France, 2017; ISBN 9782956031901.
125. Giedion, S. *Mechanization Takes Command a Contribution to Anon*, 1st ed.; Oxford University Press: Oxford, UK, 1948; ISBN-10: 0195005554.

buildings

Article

Apartment Space Planning Directions for Infectious Disease Prevention and Management: Insights Based on Residents' Experiences

Mi-Kyung Kim [1] and Eun-Jeong Kim [2,*]

[1] Department of Housing and Interior Design, Chungbuk National University, Cheong-Ju 28644, Republic of Korea; mkmkkim@chungbuk.ac.kr
[2] Dodam Design & Research, #101-904, Hangangdaero 211, Yongsan-gu, Seoul 04322, Republic of Korea
* Correspondence: dodam.design.research@gmail.com

Abstract: The COVID-19 pandemic has raised people's interest in pandemic-safe housing. This study aims to present insights into apartment housing space design to prevent and manage infectious diseases based on the actual living experiences of apartment residents. The relevant literature was reviewed, and overall satisfaction was assessed through a questionnaire targeting apartment residents in South Korea. Finally, using the photovoice method, residents' space needs were identified. By applying a mixed-use methodology and identifying the needs of residents, the following two recommendations were derived: (1) a flexible space plan to support multifunctional use while securing work efficiency and privacy in the living room, bedroom, and space for hobbies/work/learning; (2) a facility for removing contaminants at the entrance with sufficient storage space. The residents emphasized the importance of non-structural changes to the space and the convenience of use to increase actual space utilization. When planning future pandemic-safe apartment housing spaces, it is important to consider structural aspects like functionality and facilities. Convenience of use and methods of supporting family members' privacy should also be taken into account.

Keywords: residential apartment living experience; residential preference; residents' spatial needs; pandemic-safe space planning; prevention and management of infectious disease

Citation: Kim, M.-K.; Kim, E.-J. Apartment Space Planning Directions for Infectious Disease Prevention and Management: Insights Based on Residents' Experiences. *Buildings* **2023**, *13*, 2203. https://doi.org/10.3390/buildings13092203

Academic Editors: Carol Monticelli, Anna Stefańska, Daniel Mateus, Carlos C. Duarte and Nuno D. Cortiços

Received: 1 August 2023
Revised: 25 August 2023
Accepted: 28 August 2023
Published: 30 August 2023

1. Introduction

COVID-19 has brought significant changes to people's daily lives. Owing to the level of social separation it required, the pandemic has posed more social, mental, and psychological challenges to residents' lives [1]. In particular, external activities such as working, studying, and entertainment have been added to the functional needs of residential buildings [2]. Social distancing and self-quarantine have allowed residents to rethink housing design and layout. Various functional, structural, and cultural aspects of housing significantly impact resident health through the prevention and management of infectious diseases [3,4]. Many patients infected with COVID-19 overwhelmed the capacity of hospitals and quarantine institutions; therefore, households operated their homes as places of self-isolation [5].

Accordingly, efforts are being made to examine the layout and design of residential spaces from a new perspective and to suggest future design directions for a more effective response to the prevention and management of infectious diseases. Architecture has long been responsive to pandemics and health threats in innovative and adaptive ways [6]. Regarding the implications of stay-at-home requirements, many useful yet insufficient studies in multiple disciplines, including architectural design, environmental psychology, building science, engineering, urban planning, and healthcare, have been undertaken to determine the modifications necessary to make homes more adaptable to the new norm [5–10]. Scholars have introduced several factors that must be considered when designing the interiors of residential buildings. Tokazhanov et al. [2] introduced the four categories of "health and safety", "environmental resource consumption", "comfort", and "sustainable use of building services" as pandemic-resilient criteria.

Moreover, they specifically suggested various items such as natural light, temperature and humidity, greenery and gardens, activity/sports spaces, separate toilets for the infected, natural ventilation, space adjustability, personal space, acoustics, and urban farming, regarding the four main categories. AI-Qaisi [4] introduced critical considerations for post-pandemic housing characteristics, such as low-population-density areas, flexibility, pliability and adaptation, and private and removable walls and screens, based on a literature review [11–13]. Molaei et al. [14] and Aydin and Sayar [15] emphasized the importance of semi-open spaces for resting, leisure time, socializing with neighbors, and getting fresh air. Hajjar [16] argued that securing natural elements such as garden space, sunlight, air, and a view of food production helps overcome the negative psychology of residents following the spread of infectious diseases. Millán-Jiménez et al. [17] introduced findings that university students' preference for open space and flexible space use increased in terms of health and comfort after lockdown.

Bettaieb and Alsabban [3] introduced flexibility as an essential concept in housing after quarantine. They mentioned that the inside–outside relationship through windows and openings, various daily life activities, and changes in behavior in terms of privacy and comfort effected significant changes in spatial composition. To summarize the arguments in the literature, providing a flexible layout and multifunctional space using semi-open spaces and gardens to prevent and manage infectious diseases after COVID-19 effectively has become increasingly imperative.

However, this concept of flexibility requires a structurally removable wall and presents a problem regarding the accessibility of natural light, ventilation, and the outside view, according to each space division. Furthermore, although theoretical guidelines have been presented, their uniform application to individual residential spaces is limited because the lifestyles and needs of residents differ by country and region. Residential buildings play a significant role as spaces where residents mainly stay during self-isolation periods owing to infectious diseases [18]; therefore, it is critical to understand their lifestyles and needs and reflect them in the design of their space. Several authors have outlined the advantages of planning processes involving users [19–23]. In particular, Correira and Willis [24] explained that individuals have different characteristics, vulnerabilities, motivations, and ways and degrees of understanding. They asserted that people look at and respond to the present and future based on their past experiences when faced with a specific situation. For this reason, they emphasized the importance of understanding the individual's experience to improve the management of future public health issues. In addition, Hartig and Lawrence [25] argued that residence is a concept that includes people, place, activity, and time, and among them, how people act and behave in their residence has a great impact on their health. These articles suggest that understanding people's behavior-based experiences can provide critical insights to improve the quality of residential spaces in terms of health.

Heydarian et al. [26] stressed that it is critical to apply user-centered design to reflect end users' needs, requirements, and preferences in building design. Nevertheless, many building industries fail to collect accurate information from users to better understand their needs and behaviors. Consequently, residents' needs have not been detailed, and the same standards and principles have been continuously applied to building construction, ignoring regional, national, and cultural differences.

As special needs, such as the prevention and management of infectious diseases and social distancing, have emerged due to COVID-19, these characteristics should also be considered in planning residential spaces. To understand residents' needs, collecting users' opinions from various perspectives is necessary. Applying a mixed-use methodology is more effective than relying on one quantitative or qualitative research method, including surveys, interviews, observations, and site visits. Quantitative research effectively identifies the majority's general opinions and preferences. In contrast, qualitative research helps draw in-depth conclusions about latent needs by observing and analyzing the behaviors of a few representative users from various angles. For example, Cuerdo-Vilches et al. [27] described how the Spanish population perceived lockdown, how they related habitability and health aspects to housing, and how they have withstood the COVID-19 pandemic by using a survey and a photovoice method. They explained that it is useful to enrich the understanding of the topic in two aspects, quantitative and qualitative, through the mixed method.

Therefore, this study aims to identify the problems of the current residential spaces and suggest future space design directions by collecting and analyzing residents' living experiences during the COVID-19 period, in order to prevent and manage infectious diseases effectively. The participants in this study were apartment residents living in a metropolitan area of South Korea. The study examined their living experiences when quarantine was strictly enforced due to the COVID-19 pandemic. This study is divided into three stages: literature review, survey, and photovoice. It uses mixed-use methods to identify residents' vivid living experiences during the pandemic and offers practical design directions for pandemic-safe residential space plans based on user perspectives.

2. Materials and Methods

2.1. Literature Review for Identifying Planning Guides to Prevent and Control Pandemic Disease in Residential Apartments

We examined the trends and guides discussed in the literature regarding space planning in residential apartments to prevent and manage infectious diseases. A literature review was conducted using the methodology introduced by Snyder [28]. The search strategy followed the PRISMA guidelines [29] to select search terms and electronic databases, and we include an explanation of the inclusion and exclusion criteria.

2.1.1. Search Strategy

The three most used and valuable electronic databases were utilized to search academic papers comprehensively: Google Scholar, Web of Science, and Scopus [30]. For search terms, "apartment housing", "pandemic disease prevention and control", "COVID-19 pandemic", "space planning", and "residential space" were chosen as keywords. A cross-search was conducted by expanding the scope to words with meanings similar to residential apartments. The languages included both English and Korean to avoid excluding Koreans' residential experiences and needs.

2.1.2. Study Selection

From the literature search results of the three databases selected, a total of 3894 papers were searched. Among them, 251 papers were selected after reviewing the title and abstract of 2151 records, excluding redundant papers. The inclusion criteria included studies targeting apartment unit space and the exclusion criteria included studies dealing with the subjects of facilities, and architecture, building, and indoor environment requirements. As a result of assessing the relevance of the topic, 62 papers were selected, and their eligibility was closely evaluated through a full-text review. In this process, 3 papers were added from the reference list, and finally 24 papers were selected for analysis. The flow diagram of the literature review is shown in Figure 1. The final list of selected studies for review is provided in Appendix A.

2.2. Online Survey to Understand Residents' Overall Residential Experiences during the COVID-19 Pandemic

During the COVID-19 pandemic, we conducted an online survey to understand the overall residential experience of Seoul residents. The survey was conducted from 28 June to 2 July 2022 and targeted 201 adults who experienced self-quarantine during the pandemic. The participants were recruited by snowball-targeting Seoul residents. After developing the questionnaire using Google Forms, the survey link was delivered to the participants through text messages. The initial questionnaire was revised and finalized after a pilot test with researchers and ten Seoul residents who had experienced self-isolation.

The questionnaire comprised 23 items: 7 items for sociodemographic characteristics and residence type; 5 items for self-quarantine experience; 8 items for living experience during the pandemic; and 3 items for resident preference for pandemic-safe housing (Table 1 shows the items without sociodemographic characteristics). Two hundred thirty questionnaires were sent to the participants, and two hundred and one responses were received. For the data analysis, frequency analysis was conducted using SPSS version 25.0. The questionnaire obtained ethical approvement from the ethics committee of Chungbuk National University in June 2022.

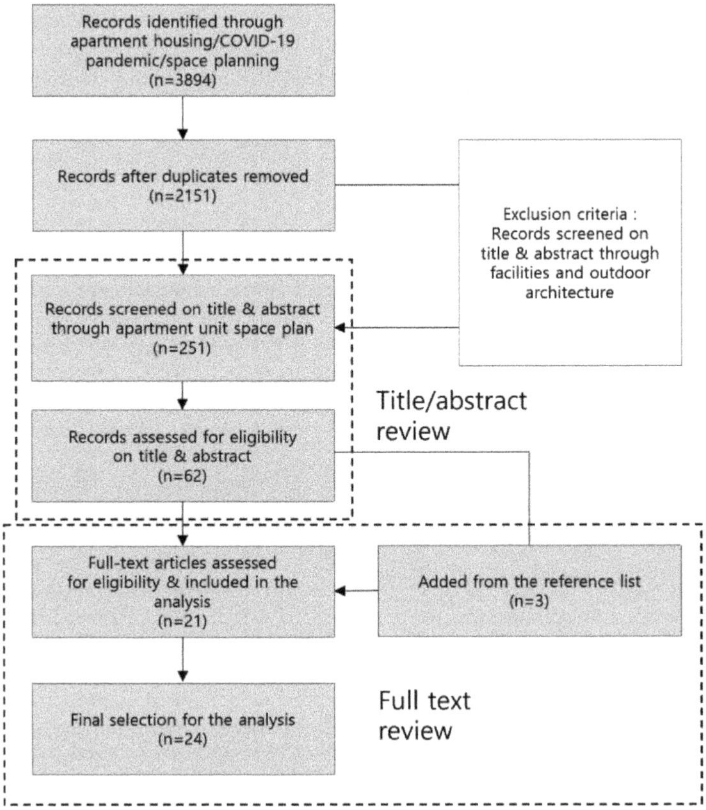

Figure 1. Flow diagram of the literature review.

2.3. *In-Depth Analysis of Apartment Residents' Living Experiences during the Pandemic*

At this stage, the photovoice technique was used to examine the living experiences of apartment residents more deeply. Photovoice is an effective method for direct interviews when dealing with sensitive or complex topics among participants by sharing one's experiences with others through photography [31]. Additionally, this method can ensure active participation in the entire process, as it can be conducted online without meeting in person [32].

In this study, to collect and analyze the residential experience during the pandemic more vividly through the voices and gazes of users, the photovoice method was divided into three stages: preliminary preparation, data collection, and data analysis.

Table 1. Questionnaire for the survey.

No.	Criteria	Questionnaire
1	Self-quarantine experience	Have you experienced self-quarantine during the COVID-19 pandemic? If yes, how long have you been in self-quarantine?
2		Where did you spend most of your time within your apartment dwelling during self-quarantine?
3		What activities did you mainly perform during self-quarantine?
4		What was the most uncomfortable aspect of living during self-quarantine?
5		During the self-quarantine period, what were you most dissatisfied with regarding the structure and function of each room?
6	Residential experience during the pandemic	Which areas of your residential space should be changed to prevent and control pandemic disease?
7		Which spaces were remodeled, restructured, or rearranged to meet the needs of residential space during the pandemic?
8		What problems do you face with your residential space in the context of the prevention and management of pandemics?
		8-1. entrance
		8-2. living room
		8-3. kitchen and dining
		8-4. bedroom
		8-5. bathroom
		8-6. balcony
14	Residents' preference for pandemic-safe housing	Where is the best place to store disinfection items such as hand sanitizer and masks?
15		What is the most preferred method of improving residential space to prevent and control pandemics?
16		What aspects should be considered most important in residential spaces to prevent and control pandemics?

First, photovoice participants were recruited in the preliminary preparation stage, and data collection and analysis methods were specified. In this study, families with elementary school children, who experienced the most significant changes in the behavior of family members owing to the social distancing policy during the pandemic, were selected as participants. The photovoice participants were housewives who interacted most actively with family members. The scope was limited to apartment residents in the metropolitan area (area less than 130 m^2), where social solid distancing policies were implemented due to the high density of residents. The participants were recruited through a purposive sampling method to select those willing to actively share their personal residence experience with others and participate in the study. As the general number of photovoice participants comprises approximately 7–10 people [33], we carefully selected seven participants after conducting brief interviews with those who expressed their willingness to participate and judging their suitability. To provide guidance and education on the photovoice method, orientation materials were prepared that introduced the purpose of the study. These materials also introduced how to take pictures and the research schedule. All the materials for photovoice obtained ethical approvement from the ethics committee of Chungbuk National University in June 2022. On 15 July 2022, we conducted a pre-education meeting for an hour on Zoom.

Second, residence experiences were collected in three stages in the data collection stage. In the first data collection stage, the participants were asked to freely take photos of spaces that revealed significant changes during the pandemic, spaces with negative living experiences, and their responses and improvements for approximately two weeks, during 15–29 July 2022. Subsequently, the participants selected approximately 20 images that were judged to adequately represent their living experience among the photos they had taken. Two representative photographs that best showed the problem were selected and submitted to the researcher with a brief description of the reason.

In the second stage of data collection, after reviewing the submitted photos, the researchers selected additional photos deemed meaningful and requested explanations from each participant. By synthesizing the primary and secondary data, images with similarities in space and function were categorized, and keywords were organized.

Moreover, a focus group interview with seven participants was conducted for approximately two hours using the classified image data (8 August 2022). When the images classified by the researcher were shown to the participants, they provided additional explanations about the photos they had taken. In-depth personal experiences were collected and shared by discussing aspects that resonated with the other participants' photos, experiences, or related opinions. Five questions from the SHOWed technique were used to present participants' opinions: (1) What do you see in this picture? (2) What is happening in this picture? (3) How does this picture relate to our lives? (4) Why do these problems, worries, and advantages exist? (5) What can we do to address the identified problems? [31].

Third, in the data analysis step, the previously collected data were divided into meaningful sentence units, assigned code numbers, and further divided into meaning units with core meanings. The interview process was recorded with prior consent, and all collected data were transcribed and documented. Categorization analysis was conducted according to context by extracting keywords that were repeatedly mentioned or conveyed meaning from the recorded documents.

2.4. Insights into Residents' Needs-Based Apartment Space Planning

Insights into residential apartment space plans were derived by synthesizing all the results obtained in the literature review, survey, and photovoice interviews. According to the guidelines identified in the literature review, the problems and experiences felt by residents were connected to the corresponding items to identify specific and practical needs. Based on the final insights, the specificity of the spatial planning needs of Korean apartment residents in response to infectious diseases is discussed.

3. Results

3.1. Literature Review for Identifying Planning Guides to Prevent and Control Pandemics in Residential Apartments

After identifying planning guides for effectively preventing and managing infectious diseases through reviews of 24 studies, 37 items were identified under five categories: general (G), entrance (EN), kitchen/dining room (KD), bedroom (BE), and balcony (BA). The general (G) category included 19 items: nine spaces (GS), two furniture (GF), three openings (GO), and five materials (GM). Two items were identified for the entrance (EN), and six items were collected for the kitchen/dining room (KD). Four items were found for the bedroom (BE), and six were identified for the balcony (BA) (Table 2).

In the general (G) category, the general space (GS) comprised variable walls, flexibility, access to outdoor space, biophilic, adjusting the opening and closing of spaces, access from the entrance to the bedroom to the living room, direct sunlight, and an improved mechanical ventilation rate. For general furniture (GF), fold/unfold flexibility/transformation and touchless motion sensor lights were emphasized. Regarding general openings (GO), ways to increase natural light and ventilation conditions were mentioned by planning multiple windows, large windows, ceiling windows, and folding doors. For general materials (GM), emphasis was placed on antibacterial materials, different finish glosses, and color control of the finish. In summary, the critical items for general apartment plans include spaces that change the structure of space and allow access to outdoor spaces; flexible and touchless furniture; different types of windows/doors for sunlight and ventilation; and variations in antibacterial and finishing materials.

Table 2. Summary of the literature review: space planning guides for pandemic-safe residential housing (* indicates items mentioned more than five times).

Category	Sub-Category	Code	Planning Guide	Keywords	Reference Code
General	Space	GS1	Minimize the placement of load-bearing walls and plan them as flexible walls. *	Flexible wall	A-1, D-1, D-4, D-5
		GS2	Plan an outdoor space of an appropriate size in addition to the indoor space for social distancing.	Access to outdoor	D-4
		GS3	Utilize biophilic design that actively introduces natural elements into the interior space.	Biophilic design	D-4, D-5
		GS4	Plan various spaces within the unit to allow easy access to the outdoors.	Access to outdoor	D-1, D-4
		GS5	Plan to open and close the space easily by utilizing movable partitions. *	Flexible wall to adjust opening/closing	A-3, A-12, D-1, D-7
		GS6	Remove the wall between the living room and the bedroom to secure variability. *	Flexible wall	A-12, A-14, A-17, D-4, D-5
		GS7	Plan to first enter each bedroom from the entrance and then connect to the living room.	Direct access from the entrance to the bedroom	D-4, D-5
		GS8	Plan to get as much direct sunlight as possible into the living space.	Sufficient sunlight	A-2
		GS9	Increase the mechanical ventilation rate to create a pleasant and hygienic living space.	Improved mechanical ventilation	D-7
	Furniture	GF1	Plan the furniture to be foldable to be unfolded and used when needed.	Folding to unfolding transformation	D-1, D-4
		GF2	Install motion lighting that reduces high-touch surfaces via motion switches.	Touchless motion sensor	D-1
	Opening	GO1	Plan at least several small windows if large windows cannot be installed.	Multiple windows	D-1
		GO2	Plan to receive as much natural light as possible through a multi-pane or large window plan.	Multiple/large windows for sufficient sunlight inflow	D-1, D-4
		GO3	Plan to install skylights or folding doors to let fresh air in.	Ceiling windows/folding door for fresh air	D-1, D-4
	Material	GM1	Use antibacterial, mold-, odor-, and stain-resistant finishing materials (including outdoor fabric) for outdoor spaces.	Antibacterial material	D-1, D-4
		GM2	Use finishes with natural antibacterial properties that eliminate germs and bacteria.	Antibacterial material	D-1, D-4
		GM3	Plan to vary the gloss level of the finish to add visual interest.	Finish variation	D-1, D-4
		GM4	Plan a comfortable living space using natural and calm colors.	Color variation	D-1, D-4
		GM5	Use bold and bright colors for wallpaper, and plan vivid colors for walls to be emphasized.	Color variation	D-1, D-4

Table 2. *Cont.*

Category	Sub-Category	Code	Planning Guide	Keywords	Reference Code
Entrance		EN1	Plan a dry wash area and an air shower zone to wash hands right after entering the front door. *	Facility for removing contaminants	A-17, D-1, D-5, D-7
		EN2	Install an inner gate between the entrance and the living space, and plan a dressing room at the entrance to remove contaminants from clothes before passing through the inner gate. *	Facility for removing contaminants	A-1, A-2, A-5, A-12, A-14, D-1, D-5, D-4
Kitchen/Dining room		KD1	Install germ-resistant countertops and flooring, touchless faucets, and appliances.	Antibacterial material, touchless appliances	D-1, D-4
		KD2	Plan a dining room that can be used as a community space between the kitchen and living room.	Multifunctional dining room	D-4, D-5
		KD3	Install home appliances using non-contact systems such as voice recognition and remote control.	Touchless appliances	A-5
		KD4	Plan to let natural light in or use natural lighting to spend more time in the kitchen.	Sufficient sunlight inflow	A-5
		KD5	Divide the kitchen into a separate space or plan a second kitchen.	Separate kitchen	D-5
		KD6	Plan an island kitchen or dining room using a semi-closed pocket door for a multifunctional space such as a home office.	Multifunctional dining room	D-4
Bedroom		BE1	Plan to have one basic bathroom for each bedroom.	Individual bathroom for each room	A-14
		BE2	The main bedroom should be planned as an isolated space among the living spaces.	A separate main bedroom	A-5, A-14
		BE3	The bedroom should be planned to be divided into multiple spaces by applying variable elements to support various activities such as work, hobbies, and relaxation.	Flexibility for multifunctional use	A-1, D-7
		BE4	The bedroom should be equipped with lighting that changes to the appropriate color and brightness to allow for various activities and supports the variability of the space.	Flexible light for multifunctional use	D-4
Balcony		BA1	Plan a balcony as a semi-public space to communicate with neighbors.	Enhanced communication with neighbors	D-1, D-4
		BA2	Plan the balcony for various purposes (e.g., home cafe, home camping, play area, vegetable garden).	Multifunctional support	A-9, D-5, D-7
		BA3	Plan a specialized floor plan with an external space for each unit.	Access to outdoors	D-6, D-7
		BA4	Connect the balcony with the garden or nature, and plan it as an emotionally comfortable space.	Garden planning	D-4
		BA5	Plan gardens to improve mental health and reduce dependence on grocery stores.	Garden planning	D-4
		BA6	Plan balconies with flexible walls or doors to be easily opened and closed.	Flexible wall to adjust opening and closing	D-4

In the entrance (EN), the installation of wash basins, air showers, dressing rooms, and foyer doors was emphasized to remove external contaminants effectively. For kitchens/dining rooms (KD), antibacterial materials, touchless appliances, adequate sunlight, and separate kitchens with multifunctional dining rooms were emphasized. For the bedroom (BE), individual bathroom plans, a separate main bedroom, and variable elements (including lighting) for multifunctional support were identified as essential guide items. The balcony (BA) highlighted enhanced communication with neighbors, multifunctional support, garden planning, and flexibility to adjust opening and closing.

Among these, five items mentioned more than five times in the literature were identified. The essential keywords can be summarized as flexible walls to adjust the opening/closing for multifunctional use (GS1, GS5, and GS6) and facility plans to remove contaminants at the entrance (EN1 and EN2). Removing contaminants at the entrance and flexible wall structures throughout the indoor space are critical for effectively managing and preventing infectious diseases.

3.2. Analysis of the Overall Residential Experiences of Apartment Housing Residents during the COVID-19 Pandemic

3.2.1. Residential Experiences of Apartment Housing Residents According to Self-Quarantine Experiences

We conducted an online survey to determine how apartment residents in Seoul felt about their living spaces while experiencing self-quarantine during the COVID-19 pandemic. The survey respondents had experienced self-quarantine within their residence for as few as 4 days and as many as 14 days. They spent most of their time in bedrooms (59.7%) and living rooms (29.9%) during quarantine. The main activities performed during self-isolation were rest (40.9%), sleep (22.0%), hobbies (12.3%), telecommuting (11.4%), and attending remote classes (7.7%). This result shows that social activities conducted outside homes before the COVID-19 outbreak were brought into residential spaces, and various activities occurred in a complex manner.

Regarding their dissatisfaction with space during the period of self-quarantine, hobbies (32.5%), rest (17.9%), telecommuting (13.2%), sleep (10.3%), and remote classes (9.8%) were identified in the order for which space and furniture did not adequately support daily activities. The result implied a functional limitation in supporting activities that were previously conducted outside when performed inside the dwelling, since it was impossible to leave (Figure 2).

3.2.2. Residential Experience during the COVID-19 Pandemic

As for the residents' dissatisfaction regarding the structure of the residential space and the function of each room, it was emphasized that there was no independent space for performing a specific function or that there was a space limit in accommodating the changed function due to the narrow area of the existing space. Further, two bathrooms were necessary for separate toilet use when family members were infected with COVID-19.

The spaces that residents felt the greatest need to change in residential housing during the pandemic were living rooms (21.7%), kitchens (12.4%), master bedrooms (9.6%), children's bedrooms (8.3%), bathrooms (8.0%), balconies (8.0%), entrances (7.6%), and study rooms or libraries (7.6%). Most of their time was spent in the bedroom and living room during self-quarantine. However, a change in the living room was found to be the most necessary because family members with different lifestyles gathered and spent time together in the living room.

When asked about their experience remodeling spaces or rearranging furniture as needed during the pandemic, living room changes (18.8%) showed the highest rate, followed by master bedrooms (10.2%), children's bedrooms (7.5%), kitchens (7.1%), and study rooms (6.0%). This result indicated that a change in the function of the living room and a consequent change in space planning were top priorities (Figure 3).

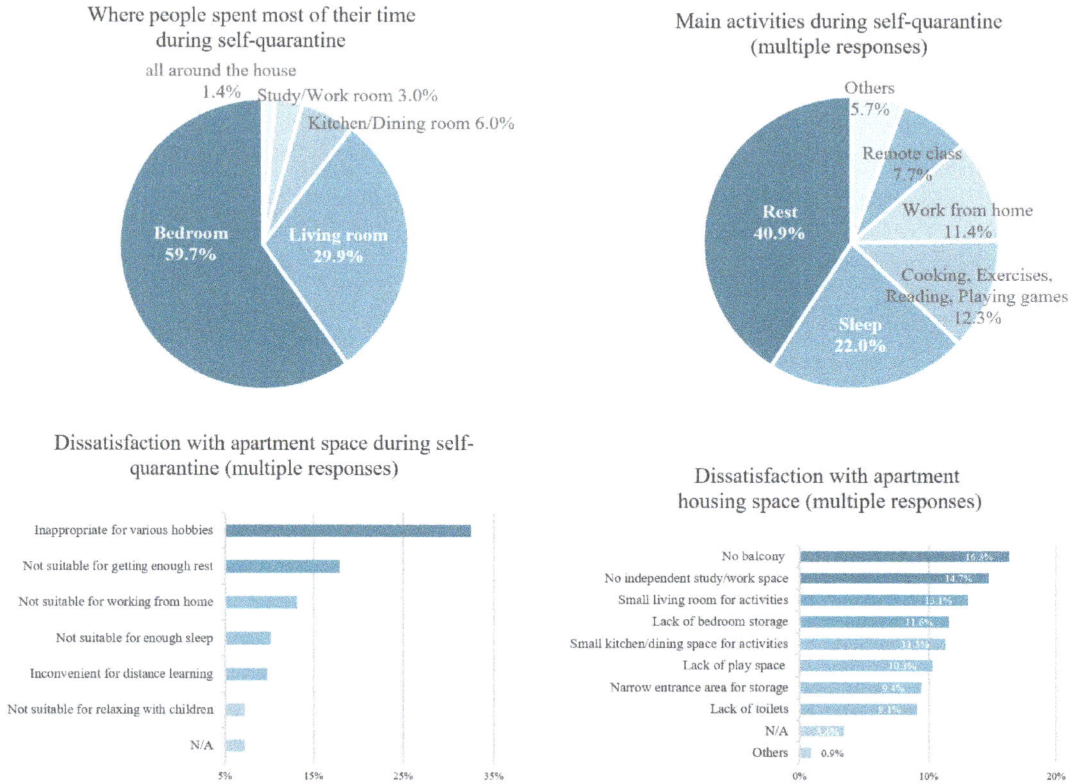

Figure 2. Summary of apartment residential experiences identified from the survey (1).

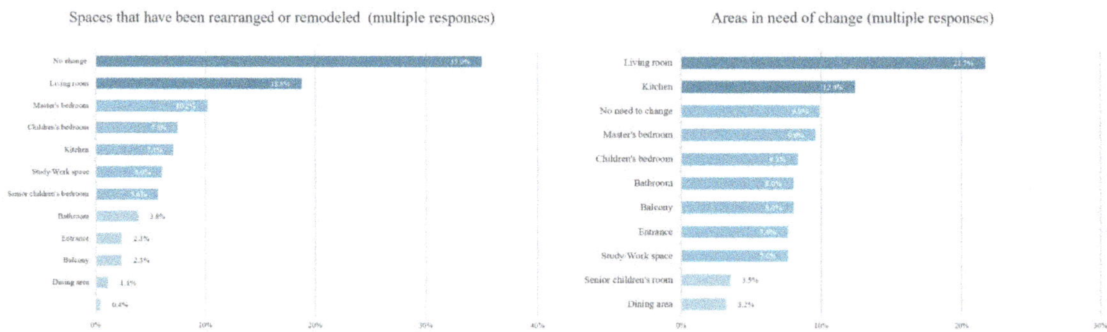

Figure 3. Summary of apartment residential experiences identified from the survey (2).

3.2.3. Resident Preference for Pandemic-Safe Housing

The preferred place to store quarantine items was the foyer room closest to the entrance (74.1%). As a space improvement method to prevent and manage infectious diseases, when considering cost, time, and realistic aspects, rather than changing the structure of the space, a simple and convenient method through furniture rearrangement (29.4%) was preferred. Subsequently, methods such as changing the function of the space (21.4%) and partial remodeling (16.4%) were preferred. It was found that the method of slightly altering the space according to the needs of residents was the most applicable.

The most relevant considerations for the effective prevention and management of infectious diseases were ventilation (52.7%), independent personal space (17.4%), and a hygienic entrance (10.9%). The emphasis on ventilation and sanitary entrances is interpreted to be due to the high rates of viral transmission and infection through the air. In addition, because family members are absorbed in several different activities indoors, securing the efficiency and privacy of their work or actions is essential (Figure 4).

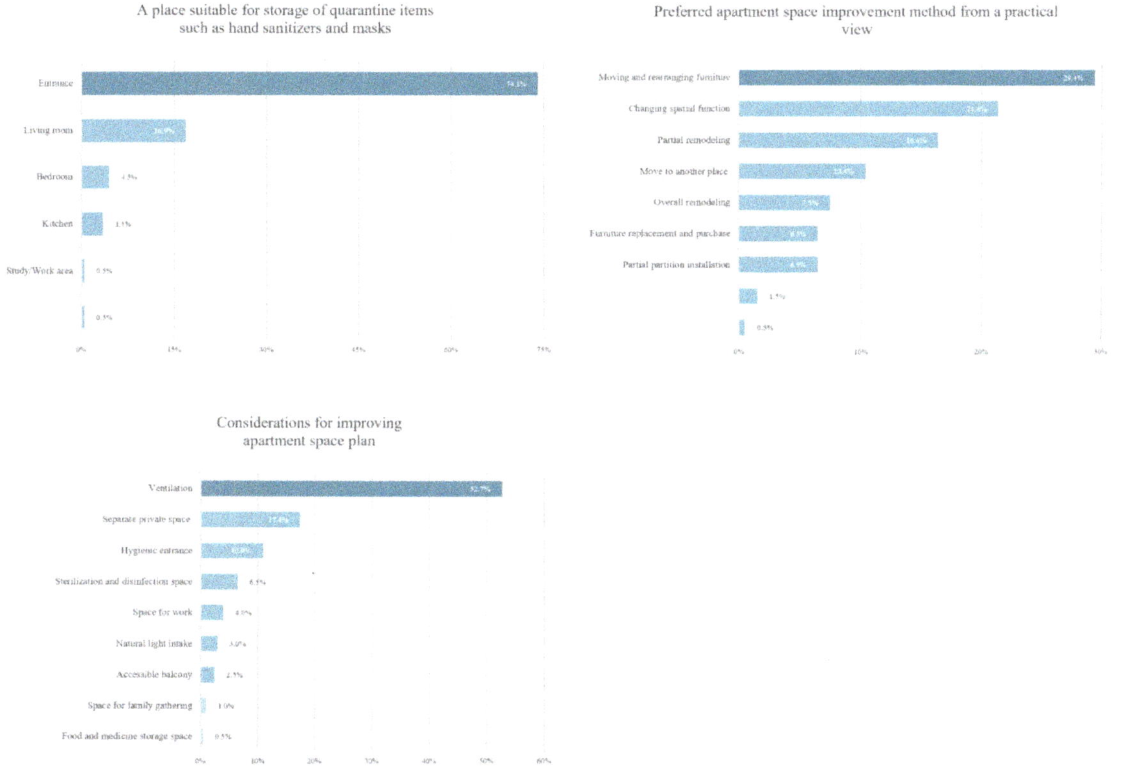

Figure 4. Summary of apartment residential experiences identified from the survey (3).

3.2.4. Summary of Apartment-Housing Residential Experience during the COVID-19 Pandemic

In summary, the pandemic caused residents to primarily occupy their bedrooms and living rooms, where they spent the majority of their time engaging in restful activities such as sleeping (62.9%), as well as pursuing their hobbies, attending remote classes, and telecommuting (31.4%). Regarding dissatisfaction with the living experience, there was no independent space for a specific function and a narrow area to accommodate the changed function, and the need for two bathrooms appeared. This revealed the need to support functions newly absorbed into the indoor space due to the pandemic. The greatest need to change spaces appeared in the living room, bedroom, and kitchen—the place where the respondents spent the maximum amount of time showed a high need for improvement. Regarding actual space improvements, living rooms and bedrooms had the most scope, indicating that residents improved spaces out of necessity.

To prevent and manage infectious diseases effectively, priority was given to minimizing the spread of viruses through ventilation and sanitary entrances. Subsequently, securing an independent personal space was emphasized to increase the efficiency of the various activities of the family members without an invasion of privacy. This suggests that it is worthwhile to increase the functionality of a private bedroom to support multiple uses rather than a living room shared by family members.

In terms of space improvement methods, furniture rearrangement, space function changes, and partial remodeling were preferred. This means that the space should be repurposed in a way that can easily change its function and ensure flexible use through furniture rearrangements.

The above discussion summarizes the insights for space planning based on their living experience during the COVID-19 pandemic that apartment residents described in their responses, as explained through the following steps: (1) reinforcing ventilation and front-door hygiene functions to prevent/manage infectious diseases; (2) supporting multifunctional use in living rooms and bedrooms; (3) securing an independent personal space that can secure hobbies, telecommuting efficiency, and privacy; (4) relaxing centered on the bedroom and enhancement of sleep function.

3.3. In-Depth Analysis of Apartment-Housing Residential Experiences during the Pandemic

The following needs were identified for each space as a result of conducting photovoice interviews with seven participants (Table 3). At the entrance, insufficient storage space was emphasized as a negative living experience. In the living room, family members' different activities and needs cause problems such as noise, distraction, and privacy issues, and the low efficiency of individual activities was highlighted as a primary problem. The dining room and kitchen are often used as temporary workspaces when they are not furnished separately. Complaints were made regarding the low work efficiency and concentration level owing to the frequent access of other family members to the workspace. The integrated bedroom, workspace, play, sleep, and study space resulted in poor concentration and poor-quality sleep. Regarding the balcony, it was pointed out that it is difficult to control the temperature according to the season and that the lack of facilities to support various activities reduces space utilization.

3.4. Insights into Residents' Needs-Based Apartment Housing Space Planning

The following summarizes spatial planning insights from apartment residents' living experiences obtained through a literature review, survey, and photovoice interviews (Table 4).

The literature review emphasized two aspects of planning a pandemic-safe residential space: (1) a flexible space plan using flexible walls, folding doors, or partitions, and (2) a facility for removing contaminants.

Regarding the flexible space plan, further considerations were derived for the living room, bedroom, hobby/telecommuting/learning space, and balcony. About the living room, support for multifunctional use was emphasized. Support for multifunctional use and reinforcement of sleeping and resting functions were highlighted in the bedroom. Securing separate, individual spaces was essential to ensure efficiency and privacy regarding hobbies, telecommuting, and learning spaces. Regarding the balcony, the need for indoor ventilation was highlighted. Regarding removing contaminants, the reinforcement of the hygiene function at the entrance was emphasized, indicating that the role of the entrance hall has become more crucial in blocking pollutants from the outside.

Additionally, residents' practical and specific needs in each space were identified. To support multifunctional living rooms, residents preferred separated areas that could be opened or closed. It was also emphasized that multifunctional support facilities, such as lighting, windows, ventilation, air conditioning, and heating, which can be individually controlled, should be built together. Regarding bedrooms, the need to prepare isolation spaces has been highlighted by suggestions of securing flexible furniture, facility planning, and exclusive toilets to support multifunctional use. Separating telecommuting spaces from study/sleep/play areas was also necessary to facilitate sleeping/resting functions.

Table 3. Barriers to the residential experience and insights for improvement identified from the photovoice.

Space	Quotes	Residential Experience	Insights
Entrance	"I bought sports equipment such as skateboards and badminton, but there was nowhere to store them, so I put them in the corner of the entrance. I had to clean it every time I opened and closed the shoe cabinet, which made my work cumbersome and the entrance even narrower." (C)	- There is insufficient storage space for sanitary items such as disinfectants and masks at the entrance. - There is insufficient space to store exercise equipment; therefore, it is stored in the corner of the entrance.	- It is necessary to secure sufficient storage space and an area to store various sanitary products and exercise equipment.
Living room	"I have two children, one on the sofa in the living room and the other on the dining table for their remote class. The kids are still young; therefore, I have to take care of them, but when they take classes simultaneously, they can hear each other; thus, it gets in the way. I could not concentrate, so I separated the older kid so he could take classes in his room." (C)	- Besides rest and family gatherings in the living room, various functions such as children's learning, couples working from home, and hobbies are introduced, all complex activities. - Activities between family members conflict at different times of day, causing discomfort such as noise, privacy issues, and activity restrictions. - When two or more children share a learning space, efficiency decreases due to noise and distraction. - When the living room area is separated, there is a darkening problem because there is no ceiling light on one side.	- Areas should be separated, or independent spaces should be secured so family members can engage in various activities within the living room. - The learning space must be clearly separated if there are two or more children. - Facilities that can support multiple functions are required to utilize the living room for various purposes. - When separating areas, lighting, windows, ventilation, air conditioning, and heating must be adjusted and considered to secure individual facilities.
Dining/Kitchen	"I work from home; however, I do not have a separate space for work, and I do it at the dining table. Even if the kids want to eat or have a snack, they cannot eat comfortably because of me." (E) "As I spend more time at home, I buy more things, such as necessities, hygiene products, and favorite foods. However, I do not have space to store them, so I just lay them outside." (F)	- It is close to households' movements and has high accessibility to their children; therefore, they work from home, mainly at the table, but their work efficiency and concentration are low. - As the number of disinfectants, sterilizers, favorite foods, and delivery foods increased, there was a shortage of space to store items and more space was needed to separate and organize garbage.	- A separate and independent space for telecommuting must be secured separately from the dining table. - Additional storage space must be secured to store hygiene products and hobbies. - Due to the increase in household waste, an efficient sanitary storage space must be secured for separate collection and organization of items.

Table 3. *Cont.*

Space	Quotes	Residential Experience	Insights
Bedroom (Main Bedroom)	"I do not have a place to work, so I am working from home with a desk and chair in the bedroom. It is inconvenient not to be able to work while my husband or children are sleeping." (G)	- If there is a home office space in the bedroom, it interferes with the sleep of other family members and restricts work hours, reducing concentration and efficiency. - The couple's bedroom has a larger area than other rooms, but it is challenging for other purposes due to fixed furniture/facilities such as a built-in wardrobe and dressing table. - The bathroom in the bedroom was used as an isolation space when COVID-19 was confirmed.	- I need an independent work-from-home space where I can concentrate on my work without disturbing the sleep of other family members. - Flexible furniture/facilities should be planned so that functions can be switched to various uses. - A dedicated toilet is secured in the room so that it can be separated into an isolated living space in the event of confirmation.
Bedroom (Children)	"When my child is taking an online class, he tries to lie down on the bed and cannot concentrate. So, I moved the child's bed to our bedroom and created a learning atmosphere in the child's room." (A)	- In the case of lower-graders, it is challenging to create a learning atmosphere because beds and toys are exposed during online learning.	- In the case of children in the lower grades, it is necessary to separate learning and sleeping functions to create a learning atmosphere.
Space for Study/Work	"Originally, the study was shared by the family, but as my husband increased his work from home, he needed an independent space due to noise and privacy issues, so we separated it." (B)	- When both husband and wife work and children learn and play in the study, efficiency is lowered due to noise and a lack of concentration.	- Spaces for work and study should be provided with independent spaces to minimize noise and increase concentration.
Balcony	"Unable to go out, I planted tomatoes and plants on the balcony and created a rest area to listen to the sound of rain or camp. However, it was too cold in winter and too hot in summer, so it was difficult to use." (B)	- It was used as a space for various leisure/hobby activities such as plant cultivation, music appreciation, and camping. - It was hot in the summer and cold in the winter, so the time to move or use the exercise equipment was limited. - There is no place to store exercise equipment, camping equipment, and toys when not in use.	- Balconies are reinforced for leisure and hobbies, requiring structures and facilities to support various activities. - A temperature control function should enable space utilization in the summer and winter. - There is a need for space to store items related to hobbies/leisure activities.

Table 4. Synthesized results of identifying user needs for a pandemic-safe housing space plan.

No.	Literature Review	Survey		Photovoice
1	Flexible space plan (flexible wall, folding door, and partition)	Living room	Support for multifunctional use	- Securing area separation or separate independent space to support various activities - Multifunctional support facility planning for multipurpose use - Individual control/security of facilities such as lighting, windows, ventilation, and air conditioning when separating areas
		Bedroom	Support for multifunctional use	- Flexible furniture/facility planning to support multifunctional spaces - Securing a separate toilet in the room in preparation for quarantine
			Strengthen sleep and rest functions	- Securing an independent homework space that does not disturb other family members' sleeping - Classification/separation of learning, sleeping, and play functions/areas
		Space for hobbies/work from home/learning	Secure an independent individual space for efficiency and privacy	- Separate learning space for two or more children - Securing an independent space for working from home, separate from the dining table (minimizing noise, considering privacy, maximizing work efficiency) - Securing storage space for related items
		Balcony	Ventilation	- Building related facilities on balconies to support leisure/hobby activities - Plan to adjust the temperature by season - Securing storage space for related items
2	Facility for removing contaminants	Entrance	Enhancement of hygiene function	- Securing sufficient storage space and space for sanitary products, delivery service, and exercise equipment - Efficient sanitary storage space planning for separate garbage collection

To secure a separate individual space for hobbies/telecommuting/learning, the separation of learning spaces for each child, ensuring independence of the home workspace, and providing storage space for related items were required. For the balcony, while the importance of ventilation was emphasized in the survey, the construction of related facilities to support leisure/hobby activities, seasonal temperature control, and securing storage space was mentioned in the photovoice. These results show that practical space utilization was considered a priority for residents. The need for storage space for sanitary products, delivery services, exercise equipment, and garbage disposal spaces was emphasized at the entrance.

4. Discussion

Considering the above results, space planning for preventing and managing infectious diseases among apartment residents in South Korea differs slightly from the results of previous studies. Previous studies [34–43] have emphasized the importance of removing contaminants by installing dry wash basins, air showers, inner gates, and entrance dressing rooms to strengthen sanitation. The opinions of apartment residents in this study indicate that securing efficient storage space for various items is more important than adding new facilities. The observed discrepancy suggests that the incorporation of supplementary amenities is practically challenging due to the financial strain and limited foyer space when implementing the optimal amenities in contemporary residential areas. Therefore, to narrow the gap between the ideal and reality, a plan to increase space utilization and share functions can be proposed by planning the circulation from the entrance to the living room through each room and linking each room, such as the entrance, laundry room, multipurpose room, and bathroom.

For living rooms, eliminating walls and flexibly adjusting space divisions using flexible walls, doors, and partitions have been emphasized in previous studies [34,37–44]. Korean apartment residents emphasized that although separating areas is necessary, establishing flexible facilities that can be individually adjusted for each space should also be considered. The reason for this is that the residents are taking into account not just the physical structure, but also the practicality and usage of the space, showing that there is a limit to the utilization of space with simple spatial division and no consideration of facilities. Previous studies [34,43] have emphasized that bedrooms should be able to support various activities such as work, rest, and hobbies. Korean residents prefer separate workspaces and study spaces for work efficiency, noise blocking, and privacy. Notably, they did not want the sleep and rest functions of the bedroom to be replaced with other functions. However, securing an individual bathroom for each bedroom and using it as an isolation space when an infectious disease is confirmed is consistent with the results of previous studies [36,38]. Taking these results together, at least one bedroom needs to be placed as far from the living room as possible and it should have an individual bathroom with the function of an isolation space. Additionally, it is desirable to strengthen the functions of sleeping and resting by separating them from learning and working.

In previous studies, antibacterial materials, touchless systems [40,41], securing natural light [32], and planning a dining room as a community space between the kitchen and living room [41,42] were emphasized to secure the independent functions of the kitchen. This aspect has the same context, as residents do not prefer the dining table as a homework and study space. Activities such as meal preparation, eating, studying, and working are unique functions that cannot be shared in one place; therefore, the study or workspace should be prepared as a separate space rather than using a dining table when considering noise, privacy, and work efficiency. In particular, because the kitchen is a place that all family members frequently access, functions that require concentration, such as studying or working, must be separated from it.

Regarding balconies, contents such as communication with neighbors, connection/expansion to external spaces, planting gardens, and vegetable garden utilization have been highlighted in previous studies [40,41]. However, for domestic residents, functions such as exercise, camping, and rest were emphasized, whereas functions such as plant cultivation, gardening, and kitchen gardens were relatively passive. Additionally, the need to build practical facilities to maximize space utilization, such as outlets, storage spaces, and temperature control functions to support multipurpose use was emphasized. Notably, while communication and networks with neighbors were considered necessary during the epidemic of infectious diseases abroad, in Korea, communication and contact with the outside were blocked entirely, and outside activities continued in a closed manner via the balcony.

In short, the guidelines of the reviewed literature (Table 2) on preventing and managing infectious diseases emphasized flexibility, communication with the outside world, and establishing facilities to block pollutants. Korean residents have expressed the importance of securing semi-enclosed flexibility within residential spaces, constructing facilities to maximize space utilization, and separating study and workspaces to increase efficiency. This difference demonstrates the practical perspective of domestic residents on maximizing multifunctionality and efficiency within a limited area. To effectively respond to future infectious diseases, the unit space of an apartment building should consider the following five critical directions: (1) enhancing sanitary function/space utilization by linking rooms around the entrance; (2) variable separation of living areas and securing flexibility of related facilities; (3) functional separation and independence of study, workspace, and kitchen; (4) considering the use of isolation space in the bedroom and strengthening the sleeping function; (5) provision of related practical facilities on the balcony to support leisure/hobby activities.

This study verified and presented the direction of spatial planning for preventing and managing infectious diseases as objectively as possible through a triangulation of literature reviews, quantitative surveys, and qualitative user interviews (photovoice). However, as the COVID-19 pandemic is a recent occurrence, it has a limitation in that the number of papers and reports considered is limited compared to other topics, and this study has focused on a specific period. Additionally, the photovoice interviews conducted to reflect the actual voice of the user in detail had seven participants. Although it is included in the range of adequate numbers required to identify needs, the additional needs of many users still need to be discovered. When domestic residents' experience-based space planning needs are continuously met, a residential space design with high resident satisfaction is proposed. In addition, since the main purpose of this study was to understand practical experiences such as behavior, emotion, and interactions of residents, indoor environment requirements such as ventilation, windward flow, and fresh air were not directly addressed within the scope of this study. Experts were not included as participants because the focus of this study was to collect and analyze the vivid voices of urban residents. For this reason, rather than presenting a detailed and professional-level planning guideline, the results were limited to presenting critical insights that can be referred to in setting the criteria and direction of future residential space plans.

5. Conclusions

The COVID-19 pandemic has become endemic; however, many researchers are still considering the potential occurrence and spread of different types of infectious diseases. Continuous study of space planning is essential for residential spaces to function as comfortable and safe shelters for humans, especially regarding the effective prevention and management of infectious diseases, in terms of public health. This is due to the fact that residents consider not only the physical structure but also the practicality and usage of the space. In terms of flexibility, Korean urban residents showed greater interest in efficient space utilization such as storage and access to outlets, switches, and windows when dividing the space. In addition, it was remarkable that spatial support for outdoor physical activities such as exercise and camping was more important than plants and gardening for Korean residents.

Understanding residents' potential and practical needs is usually limited to presenting a theoretical guide or collecting quantitative opinions. Thus, we used a mixed-use method, as it is essential to understand user needs from multiple perspectives. This study is significant because it deeply grasps the living experiences of Korean residents during the COVID-19 period and derives practical design directions based on these experiences. The reason why we still need such research at the end of the pandemic is that public health issues including infectious diseases are critically related to human safety and must be jointly addressed worldwide. Recalling that the level of quarantine among countries was not the same during the COVID-19 pandemic, in order for the world to properly respond to infectious diseases in the future, it is critical to systematically understand the experiences we have gone through. The results of this study can be used as a framework for designing user research methods for user-centered residential space designs in the future. Finally, we believe that the insights for space planning presented in this study can systematically respond to future public health issues including infectious diseases, and serve as a guide to reflect the practical needs of urban residents.

Author Contributions: Conceptualization, M.-K.K.; methodology, M.-K.K. and E.-J.K.; validation, M.-K.K. and E.-J.K.; formal analysis, E.-J.K.; investigation, E.-J.K.; writing—original draft preparation, E.-J.K.; writing—review and editing, M.-K.K.; supervision, M.-K.K.; project administration, M.-K.K. All authors have read and agreed to the published version of the manuscript.

Funding: This research was funded by the National Research Foundation of Korea (NRF), grant number NRF-2021R1F1A1062673.

Institutional Review Board Statement: The study was conducted in accordance with the Declaration of Helsinki, and approved by the Ethics committee of Chungbuk National University (CBNU-202206-HR-0138, 24 June 2022).

Data Availability Statement: The datasets used and analyzed during the current study are available from the corresponding author upon reasonable request.

Acknowledgments: We thank Seong-Ju Han, Ye-Bin Jeon, and Ji-Eun Park of Chungbuk National University for supporting data collection for this study.

Conflicts of Interest: The authors declare no conflict of interest.

Appendix A

Table A1. Reviewed literature.

Code.	Authors (Year)	Title
A-1	AIA (2020 [34]	Strategies for Safer Multifamily Housing
A-2	AIA (2020) [35]	Re-occupancy Assessment Tool V3.0
A-3	Morgan, B. (2020) [44]	4 Ways COVID Has Changed Home Design
A-4	Peters, T., & Halleran, A (2020) [7]	How Our Homes Impact Our Health: Using a COVID-19 Informed Approach to Examine Urban Apartment Housing
A-5	Lee, H. K. (2021) [36]	Hyundai Engineering Develops Specialized Flats Tailored to the Era of COVID-19
A-6	Lee, S. W. (2021) [45]	Synergy between ePyeonhansesang C2 House and 4Bay innovative flat.
A-7	Massenburg, D. R. (2021) [46]	The house that quarantine built: Post-pandemic home design trends
A-8	Scileppi, T. (2021) [47]	The Latest Home Design Trends of 2021 as told by Debra Design Group
A-9	An, S. Y. (2022) [48]	Catch end-users who have become strict... Constructor plane specialization war.
A-10	Spennemann, D.H. (2022) [49]	Designing for COVID-2x: Reflecting on Future-Proofing

Table A1. *Cont.*

Code.	Authors (Year)	Title
A-11	Elrayies, G. M. (2022) [50]	Prophylactic Architecture: Formulating the Concept of Pandemic-Resilient Home
A-12	Lee, K. W. (2022) [37]	LH, Proposes an Integrated Public Rental Housing Plan Tailored to the Life Cycle
A-13	Glavan et al. (2022) [51]	COVID-19 and City Space: Impact and Perspectives
A-14	Tanamas et al. (2022) [38]	Space Adaptations During Pandemic in Apartments
A-15	Benbow, W. (2022) [52]	COVID-19 in Long-Term Care: The Built Environment Impact on Infection Control
A-16	Ching & Rani (2023) [53]	The Impact of COVID-19 Pandemic on Home Spatial Design
A-17	Chen et al. (2023) [39]	Effects of COVID-19 on Residential Planning and Design: A Scientometric Analysis
D-1	Kim, M. S. (2021) [40]	A study on Architectural Approaches Corresponding to the Post-COVID Era
D-2	Kim et al. (2022) [54]	Analysis of Prior Research Studies to Develop Guidelines for Apartment Housing Planning in Response to Infectious Disease Disasters
D-3	Shin et al. (2022) [55]	The Architectural Planning Strategies for Multifamily Housing in a Post-Pandemic World
D-4	Lee et al. (2022) [41]	Changes in Housing Function and Space Preference in the Post COVID Era
D-5	Jeon et al. (2022) [42]	Reconstruction of Residential Space in the Post-COVID Era
D-6	Kim et al. (2023) [56]	Analysis of User Experience Using Photo Voice Technique for Apartment Residential Space Planning in Responding to Infectious Diseases
D-7	Park et al. (2023) [43]	Analysis of Residents' Interest Areas and Preference of Based on Eye Tracking for Apartment Planning in Response to Infectious Disease

References

1. Saied, A.A.; Metwally, A.A.; Madkhali, N.A.B.; Haque, S.; Dhama, K. Egypt's COVID-19 Recent Happenings and Perspectives: A Mini-Review. *Front. Public Health* **2021**, *9*, 696082. [CrossRef]
2. Tokazhanov, G.; Tleuken, A.; Guney, M.; Turkyilmaz, A.; Karaca, F. Assessment Method for New Sustainability Indicators Providing Pandemic Resilience for Residential Buildings. *MethodsX* **2021**, *8*, 101577. [CrossRef] [PubMed]
3. Bettaieb, D.M.; Alsabban, R. Emerging Living Styles Post-COVID-19: Housing Flexibility as a Fundamental Requirement for Apartments in Jeddah. *Archnet-IJAR Int. J. Archit. Res.* **2020**, *15*, 28–50. [CrossRef]
4. Al-Qaisi, M.O.M. An overview of post-pandemic housing: Through stay-at-home experience. *Arab. J. Sci. Publ.* **2021**, *2663*, 5798.
5. Alhadedy, N.H.; Gabr, H.S. Home Design Features Post-COVID-19. *J. Eng. Appl. Sci.* **2022**, *69*, 87. [CrossRef]
6. Aqarmap Guide. Available online: https://advice.aqarmap.com.eg/en/aqarmap-guide-to-the-most-important-real-estate-exhibitions-in-egypt-and-the-arab-world (accessed on 5 March 2022).
7. Peters, T.; Halleran, A. How Our Homes Impact Our Health: Using a COVID-19 Informed Approach to Examine Urban Apartment Housing. *Archnet-IJAR Int. J. Archit. Res.* **2020**, *15*, 10–27. [CrossRef]
8. Alhusban, A.A.; Alhusban, S.A.; Alhusban, M.A. How the COVID 19 Pandemic Would Change the Future of Architectural Design. *J. Eng. Des. Technol.* **2022**, *20*, 339–357, *ahead-of-print*. [CrossRef]
9. Erti Idawati, D. Sustainable Housing as an Approach to Adaptive Built Environment to COVID-19 (Case Study: Keudah Walk-up Flat in Banda Aceh). *IOP Conf. Ser. Earth Environ. Sci.* **2021**, *881*, 012012. [CrossRef]
10. Xu, J. *The Impact of Pandemics on Future Residential Buildings in China*; Rochester Institute of Technology: Rochester, NY, USA, 2020.
11. Liu, S.; Su, Y. The impact of the COVID-19 pandemic on the demand for density: Evidence from the US housing market. *Econ. Lett.* **2021**, *207*, 110010. [CrossRef]
12. Hipwood, T. *Coronavirus: An Architect on How the Pandemic Could Change Our Homes Forever*; The Conversation: London, UK, 2020.
13. Bahadursingh, N. *Ways COVID-19 Will Change Architecture*; Architizer: New York, NY, USA, 2020.
14. Molaei, P.; Hashempour, P.; Tang, L.M. Semi-open spaces of apartments considering COVID-19 pandemic: General expectations of balcony design in the post-pandemic world. *Archit. Eng. Des. Manag.* **2022**, *18*, 705–722. [CrossRef]
15. Aydin, D.; Sayar, G. Questioning the use of the balcony in apartments during the COVID-19 pandemic process. *Archnet-IJAR Int. J. Archit. Research* **2021**, *15*, 51–63. [CrossRef]
16. Hajjar, R.M. Exploring a new housing design paradigm for post pandemic multi-story buildings in Lebanon. *Archit. Plan. J.* **2021**, *27*, 1. [CrossRef]

17. Millán-Jiménez, A.; Herrera-Limones, R.; López-Escamilla, Á.; López-Rubio, E.; Torres-García, M. Confinement, comfort and health: Analysis of the real influence of lockdown on university students during the COVID-19 pandemic. *Int. J. Environ. Res. Public Health* **2021**, *18*, 5572. [CrossRef] [PubMed]
18. Garber, M. The Atlantic. Homes Actually Need to be Practical Now: One of the Ironies of Social Distancing Is That It Can Put Privacy in Short Supply. Available online: https://www.theatlantic.com/culture/archive/2020/03/finding-privacy-during-pandemic/608944/ (accessed on 19 August 2022).
19. Lerup, L. *Building the Unfinished: Architecture and Human Action*; Sage Publications: Beverly Hills, CA, USA, 1977.
20. Albrecht, J. Towards a theory of participation in architecture—An examination of humanistic planning theories. *J. Archit. Educ.* **1988**, *42*, 24–31.
21. Tzortzopoulos, P.; Cooper, R.; Chan, P.; Kagioglou, M. Clients' activities at the design front-end. *Des. Stud.* **2006**, *27*, 657–683. [CrossRef]
22. Nilsson, B.; Peterson, B.; Holden, G.; Eckert, C. Design Med Omtanke: Participation and sustainability in the design of public sector buildings. *Des. Stud.* **2011**, *32*, 235–254. [CrossRef]
23. Eriksson, J.; Glad, W.; Johansson, M. User involvement in Swedish residential building projects: A stakeholder perspective. *J. Hous. Built Environ.* **2015**, *30*, 313–329. [CrossRef]
24. Correia, T.; Willis, K. Applying critical realism to the COVID-19 pandemic to improve management of future public health crises. *Int. J. Health Plan. Manag.* **2022**, *37*, 599. [CrossRef]
25. Hartig, T.; Lawrence, R.J. Introduction: The residential context of health. *J. Soc. Issues* **2003**, *59*, 455–473. [CrossRef]
26. Bullinger, H.J.; Bauer, W.; Wenzel, G.; Blach, R. Towards user centred design (UCD) in architecture based on immersive virtual environments. *Comput. Ind.* **2010**, *61*, 372–379. [CrossRef]
27. Cuerdo-Vilches, T.; Navas-Martín, M.Á.; Oteiza, I. A mixed approach on resilience of Spanish dwellings and households during COVID-19 lockdown. *Sustainability* **2020**, *12*, 10198. [CrossRef]
28. Snyder, H. Literature review as a research methodology: An overview and guidelines. *J. Bus. Res.* **2019**, *104*, 333–339. [CrossRef]
29. Page, M.J.; McKenzie, J.E.; Bossuyt, P.M.; Boutron, I.; Hoffmann, T.C.; Mulrow, C.D.; Shamseer, L.; Tetzlaff, J.M.; Akl, E.A.; Brennan, S.E.; et al. The PRISMA 2020 statement: An updated guideline for reporting systematic reviews. *Int. J. Surg.* **2021**, *88*, 105906. [CrossRef]
30. Birkle, C.; Pendlebury, D.A.; Schnell, J.; Adams, J. Web of Science as a data source for research on scientific and scholarly activity. *Quant. Sci. Stud.* **2020**, *1*, 363–376. [CrossRef]
31. Catalani, C.; Minkler, M. Photovoice: A review of the literature in health and public health. *Health Educ. Behav.* **2010**, *37*, 424–451. [CrossRef]
32. Call-Cummings, M.; Hauber-Özer, M. Virtual photovoice: Methodological lessons and cautions. *Qual. Rep.* **2021**, *26*, 12. [CrossRef]
33. Wang, C.C. Photovoice: A participatory action research strategy applied to women's health. *J. Women's Health* **1999**, *8*, 185–192. [CrossRef]
34. The American Institute of Architects. AIA, Strategies for Safer Multifamily Housing. Available online: https://content.aia.org/sites/default/files/2020-08/AIA_COVID_Building_type_Multifamily_Housing.pdf (accessed on 21 July 2023).
35. The American Institute of Architects. AIA, Re-occupancy Assessment Tool V3.0. Available online: https://content.aia.org/sites/default/files/2020-08/ReOccupancy_Assessment_Tool_v3.pdf (accessed on 21 July 2023).
36. Construction Times. Hyundai Engineering, Hyundai Engineering Develops Specialized Flats Tailored to the Era of COVID-19. Available online: http://www.constimes.co.kr/news/articleView.html?idxno=220285 (accessed on 21 July 2023).
37. Lee, K.W. Kukto Ilbo. Available online: http://www.ikld.kr/news/articleView.html?idxno=255077 (accessed on 17 August 2022).
38. Tanamas, K.; Widayati, N.; Lianto, F. Space Adaptations During Pandemic in Apartments. In Proceedings of the 3rd Tarumanagara International Conference on the Applications of Social Sciences and Humanities (TICASH 2021), Jakarta, Indonesia, 25 August 2021; Atlantis Press: Dordrecht, The Netherlands, 2021; pp. 276–282.
39. Chen, Q.; Sun, Z.; Li, W. Effects of COVID-19 on Residential Planning and Design: A Scientometric Analysis. *Sustainability* **2023**, *15*, 2823. [CrossRef]
40. Kim, M.S. A Study on Architectural Approaches Corresponding to the Post-COVID Era—Proposal of Prevention of Infectious Disease Through Environmental Design. *J. Archit. Inst. Korea* **2021**, *37*, 67–75.
41. Lee, D.Y.; Kwon, O.J.; Kim, J.Y. Changes in Housing Function and Space Preference in the Post COVID Era—Plan Proposal for the Echo Boomer Generation. *J. Korean Hous. Assoc.* **2022**, *33*, 43–53. [CrossRef]
42. Jeon, K.S.; Lee, M.S.; Choi, K.B. Reconstruction of Residential Space in the Post-COVID era. *J. Korean Inst. Cult. Archit.* **2022**, *78*, 94–101.
43. Park, E.C.; Kim, E.J.; Kim, M.K. Analysis of Residents' Interest Areas and Preference of based on Eye Tracking for Apartment Planning in Response to Infectious Disease. *J. Korean Inst. Inter. Des.* **2023**, *32*, 96–106. [CrossRef]
44. Morgan, B.; Forbes. 4 Ways COVID Has Changed Home Design. Available online: https://www.forbes.com/sites/blakemorgan/2020/09/28/4-ways-covid-has-changed-home-design/?sh=4713b51568433 (accessed on 21 July 2023).
45. Global Economy Newspaper. Synergy between ePyeonhansesang C2 House and 4Bay Innovative Flat. Available online: https://www.getnews.co.kr/ (accessed on 21 July 2023).

46. WRAL. The House That Quarantine Built: Post-Pandemic Home Design Trends. Available online: https://www.wral.com/the-housethat-quarantine-built-post-pandemic-home-designtrends/19621460/ (accessed on 21 July 2023).
47. 2021—LI Press. Available online: https://www.longislandpress.com/2021/ (accessed on 7 September 2022).
48. An, S.Y. Aju Business Daily. Available online: https://www.ajunews.com/view/2021 (accessed on 19 August 2022).
49. Spennemann, D.H.R. Designing for COVID-2x: Reflecting on Future-Proofing Human Habitation for the Inevitable next Pandemic. *Buildings* **2022**, *12*, 976. [CrossRef]
50. Elrayies, G.M. Prophylactic Architecture: Formulating the Concept of Pandemic-Resilient Homes. *Buildings* **2022**, *12*, 927. [CrossRef]
51. Lekić Glavan, O.; Nikolić, N.; Folić, B.; Vitošević, B.; Mitrović, A.; Kosanović, S. COVID-19 and City Space: Impact and Perspectives. *Sustainability* **2022**, *14*, 1885. [CrossRef]
52. Benbow, W. COVID-19 in Long-Term Care: The Built Environment Impact on Infection Control. *Health Environ. Res. Des. J.* **2022**, *15*, 287–298. [CrossRef]
53. Ching, L.S.; Rani, M.F.A. The impact of COVID-19 pandemic on home spatial design. *Malays. J. Sustain. Environ.* **2023**, *10*, 207–224. [CrossRef]
54. Kim, M.K.; Park, E.C.; Kim, E.J. Analysis of prior research studies to develop guidelines for apartment housing planning in response to infectious disease disasters. *Korean J. Hum. Ecol.* **2022**, *31*, 275–288. [CrossRef]
55. Shin, D.S.; Lee, M.S.; Yoo, Y.H. The Architectural Planning Strategies for Multifamily Housing in a Post-Pandemic World. *J. Archit. Inst. Korea* **2022**, *38*, 73–83.
56. Kim, M.K.; Han, S.J.; Kim, E.J. Analysis of User Experience Using Photo Voice Techniques for Apartment Residential Space Planning in Responding to Infectious Diseases. *J. Korean Inst. Inter. Des.* **2023**, *32*, 66–75. [CrossRef]

buildings

MDPI

Article

Architecture towards Technology—A Prototype Design of a Smart Home

Pedro Racha-Pacheco [1], Jorge T. Ribeiro [1,2,3,*] and José Afonso [3]

1 Lisbon School of Architecture, University of Lisbon, 1349-063 Lisboa, Portugal; prachapacheco@gmail.com
2 CERENA, Centre of Natural Resources and Environment, 1049-001 Lisboa, Portugal
3 CIAUD, Research Centre for Architecture, Urbanism and Design, 1349-063 Lisboa, Portugal; jafonso@fa.ulisboa.pt
* Correspondence: jribeiro@fa.ulisboa.pt

Abstract: Humanity's way of life has been irreversibly transformed by new technological advancements during the past decades. Although such technological innovations have been gradually transposed into architecture, their full integration is not yet achieved. This article addresses the issue of incorporating cutting-edge technologies (such as smart thermostats, lighting sensors, security cameras, remote commands, graphic user interfaces, smartphones, mobile apps, gestures, voice commands, etc.) into urban small-scale residential architecture, in the future evolution context. For this purpose, a methodology was conceived that the main concepts regarding automation and information networks were researched, as well as their practice in some reference architecture cases. The guidelines for the prototype architectonic design were defined based on the previous knowledge acquired. Then, a prototype design of an intelligent home was iteratively developed as a machine for living in constant change. It was expected to contribute to increasing and disseminating knowledge in these fields, explaining their benefits and limitations. The prototype design presented in this article contributes to sensitizing architecture professionals to the importance of integrated and systematized thinking in all procedures of a smart home design.

Keywords: building automation; network components and devices; housing prototype; architectural design; smart home

Citation: Racha-Pacheco, P.; Ribeiro, J.T.; Afonso, J. Architecture towards Technology—A Prototype Design of a Smart Home. *Buildings* **2023**, *13*, 1859. https://doi.org/10.3390/buildings13071859

Academic Editor: Wolfgang Kastner

Received: 16 June 2023
Revised: 5 July 2023
Accepted: 20 July 2023
Published: 22 July 2023

1. Introduction

The technological development provided by the first two Industrial Revolutions known, especially since the second half of the 20th century, were an extraordinary and fast acceleration [1], which was translated into the Electronic Revolution (i.e., 3rd Industrial Revolution) and the Communication Revolution (i.e., 4th Industrial Revolution).

The Electronic Revolution occurred at the end of 1960s and drastically changed the technology for the controls of both electrical and thermal systems by introducing Programmable Logic Controllers (PLC), which were inexpensive and easy to program. Later, in the 1990s, Direct Digital Control (DDC)—a microprocessor that updates in real-time an internal information database by monitoring information of an environment and continuously produces corrective output commands as an answer to changing control conditions [2]—was used for the first time in heating, ventilation and air-conditioning systems (HVAC) [3] of commercial buildings to enable functions to run automatically. The core of the modern building automation and control systems (BACS) is based on PLCs and solid-state devices—an electrical component/system that is based largely or entirely on a semiconductor [4]. Although the BACS are often associated with commercial and service buildings, they can also be defined as home and building electronic systems (HBES). The BACS are one of the most critical enabling technologies for creating microgrids for smart buildings and energy communities [5].

The arising (the early 1960s [6]) of the internet and its exponential development (commercial phase 1984–1989 [6]) are the geneses of the 2000s Communication Revolution. By 2020, more than half of the world's population was estimated to have access to the Internet [7]. Further, that number is growing, largely due to the prevalence of smart technology and the Internet of Things (IoT), where computer-like devices connect with the Internet or interact via wireless networks [7]. These "things", particularly associated with homes, include smartphones, apps, thermostats, lighting systems, irrigation systems, security cameras, vehicles, and even cities [7]. Simultaneously, the Communication Revolution has further changed the BACS/HBES.

Although the first imagined visions of a Smart Home came from the early 20th century [8], and the first consistent references came from the 1960s [9], the first Smart House was designed by Pierre Sarda in 1974 (https://youtu.be/cqPsI1YBSgc (accessed on 21 March 2023). A Smart House is actually a dwelling where an organized automation system connects all the electrical devices by sensors and telecommunication systems (buttons, touchscreens, keyboards, and voice and gesture recognition) for remote control or assistance. It manages lighting, heating, air conditioning, ventilation, security alarm system, audio and video system, calls devices, energy control equipment, presence, automation (doors, windows, blinds, gates), technical alarms (e.g., in case of unwanted water spillage), household appliances, etc. [10].

Currently, the term Domotics—which comes originally from the Latin 'domus' which means house, and 'tics' which includes robotics, telematics, and computational science [11]—is also linked to building automation. It is a technology for private home automation, providing several services for the resident's comfort and security. Although many automation systems were also used in BACS—or Smart Homes—and Smart Houses, the Domotics also refers to additional functions such as multimedia home entertainment systems, automatic plant watering and pet feeding, and automatic scenes for dinners and parties.

Nowadays, smart systems are seen as context-aware systems for increasing building skills. They should be aware when something is not happening according to the user specifications or through environmental information—i.e., aimed to know what is happening and what are the users' intentions and needs in order to anticipate an action or an undesirable condition [12]. The future expectation regarding home automation is linked to contextual information that can be adapted to each user, which is essential to conceive user behavioral models. Architecture should promote a space design that collects the greatest amount of information, without compromising the inhabitants' comfort, and simultaneously promotes it through more efficient use.

Despite those technological advances, the mainly marketed building automation systems concern energy consumption management and thermal comfort control [13]. Thus, there is still a long path to the generalized application of BACS new features, and this article is a small step in that direction.

In fact, the above-mentioned advances established the main purposes of this paper: (a) systematizing and disseminate building automation knowledge, especially for traditional architecture professionals, highlighting the growing importance of automation for the contemporary world and for the new ways of living; (b) show the design of a smart home—housing prototype—defining the main technological concepts useful for architectural design, and the integrated and systematic thinking in all process is another contribution; and (c) sensitize architecture professionals to the new challenges of the near future in the architecture field.

This paper has five sections. After the Introduction, Section 2 describes the methodology conceived to develop the architectural design of the housing prototype. The main components of a building automation system were presented in Section 3. Section 4 describes, in detail, the housing prototype developed and discusses it in the contemporary context. Section 5 presents the main conclusions.

2. Methodology

The methodology conceived for the housing prototype design integrating the smart systems was adapted from Gomes et al. [14].

It starts with literature research on the concepts and smart systems applied to architecture. While intense literature research was carried out, several reference cases were also studied to consolidate the acquired knowledge. An empirical program that integrated smart systems was defined, balancing the two previous tasks. Then, iteratively, a housing prototype was developed until matching all requirements. All iterative models are performed in the Blender 2.7 software [15], including the final result presented in Section 4 as a housing prototype for a single inhabitant. Figure 1 outlines the methodological process to achieving the main objective—housing architectural design with home automation integration.

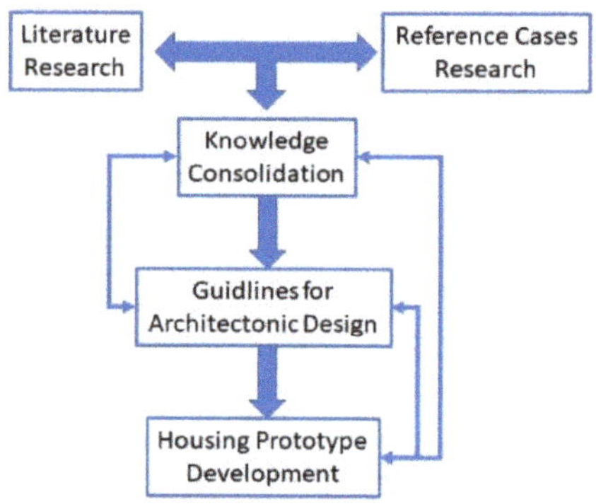

Figure 1. Methodology adapted from Gomes et al. [14].

3. Building Automation System

A building automation system mainly needs devices that communicate between them, with the user, and communication protocols. It is typically organized into three levels (Figure 2): field, automation, and management [3,16]. The field level interacts with the physical world, usually with sensors and actuators. At the automation level are developed control logic operations, by controllers, to execute appropriate actions. The management level is used by operators to monitor, configure, and control the whole system.

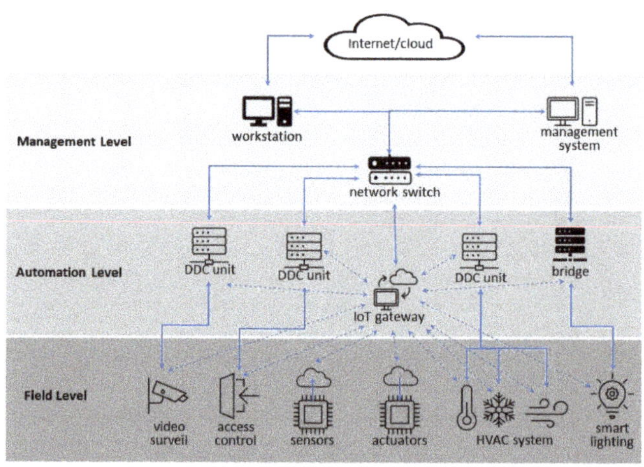

Figure 2. Three levels of the functional hierarchy of a building automation system. Source: second author.

3.1. Sensors and Actuators

Sensors and actuators are often contained in the same equipment. Sensors are devices that convert a physical reality into a signal that can be measured [17]. The most common are motion, pressure, or contact sensors, widely used in space monitoring and security. Actuators are also devices, such as window blinds or ceiling lamps [17], that react to signals by closing circuits or varying the intensity of electric loads. Actuators complement sensors by mechanically regulating a given system—e.g., the motorized shading system, removable walls, height-adjustable workstations, etc. In a simple way, the sensor captures and the actuator reacts by changing the environment by automatic or manual instruction.

3.2. Controllers

Controllers operate joined with actuators to regulate the operation of valves or other devices—e.g., an intelligent lamp that knows to turn on when the level of natural light justifies it through a brightness sensor. This control is done through a network that connects all electrical or mechanical instruments by DDC [3].

To establish the connections, each vendor develops specific drivers for different protocol systems and applications, such as Konnexbus (KNX), LonWorks, BACnet, etc. [17].

3.3. Interfaces

Interfaces can be understood as devices that allow communication between two different systems. User interfaces establish communication between the user and the system, which can occur with remote commands, mobile applications, gestures, and voice commands, depending on the complexity. These devices often use GUI (Graphic User Interface) to translate instructions into graphical and user-friendly support.

System interfaces, in addition to interacting directly with the system, allow diagnosis and maintenance [3,12].

3.4. Communication Protocols

Communication between devices is carried out through an Application Programming Interface (API), which translates the "language spoken" by each of them. In fact, one of the biggest challenges is the need to articulate many communication protocols and standards without compromising compatibility [12,18]. Despite the experts' awareness, the scarcity of authoritative literature and international rules regarding the subject remains a surprising aspect of building automation. This situation disturbs communication between developers and contributes to worsening the heterogeneity problem, making it difficult for the articulation and integration of different manufacturers' systems [17].

Until the creation of international standards, building automation systems were mainly guided by the technical documentation of the suppliers of each automation system. KNX, LonWorks, and BACnet systems are examples of open standardized protocols [17,19].

In the European context, the Konnexbus (KNX) protocol stands out—standardized through the EN 50090 standard—and is currently recognized internationally as ISO/IEC 14543-3 [17]. It is more field- and automation-levels oriented and can be used for the automation of every building type and size—from family houses to office complexes and airports [19]. Each device's own "intelligence" allows it to know what to receive/send from/to the bus and how to process the received data [19]. That skill makes that complex home and building automation system fully decentralized.

LonWorks, or Local Operating Network (LON), is also available as the European standard EN 14908 and as the international standard ISO/EIC-14908 [17]. More popular in North America than in Europe, LonWorks is mainly focused on automation and field levels. Currently, it is not only used for building automation, given its great flexibility and complexity [19].

Building automation and control networking protocol (BACnet) was published as ANSI/ASHRAE 135 standard and later became a CEN and ISO standard within the ISO 16484 series [17]. It is focused on the management and automation levels and is used mainly in non-residence buildings as a Building Management System (BMS). The information exchanged between devices on BACnet uses objects and services, basically. According to Nývlt [19], besides traditional tasks (HVAC and light control), BACnet is able to supervise BMS and complex building tasks (e.g., security and fire safety systems, access control systems, vertical transport systems, elevator control, maintenance or waste management).

3.5. Network Topologies

The communication between devices, via wired (e.g., coaxial cables, electrical network, Ethernet cables, etc.) or wireless (e.g., Bluetooth, etc.), establishes a network, whose main topologies [20] are summarized in Table 1.

Table 1. Network topologies.

Network Topology	Description	Advantages	Disadvantages
Bus	a single cable is used to connect all the devices on the network	easy installation and minimal cabling required;	performance limitations on the number of network nodes;
		failure of one node does not affect other nodes;	network connectivity shut down when the cable fails.
		messages from one node can be seen simultaneously by all other nodes.	
Ring	each node is connected to two other nodes, and the last is connected to the first		relatively long transmission time between nodes;
		messages sent in one direction or both directions (using two cables between each connected node);	cabling failure between two nodes has a broader effect on the entire network;
		commonly used in networks with low inter-node traffic	relative communication delays between nodes;
			network expansion or reconfiguration requires new cabling
Star	requires the use of a top-level central node to which all other nodes are connected	reduced messaging delay between nodes;	need for more wires;
		a connection failure between a higher-level node and any of its subordinate nodes, or failure of one subordinate node, will not affect the entire network;	a failure of the top-level node will disrupt all communication on the network
		cabling failures location;	a limited number of higher-level node connections
		often used in LANs with larger geometrical area	
Tree	built by creating a set of star topologies below a central node or by connecting a set of star topologies directly through a bus	easy network scaled;	
		cabling fault location;	a higher-level node failure or the cable connection failure, then the partial network will be lost to communication.
		expansion can be as simple as attaching an additional star network topology to the bus	
Mesh	full—each node is directly connected to all other nodes; section—has a number of network nodes that are indirectly connected to other nodes	path redundancy;	
		for high traffic between nodes;	high cost of setting up the network;
		reduced probability of single-point network failure;	each node needs a routing algorithm to compute the path
		source nodes define the best route from sender to destination	
		wireless suitable	

Many networks are described as hybrid topologies, which combine the features of two or more of the above networks [20].

3.6. Network Components and Devices

Networks use several components and devices, which are described below, based on Bird & Hartwood [21].

Repeaters, mostly linked to coaxial network configurations, were used in the past to increase the usable cable length. As coaxial networks are less used currently and repeaters' functionality was included in other devices, such as hubs and switches, repeaters are now rarely used.

A hub is a multiport repeater. It means that the hub has the basic function of collecting data from one of the connected devices and forwarding them to all the other ports. It is a layer-one device that regenerates signals. Currently, hubs are slowly being replaced with switches, mainly due to the continuously increasing demand for more bandwidth and also due to their inefficiencies.

Bridges are data link layer devices that are used to connect subnets and manage the traffic flow between them. In fact, sometimes networks must be divided into subnets to reduce the amount of traffic or for security reasons. Driving data frames from one segment to the other are decided based on the MAC (Media Access Control) address, a unique number that is stamped into each data frame. Today, network switches have largely replaced bridges.

A switch is a multiport bridge and is a two devices layer. It receives incoming frames and, in contrast to a hub, uses the devices-connected MAC addresses to determine specific addresses and forwards the frames to the correct port. Like a bridge, a switch is used to increase the capability of an Ethernet LAN (Local Area Network) by dividing the network into several collision domains. This reduces the data traffic in each subnet and increases the usable bandwidth. When larger networks are needed, they can be created through the multiple switches' interconnection.

Routers are network devices, as the name suggests, that route data around the network. The router can define the data destination address, examining it upon arrival. It uses defined route tables (software-configured) to make decisions—that is, to determine the best way for the data to follow its tour. This approach, contrary to bridges and switches which use the MAC address (hardware-configured) to determine the data destination, makes routers more functional. It also makes them more complex because they have to work harder to determine the information destination.

Gateway is any device, system, or software application that can perform the function of translating data from one format to another. The key feature of a gateway is that it converts the data format, not the data itself.

A Channel Service Unit/Data Service Unit (CSU/DSU) changes the signals' digital format—computer signals to communication signals. It is used as a translator between the different technologies of LAN and WAN (Wide Area Network) links. The increasing use of WAN links means that some router manufacturers are now including the CSU/DSU functionality in routers or are providing the expansion capability to do so.

Network Interface Card (NIC) is the mechanism by which computers connect to a network.

ISDN (Integrated Services Digital Network) terminal adapter is an external device that allows this type of digital communication, using a conventional phone line. It is available as add-in expansion cards fitted into computers, which connect to the serial interfaces of PCs—or modules—in a router.

Wireless Access Point (WAP) is a transceiver (see brief description below) network device used for wireless LAN (WLAN) radio signals. A WAP is typically a separate device with a built-in antenna, transmitter, and adapter. It can operate as a bridge connecting a standard wired network to wireless devices or as a router for data transmissions. In wireless networks, there may be multiple access points to cover a large area or only a single access point for a small area, such as a single home or small building.

A modem is a contraction of the terms modulator and demodulator. Modems translate digital signals from a computer into analog signals that can travel across conventional phone lines. However, due to the relatively slow communication, other remote access types (e.g., ISDN) are commonly preferred.

A transceiver (a contraction of the terms transmitter and receiver) has the function of transmitting and receiving analog or digital signals. It can be found embedded in devices such as network cards, as well as external devices.

A firewall is a networking device that controls (either hardware or software-based) its access, protecting data and resources from outside threats, and typically protects internal networks from public networks. For that reason, it is typically located at the entry/exit points of a network. In small offices and homes, a firewall is commonly installed on the local system and configured to control traffic.

4. Results and Discussion

4.1. Housing Prototype

The housing prototype that is described below, scaled for a single user, includes automation systems, but is not an executable architecture design. It intended to merge the architectural shape into its canonical elements, combined with the technological components, creating an automated capsule with four functional areas: bedroom, personal hygiene, cooking, and work [22]. The architecture design fit into prefabricated modular construction and was influenced by metabolism, a movement driven by Kisho Kurokawa, advocating the architectural idea as a living organism in constant change and adaptation. The strong reference to Kurokawa's capsules was due to their status as the house of the future (designed around 50 years ago), and because they were a synthesis considered useful for simplifying the house concept.

The capsule typology justified the mention of the minimum habitable space concept addressed by authors such as Walter Gropius and Le Corbusier, favoring the functionality of the space [23]. This line of thought defended that Man should adapt to Architecture—the housing prototype intended to show a hypothetical solution to the needs and expectations of interaction with the digital environment, based on the conceptual axioms of a limited architectural space.

To cope with the mass housing needs, authors such as Buckminster Fuller, Jean Prouvé or Pierre Koenig defended modular construction strategies, inspired by the industrial processes of prefabrication and mass production, providing themselves with an ethical conscience in the establishment of more sustainable, affordable, and adaptable.

In general, the influential modernist authors defend, in the light of reason and logic, the reduction of architecture to the essential, proposing an architecture that promoted minimum conditions and defining a behavioral criterion in the inhabiting act. However, the scarcity of space and resources can also be an engine of creativity, establishing constraints, but guiding the architectural design of which Ernst May's experience of social architecture in Frankfurt—which is at the origin of the concept of *Existenzminimum*—which is an example [24]. Le Corbusier proposed *a contrario,* the *maison maximum* concept as an analogously inverse answer to the same problem, proposing the maximum space that should be inhabited [22].

Considering the current and growing importance of technology in society and its digital development, the housing prototype attributed greater relevance to the technology, than to the construction process or to the comfort, which depended on the context—cultural, temporal, spatial, etc. In this sense, the capsule proposed with contemporary devices, expands on the type of relationships between the inhabitant and his space, making it progressively more reactive and interactive in a continuous adaptation.

The shape of the capsule was designed according to the devices that were part of the daily experience of human beings in the 21st century. Regarding the intelligence of a technological environment as its contextual information, it could be stated that the inhabiting machine creation was analogous to the living being creation, according to the metabolist principles. The architectural environment design, fed by the information that was introduced and captured, is able to react and adapt to human variables. The individual is an extension of his home and vice versa. Taking into account the smartphone and personal computer interoperability potentials, a way of life characterized by the synchrony between all the technological devices of an individual, including his house and the respective components, is hypothesized in the housing prototype.

The capsule, inspired by the Kurokawa's Nakagin Tower [25], is a prefabricated steel structure, clad in galvanized steel panels with an exterior waterproofing treatment. Conceived by injection molds, in order to reduce the joints, the manufacturing process involves the structural envelope design for a molded parts aggregation (Figure 3).

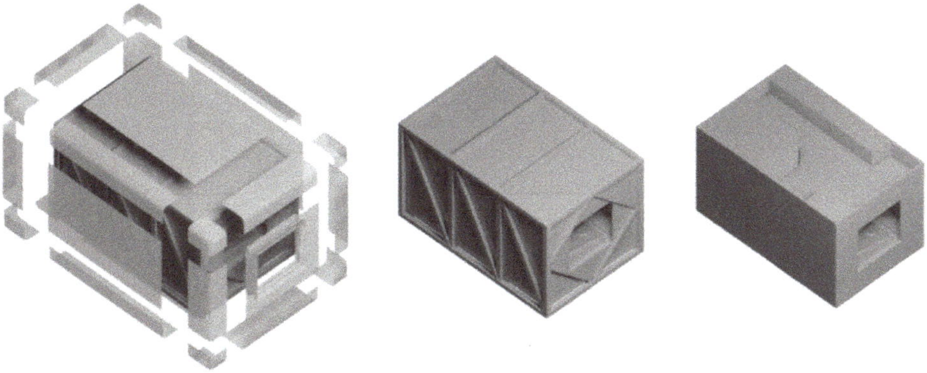

Figure 3. Structure scheme. Source: first author.

The technical systems—i.e., water, air quality, interfaces, sensors/actuators, lighting, security/privacy, and sound (Figure 4)—which will be further developed by experts, were accessed from the inside and/or outside and located in specific places designated by the technical team. Technical assistance would be carried out from the inside. Photovoltaic panels on the capsule roof would be fitted. The capsule joints in a tower were also foreseen, which required adaptations in some capsules, such as their individual removal for technical assistance and photovoltaic panels setting only in the top capsules.

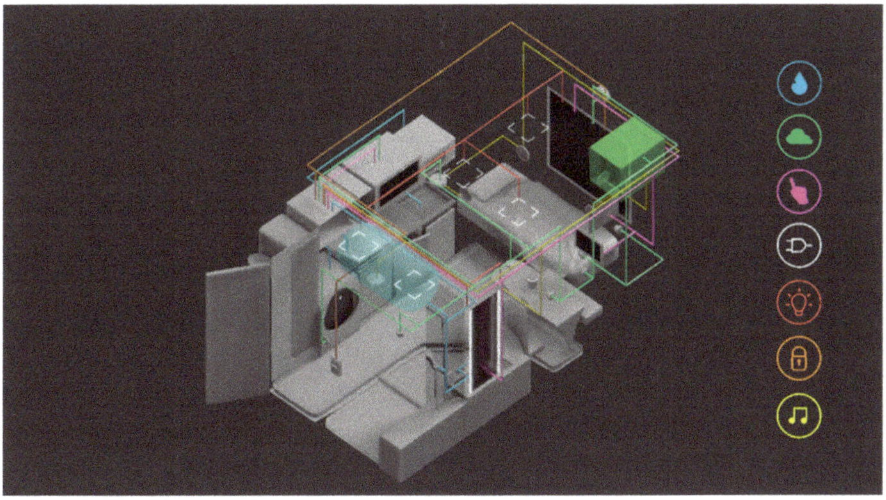

Figure 4. Technical systems. Source: first author.

As in the Welfare Techno House (WTH) [12,26], the Intelligent Room Project [27], and MavHome [28], the capsule will have a set of sensors (e.g., on the floor [29] or on the smart mattress), microphones and cameras to information capture at several levels, e.g., ECG (electrocardiogram) and body mass recording. This information is used, through Artificial Intelligence and Machine Learning systems, to build the inhabitant's behavioral model [27] and for accident prevention, allowing also for the monitoring and connecting to emergency systems. Although, the images and sounds captured are sensitive matters in terms of privacy preservation and security of inhabitants. So, those issues should be solved at the individual level by the owner and technical team.

The weather information is monitored by a sensor located on the outside of the capsule.

At the entrance there will be access control through a smart lock with the biometric reader and an intelligent thermostat will be implemented for the direct control of the HVAC system. The atmospheric sensor on the ceiling records and monitors air quality and detects fires.

A signal diffusion strategy adopted is wired combined with wireless, organized in a mesh to avoid signal collisions (interference). A network of KNX devices and its gateway will also be foreseen to enable control by external devices—e.g., smartphone, smart assistant, etc. The network includes a set of touch sensors to provide direct control of the lighting system. Devices without native intelligence will be used in the system via smart plugs.

As in Nakagin [25], water is drained from the collectors at floor level. If the capsule can be connected to an external sewage network, the system would be similar to the traditional one. If this does not happen, a dry system was foreseen. Additionally, a cistern would be implemented to collect rainwater, similar to the EcoCapsule [30].

The plan design was based on Kurokawa's Nakagin Tower capsule, as it is an architectural reference previously tested in terms of efficiency and is focused on house intelligence (Figures 5 and 6).

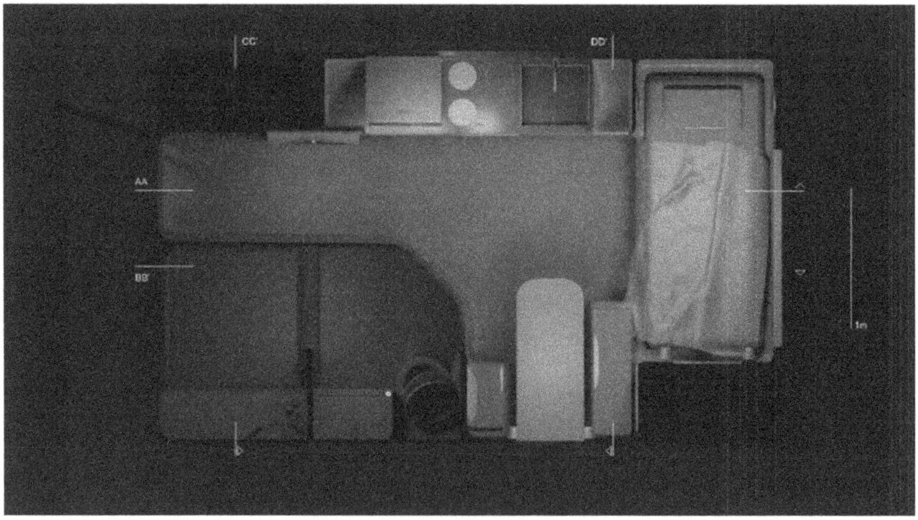

Figure 5. Capsule plan. Source: first author.

The negative space, related to the structure and technical infrastructure, was understood as the entire interstitial space between the interior and exterior cladding, which assumed a prominent role in the design of this capsule. Flexibility between storage and technical infrastructures spaces was considered, depending on needs, allowing different sizes for each space and, consequently, the type and diversity of technological systems. The access point to the automation system infrastructure was located in a technical cabinet located at the upper limit of the capsule (Figure 6a). This location was chosen for its relevance to the geometric volume design, corresponding to the negative space of the capsule. This extends inside the capsule, tangent to its perimeter so that the useful space for circulation and permanence was free.

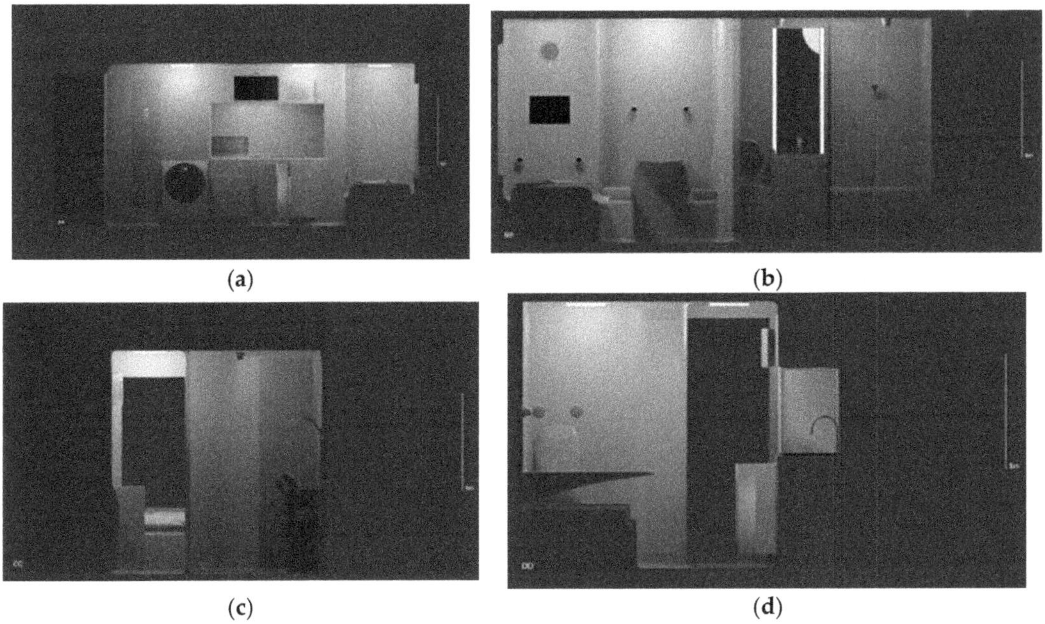

Figure 6. Cross sections located in Figure 5: (**a**) cross section AA′; (**b**) cross section BB′; (**c**) cross section CC′ and (**d**) cross section DD′. Source: first author.

4.1.1. Bedroom

The bedroom was designed for rest and entertainment purposes (Figure 7a). The mattress (#1 in Figure 7a), with built-in heating, has ergonomic articulation, allowing it to be adapted to the user's physiognomy and stiffness configuration according to preference. Two speakers located laterally close to the cushion (#2 in Figure 7a) and a screen (#4 in Figure 7a) was the entertainment system. Behind the screen, the technical column (negative space) included the necessary connections and the ventilation system (#3 in Figure 7a). There was also a built-in induction charger, close to the rest area, for charging mobile devices. Additionally, an intelligent compartment (#5 in Figure 7a) was proposed to regulate the transparency through the electrochromic glass and display information regarding the capsule. This was a transparent LCD (Liquid Crystal Display) touchscreen that served the entire capsule, excluding the sanitary room, and can be considered as an interface of the automation system.

4.1.2. Textile and Laundry Area

At the entrance was located the textile and laundry area (Figure 7b). It was characterized by four compartments: clothes storage (#6 in Figure 7b), a laundry system built into a closet, using steam and gravity to wash and iron clothes (#7 in Figure 7b), a washing and drying machine (#8 in Figure 7b), and the access point to the central domestic control system (#9 in Figure 7b) from which the technical installation (#10 in Figure 7b) is developed. The size of each of these compartments could be adjusted according to the specific needs of each user, e.g., #7 and #8 could have been neglected to increase the storage area (#6).

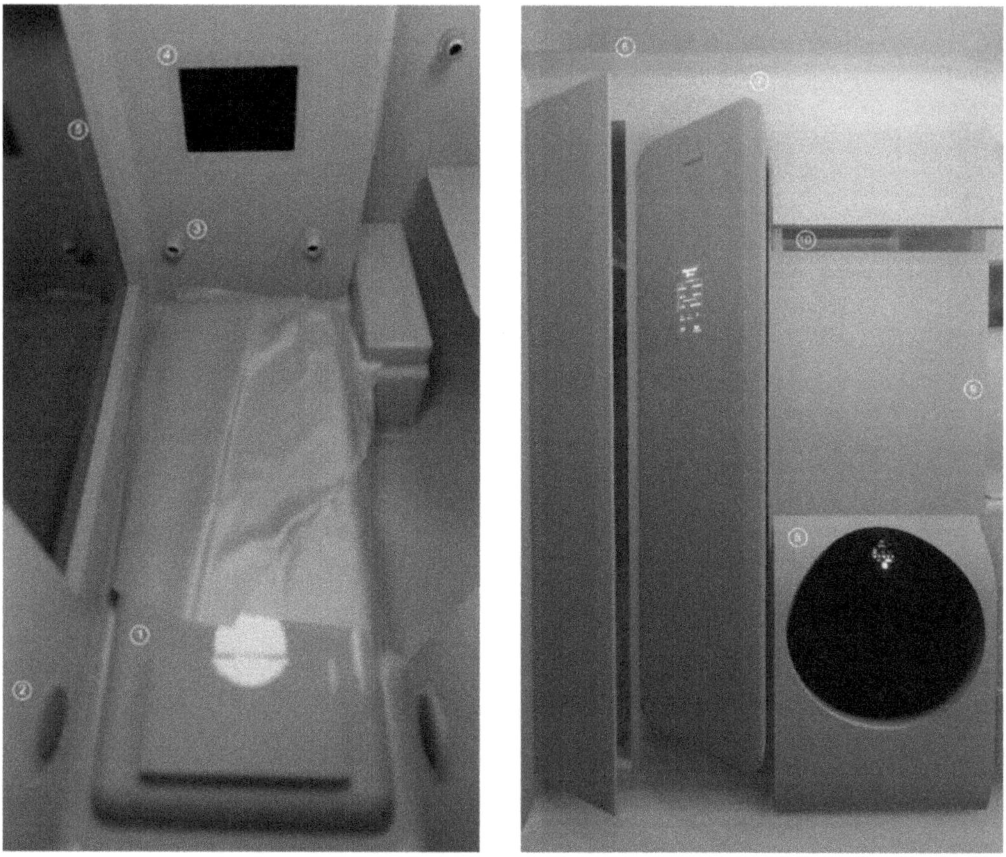

(a) (b)

Figure 7. (**a**) Bedroom [20]; (**b**) Textile and laundry area. Source: first author.

4.1.3. Personal Hygiene Area

The personal hygiene area (Figure 8a) was protected by the only visual barrier of the capsule, which could be replaced by electrochromic glass. The area had a smart toilet (#11 in Figure 8a) and a shower area. The washbasin (#12 in Figure 8a) was equipped with a gesture-controlled mechanism. The basin base moved vertically between two positions: aligned with the height of the washbasin, or making the cavity that confined the basin. The smart mirror (#13 in Figure 8a) was equipped with technology identical to the capsule window, as described above, enabling tactile interaction with the control system and also providing information about it. The shower area (#14 in Figure 8a) had a water-resistant panel built into the shower wall, which interacted with the control system, replacing the faucet mechanism. This device could be centrally controlled from a smartphone or smart assistant, programming different types of showers and interacting with the water heater and smart valves in the water network.

(a) (b)

Figure 8. (**a**) Personal hygiene area; (**b**) Cooking area. Source: first author.

4.1.4. Cooking Area

The cooking area (Figure 8b) was for cooking meals and the cook robots were excluded. A solution was favored that attributes centrality to a kitchen hub (#16 in Figure 8b), which served as the main interface to the capsule. Meals could be cooked in a combined oven/microwave (#17 in Figure 8b) or on an induction cooker (#20 in Figure 8b). The stove had two surfaces with circular cutouts used as induction bases for cooking, another cutout used as a weighing scale, and a fourth as an induction charger. The adjacent area (#21 in Figure 8b) was occupied by a basin with an identical mechanism to the basin in the personal hygiene area. The dishwasher (#19 in Figure 8b) is built-in and accessible from the top of the workbench. The lower space of the bench contained storage for kitchen utensils (#22 in Figure 8b), and a smart fridge (#23 in Figure 8b) with the functionality of inventorying the items contained and communication with ordering services. There was also a space for waste disposal (#24 in Figure 8b). The side spaces of the workbench (#18 in Figure 8b) were equipped with smart electrical sockets for not foreseen equipment. These, as previously mentioned, allowed for the basic control of additional equipment that was not natively smart.

4.1.5. Work Area

The work area (Figure 9) was multipurpose, as it fulfilled work, socializing, and meal functions. It was equipped with a screen (#25 in Figure 9) that served as a second monitor for the laptop used on the desktop. There were three sound columns (#26 in Figure 9) in the work area. One was located at the upper limit of the capsule, working together with the bedroom. The remaining two speakers were positioned on either side of the workstation user. There were also two HVAC system diffusers (#27 in Figure 9) on the side wall of the work area in order to contribute to general or zonal thermal control of the capsule.

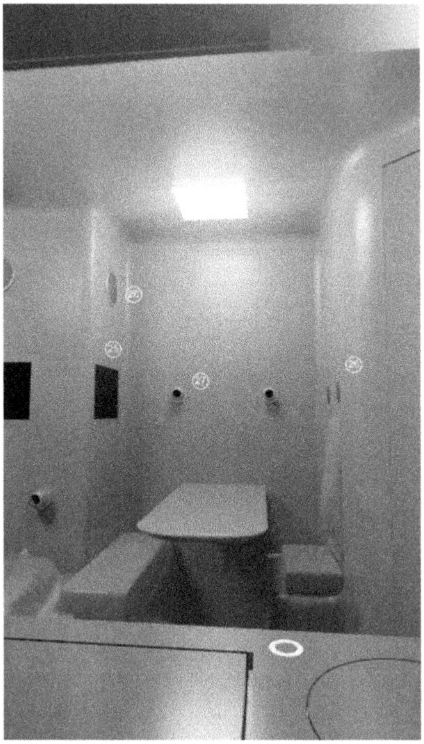

Figure 9. Work area. Source: first author.

4.2. Discussion

According to [3], it was expected that an intelligent building would be able to satisfy the needs of its users in four domains: security, multimedia, temperature, and lighting control, which could be manually or automatically controlled.

Nevertheless, the automation systems do not yet significantly add value and present solutions for problems that do not need to be solved (most people do not need a new thermostat, a Wi-Fi fridge, or a smart lock to replace their manual and/or mechanical devices, which in most cases were not yet obsolete). Changing these devices for more expensive and potentially more complicated devices was not something that most people were willing to do, and instead were happy with their current situation. Beyond the high equipment costs, concerns regarding privacy and security, lack of standardized equipment and compatibility issues were also limitations to smart systems market expansion [8,12]. Nowadays, interoperability is being put forward as the main challenge not only for building automation, but also for IoT technologies [17]. Thus, perhaps installing smart systems in pre-existing buildings was still not recommended.

However, the BACS allows for optimizing the operational energy costs of a building, maintaining the comfort and accessibility of its users [17], and will be the future for new buildings.

The smart home can also play an active and decisive role in preventing domestic accidents and supporting weak (mental or physical) individuals. Contextual information, combined with the monitoring and forecasting of impaired individuals' activity, can be the key to formalizing a safe and accessible space.

5. Conclusions

Krueger [31] believed, in 2006, that architecture had neglected the possibilities offered by new technologies. To cope with the previous statement, increasing and disseminating knowledge in these fields was needed. This article brought those areas of knowledge closer together, encouraging their relationship to provide mutual benefits and developments in the near future.

The potential provided by the new technologies was explored by architecture in this paper through summarizing technical knowledge, which was useful for the building automation architectural project. Far from exhausting the subject, it intended to be a starting point for the design of a living and constantly evolving intelligent system—the house as an information cell that integrates the urban fabric, extending to the city and the world scales.

The housing prototype developed, not being an executable project design, should be seen as an exploratory and hypothetical architectural solution of what the future could be. The project design will be further developed, in a collaborative way, with experts in several technical aspects.

The housing prototype intended to show the house automation architectural design as a technological product at the service of the user. Understanding that the house can be one more product that can integrate (or coordinate) a smart grid, can decisively change the field of architecture and challenge its limits. The capsule can be used as assisted living, however, presenting challenges to be overcome in terms of interior mobility.

It was foreseeable that the current competitiveness of the BACS market generated integrated operating systems that provided the connection between all the smart devices in the house. Architecture, by its nature and through its design, may have an increased responsibility in the adoption and dissemination of these systems. The house of the future is repeatedly reinvented, and architects must know how to position themselves in a volatile market.

A business model for architecture could be created, similar to that used by large technological companies—e.g., Google, Facebook, etc.—that promote a free service linked to monitored user data. These data and information have intrinsic value for entities and companies interested in the targeted promotion of their products or services. It is possible that this data trade could represent a revolution in today's housing market, transforming the value and ownership notions.

Author Contributions: Conceptualization, J.T.R. and J.A.; methodology, J.T.R.; formal analysis, J.T.R. and J.A.; investigation, P.R.-P. and J.T.R.; resources, P.R.-P. and J.T.R.; writing—original draft preparation, J.T.R.; writing—review and editing, J.A., P.R.-P. and J.T.R. All authors have read and agreed to the published version of the manuscript.

Funding: This research received no external funding.

Data Availability Statement: Data sharing is not applicable to this article.

Conflicts of Interest: The authors declare no conflict of interest.

References

1. Roser, M. Technology over the Long Run: Zoom out to See How Dramatically the World Can Change within a Lifetime. Available online: https://ourworldindata.org/technology-long-run (accessed on 31 March 2023).
2. Coffin, M.J. Introduction to Direct Digital Control Systems. In *Direct Digital Control for Building HVAC Systems*; Springer: Boston, MA, USA, 1999. [CrossRef]
3. Merz, H.; Hansemann, T.; Hübner, C. *Building Automation—Communication Systems with EIB/KNX, LON and BACnet*, 2nd ed.; Springer Nature: Mannheim, Germany, 2018.
4. IEEE (Institute of Electrical and Electronics Engineers) Technology Navigator 2023. Solid State Devices. Available online: https://technav.ieee.org/topic/solid-state-devices (accessed on 31 March 2023).

5. Martirano, L.; Mitolo, M. Building Automation and Control Systems (BACS): A Review. In Proceedings of the 2020 IEEE International Conference on Environment and Electrical Engineering and 2020 IEEE Industrial and Commercial Power Systems Europe (EEEIC/I&CPS Europe), Madrid, Spain, 9–12 June 2020. [CrossRef]
6. Cohen-Almagor, R. Internet History. *Int. J. Technoethics* **2011**, *2*, 45–64. [CrossRef]
7. Dennis, M.A.; Kahn, R. Internet Computer Network. 2023. Available online: https://www.britannica.com/technology/Internet (accessed on 3 April 2023).
8. Serrenho, T.; Bertoldi, P. *EUR 29750 EN*; Smart Home and Appliances: State of the Art—Energy, Communications, Protocols, Standards. Publications Office of the European Union: Luxembourg, 2019. [CrossRef]
9. Harper, R. Inside the smart home: Ideas, possibilities and methods. In *Inside the Smart Home*; Harper, R., Ed.; Springer: London, UK, 2003; pp. 1–13. [CrossRef]
10. Dvoršak, B.; Havelka, J.; Mainardi, E.; Pandžić, H.; Selič, T.; Tretinjak, M. Smart Home Systems. This Publication is Part of the SHVET Project, Funded with Support from the European Commission. Available online: https://ec.europa.eu/programmes/erasmus-plus/project-result-content/4df4e928-8958-4552-80da-146977e666b9/Smart_Home_systems_FINAL.pdf (accessed on 22 April 2023).
11. Simonet, C.; Noyce, A.J. Domotics, Smart Homes, and Parkinson's Disease. *J. Park. Dis.* **2021**, *11*, S55–S63. [CrossRef] [PubMed]
12. Bejarano, A.; Fernandez, B.; Jimeno, M.; Salazar, A.; Wightman, P. Towards the Evolution of Smart Home Environments: A Survey. *Int. J. Autom. Smart Technol.* **2016**, *6*, 105–136. [CrossRef]
13. De Groote, M.; Volt, J.; Bean, F. *Is Europe Ready for the Smart Buildings Revolution*; Buildings Performance Institute Europe (BPIE): Brussels, Belgium, 2017.
14. Gomes, I.B.; Ribeiro, J.T.; Macedo, C.L. Lisbon's Águias Villa's conversion into a hotel: Merging sustainability, architectural rehabilitation, and tourism concepts. *Int. J. Architecton. Spat. Environ. Des.* **2021**, *15*, 85–108. [CrossRef]
15. Blender Foundation. Software Blender 2.7. Available online: https://www.blender.org/ (accessed on 22 November 2020).
16. Kastner, W.; Neugschwandtner, G.; Soucek, S.; Newman, H.M. Communication systems for building automation and control. *Proc. IEEE* **2005**, *93*, 1178–1203. [CrossRef]
17. Domingues, P.; Carreira, P.; Vieira, R.; Kastner, W. Building automation systems: Concepts and technology review. *Comput. Stand. Interfaces* **2016**, *45*, 12. [CrossRef]
18. Miori, V.; Russo, D. Domotic evolution towards the IoT. In *International Conference on Advanced Information Networking and Applications Workshops*; IEEE: Victoria, BC, Canada, 2014; pp. 809–814.
19. Nývlt, O. *Buses, Protocols and Systems for Home and Building Automation*; Department of Control Engineering Faculty of Electrical Engineering Czech Technical University in Prague: Prague, Czech Republic, 2009–2011.
20. Thakur, M.; Verma, P. A Review of Computer Network Topology and Analysis Examples. *Int. J. Adv. Eng. Manag.* **2022**, *4*, 1523–1530. [CrossRef]
21. Bird, D.; Harwood, M. *Network+ Training Guide*; Que Publishing: Indianapolis, IN, Canada, 2003.
22. Corbusier, L. *Towards A New Architecture*; Dover Publications Inc.: New York, NY, USA, 1986.
23. Seelow, A.M. Function and Form: Shifts in Modernist Architects' Design Thinking. *Arts* **2017**, *6*, 1. [CrossRef]
24. Brysch, S. Reinterpreting *Existenzminimum* in Contemporary Affordable Housing Solutions. *Urban Plan.* **2019**, *4*, 1–20. [CrossRef]
25. Ishida, A. Metabolic impermanence: The Nakagin capsule tower. *Inflection J.* **2017**, *4*, 32–43.
26. Tamura, T.; Kawarada, A.; Nambu, M.; Tsukada, A.; Sasaki, K.; Yamakoshi, K.-I. E-Healthcare at an Experimental Welfare Techno House in Japan. *Open Med. Inform. J.* **2007**, *1*, 1–7. [CrossRef] [PubMed]
27. Brooks, R.A. The Intelligent Room Project. In Proceedings of the Second International Conference on Cognitive Technology Humanizing the Information Age, Aizu-Wakamatsu, Japan, 25–28 August 1997; IEEE: Piscataway, NJ, USA, 1999. [CrossRef]
28. Cook, D.J.; Youngblood, M.; Heierman, E.O.; Gopalratnam, K.; Rao, S.; Litvin, A.; Khawaja, F. MavHome: An agent-based smart home. In Proceedings of the First IEEE International Conference on Pervasive Computing and Communications, Fort Worth, TX, USA, 26 March 2003; IEEE: Piscataway, NJ, USA, 2003. [CrossRef]
29. Yamazaki, T. The Ubiquitous Home. *Int. J. Smart Home* **2007**, *1*, 17–22.
30. Pospisil, M. EcoCapsule, self-sustainable microhome. In Proceedings of the DARCH 2021 1st International Conference on Architecture & Design. International Organization Center of Academic Research, Online, 12–13 July 2021; pp. 85–91.
31. Krueger, T. Metadermis: Like a second skin. In *Architecture—Machine and Body: Notes on the New Technology in Architecture*; Afonso, R.B., Furtado, G., Eds.; Carneiro, A.; Peck, B.; Alves, I., Translators; FAUP: Porto, Portugal, 2006; pp. 87–98.

buildings

MDPI

Article

The Impact of High-Density Urban Wind Environments on the Distribution of COVID-19 Based on Machine Learning: A Case Study of Macau

Liang Zheng †, Yile Chen †, Lina Yan † and Jianyi Zheng *

Faculty of Humanities and Arts, Macau University of Science and Technology, Macau 999078, China
* Correspondence: jyzheng@must.edu.mo
† These authors contributed equally to this work.

Abstract: The COVID-19 epidemic has become a global challenge, and the urban wind environment, as an important part of urban spaces, may play a key role in the spread of the virus. Therefore, an in-depth understanding of the impact of urban wind environments on the spread of COVID-19 is of great significance for formulating effective prevention and control strategies. This paper adopts the conditional generative confrontation network (CGAN) method, uses simulated urban wind environment data and COVID-19 distribution data for machine training, and trains a model to predict the distribution probability of COVID-19 under different wind environments. Through the application of this model, the relationship between the urban wind environment and the spread of COVID-19 can be studied in depth. This study found that: (1) there are significant differences in the different types of wind environments and COVID-19, and areas with high building density are more susceptible to COVID-19 hotspots; (2) the distribution of COVID-19 hotspots in building complexes and the characteristics of the building itself are correlated; and (3) similarly, the building area influences the spread of COVID-19. In response to long COVID-19 or residential area planning in the post-epidemic era, three principles can be considered for high-density cities such as Macau: building houses on the northeast side of the mountain; making residential building layouts of "strip" or "rectangular" design; and ensuring that the long side of the building faces southeast (the windward side). (4) It is recommended that the overall wind speed around the building be greater than 2.91 m/s, and the optimal wind speed is between 4.85 and 8.73 m/s. This finding provides valuable information for urban planning and public health departments to help formulate more effective epidemic prevention and control strategies. This study uses machine learning methods to reveal the impact of urban wind environments on the distribution of COVID-19 and provides important insights into urban planning and public health strategy development.

Keywords: machine learning; COVID-19; urban wind environment; high-density city; urban planning; urban public health

Citation: Zheng, L.; Chen, Y.; Yan, L.; Zheng, J. The Impact of High-Density Urban Wind Environments on the Distribution of COVID-19 Based on Machine Learning: A Case Study of Macau. *Buildings* **2023**, *13*, 1711. https://doi.org/10.3390/buildings13071711

Academic Editors: Carol Monticelli, Anna Stefańska, Daniel Mateus, Carlos C. Duarte and Nuno D. Cortiços

Received: 14 June 2023
Revised: 29 June 2023
Accepted: 3 July 2023
Published: 4 July 2023

1. Introduction

1.1. Research Background

According to the World Health Organization (WHO), approximately 10–20% of people infected with COVID-19 may continue to experience medium- and long-term effects of COVID-19 [1–3]. These effects are collectively referred to as the "long-term effects of COVID-19" or "long-term COVID-19 syndrome" (long COVID-19). Common symptoms of "long COVID-19" include fatigue, shortness of breath, cognitive impairment, headache, chest pain, joint pain, etc., which may affect daily life [4]. Taking cities in 31 provinces in the Chinese mainland as an example, the number of people infected with COVID-19 from 1 May to 31 this year was 2777, and 164 died. Many netizens claim on social platforms that they have been infected with COVID-19 for the second or even third time, and it is evident that the risk of COVID-19 still exists and cannot be ignored. In the face of this sudden public health incident, we sincerely appreciate the government's timely and powerful management and control capabilities and admire the silently contributing medical staff. At the same time, in our roles, as an architect, urban design worker, or scholar of urban architecture research, we also deeply understand the inadequacy of existing cities in responding to public health emergencies. The global pandemic has given the scientific community a lot to think about, but the recurrence of the pandemic or the outbreak of similar diseases is unknown in the future. Unknown things are difficult to predict, but it is possible to work hard for the well-being of mankind and improve housing conditions objectively. How can we prevent problems before they happen; be prepared when other unknown pandemic diseases are introduced; and provide people with a safe, secure, healthy, and comfortable living space?

Because of regional differences and differences between cities, different urban characteristics and urban environments also have differences. Macau is a city on the southeast coast of China and the west bank of the Pearl River (Figure 1). In the north, it is 145 km from Guangzhou, and in the east, it is 70 km from Hong Kong, across the Lingding Ocean. In the past, it included the Macau Peninsula, Taipa, and Coloane. However, now, the two islands of Taipa and Coloane have been reclaimed as one city. The Macau Peninsula was only 2.78 square kilometers in area in 1840. After extensive land reclamation, Macau's total area reached 33.3 square kilometers [5]. Macau has an estimated population of 673,600, with a population density of 20,745 people per square kilometer, according to official data. The northern portion of the Macau Peninsula is one of the world's most densely populated regions [5]. This is a veritable ultra-high-density city. At the same time, this aspect also introduces challenges that differ from other cities in terms of pandemic diseases and residential area planning. The impact of the COVID-19 epidemic on Macau began in early 2020 and has been divided into six waves. On 8 January 2023, the transition period of the epidemic ended, and severe special infectious pneumonia was listed as an "endemic disease" [6]. In February of the same year, the influenza A virus also broke out, and most outbreaks were related to mass gathering incidents. Controlling epidemic diseases requires the joint efforts of all parties, but in the urban environmental space, it is also worthy of attention and research.

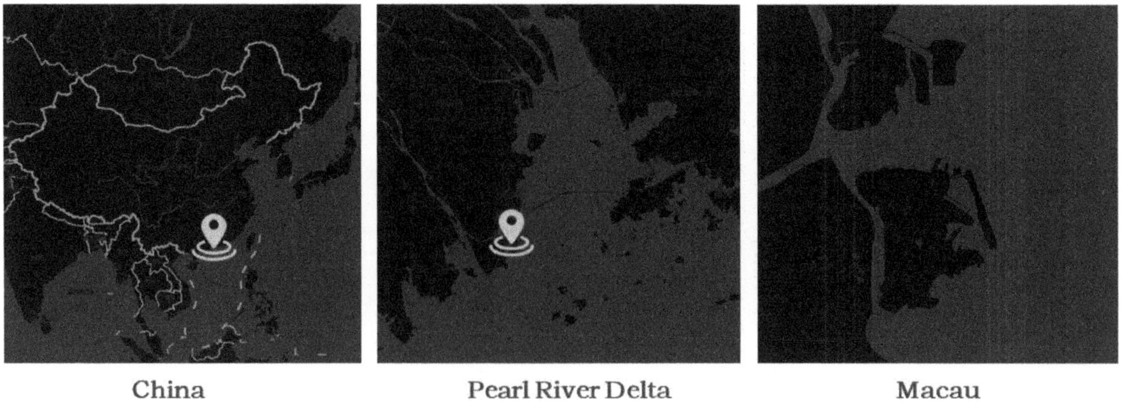

Figure 1. Macau's location. The three small pictures are, respectively, the geographic location of Macau in China, the location of Macau in the Pearl River Delta, and the Macau Special Administrative Region.

1.2. Literature Review

1.2.1. The Pandemic and the Urban Wind Environment

Safe urban public spaces are very important to public health. Whether during a pandemic or in the post-epidemic era of COVID-19, urban design inevitably needs to consider the risk of virus transmission. Therefore, the design of the urban and architectural wind environment has also become an important consideration. At present, some scholars have explored this. For example, the urban wind environment is closely related to the pollutant index, which will increase the risk of disease transmission [7–11]. Researchers have also explored geometric design parameters and the infection possibility of courtyards under calm wind conditions [12]; natural ventilation conditions have been shown to effectively reduce the risk of COVID-19 spreading in open offices [13]; a semantic similarity model based on point of interest (POI) selection was used to explore the natural ventilation potential (NVP) of four basic residential layouts (point layout, parallel layout, central layout, and mixed layout) in Wuhan to evaluate measures to improve urban ventilation [14]; using Design-Builder (licensed version) and Autodesk CFD software (student version), researchers performed simulations to compare the amount of natural ventilation and lighting before and after modifying the shape of the building, and then they obtained a multi-story residential design adapted to COVID-19 [15].

1.2.2. Application Areas of Machine Learning in COVID-19

On the other hand, with the rise and vigorous promotion of artificial intelligence, machine learning, as one of its core technologies, has also been applied in the research of urban environments and COVID-19 [16–21]. For example, some scholars incorporated the parameters of the urban environment into the effect of daily COVID-19 case prediction and used long short-term memory (LSTM) models to predict India's (New Delhi and Nagpur), the United States's (Yuma and Los Angeles), and Sweden's (Stockholm, Skne, Uppsala, and Vastra Götaland) daily cases (tropical, subtropical, and boreal) in nine cities in different climatic zones of these three countries [22]. However, more models are still being applied to the prediction of disease data models in the field of medical care, ranging from the prediction of disease transmission trajectory to the development of diagnosis and prognosis models [23–25]; the identification and diagnosis of radiological images and their use in treatment [26–32]; and the discovery of potential drugs and the prediction of their structures [32–34]. In addition, there is a significant amount of research on forecasting for economic recovery and financial calculation. However, in the field of urban design, there is still a lot of room for exploration.

1.2.3. COVID-19 and Housing Conditions

In addition, in the past three years, some scholars have researched COVID-19 and housing conditions. The relationship between housing conditions and the health of occupants is direct, affecting the quality of life and life expectancy as well as predisposing to the development or exacerbation of different pathologies [35]. A lack of open space, the environment, and the inability to use the most commonly used spaces in the house due to COVID-19 lockdowns may have negatively impacted the health and comfort of college students during quarantine. Therefore, increasing the size of bedrooms and having a balcony or terrace have become elements worth considering [36]. Others have investigated how residential properties in Spain recovered resiliently during the COVID-19 lockdown. Additionally, the authors pointed out that improved designs to achieve healthier and happier housing need to consider the size and characteristics of the built environment (housing and surrounding urban spaces), independent energy and equipment supply for each space, natural ventilation and lighting, and open space contact with the outside world [37]. However, in North America, people who use drugs in various housing settings, including supportive housing buildings, may be particularly at risk for drug-related injuries. Therefore, in the face of such groups, housing conditions also need to consider nursing space support under COVID-19 [38]. There are also rich empirical cases in different countries and regions showing that poor housing conditions lead to further mortality from COVID-19 [39–43]. Nonetheless, the correlation between housing planning and the pandemic at the macro level is worth exploring further. This helps urban designers consider more risk factors when making decisions, thereby contributing to the well-being of human settlements.

1.3. Problem Statement and Objectives

According to a study conducted in Macau, maintaining social distance, wearing masks [44], and voluntarily isolating oneself [45] are effective methods for preventing the community transmission of COVID-19. COVID-19 has disrupted the rhythm of daily life, which has significantly increased the likelihood of anxiety and isolation [46]. The decline in quality of life increases the incidence of insomnia (27.6% of Macau residents suffer from insomnia) [47], and chronic insomnia can lead to depressive symptoms (38.5%) [47]. The main psychological symptoms that Macau residents display during the epidemic are fatigue, mental disorders, and guilt [48], and a lack of exercise increases the likelihood of mental illness. However, in high-density cities with limited land area, it is difficult for many residential areas to provide outdoor rehabilitation sports facilities. At the same time, the current transmission route of epidemic diseases is mainly through the air, and the correlation with the urban wind environment is particularly significant. In the post-epidemic era, how to improve the planning of residential areas has also become a subject of ongoing research. In this paper, the researchers explored the following six questions:

(1) Taking Macau, China, as an example, what is the impact of the high-density urban space and the urban wind environment it shapes on the distribution of COVID-19?
(2) How does machine learning technology assist in analyzing the distribution of COVID-19?
(3) Further, based on the footprint data of 500 cases of the outbreak in Macau, China, in June 2022, what is the correlation between COVID-19 and the urban wind environment?
(4) How do urban wind environments promote or inhibit the distribution of COVID-19 under different morphological layouts?
(5) Using the layout planning and design of sustainable residential areas, which type of form is more conducive to adapting to the environment of pandemic diseases?
(6) In the post-pandemic era, what reflections can this research provide for other similar epidemics?

2. Materials and Methods

2.1. Data Collection

As mentioned above, Macau is a coastal city located in the south of China, near the mouth of the Pearl River. Its characteristics in the wind environment are closely related to its geographical location and subtropical monsoon climate (Figure 2). In summer, Macau is usually affected by the southeast monsoon, with relatively strong winds and an average wind speed of 5 to 7 m per second, often accompanied by brief gusts. The wind direction is mainly from the southeast and blows inland, but it also brings marine air currents and humid weather conditions. This monsoon wind environment may have an impact on the spread of COVID-19. On the one hand, strong winds help air flow and dilute virus particles, reducing their residence time in the air and thereby reducing the risk of transmission. On the other hand, southeasterly winds may blow from densely populated areas, carrying potential virus particles and increasing the possibility of infection. In winter, Macau is mainly affected by the northern monsoon. At this time, the wind is relatively weak, with an average wind speed of about 2 to 4 m per second, and the wind direction is mainly northwest, blowing towards the sea. Under the winter monsoon environment, the air is relatively stable, which may cause virus particles to stay in the air for a long time, increasing the risk of transmission. In addition, Macau is also affected by climate change, such as frequent typhoons, but due to the complexity of extreme climates, the impact of the wind environment in extreme climates on the epidemic is not considered in this study.

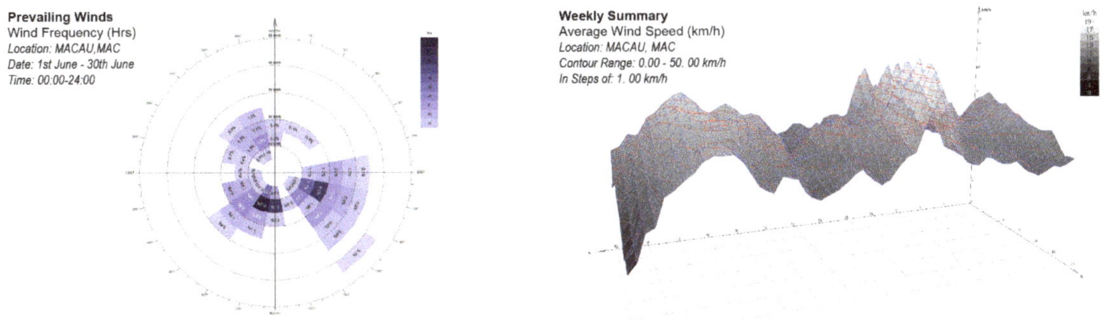

Figure 2. Analysis of prevailing winds and annual wind speed in Macau.

To simulate the wind environment in Macau, the researchers collected wind speed and direction data published by the Macau Meteorological Bureau. These data include changes in wind strength and wind direction at different times of the day. These data reflect the real wind conditions in Macau. Second, the Ecotect building environment analysis software (version 2011) was used (refer to Appendix A for the operating conditions of the software when performing wind environment simulation using Ecotect), which provides a scientific and accurate method to simulate the urban wind environment and provides researchers with a tool to quantitatively analyze the urban wind [49,50]. Ecotect is a commonly used environmental simulation software; some previous studies can provide a basis for its reliability [51]. Finally, the researchers applied these meteorological data to the model after building a three-dimensional model of Macau city to further simulate the wind environment in different regions of Macau and achieve an accurate simulation of the urban wind environment (Figure 3a).

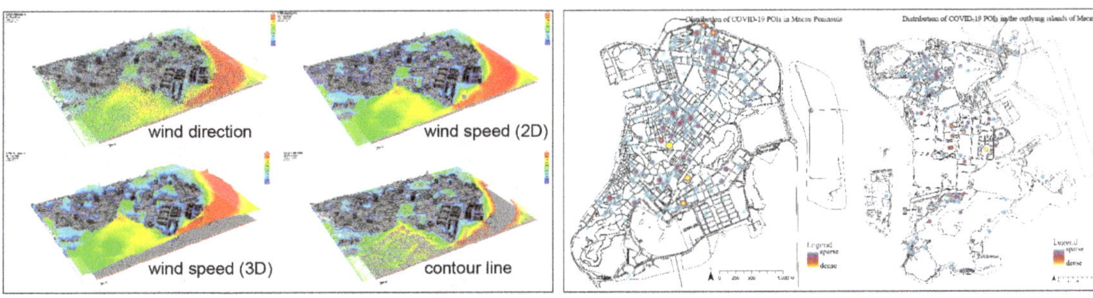

(**a**) (**b**)

Figure 3. Macau wind environment simulation map and COVID-19 heat map. (**a**) Urban wind environment data simulation; (**b**) COVID-19 data collection and aggregation.

In this study, the use of Ecotect software (version 2011) has the following advantages:

(1) Building performance simulation: Ecotect can simulate the performance of buildings under different wind conditions. With the software, researchers can accurately assess the impact of different wind environments on the spread of COVID-19 and gain insight into the relationship between the built environment and virus spread.
(2) Data collection and analysis: Ecotect provides a wealth of data collection tools to obtain the environmental parameters required by the building, such as wind speed, wind direction, and indoor and outdoor temperatures. These data are crucial for studying the association of urban wind environments with the spread of COVID-19.

(3) Result visualization: Ecotect provides an intuitive result visualization function. Through charts and images, researchers can clearly display and present the analysis results. These visualization tools are important to convey research findings and conclusions to judges and readers and contribute to a better understanding of the link between urban wind environments and the spread of COVID-19.

To summarize, Ecotect software (version 2011) played an important role in this study. It provides functions such as building performance simulation, data collection and analysis, and result visualization, providing strong support for researchers to explore the relationship between urban wind environments and COVID-19 transmission.

At the same time, this study also collected the footprint data of COVID-19 patients provided by the Macau Health Bureau in June 2022. These data included the location and range of activities of 500 COVID-19 patients before onset, totaling 3982 footprints [52]. In order to convert these footprint points into geographic coordinates and analyze them, this research uses Google Maps API Web Services to convert the address information of footprint points into accurate geographic coordinates. After the footprint points are converted into geographic coordinates, these data are input into the ArcGIS Pro software (version 3.1) for processing. To create a heat map of the epidemic footprint, the researchers used the hotspot analysis feature in ArcGIS Pro. This function can generate a heat map according to the density of footprints, visually showing the concentrated areas of epidemic activity and hotspots of transmission. In order to ensure the accuracy and consistency of the data, the researchers uniformly transformed all geographic coordinates into the Observatorio Meteorologico 1965 Macau Grid coordinate system for correction. This ensures that the data used in producing the heatmap have a consistent reference frame, making the research results more reliable and comparable (Figure 3b).

In addition, since the epidemic data collected in this study are from June 2022, it is necessary to set the time of the wind simulation to the average wind force and direction of that month. This can ensure that the simulation results are consistent with the actual situation and provide an accurate data basis for this study on the relationship between the urban wind environment and the spread of COVID-19. Therefore, this study mainly focuses on the relationship between the summer monsoon environment, the distribution footprint of the COVID-19 epidemic, and the layout of buildings.

2.2. Data Processing

Using the above method, the researchers obtained the wind environment simulation results for the Macau Peninsula and outlying islands and generated the corresponding COVID-19 heat map (Figure 4). When simulating the wind environment, considering the difference in the height of Macau's topography, it is necessary to set the analysis height of the wind environment to 5 m above sea level so as to reflect the wind environment conditions in most parts of Macau. However, it should be noted that some buildings in low-lying areas may be ignored in the simulation. When generating the COVID-19 heat map, this study used black-and-white image drawing to simplify data presentation. Black areas in the image indicate areas without COVID-19 cases, while white areas indicate COVID-19 cases. In addition, the depth of the image represents the density of hotspots, and the darker the color, the higher the density of hotspots. This drawing method can visually illustrate and compare the COVID-19 epidemic situation in different regions, identify areas where COVID-19 hotspots are concentrated, and help to understand the spread of the epidemic and the density distribution of hotspots.

In addition, since the training of the CGAN model requires a large amount of sample data (refer to Appendix B for the operating conditions of the machine learning environment), the above-mentioned images need to be cropped uniformly to generate a picture with a size of 512 $\times$ 512 pixels (Figure 5). In this study, pixels refer to the basic display unit in an image. In the wind simulation, the pixels of different colors represent the numerical value of the wind speed. In the COVID-19 heat map, white to gray pixels represent the extent of an outbreak, and black pixels represent no outbreaks. Each cropped image covers an area of 4 hectares. The specific cropping area depends on the distribution of COVID-19, which is mainly concentrated in the area containing COVID-19. The selection of these regions can help the CGAN model learn the correlation between urban wind environments and COVID-19. Regions that do not contain COVID-19 may interfere with the training of CGAN and cause the model to not fit the data well, so they are excluded during the cropping process. In this study, a total of 54 sets of images were cropped from the Macau Peninsula and used for the training phase of the model. At the same time, 35 sets of pictures were cropped from the outlying islands of Macau (mainly Taipa and Coloane), and these data were used to test the robustness of the trained model and its accuracy and generalization.

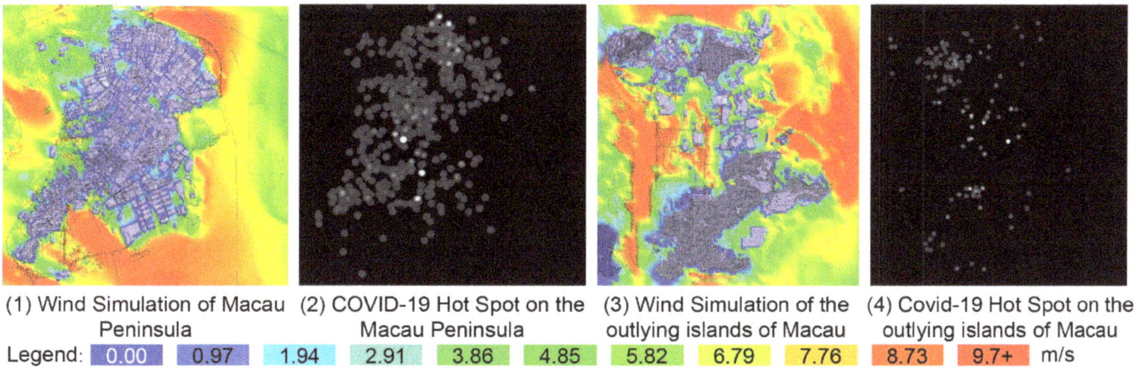

Figure 4. Wind environment simulation map of Macau Special Administrative Region (including Macau Peninsula, Taipa, and Coloane) and its distribution map of COVID-19 hotspots. Among them, Taipa and Coloane belong to the outlying islands of the Macau Special Administrative Region.

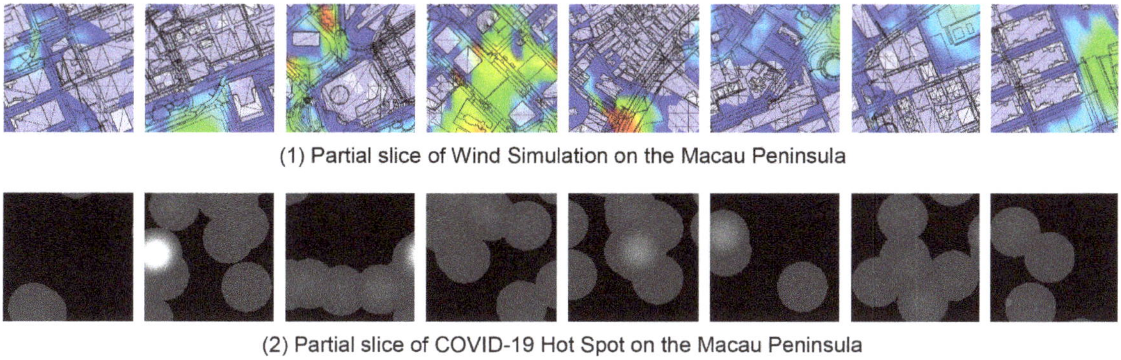

Figure 5. A slice of the Macau Peninsula wind environment simulation map and COVID-19 hotspot distribution map. (1) Partial slice of wind simulation on the Macau Peninsula; (2) Partial slice of COVID-19 hot spot on the Macau Peninsula.

In the processing of materials, by clipping and selecting areas containing COVID-19 as samples, it is possible to ensure that the trained CGAN model can accurately reflect the relationship between the urban wind environment and COVID-19, thereby improving the predictive ability of the model. At the same time, by using the data of Macau's outlying islands (mainly Taipa and Coloane) for testing, the model's reliability can be verified, and its adaptability and promotion in different geographical situations can be evaluated. Such a design can ensure that the research conclusions have certain scientific credibility and provide valuable references for further research and practical application.

2.3. CGAN Method

CGAN, or conditional generative adversarial network, is the machine learning method used in this study. The characteristic of CGAN is to generate synthetic data that meet the given conditions through the game of the generator and the discriminator (Figure 6). CGAN consists of two key components: a generator and a discriminator. The generator's goal is to generate corresponding synthetic data (COVID-19 distribution data) according to the given conditions (in this study, the conditions are urban wind environment data). The discriminator is responsible for judging whether the given data are real or generated by the generator. The generator and the discriminator compete with each other through adversarial training to continuously optimize their respective capabilities.

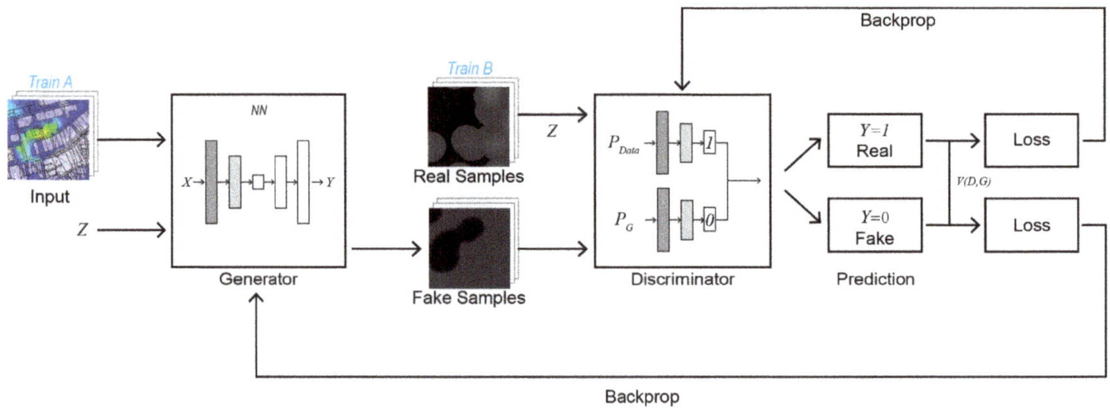

Figure 6. CGAN principle flow chart.

During training, the generator takes input conditional data (urban wind environment data) and transforms it into synthetic data similar to real COVID-19 distribution data. In contrast, the discriminator receives both the real COVID-19 distribution data and the synthetic data generated by the generator and tries to distinguish their authenticity. The goal of the generator is to generate data that can fool the discriminator into being unable to accurately distinguish real data from synthetic data, while the goal of the discriminator is to distinguish real data from generated synthetic data as accurately as possible. By repeatedly performing confrontational training between the generator and the discriminator, the network framework of CGAN can gradually optimize the generator's ability. The synthetic data generated by it are closer to the distribution of real data, and the accuracy of the discriminator is improved to make it more effective in distinguishing real data from synthetic data. Finally, the trained generator can generate a probability model that predicts the distribution of COVID-19 under different urban wind environments based on the given urban wind environment data, thereby revealing the impact of urban wind environments on the distribution of COVID-19.

3. Model Training

3.1. Model Training Process and Verification

The model training in this study involves two models, namely, Model 1, for generating COVID-19 heat maps from wind environment images, and Model 2, for generating wind environment images from COVID-19 heat maps. The training goal of these two models is to reveal the association between the urban wind environment and the spread of COVID-19 and achieve data generation and prediction under given conditions.

The input in Model 1 is wind environment image data, and the goal is to generate the corresponding COVID-19 heat map. The model's training uses the wind environment images of the Macau Peninsula and the corresponding COVID-19 heat map as training samples. By inputting the wind environment image into Model 1, the synthetic COVID-19 heat map generated by Model 1 is compared with the real COVID-19 heat map, and the difference between the two is calculated as the loss function of Model 1.

The training process of Model 2 is similar to that of Model 1 but in the opposite direction. The input of Model 2 is the COVID-19 heat map, and the goal is to generate the corresponding wind environment image. Additionally, using the same training samples, the COVID-19 heat map is input into Model 2, the synthetic wind environment image generated by Model 2 is compared with the real wind environment image, and the difference between the two is calculated as the loss function of Model 2.

Model 1 and Model 2 were trained for 200 epochs, respectively, and statistical analysis was performed by drawing a line graph of the training log (Figure 7). Observing the training log graph leads to the following conclusions:

(1) In the early stages of training, Model 1 and Model 2 rapidly reduce the loss value, showing a preliminary understanding and grasp of the characteristics of the data. As the training progresses, they gradually optimize the parameters and network structure, and their ability to adapt to the data distribution continues to improve. This leads to a continuous decrease in the loss value, which reaches a lower level at a certain stage. With the deepening of training, the loss value gradually tends to be stable, indicating that the model has learned the key features of the data and reached an equilibrium state during the optimization process. The model has reached a good level of performance, and further training may not improve performance significantly. Therefore, the model goes through an initial stage of learning and progressive optimization, and finally it reaches a state where learning converges.

(2) In the training log of statistical CGAN, it is observed that the minimum values of different indicators appear at different iteration numbers. This shows that the optimization process of the model has achieved a balance among various indicators, and the performance of each indicator is considered comprehensively rather than paying too much attention to a specific indicator. At the same time, the minimum values of different indicators are distributed on different iteration numbers, which may indicate mutual influence and interaction between them. The optimization of the model requires trade-offs and adjustments between various indicators to achieve a balanced and optimal training result. In addition, the minimum values of different indicators appear at different iteration numbers, which may also mean that the model gradually optimizes different aspects at different stages. A model may first focus on the optimization of a certain metric and then gradually shift to the optimization of other metrics to achieve better overall performance. These observations thus reveal important features such as balanced optimization, inter-metric interactions, and stepwise optimization during model training.

(3) The lower Mean G_GAN, Mean D_fake, and Mean D_real indicate that the adversarial training between the generator (G) and the discriminator (D) has reached a balanced state. The generator can successfully fool the discriminator into generating realistic samples. At the same time, the discriminator can effectively distinguish generated samples from real samples and give accurate discriminative results. This shows that the model has stability, and the generator has effectively learned the characteristics of the data distribution during the training process and can generate high-quality samples similar to real samples. The discriminator has a high discrimination ability and can accurately judge the authenticity of the generated samples. Therefore, these observations reflect the model's stability, efficient generative ability, and high-quality discriminative ability.

In addition to parsing the training log of the model, the quality of the model generation can also be evaluated by observing the test pictures of each generation of the trained model. In this study, the pictures generated by Models 1 and 2 were observed and compared with real pictures to verify the training effect and generation quality of the models.

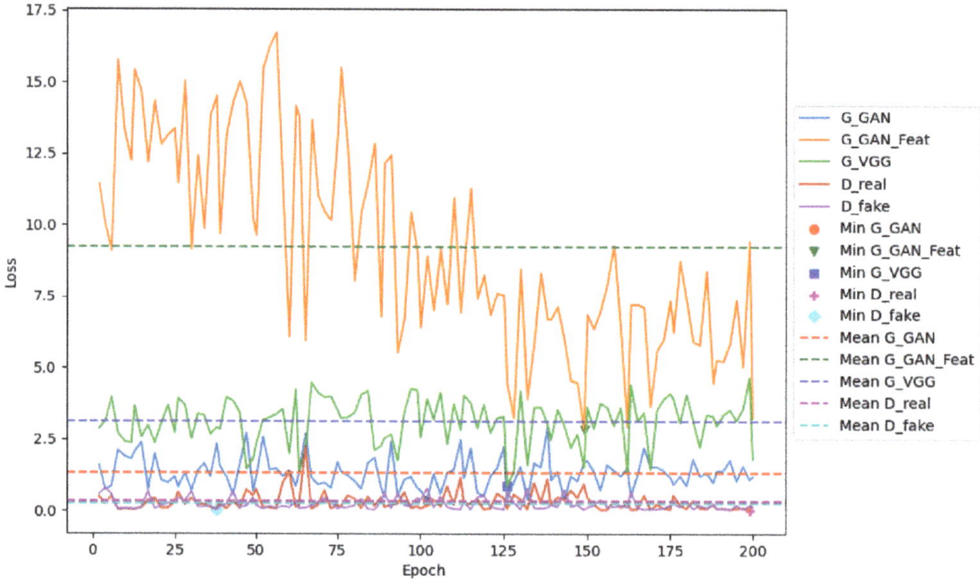

(1) The loss curve of model 1

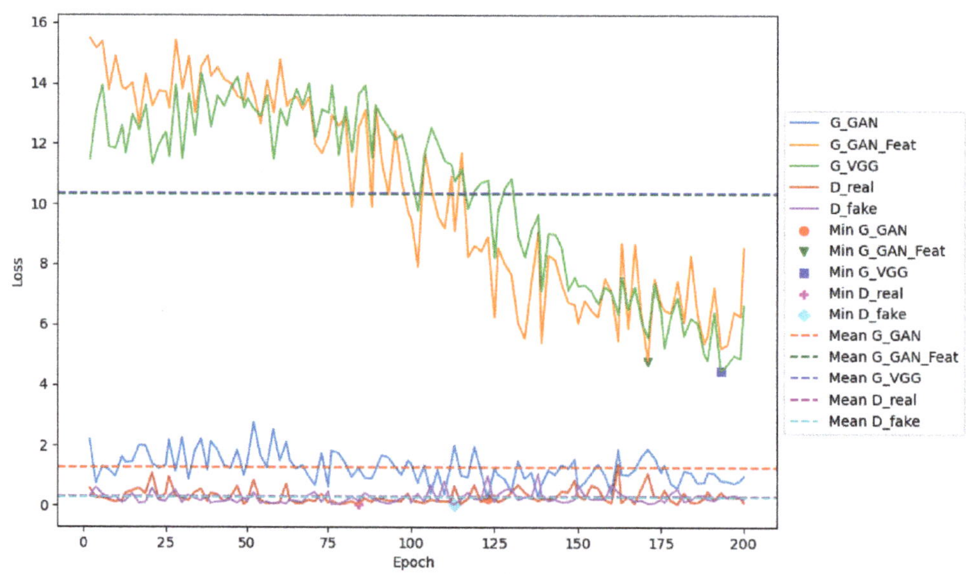

(2) The loss curve of model 2

Figure 7. Line chart of model training log.

It can be observed in the training process pictures of Model 1 that in the first 50 epochs, there is a large difference between the generated pictures and the real pictures (Figure 8). However, as training progresses, the model gradually improves the quality of the generated images as the number of epochs increases. After about 100 and 150 training epochs, the generated pictures gradually approach the reality of real ones. When the training reaches about 200 epochs, the generated images are almost indistinguishable from the real ones. Therefore, Model 1 has completed the training and has good epoch quality.

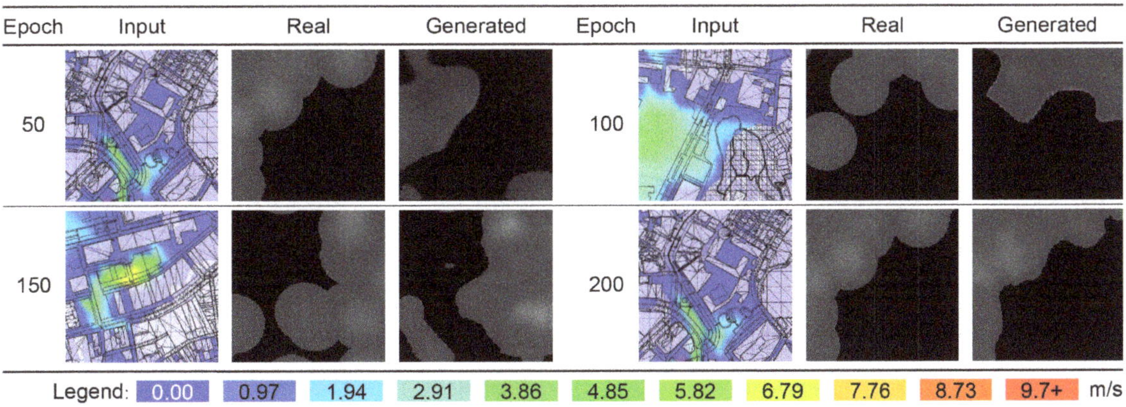

Figure 8. Comparison of the images generated during the training process of Model 1 with the original images. "Epoch" indicates the number of training iterations. This model has been trained for 200 epochs in total, so the test images of the 50th, 100th, 150th, and 200th epochs during the model training process are selected for analysis. "Input" represents the input wind environment simulation. The slice material of the result "Real" represents the actual COVID-19 hotspot image corresponding to the slice. "Generated" represents the image generated by the model through the material of "input".

Similarly, a situation similar to that of Model 1 is also observed in the pictures of the training process of Model 2 (Figure 9). There is a large difference between the pictures generated in the initial stage and the real pictures, but as the training progresses, the model gradually improves the quality of the generated pictures. After about 100 and 150 training epochs, the generated pictures gradually approach the reality of real ones. Eventually, after about 200 training epochs, the generated images are nearly indistinguishable from real ones. This further verifies the completion of the training of Model 2 and the improvement in the epoch quality.

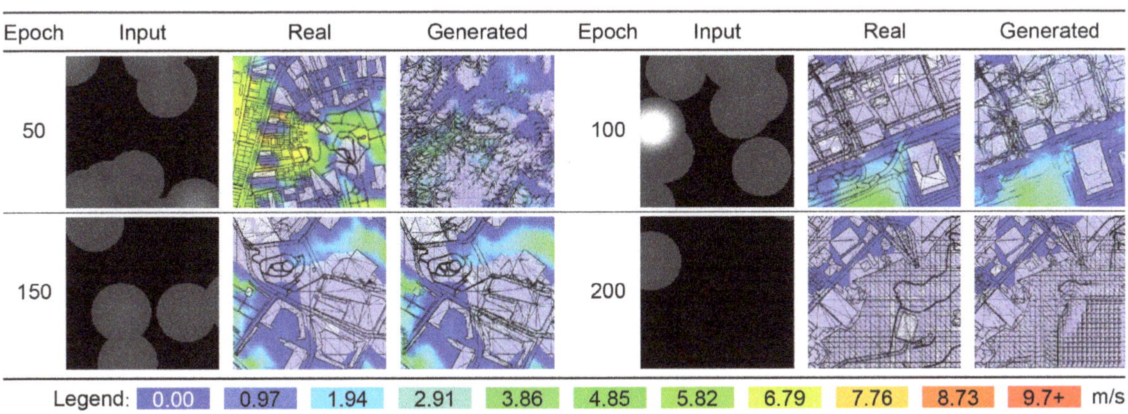

Figure 9. Comparison of the images generated during the training process of Model 2 with the original images. "Epoch" indicates the number of training iterations. This model has been trained for 200 epochs in total, so the test images of the 50th, 100th, 150th, and 200th epochs during the model training process are selected for analysis. "Input" represents the input wind environment simulation. The slice material of the result "Real" represents the actual COVID-19 hotspot image corresponding to the slice. "Generated" represents the image generated by the model through the material of "input".

Through the above observations, it can be concluded that after a certain number of training algebras, the epoch quality of Models 1 and 2 has been significantly improved, and the difference between them and the real picture is gradually reduced. This shows that the model training process in this study is effective, successfully learns the association between the urban wind environment and the COVID-19 heat map, and can generate images of high quality. This provides a reliable basis for further analysis of the relationship between urban wind environments and the spread of COVID-19.

3.2. *Correlation Analysis of the Wind Environment and COVID-19 in Different Building Layout Types*

In order to understand the correlation between the wind environment of different building layout types and COVID-19, this study takes the materials of wind simulation and COVID-19 hotspots as the input of Model 1 and Model 2 and performs image generation of COVID-19 hotspots and wind simulations. The following characteristics can be observed through the results of generating 54 groups of picture materials on the Macau Peninsula:

In the generated results of Model 1, it can be observed that the higher the building density, the higher the distribution of COVID-19 hotspots, which aligns with common cognition (Figure 10). Among them, Figure 10A shows the situation of modern buildings, and Figure 10B shows the situation of historical buildings. In modern complexes, the distribution of COVID-19 hotspots is relatively uniform, while in historic complexes, the area in the square has the highest density of COVID-19 hotspots. This shows that the building density of the modern building complex is relatively reasonable, and there are many public spaces in it. However, the distance between buildings in the historical building complex is relatively small, and there is only one square for activities, which leads to a concentration of crowds in this square, leading in turn to a high COVID-19 hotspot density.

In addition, in Figure 10C,D, it can be observed that the distribution of COVID-19 hotspots basically matches the shape of the building complex. Especially in public spaces with high wind speeds, the distribution of COVID-19 was not observed, which indicates that open spaces with high wind speeds can effectively alleviate and inhibit the occurrence of COVID-19. Given the above observations, it can be concluded that there are obvious differences in the correlation between different types of wind environments and COVID-19. Areas with a high building density are more prone to COVID-19 hotspots. Modern buildings have relatively better evacuation and public space layouts, which can reduce the concentration of hotspots, while in open spaces with high wind speeds, the spread of COVID-19 is effectively suppressed.

In the generated results of Model 2, it can be observed that more COVID-19 hotspots are typically distributed in building complexes, while less COVID-19 is typically distributed in natural mountain landscapes, which is consistent with our assumption (Figure 11). Figure 11A,B show the distribution of many COVID-19 hotspots, but their distribution patterns are not the same, and they are also consistent with the distribution of building groups in the generated wind environment picture. This shows that the building arrangement is the main driver of the distribution of COVID-19, and the scope of the buildings largely determines the spread of COVID-19.

In addition, in Figure 11C,D, it can be observed that the distribution of COVID-19 hotspots is reduced. Under this condition, the final generated pictures all point to the natural mountain landscape, which shows that the spread of the epidemic can be effectively suppressed in the natural outdoor environment. The specific reason for this may be that the natural outdoor environment has better ventilation conditions, and some plant species also play a role in inhibiting the spread of COVID-19, thereby reducing the formation of hotspots.

In summary, there are obvious differences in the correlation between different types of wind environments and COVID-19. The distribution of more COVID-19 hotspots in the building complex relates to the shape or arrangement of the building itself, and the size of the building determines the spread of COVID-19. In contrast, the lower COVID-19 distribution in natural mountain landscapes may be due to the advantages of the outdoor environment, such as the combined effects of better ventilation conditions and plant suppression.

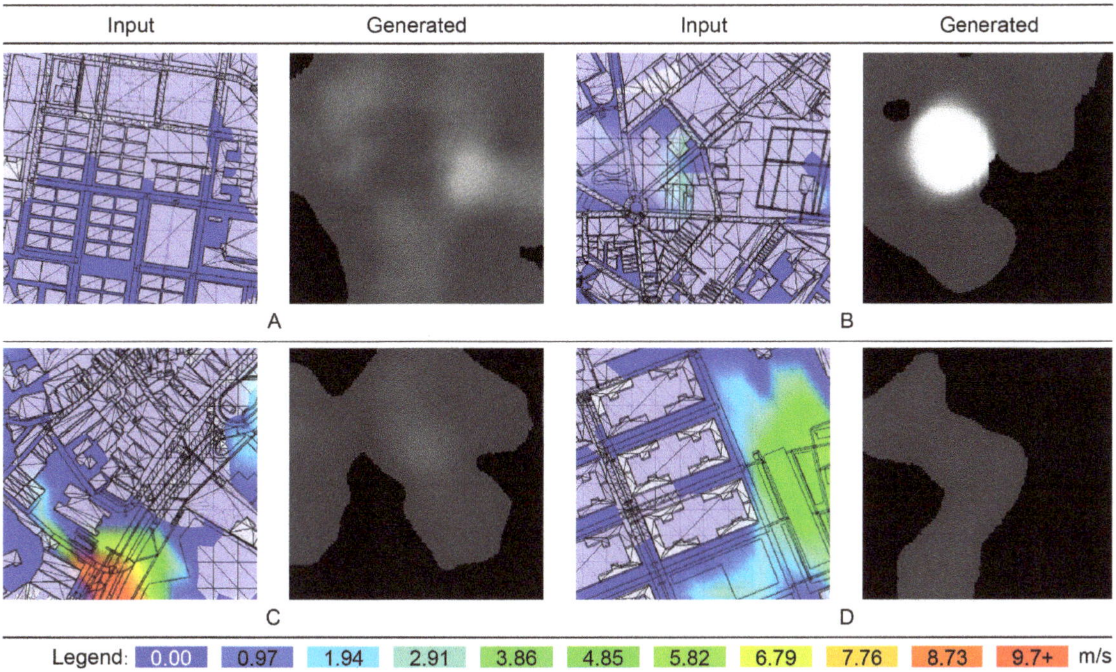

Figure 10. Different wind environments compared to COVID-19 hotspots. "Input" represents the input wind environment simulation. "Generated" represents the image generated by the model through the material of "input". (**A–D**) represent the group number.

3.3. Robustness Test of the Model

In order to test the robustness of the model, it is necessary to select some new materials that were not used in the model training process and ensure that there is a certain logical relationship between these materials and the training materials. Therefore, in this study, some wind environment simulations and slices of COVID-19 hotspots in Macau's outlying islands (mainly Taipa and Coloane) were selected for this study, and a total of 35 slices were acquired. As part of the Macau Special Administrative Region, the Macau Islands share climatic conditions and urban planning with the Macau Peninsula. By evaluating the results of model generation with these new materials, the robustness of the model can be further judged.

Figure 11 shows the generation results of Model 1 for the wind environment simulation materials on Macau's outlying islands. Overall, there is a certain similarity between the generated COVID-19 hotspot distribution map and the actual results. However, there is a certain difference in percentage, ranging from 22.76% to 56.70%, and the fluctuation in accuracy is more obvious.

(1) In Figure 12 group number 1,2, the difference percentages are small (26.81% and 22.76%). This is because the two illustrations have a clear separation of building groups and public spaces, as well as higher wind speeds, factors that allow the model to generate a more accurate map of the distribution of COVID-19 hotspots.
(2) In Figure 12 group number 3,4, the difference percentages are higher (56.70% and 49.47%). The reason for this is that some of these buildings are not residential or have been out of use, in which case the model cannot distinguish well, resulting in

generated COVID-19 hotspot distribution results that deviate significantly from the actual situation. In addition, the model avoids the generation of COVID-19 hotspots in the position of wind flow so that the overall generation results still have a certain degree of accuracy.

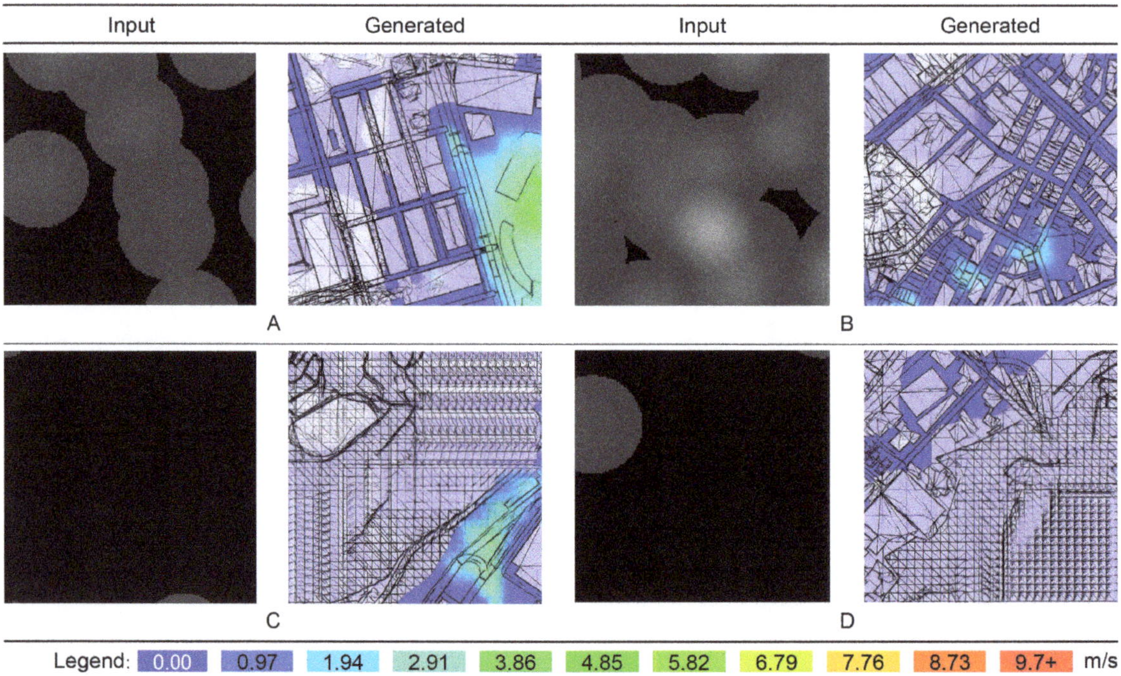

Figure 11. Different COVID-19 hotspots compared to wind environments. "Input" represents the input wind environment simulation. "Generated" represents the image generated by the model through the material of "input". (**A–D**) represent the group number.

In summary, the robustness and accuracy of Model 1 can be preliminarily understood by evaluating the results of model generation on materials from Macau's outlying islands. Despite some differences, Model 1 can still generate a COVID-19 hotspot distribution map similar to the actual results. Especially in the case of obvious separation of building groups and public spaces and high wind speeds, it performs better. However, for some non-residential buildings or areas that have been discontinued, the accuracy of the model drops. These findings provide valuable guidance for researchers to further optimize and improve Model 1.

Figure 13 shows the analysis of the generation results of Model 2 on the COVID-19 hotspot materials on the outlying islands of Macau. Overall, Model 2 has poor quality in generating architectural forms and can barely restore the actual architectural forms, presenting a chaotic image. However, the model showed some accuracy in generating the main locations of buildings and wind environments, but the overall error value was large (ranging from 75.83% to 87.83%). This is because the input material is the distribution map of COVID-19 hotspots, and this type of material contains much less information than the picture of the wind environment simulation. It is difficult to make the model accurately restore the actual urban architecture and wind environment based on the COVID-19 heat map alone.

Nevertheless, the experiment can still observe a certain regularity.

(1) Figure 13 group number 1 features the smallest error value among the other pictures (75.83%). This is because the input COVID-19 heat map has only a small number of hotspot distributions, and the hotspot density is low. The model considers these areas to be connected to the mountain landscape, so relatively accurate results are obtained, which shows that there is a high correlation between the distribution of COVID-19 and the mountain landscape.

(2) In Figure 13 group number 2, the results generated by the model show that strong wind flow is located in the middle of the building complex. This is because there is no COVID-19 hotspot distribution in the middle of Figure 13 group number 2, and the model believes this area should have good ventilation. The results of this generation are consistent with the actual situation, showing that a better ventilation environment helps to suppress the spread of COVID-19.

(3) The generated results in Figure 13 group number 3 are completely different from the actual situation, and the error percentage is the highest (87.83%). This is because this location is the main commercial area of Macau's outlying islands, and the spread of COVID-19 is mainly caused by contact between people in commercial activities, such as people smoking in this area, and COVID-19 spreads through smog. This particular case was not anticipated by the model and thus generated completely inaccurate results.

(4) The generated result of Figure 12 group number 4 is similar to that of Figure 13 group number 2. Likewise, in areas where there is no distribution of COVID-19, the model considers them to be better ventilated. Therefore, there is a clear boundary between the building and the wind environment in the generated results.

However, for large buildings with medical functions or related diseases, such as hospitals, when the model generates a COVID-19 hotspot distribution map, the accuracy of the hospital area may be affected to a certain extent. As important public facilities, hospitals have different personnel flow and activity patterns from residential areas and may face higher epidemic risks and transmission challenges. Therefore, when assessing and predicting the spread of COVID-19 in hospital areas, more detailed and in-depth research is required, combined with hospital-specific environmental factors and control measures for analysis.

In summary, Model 2 is less accurate in generating COVID-19 hotspot material but can still show some accuracy in specific cases. This emphasizes the importance of considering the type of input material in the quality of model generation, and it is difficult to accurately restore the form of urban buildings and wind environments by only relying on the COVID-19 heat map. In future research, the robustness of Model 2 can be further improved to improve its adaptability to complex inputs.

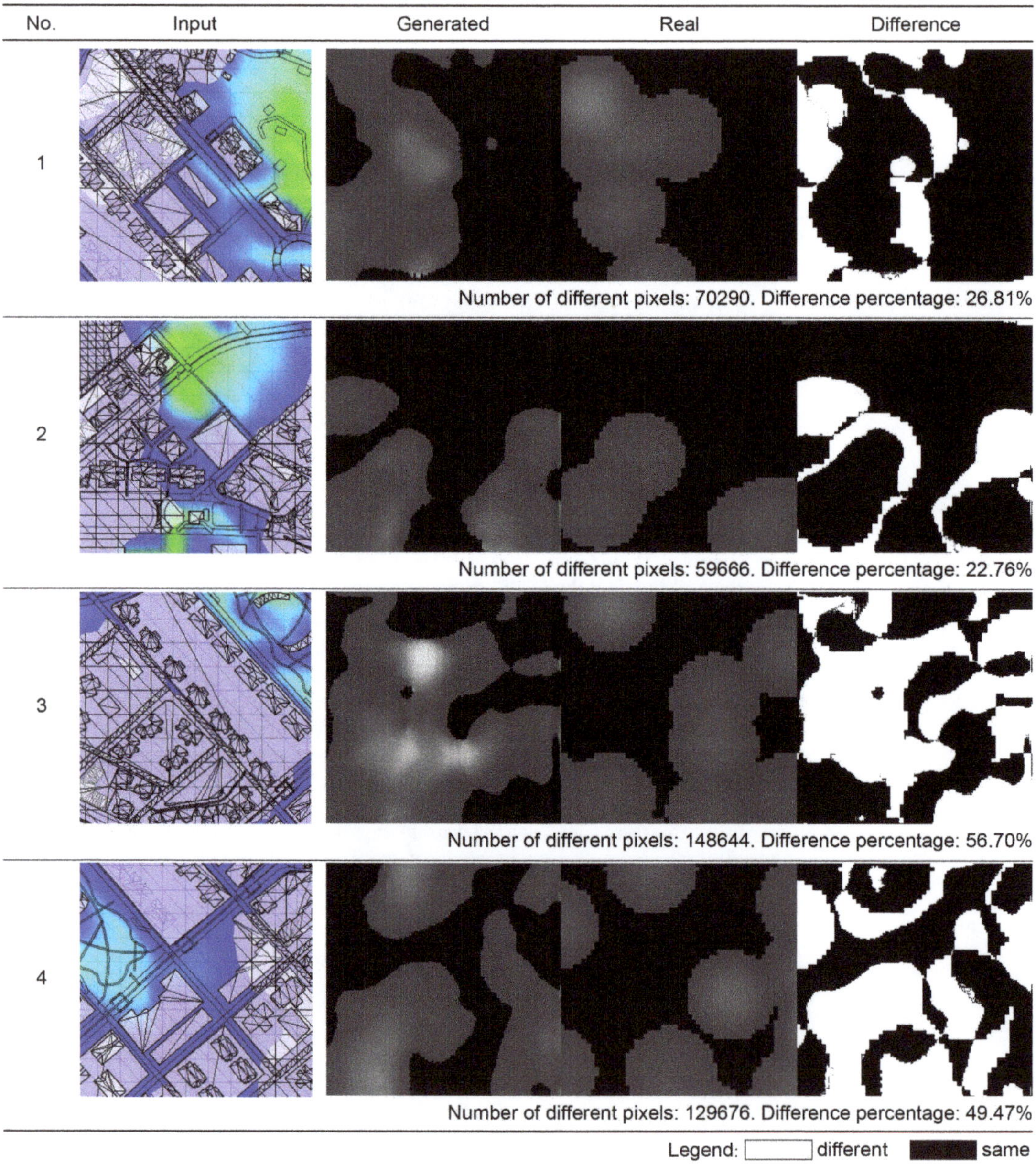

Figure 12. Results of model testing using wind simulation data from Macau's outlying islands. "Input" represents the input wind environment simulation. "Generated" represents the image generated by the model through the material of "input". The slice material of the result "Real" represents the actual COVID-19 hotspot image corresponding to the slice. "Difference" represents an overlay of "Generated" and "Input" materials. Parts that are the same are black, and parts that are different are white to better analyze the accuracy difference between the results generated by the model and the actual results. "1–4" represents the group number of the simulation experiment control.

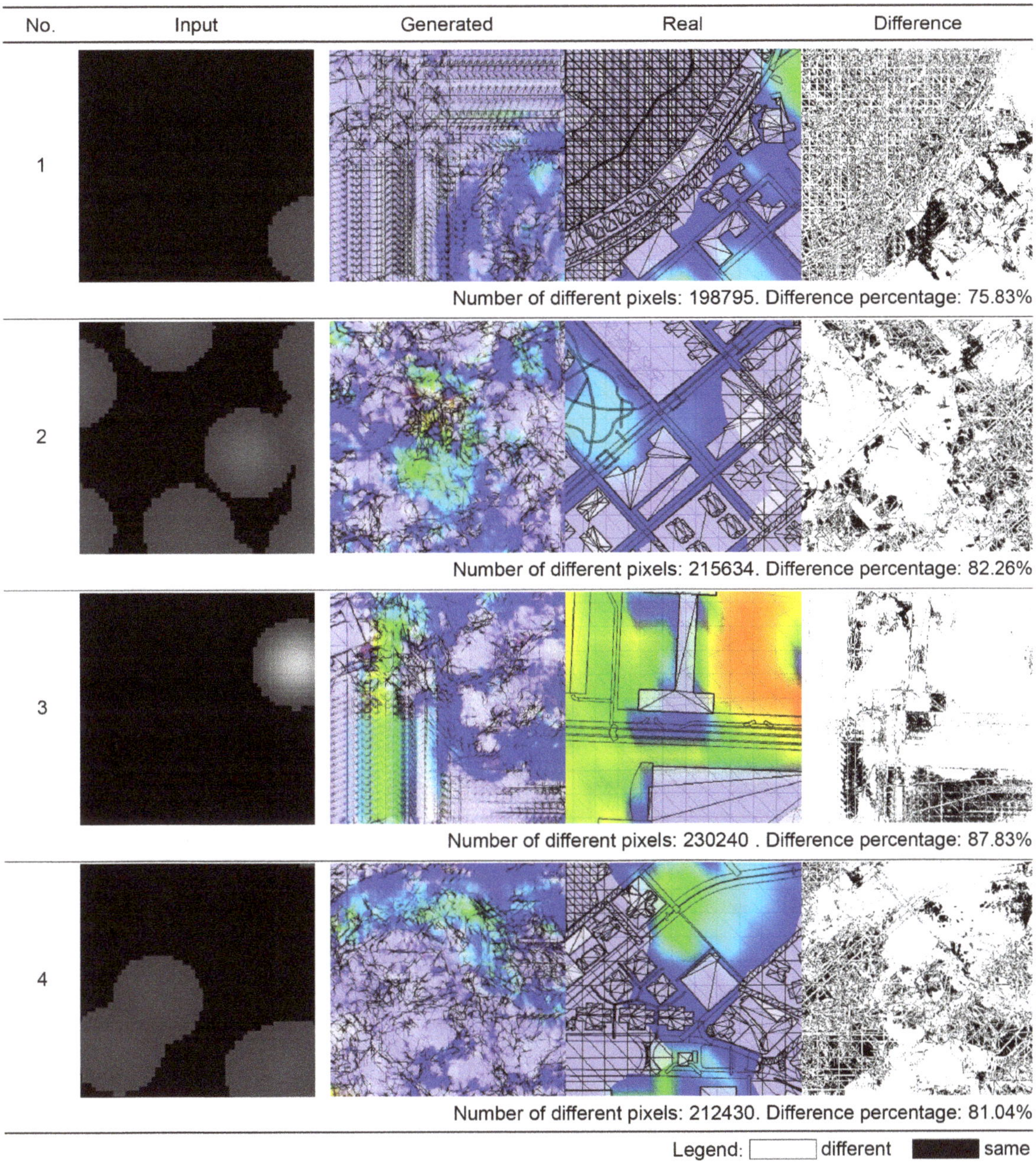

Figure 13. Results of model testing using COVID-19 data from Macau's outlying islands. "Input" represents the input wind environment simulation. "Generated" represents the image generated by the model through the material of "input". The slice material of the result "Real" represents the actual COVID-19 hotspot image corresponding to the slice. "Difference" represents an overlay of "Generated" and "Input" materials. Parts that are the same are black, and parts that are different are white so as to better analyze the accuracy difference between the results generated by the model and the actual results. "1–4" represents the group number of the simulation experiment control.

4. Discussion: Residential Planning under Long COVID-19

In this study, it was found that there is a certain correlation between wind environment factors and the distribution of hotspots of the new coronavirus, and different types of wind environments have different effects on the distribution of hotspots. This means that urban planning and public health departments can reduce the risk of epidemic transmission through reasonable urban design and wind environment control. Based on this, the following four sections—typical residential building types in low-epidemic-risk areas, wind environment simulation, epidemiological situation analysis and verification, and design principles—discuss the relationship between the daily urban wind environment, the distribution of COVID-19 hotspots, and the shape of building layouts.

4.1. Typical Residential Building Types in Low-Epidemic-Risk Areas

How can we further determine which type of architectural form is more suitable for the future long COVID-19 environment? Based on the above-mentioned relationship between the wind environment of residential buildings in Macau and the impact of the epidemic, the researchers examined specific architectural space forms in areas less affected by the epidemic. First of all, taking Macau Peninsula as a typical research object, the wind speed of the epidemic-affected area and the unaffected area was analyzed to obtain the relationship between the epidemic and wind speed. To obtain the wind speed map of the epidemic-affected area and the wind speed map of the epidemic-unaffected area, the researchers superimposed the epidemic distribution map and the wind speed map (Figure 14). Second, the wind speeds of the above two epidemic distributions were counted separately to obtain the overall wind speed of the affected area and the wind speed of the area unaffected by the epidemic (Figure 15).

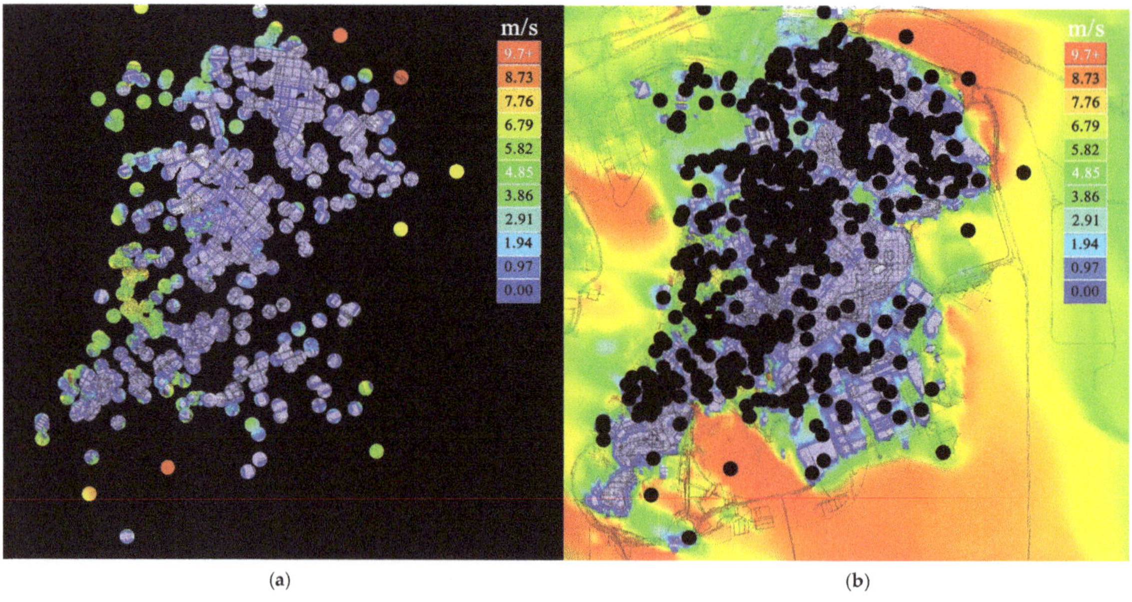

Figure 14. Wind speed distribution map of the affected area and the unaffected area. (**a**) Wind speed map of the area affected by the epidemic; (**b**) wind speed map for areas not affected by the epidemic.

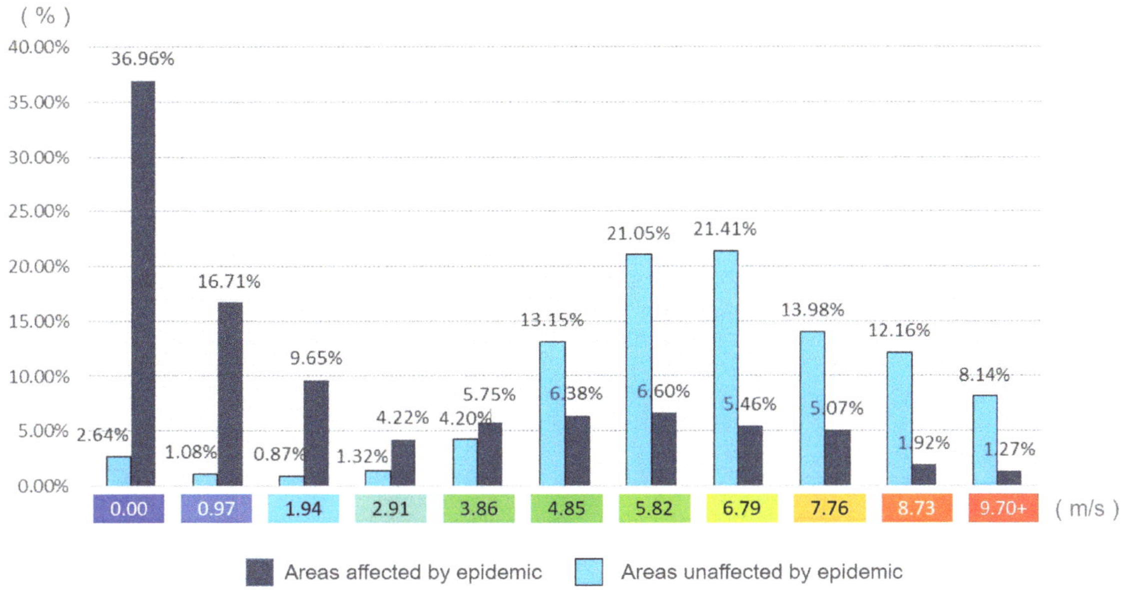

Figure 15. A statistical map of wind speeds in areas affected by the epidemic and those unaffected.

Given the wind speed statistics in Figure 14, it can be seen that in the areas affected by the epidemic, 63.32% of the areas affected by the epidemic have a wind speed between 0.00 and 1.94 m/s (the wind speed is 0.00 m/s, accounting for 36.96%; the wind speed is 0.97 m/s, accounting for 16.71%; and the wind speed is 1.94 m/s, accounting for 9.65%). Wind speeds between 2.91 and 9.7 m/s are relatively less affected by the epidemic, accounting for between 1.27% and 6.60%. In the areas not affected by the epidemic, 94.09% of the wind speeds are between 3.86 and 9.70 m/s, and most of them are between 4.85 and 8.73 m/s (accounting for 68.61%). Comparing the superposition of the two situations, it is found that the proportion of wind speed at 3.86 m/s is relatively similar. When the wind speed is higher than 2.91 m/s, the proportion of the overall epidemic distribution is small, indicating that most areas can reduce the spread of the epidemic to a large extent when the wind speed is higher than 2.91 m/s.

Therefore, for the areas not affected by the epidemic, the researchers selected the areas with wind speeds greater than 2.91 m/s around the buildings, combined with the building density and the surrounding topography, to further study the architectural spatial form. According to the characteristics of urban residential buildings and environmental spaces in Macau, they can be divided into the following four categories based on building density and surrounding terrain conditions:

(1) High-density residential buildings without mountains: the building density is relatively high, and there is no obvious mountainous terrain around (the surrounding wind speed is 0.00~6.79 m/s). Its characteristics are: the form of a single residential building presents a "C" shape. Among them, the building is a semi-enclosed "C"-shaped building, and the inwardly enclosed frontal inner part of the building faces the east or south side (windward side) (Figure 16).

(2) High-density residential buildings with mountainous types: The surrounding mountainous terrain has a relatively high building density (the surrounding wind speed is 0.97~8.73 m/s). Residential buildings are built at the foot of the mountain or on top of the mountainous terrain. Its characteristics are that most of the buildings are located on the southeast side of the mountain, and the building form is "strip" or "rectangular". Further research found that the east and south sides of the mountain

are the main windward sides, and residential buildings on the southeast side of the mountain are better ventilated and less affected by the epidemic. The buildings on the west side of the mountain have poor ventilation, and the airflow is blocked by the mountain, which means the buildings are greatly affected by the epidemic. Therefore, residential buildings should be avoided on the west side of the mountain. The single building built at the foot of the mountain has a "strip" shape, with its back against the mountain and its long side facing the southeast, maximizing the windward area. Most buildings built on the mountain are "rectangular", and the southeast side is less affected by the epidemic (Figure 17).

(3) Low-density residential buildings without mountains: the building density is relatively low, and there is no mountainous terrain around (the surrounding wind speed is 0.00~5.82 m/s). Its characteristics are that the residential building presents a single shape, mostly an "L" or "+". Among them, the short side of the high-density strip, or "L"-shaped building unit, is the windward side, and the long side faces the northeast or southwest side. The "+"-shaped building layout faces the southeast side, and the "+" is also rotated by 45° to increase the windward area (Figure 18).

(4) Low-density residential buildings with mountainous types: with mountainous terrain all around, the building density is relatively low (the surrounding wind speed is 0.00~4.85 m/s). Its characteristics are that most of the buildings are located on the northwest side of the mountain. The residential buildings at the foot of the mountain are more affected by the wind in the northwest, so the epidemic's impact is lessened. The single residential building on the northwest side at the foot of the mountain presents a "C" shape, or "rectangle". Among them, the building is a semi-enclosed "C"-shaped building, and the inwardly enclosed frontal inner part faces the southeast (windward side) or northeast side. The single building located on the northwest side of the mountain is mainly "rectangular", with its back against the mountain and its long side facing the southeast side (windward side). Compared with buildings located on the northwest side of the mountain, they are more susceptible to the impact of the epidemic. It is recommended that residential buildings be built on the southeast side of the mountain (Figure 19).

To summarize, in residential areas with high density, residential buildings that are less affected by the epidemic are mainly elongated, "L", "C," and "+" types. However, low-density residential areas generally have higher wind speeds and are less affected by the epidemic. In the case of a mountainous environment, the wind environment of the mountain is obvious. Buildings built on the windward side of the southeast side of the mountain are generally less affected by the epidemic. The mountain blocks residential buildings built on the northwest side of the mountain from most of the wind and air flow, and the buildings are more affected by the epidemic. Typical residential buildings can be summarized as follows (Figure 20):

(1) "C"-type, high-density residential buildings without mountains (wind speed 2.91~6.79 m/s);
(2) "L" type or "+" type, low-density residential buildings without mountains (wind speed 2.91~5.82 m/s);
(3) "Strip"-type, high-density residential buildings, mostly located on the northeast side of the mountain (wind speed 2.91~8.73 m/s).
(4) "Rectangular"-type high-density residential buildings with mountains, mostly located on the southeast side of the mountain (wind speed 2.91~8.73 m/s).
(5) "C"-type, low-density residential buildings with mountains, mostly located at the foot of the mountain in the northwest (wind speed 2.91~4.85 m/s).
(6) "Rectangular"-type, low-density residential buildings with mountains, mostly located at the foot of the mountain and on the northwest side (wind speed 2.91~4.85 m/s).

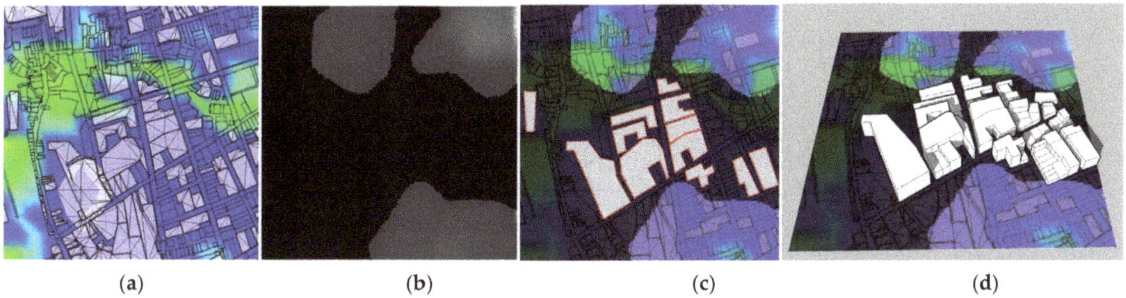

Figure 16. High-density residential buildings without mountains. (**a**) Wind environment map; (**b**) epidemic distribution map; (**c**) "C" layout of typical residential buildings; (**d**) building space model.

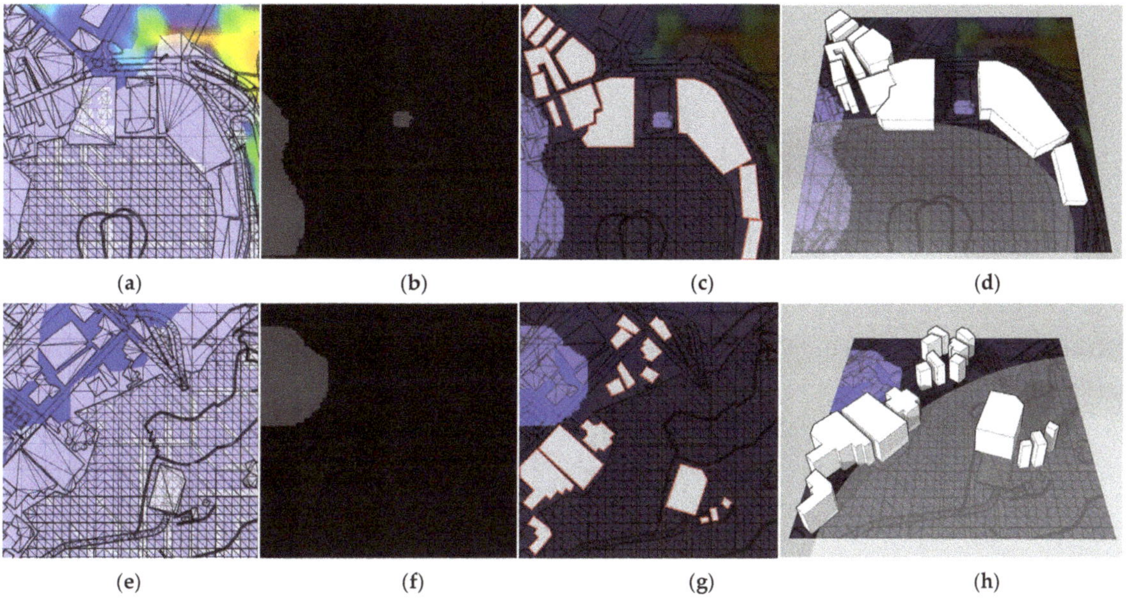

Figure 17. High-density residential buildings without mountains. (**a**,**e**) Wind environment map; (**b**,**f**) epidemic distribution map; (**c**) "long strip" layout of typical residential buildings; (**g**) "rectangular" layout of typical residential buildings; (**d**,**h**) building space model.

4.2. Wind Environment Simulation and Epidemic Situation Analysis and Verification

In order to further verify that the typical residential space summarized above has the characteristics of a low-risk downwind environment, the researchers conducted wind environment simulation and epidemic prediction verification for several typical residential building forms summarized above. First, they simulated the wind environment of the six typical residential types obtained above (the wind simulation parameters are the same as before). Second, the typical residential layout type map was imported into the machine learning model to predict the epidemic situation under this form. Third, the wind environment simulation results were superimposed and compared with the epidemic prediction results, and the location of residential buildings was compared with the COVID-19 outbreak prediction map. In this way, it was verified that the numerous typical residential building forms summarized in this paper could effectively reduce the impact of the COVID-19 outbreak due to the influence of the wind environment.

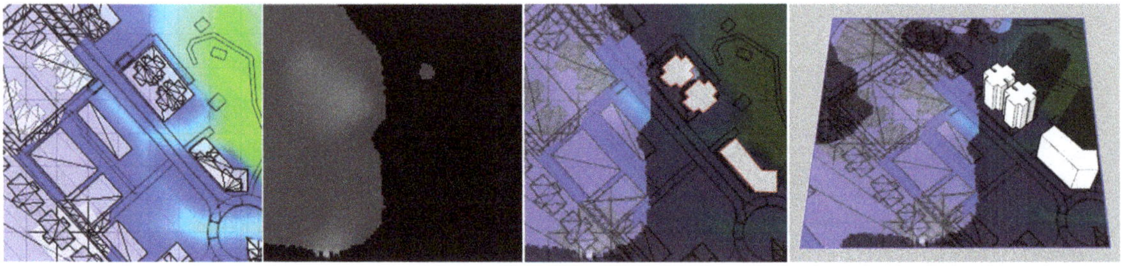

(a) (b) (c) (d)

Figure 18. Low-density residential buildings without mountains. (**a**) Wind environment map; (**b**) epidemic distribution map; (**c**) "L" or "+" layout of typical residential buildings; (**d**) building space model.

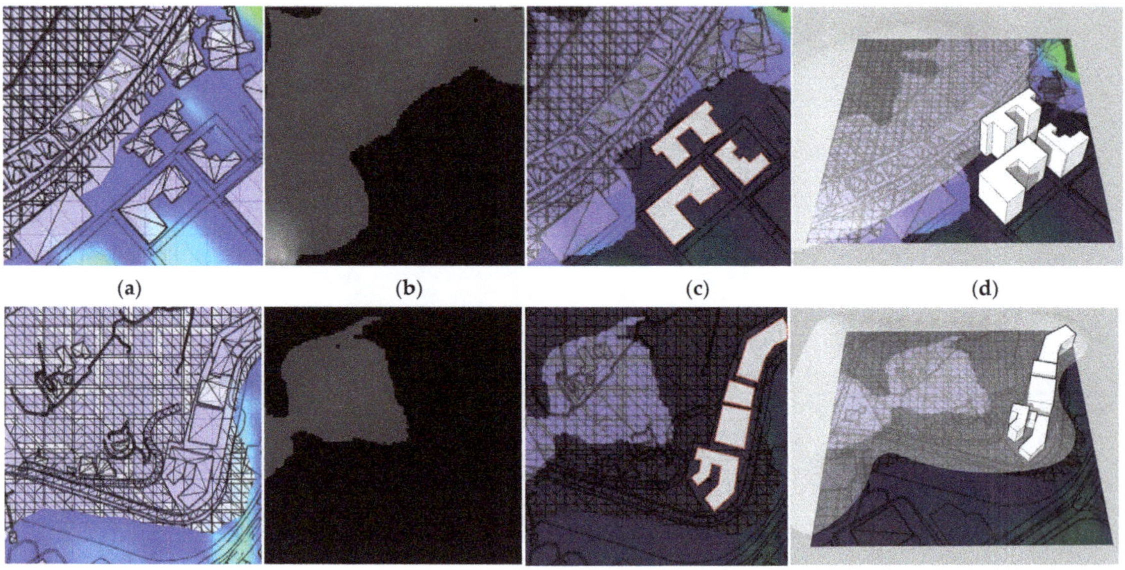

(a) (b) (c) (d)

(e) (f) (g) (h)

Figure 19. High-density residential buildings without mountains. (**a**,**e**) Wind environment map; (**b**,**f**) epidemic distribution map; (**c**) "C" shape layout of typical residential buildings; (**g**) "rectangular" layout of typical residential buildings; (**d**,**h**) building space model.

It can be seen in Figure 18 that the six typical residential building forms are all affected by the wind environment to a certain extent, but they are less affected by the outbreak of the epidemic (the test wind speed was 2.91~9.7 m/s). Among them, the best performance is in the situation in Figure 21 group number 3: "strip"-type, high-density residential buildings with mountains (average wind speed of 5.82 m/s). The second-best performance is the situation in Figure 21 group number 6: "rectangular"-type, low-density residential buildings with mountains (average wind speed of 4.85 m/s). The typical forms of Figure 21 group number 3,6 both include mountains, and the buildings are located on the east side of the mountain terrain. Observing Figure 21 group number 3, the simulation diagram of the wind environment of the "strip" high-density residential building with mountains, it can be seen that the length of the strip and rectangular buildings is located on the windward side of the direction, and the wind passes through the building on the side of the mountain. After the height changes, it continues to rise, and an upward airflow is generated between the building and the mountain to quickly dredge the airflow between the buildings. Figure 21 group number 4 is similar to this situation, but the building is located on the northeast side of the mountain, and the mountain blocks part of the wind, so the effect is not as good as the situation in Figure 21 group number 3. Among the remaining types, most of the buildings are included in the non-outbreak area under the influence of the wind environment, but some buildings are in the predicted outbreak area.

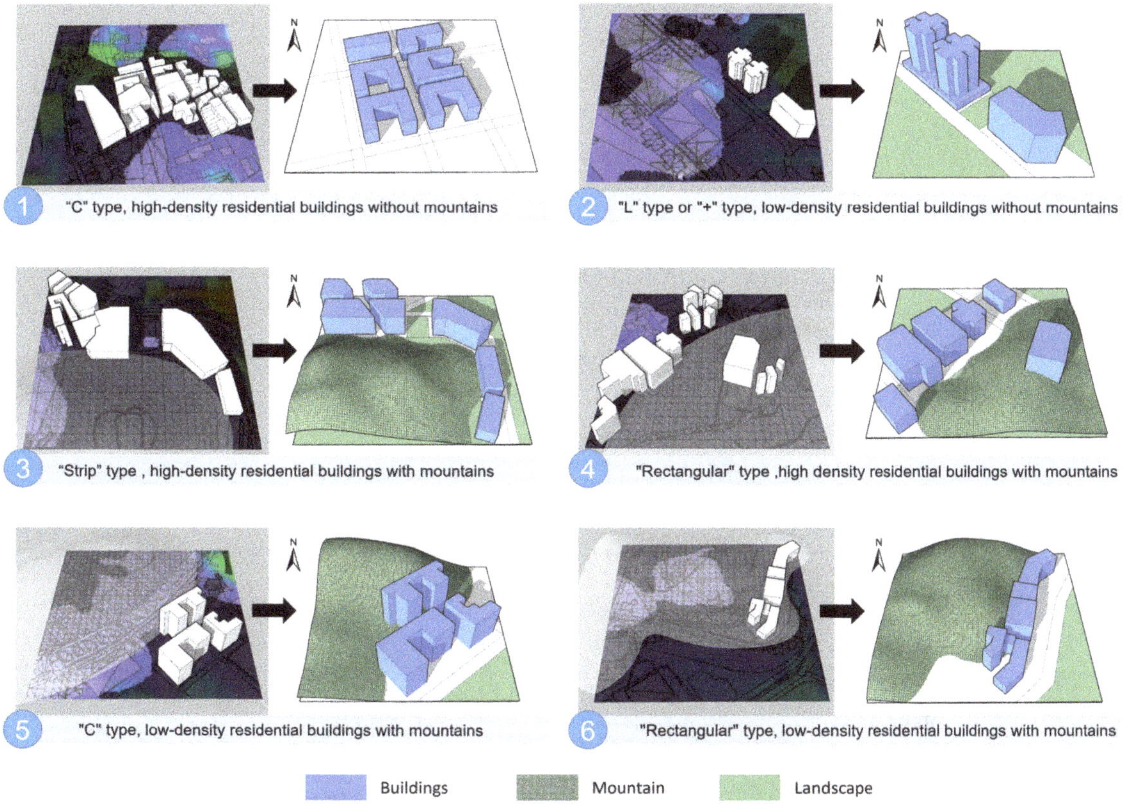

Figure 20. Analysis of typical residential building types in areas with low incidence of COVID-19. "1–6" represents the group number of the simulation experiment control.

In general, most of these six typical residential building forms are in areas without epidemics under the influence of wind environments, have the characteristics of low epidemic risk, and can be used as an important reference layout for residential buildings. Against the background of the development of long COVID-19, it can be used as a reference for future residential building planning.

4.3. Design Principles

Based on the above analysis of Macao's wind environment and the outbreak of the epidemic, especially in high-density cities, we can summarize three principles and suggestions for the future long COVID-19 residential building form planning:

(1) Select a site with mountains surrounding it for the residential buildings. Mountain scenery can effectively suppress the epidemic. The northeast side of the mountain is the best place to put a building, followed by the southeast and northwest sides.

Keep a certain distance between the building and the mountain to increase the air flow between the building and the mountain.

(2) Residential building form: adopt a "strip" or "rectangular shape", and use the long side of the building as the windward side as much as possible. Guide the wind into the building to the greatest extent and renew the air in the residential building. The six typical architectural forms summarized above can be used as planning references.

(3) Residential building orientation: Orient the long side of the building towards the southeast side (windward side). Most of the six typical residential buildings summarized above are inclined towards the southeast, and the purpose is to increase the airflow on the windy side as much as possible.

(4) Surrounding wind speed control: It is recommended that the overall wind speed around the building be above 2.91 m/s. The wind speed can be controlled by referring to the wind speed values corresponding to the six typical residential building types obtained in the experiments in this paper. The best wind speed is between 4.85 and 8.73 m/s.

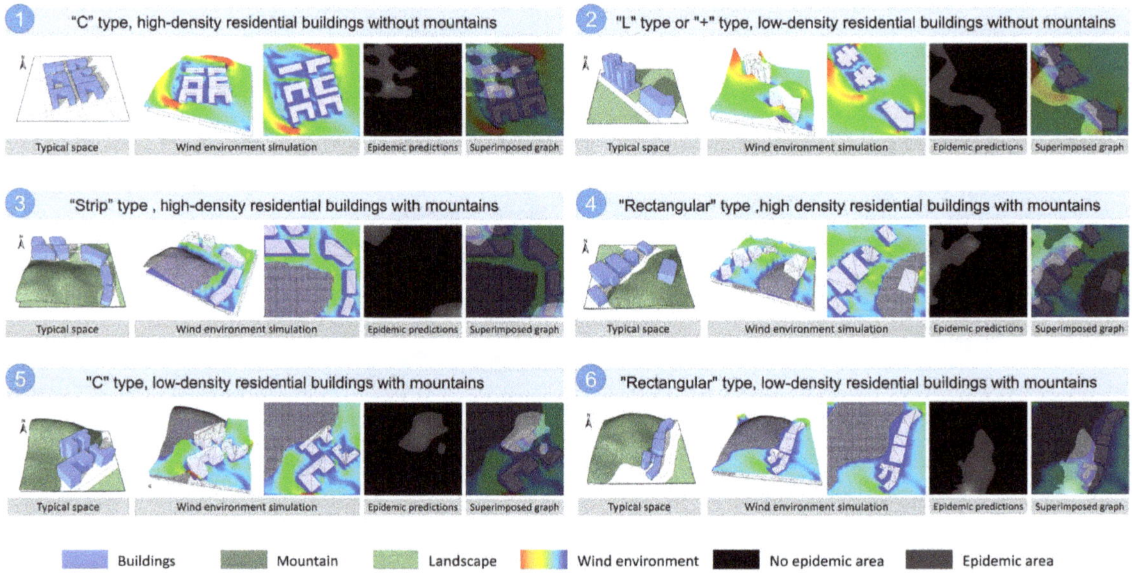

Figure 21. Wind environment simulation and COVID-19 prediction of typical layout of residential buildings. "1–6" represents the group number of the simulation experiment control.

It is undeniable that the design principles in this study belong to the stage of theoretical exploration. However, the application of these principles in practice may face some challenges. Especially in the context of the current tapering of the pandemic, architectural and urban planning decisions need to consider long-term sustainability and variability. The lifespan of a building may span multiple pandemic cycles, so design decisions need to take into account a variety of factors, including special needs during the pandemic and future sustainability goals. In addition, the COVID-19 virus is not the only disease; other diseases, such as asthma, are associated with factors such as humidity and air quality. Therefore, the impact and needs of other diseases should also be considered in implementing the design principles. At the same time, it is important to actively cooperate with professionals in related fields, such as urban planners, architects, and public health experts. Further, design principles and guidelines should be validated and improved through field observations, simulation experiments, and case studies to ensure their feasibility and effectiveness in future practice.

5. Conclusions

This study aims to explore the relationship between the urban wind environment and the spread of COVID-19 to reveal the importance of the wind environment in epidemic prevention and control. Additionally, we aimed to provide new perspectives for the urban planning and public health sectors to develop effective strategies. Additionally, the study results are universal to a certain extent. This study strives to ensure the scientificity and credibility of the research by adopting systematic research methods (Ecotect wind environment simulation, CGAN) and reliable data sources (climate statistics published by the Macau Meteorological and Geophysical Bureau). The researchers used Generative Adversarial Networks (CGAN) as a main method to generate urban wind environment data and COVID-19 hotspot distribution maps. As mentioned in the literature review, GAN is a machine learning-based tool widely used in several fields and has a solid scientific foundation. The method achieves accurate simulation and generation of real data by training a generator network and a discriminator network. Generative adversarial networks have shown outstanding performance in tasks such as image generation, data augmentation, and simulation and have been extensively researched and validated. Based on the existing urban wind environment data and COVID-19 epidemic data, the researchers generated reliable and accurate wind environment data and COVID-19 hotspot distribution maps through generative adversarial networks. In this research, the training and verification of the model were carried out, and a comparison with and analysis of the actual data were conducted. By comparing the data generated by the model and the real data, the study found that the difference between the two was not very significant (Figures 7 and 8). This showed that the method used was scientific and useful. The following conclusions can be drawn from this study:

(1) There are significant differences in the correlation between different types of wind environments and COVID-19. Areas with high building density are more prone to COVID-19 hotspots. Modern building complexes have better evacuation and public space layouts, which can reduce the concentration of hotspots. In addition, open spaces with high wind speeds can effectively suppress the spread of COVID-19.

(2) The distribution of COVID-19 hotspots in the building complex relates to the characteristics of the building itself. The size of the building area determines the spread of COVID-19. In contrast, the lower COVID-19 distribution in natural mountain landscapes may be aided by the outdoor environment, such as better ventilation conditions and the inhibitory effect of plants.

(3) The model (Model 1) that generates the distribution of COVID-19 hotspots from wind environment data performs well in generating the distribution of COVID-19 hotspots, especially when there is a clear separation between building groups and public spaces, and the wind speed is high. However, in non-residential buildings or areas that have been discontinued, the model's accuracy drops, requiring further optimization and improvement.

(4) The model (Model 2) that generates wind environment data through the distribution of COVID-19 hotspots is less accurate in generating building and wind patterns but can still show some accuracy in specific cases. This emphasizes the importance of considering the type of input material on the quality of model generation, and the robustness of Model 2 needs to be further improved for complex generation tasks.

(5) For high-density cities such as Macau, in response to long COVID-19 or residential area planning in the post-epidemic era, three principles can be considered: building houses on the northeast side of the mountain; ensuring residential building layouts are of "strip" or "rectangular" design; and orienting the long side of the building towards the southeast (windward side). At the same time, it is recommended that the overall wind speed around the building be above 2.91 m/s, and the best wind speed is between 4.85 and 8.73 m/s.

(6) There is a close relationship between the urban wind environment and the spread of COVID-19, and the specific form of the urban wind environment may have a

significant impact on the speed and scope of virus transmission. Therefore, urban planning and public health departments must consider urban wind environments in epidemic prevention and control strategies.

In addition, there are some limitations to this study:

(1) This study only takes Macau as an example, and the research results may be affected by geographical specificity. Therefore, similar studies need to be carried out in more areas (such as low-density cities) to verify the generalizability of the results.
(2) This study only focuses on the impact of urban wind environments on the spread of COVID-19 without considering the combined effects of other environmental factors. Future research can continue to explore the comprehensive impact of different environmental factors (such as light and thermal environments) on the spread of the epidemic.
(3) The design principles and guidelines proposed in this study are based on the exploration of the relationship between the urban wind environment and the spread of COVID-19. This provides new perspectives for urban planning and public health authorities to develop strategies to help reduce the risk of spreading the disease. However, design principles belong to the stage of theoretical exploration. This may limit its direct application in practice. The impact and needs of other diseases should be considered in implementing subsequent design principles to respond to the ever-changing urban environment. At the same time, builders should actively cooperate with professionals in related fields, such as urban planners, architects, and public health experts. Architects can verify and improve design principles and guidelines through methods such as field observations, simulation experiments, and case studies to ensure their feasibility and effectiveness in practice.
(4) This study focuses on exploring the relationship between the urban wind environment and the spread of COVID-19, and the conclusions mainly apply to COVID-19. From now on, designing cities around COVID-19 may not be enough, because outbreaks and virus characteristics may change over time. However, such a method can still be used as a reference in the future. At the same time, this study also has certain reference values for improving the control of other diseases. Although the transmission mechanism and characteristics of each disease are different, there may be some commonalities in the influencing factors of the urban wind environment on disease transmission. For example, the spread of diseases relating to humidity or air quality, such as asthma, may also be affected by the urban wind environment [53].

Nonetheless, this study provides an analytical approach focusing on the urban wind environment and COVID-19. This approach could provide useful guidance for urban planners and public health authorities in designing cities with the risk of transmission of different diseases in mind. However, since transmission mechanisms and risk factors may differ for each disease, further research and context-specific analyses are needed when applying this study to improve the control of other diseases. This involves an in-depth study and evaluation of the transmission pathways, environmental factors, and associated data for a particular disease. Therefore, although the conclusions of this study are mainly applicable to COVID-19, the research methods and analytical framework can provide useful references for improving the control of other diseases and inspire future related research in the fields of urban planning and public health.

For buildings and complexes that generate large numbers of users, such as so-called "urban interchanges," the challenge may be even greater. These areas often have a higher turnover and density of people, which may increase the risk of COVID-19 transmission. Therefore, studying the relationship between the wind environment at urban intersections and the spread of COVID-19 is a direction worthy of further research. For these special areas, more factors need to be considered, such as the concentration of people, building layout and usage patterns, and ventilation systems, in order to more comprehensively assess their relevance to the spread of the epidemic. Although this study only focuses on residential areas, public facilities, buildings, and urban intersections have an important impact on the spread of the epidemic. Therefore, future research can further expand the scope, deeply explore the relationship between the characteristics of the wind environment and the spread of the epidemic in these special areas, and formulate corresponding prevention and control strategies.

In order to further improve future research, researchers can deeply explore the comprehensive influence of different environmental factors on the spread of the epidemic and optimize the model's performance to improve the accuracy of complex inputs and non-residential building areas. At the same time, studies could focus on conducting broader research to expand the understanding of the relationship between urban wind environments and COVID-19 transmission and provide more specific guidance for urban planning and public health departments.

Author Contributions: Conceptualization, L.Z. and Y.C.; methodology, L.Z. and L.Y.; software, L.Z. and L.Y.; validation, L.Z. and L.Y.; formal analysis, L.Z. and Y.C.; investigation, L.Z. and Y.C.; resources, L.Z. and Y.C.; data curation, L.Z. and Y.C.; writing—original draft preparation, L.Z., Y.C. and L.Y.; writing—review and editing, L.Z. and Y.C.; visualization, L.Z. and Y.C.; supervision, J.Z.; project administration, J.Z.; funding acquisition, J.Z. All authors have read and agreed to the published version of the manuscript.

Funding: This research was funded by the National Social Science Foundation's special academic team project for unpopular research (21VJXT011).

Institutional Review Board Statement: Not applicable for studies not involving humans or animals.

Informed Consent Statement: Not applicable for studies not involving humans.

Data Availability Statement: The datasets used and analyzed during the current study are available from the corresponding author upon reasonable request.

Conflicts of Interest: The authors declare no conflict of interest.

Appendix A

Ecotect's operating environment: the operating system was Windows 10 (X64), the Autodesk Ecotect Analysis version was 2011, the graphics card was NVIDIA Quadro RTX 5000 (16G), and the processor was Intel(R) Xeon(R) Gold 6230R (2.1 GHz).

Appendix B

Machine learning environment configuration: the operating system was Windows 11 (X64), the Cuda version was 11.5, the deep-learning framework was Pytorch, the graphics card was GeForce GTX 3070 (16G), and the processor was AMD Ryzen 9 5900HX (3.30 GHz).

References

1. Post COVID-19 Condition. Available online: https://www.who.int/emergencies/diseases/novel-coronavirus-2019/media-resources/science-in-5/episode-47{-}{-}-post-covid-19-condition?gclid=CjwKCAjwg-GjBhBnEiwAMUvNW6KAnNghamKfuAZgrMh56QRZbpESvzBoupjo7NNoe0USPARjdcFkSBoCerQQAvD_BwE (accessed on 1 June 2023).
2. Post COVID-19 Condition (Long COVID). Available online: https://www.who.int/europe/news-room/fact-sheets/item/post-covid-19-condition (accessed on 1 June 2023).
3. Davis, H.E.; McCorkell, L.; Vogel, J.M.; Topol, E.J. Long COVID: Major findings, mechanisms and recommendations. *Nat. Rev. Microbiol.* **2023**, *21*, 133–146. [CrossRef] [PubMed]
4. Raveendran, A.V.; Jayadevan, R.; Sashidharan, S. Long COVID: An overview. *Diabetes Metab. Syndr. Clin. Res. Rev.* **2021**, *15*, 869–875. [CrossRef] [PubMed]
5. Basic Information of Macau. Available online: https://www.ipim.gov.mo/zh-hant/Macau-exhibition-and-conference/information-on-exhibition-and-conference/introduction/ (accessed on 1 June 2023).
6. Macau Ended the Transitional Period of the Epidemic on SUNDAY, and the Government Estimates that 60% to 70% of the Infected Population. Available online: https://www.tdm.com.mo/zh-hant/news-detail/788659?isvideo=false&lang=zh-hant&category=all (accessed on 1 June 2023).
7. Coccia, M. How (Un)sustainable Environments Are Related to the Diffusion of COVID-19: The Relation between Coronavirus Disease 2019, Air Pollution, Wind Resource and Energy. *Sustainability* **2020**, *12*, 9709. [CrossRef]

8. Amnuaylojaroen, T.; Parasin, N. The association between COVID-19, air pollution, and climate change. *Front. Public Health* **2021**, *9*, 662499. [CrossRef] [PubMed]
9. Purwanto, P.; Astuti, I.S.; Rohman, F.; Utomo, K.S.B.; Aldianto, Y.E. Assessment of the dynamics of urban surface temperatures and air pollution related to COVID-19 in a densely populated City environment in East Java. *Ecol. Inform.* **2022**, *71*, 101809. [CrossRef]
10. Fadaei, A. Ventilation systems and COVID-19 spread: Evidence from a systematic review study. *Eur. J. Sustain. Dev. Res.* **2021**, *5*, em0158. [CrossRef]
11. Leng, P.C.; Aw, S.B.; Ali, N.E.H.; Ling, G.H.T.; Lee, Y.L.; Ahmad, M.H. Solar Chimneys as an Effective Ventilation Strategy in Multi-Storey Public Housing in the Post-COVID-19 Era. *Buildings* **2022**, *12*, 820. [CrossRef]
12. Leng, J.; Wang, Q.; Liu, K. Sustainable design of courtyard environment: From the perspectives of airborne diseases control and human health. *Sustain. Cities Soc.* **2020**, *62*, 102405. [CrossRef]
13. Abbas, G.M.; Gursel Dino, I. The impact of natural ventilation on airborne biocontaminants: A study on COVID-19 dispersion in an open office. *Eng. Constr. Archit. Manag.* **2022**, *29*, 1609–1641. [CrossRef]
14. Lu, H.; Xia, M.; Qin, Z.; Lu, S.; Guan, R.; Yang, Y.; Miao, C.; Chen, T. The Built Environment Assessment of Residential Areas in Wuhan during the Coronavirus Disease (COVID-19) Outbreak. *Int. J. Environ. Res. Public Health* **2022**, *19*, 7814. [CrossRef]
15. Waheeb, M.I.; Hemeida, F.A. Study of natural ventilation and daylight in a multi-storey residential building to address the problems of COVID-19. *Energy Rep.* **2022**, *8*, 863–880. [CrossRef]
16. Moulaei, K.; Shanbehzadeh, M.; Mohammadi-Taghiabad, Z.; Kazemi-Arpanahi, H. Comparing machine learning algorithms for predicting COVID-19 mortality. *BMC Med. Inform. Decis. Mak.* **2022**, *22*, 2. [CrossRef]
17. Dairi, A.; Harrou, F.; Zeroual, A.; Hittawe, M.M.; Sun, Y. Comparative study of machine learning methods for COVID-19 transmission forecasting. *J. Biomed. Inform.* **2021**, *118*, 103791. [CrossRef] [PubMed]
18. Prakash, K.B.; Imambi, S.S.; Ismail, M.; Kumar, T.P.; Pawan, Y.V.R.N. Analysis, prediction and evaluation of COVID-19 datasets using machine learning algorithms. *Int. J.* **2020**, *8*, 2199–2204.
19. Wathore, R.; Rawlekar, S.; Anjum, S.; Gupta, A.; Bherwani, H.; Labhasetwar, N.; Kumar, R. Improving performance of deep learning predictive models for COVID-19 by incorporating environmental parameters. *Gondwana Res.* **2023**, *114*, 69–77. [CrossRef] [PubMed]
20. Pinter, G.; Felde, I.; Mosavi, A.; Ghamisi, P. Richard Gloaguen, COVID-19 pandemic prediction for Hungary; a hybrid machine learning approach. *Mathematics* **2020**, *8*, 890. [CrossRef]
21. Kushwaha, S.; Bahl, S.; Bagha, A.K.; Parmar, K.S.; Javaid, M.; Haleem, A.; Singh, R.P. Significant applications of machine learning for COVID-19 pandemic. *J. Ind. Integr. Manag.* **2020**, *5*, 453–479. [CrossRef]
22. Lalmuanawma, S.; Hussain, J.; Chhakchhuak, L. Applications of machine learning and artificial intelligence for COVID-19 (SARS-CoV-2) pandemic: A review. *Chaos Solitons Fractals* **2020**, *139*, 110059. [CrossRef]
23. Wynants, L.; Van Calster, B.; Collins, G.S.; Riley, R.D.; Heinze, G.; Schuit, E.; Bonten, M.M.J.; Dahly, D.L.; Damen, J.A.A.; Debray, T.P.A.; et al. Prediction models for diagnosis and prognosis of COVID-19: Systematic review and critical appraisal. *BMJ* **2020**, *369*, m1328. [CrossRef]
24. Sear, R.F.; Velasquez, N.; Leahy, R.; Restrepo, N.J.; El Oud, S.; Gabriel, N.; Lupu, Y.; Johnson, N.F. Quantifying COVID-19 content in the online health opinion war using machine learning. *IEEE Access* **2020**, *8*, 91886–91893. [CrossRef]
25. Apostolopoulos, I.D.; Mpesiana, T.A. COVID-19: Automatic detection from x-ray images utilizing transfer learning with convolutional neural networks. *Phys. Eng. Sci. Med.* **2020**, *43*, 635–640. [CrossRef]
26. Apostolopoulos, I.D.; Aznaouridis, S.I.; Tzani, M.A. Extracting possibly representative COVID-19 biomarkers from X-ray images with deep learning approach and image data related to pulmonary diseases. *J. Med. Biol. Eng.* **2020**, *40*, 462–469. [CrossRef] [PubMed]
27. Ardakani, A.A.; Kanafi, A.R.; Acharya, U.R.; Khadem, N.; Mohammadi, A. Application of deep learning technique to manage COVID-19 in routine clinical practice using CT images: Results of 10 convolutional neural networks. *Comput. Biol. Med.* **2020**, *121*, 103795. [CrossRef]
28. El Asnaoui, K.; Chawki, Y. Using X-ray images and deep learning for automated detection of coronavirus disease. *J. Biomol. Struct. Dyn.* **2021**, *39*, 3615–3626. [CrossRef]
29. Paul, H.Y.; Kim, T.K.; Lin, C.T. Generalizability of deep learning tuberculosis classifier to COVID-19 chest radiographs: New tricks for an old algorithm? *J. Thorac. Imaging* **2020**, *35*, W102–W104.
30. Zargari Khuzani, A.; Heidari, M.; Shariati, S.A. COVID-Classifier: An automated machine learning model to assist in the diagnosis of COVID-19 infection in chest X-ray images. *Sci. Rep.* **2021**, *11*, 9887. [CrossRef]
31. Oh, Y.; Park, S.; Ye, J.C. Deep learning COVID-19 features on CXR using limited training data sets. *IEEE Trans. Med. Imaging* **2020**, *39*, 2688–2700. [CrossRef]
32. Heo, L.; Feig, M. Modeling of severe acute respiratory syndrome coronavirus 2 (SARS-CoV-2) proteins by machine learning and physics-based refinement. *BioRxiv* **2020**. [CrossRef]
33. Ke, Y.-Y.; Peng, T.-T.; Yeh, T.-K.; Huang, W.-Z.; Chang, S.-E.; Wu, S.-H.; Hung, H.-C.; Hsu, T.-A.; Lee, S.-J.; Song, J.-S.; et al. Artificial intelligence approach fighting COVID-19 with repurposing drugs. *Biomed. J.* **2020**, *43*, 355–362. [CrossRef] [PubMed]
34. Ong, E.; Wong, M.U.; Huffman, A.; He, Y. COVID-19 coronavirus vaccine design using reverse vaccinology and machine learning. *Front. Immunol.* **2020**, *11*, 1581. [CrossRef] [PubMed]

35. World Health Organization. Directrices de la OMS Sobre Vivienda y Salud: Resumen de Orientación. Organización Mundial de la Salud, 2018. Available online: https://apps.who.int/iris/handle/10665/279743 (accessed on 27 June 2023).
36. Millán-Jiménez, A.; Herrera-Limones, R.; López-Escamilla, Á.; López-Rubio, E.; Torres-García, M. Confinement, Comfort and Health: Analysis of the Real Influence of Lockdown on University Students during the COVID-19 Pandemic. *Int. J. Environ. Res. Public Health* **2021**, *18*, 5572. [CrossRef] [PubMed]
37. Cuerdo-Vilches, T.; Navas-Martín, M.Á.; Oteiza, I. A Mixed Approach on Resilience of Spanish Dwellings and Households during COVID-19 Lockdown. *Sustainability* **2020**, *12*, 10198. [CrossRef]
38. MacKinnon, L.; Socías, M.E.; Bardwell, G. COVID-19 and overdose prevention: Challenges and opportunities for clinical practice in housing settings. *J. Subst. Abus. Treat.* **2020**, *119*, 108153. [CrossRef] [PubMed]
39. Ahmad, K.; Erqou, S.; Shah, N.; Nazir, U.; Morrison, A.R.; Choudhary, G.; Wu, W.-C. Association of poor housing conditions with COVID-19 incidence and mortality across US counties. *PLoS ONE* **2020**, *15*, e0241327. [CrossRef]
40. Ralli, M.; Cedola, C.; Urbano, S.; Morrone, A.; Ercoli, L. Homeless persons and migrants in precarious housing conditions and COVID-19 pandemic: Peculiarities and prevention strategies. *Eur. Rev. Med. Pharmacol. Sci.* **2020**, *24*, 9765–9767. [PubMed]
41. Tinson, A.; Clair, A. Better housing is crucial for our health and the COVID-19 recovery. *Health Found.* **2020**, *20*, 1–25.
42. Zimmermann, R.; Sarma, N.; Thieme-Thörel, D.; Alpers, K.; Artelt, T.; Azouagh, K.; Bremer, V.; Broistedt, P.; Eckmanns, T.; Feltgen, N.; et al. COVID-19 outbreaks in settings with precarious housing conditions in Germany: Challenges and lessons learned. *Front. Public Health* **2021**, *9*, 708694. [CrossRef]
43. Hu, M.; Roberts, J.D.; Azevedo, G.P.; Milner, D. The role of built and social environmental factors in COVID-19 transmission: A look at America's capital city. *Sustain. Cities Soc.* **2021**, *65*, 102580. [CrossRef]
44. Ieng, S.M.; Cheong, I.H. An overview of epidemiology of COVID-19 in Macau SAR. *Front. Public Health* **2020**, *8*, 550057. [CrossRef]
45. Raposo, V.L. Macau, a Luta contra a COVID-19 no Olho do Furação. *Cad. Ibero-Am. De Direito Sanitário* **2020**, *9*, 12–28. [CrossRef]
46. Zuev, D.; Hannam, K. Anxious immobilities: An ethnography of coping with contagion (COVID-19) in Macau. *Mobilities* **2021**, *16*, 35–50. [CrossRef]
47. Bai, W.; Zhao, Y.-J.; Cai, H.; Sha, S.; Zhang, Q.; Lei, S.-M.; Lok, G.K.I.; Chow, I.H.I.; Cheung, T.; Su, Z.; et al. Network analysis of depression, anxiety, insomnia and quality of life among Macau residents during the COVID-19 pandemic. *J. Affect. Disord.* **2022**, *311*, 181–188. [CrossRef] [PubMed]
48. Zhao, Y.-J.; Bai, W.; Cai, H.; Sha, S.; Zhang, Q.; Lei, S.M.; Lok, K.-I.; Chow, I.H.I.; Cheung, T.; Su, Z.; et al. The backbone symptoms of depression: A network analysis after the initial wave of the COVID-19 pandemic in Macau. *PeerJ* **2022**, *10*, e13840. [CrossRef]
49. Cheshmehzangi, A.; Zhu, Y.; Li, B. Application of environmental performance analysis for urban design with Computational Fluid Dynamics (CFD) and EcoTect tools: The case of Cao Fei Dian eco-city, China. *Int. J. Sustain. Built Environ.* **2017**, *6*, 102–112. [CrossRef]
50. Yang, L.; He, B.J.; Ye, M. Application research of ECOTECT in residential estate planning. *Energy Build.* **2014**, *72*, 195–202. [CrossRef]
51. Thiodore, J.; Srinaga, F. Open Space Scenario on Riverside Settlement to Access Comfortable Wind Environment. In *IOP Conference Series: Earth and Environmental Science*; NIOP Publishing: Bristol, UK, 2021; Volume 764, p. 012022.
52. Chen, Y.; Zheng, L.; Song, J.; Huang, L.; Zheng, J. Revealing the Impact of Urban Form on COVID-19 Based on Machine Learning: Taking Macau as an Example. *Sustainability* **2022**, *14*, 14341. [CrossRef]
53. Etzel, R.A. How environmental exposures influence the development and exacerbation of asthma. *Pediatrics* **2003**, *112* (Suppl. S1), 233–239. [CrossRef]

buildings

Article

Supply–Demand Matching of Smart Parcel Lockers in a Residential Area: Insights from Tianjin

Mengyue Ding [1,2], Yan Song [2] and Yike Hu [1,*]

[1] School of Architecture, Tianjin University, Tianjin 300072, China; dmengyue@unc.edu
[2] Department of City and Regional Planning, University of North Carolina at Chapel Hill, Chapel Hill, NC 27599, USA; ys@email.unc.edu
* Correspondence: huyike11@tju.edu.cn

Abstract: The COVID-19 pandemic has led to a surge in the use of self-service facilities (SSFs) in residential areas worldwide. Previous studies on SSFs mainly focused on their application in commercial or other scenarios. However, SSFs in residential areas have not been thoroughly studied. This study develops an analytical framework for assessing both the supply and demand for SSFs in residential areas. The study evaluates 2693 residential communities and 479 smart parcel lockers (SPLs) in Tianjin, China. The results show that the high-demand area for SPLs is within 300 m of home, while the high-supply area is 300–600 m from home. Further analysis using the Gini coefficient and location quotient shows that the top 20% of the population have access to 80% of SPLs, and most residential communities experience an oversupply. Our study suggests that a mismatch between the supply and demand of SPLs may result in massive public space waste, resource waste, and inequity. Given the many uncertainties of the future, this study highlights the need to consider the dynamic supply–demand relationship of SSFs. This may encourage urban planners, policymakers, and experts in other related disciplines to work towards a more service-efficient and equitable utilization of SSFs in residential areas.

Keywords: smart parcel lockers; self-service technology; self-service facilities; supply–demand matching; residential communities; spatial assessment; Tianjin

Citation: Ding, M.; Song, Y.; Hu, Y. Supply–Demand Matching of Smart Parcel Lockers in a Residential Area: Insights from Tianjin. *Buildings* **2023**, *13*, 2088. https://doi.org/10.3390/buildings13082088

Academic Editors: Nuno D. Cortiços, Anna Stefańska, Daniel Mateus, Carol Monticelli and Carlos C. Duarte

Received: 29 June 2023
Revised: 8 August 2023
Accepted: 15 August 2023
Published: 17 August 2023

1. Introduction

The COVID-19 pandemic has had a significant impact on people's lifestyles in two ways [1,2]. First, the lockdown policy has indirectly limited people's lives to their living space, which has encouraged a shift from offline to online shopping. Second, the no-contact policy has made self-service facilities (SSFs) an effective COVID-19 prevention and control measure, as well as an unavoidable choice for people's physical and mental health. As a result of these changes, the rapid expansion of SSFs in residential areas has become a concern.

SSFs are pioneering the shift of traditional manual services from "high-touch" to "high-tech" [3,4]. With the advantage of reducing or eliminating human interaction, SSFs have been widely used in the service industry since the pandemic's outbreak [5]. Smart parcel lockers (SPLs) are one of the most widely used self-service solutions in recent years. They are created to enable contactless deliveries in response to e-commerce orders [4,6,7]. SPLs are typically designed as collection-and-delivery machines that can be easily placed in a wide variety of living and commercial environments, such as transportation hubs, office buildings, shopping malls, and residential communities [8]. Unlike traditional home delivery services and collection-and-delivery service points, SPLs provide consumers with the ease of 24 h access. They also significantly minimize delivery failures and are considered beneficial from both sociological and operational perspectives [9–11].

As both supply and demand are increasing, SPLs are applied worldwide. Prominent examples include ByBox in the UK, PackStation in Germany, InPost in Poland, SingPost and Ninjia Van in Singapore, and Hive Box in China [8,10,12–14]. While most studies have focused on techniques to refine the model or increase the accuracy of the algorithm to optimize logistical performance and better understand consumer demands, some researchers have recognized the significance of SPL location selection. They have summarized a variety of SPL location issues, such as accessibility, availability, security, environmental impact/land occupation, costs, methods of use, and regulations [15–18]. In fact, the location of SPLs is where the conflict between supply and demand occurs. Addressing location issues is critical to long-term supply and demand for SPLs. It is important to investigate a balanced path for the development of SPLs under the new type of supply–demand interaction between logistics businesses, customers, and governments. In this regard, SPLs could be more service-efficient and sustainable.

In this paper, we conduct a case study on the assessment of the supply–demand match of SPLs on a residential community level in the central metropolitan region of Tianjin, China. Most urban residential communities in China have distinct physical boundaries [19], which divide the inside from the outside. This means that residents are more likely to use SPLs in their own community, rather than in another community. As a result, the acceptable distance for Chinese residents to use SPLs may be shorter than in other countries. Tianjin is one of China's four municipalities, and it is also a notable e-commerce powerhouse. The express delivery sector in Tianjin has witnessed tremendous growth, and it consistently outpaced the national average growth [8]. In 2016, there were only five SPLs in Tianjin. From 2019 to 2022, the number of SPLs climbed from 51 to 479, reflecting an average annual growth rate of roughly 279.74% during the pandemic period [8]. The increase in supply has led to an increase in complaints from residents. This may be because the policy lacks limits on where to install and where not to place SPLs. As a result, some SPLs are located in inconvenient locations, such as in residential buildings, in a remote corner of a community, or in the middle of a busy street outside the boundary of the community. The government of Tianjin intends to raise the number of SPLs to 5000 [20]. However, installing SPLs blindly may result in a massive waste of resources. Therefore, research on the supply and demand matching of SPLs can provide valuable reference for policy formulation, urban planning and management, and city logistics layout.

We first obtained basic data of 2693 residential communities and 479 SPLs in the study area from open access websites. Using the analytical framework we developed, we examined the supply and demand of SPLs. We found that both supply and demand of SPLs are positively correlated with road density, population density, and per capita income. We then assessed the supply–demand matching state of SPLs. The results showed that the supply of SPLs is concentrated in the central urban area, and the top 20% of the population has access to 80% of the SPLs. This could lead to significant improper use of public spaces, resource waste, and inequity. Our extensive analysis indicates the critical importance of research on the supply–demand matching of new emergent facilities such as SPLs. To the best of our knowledge, this is the first study to address location difficulties by integrating both supply and demand for SPLs. Our findings indicate that location, particularly the area within 600 m of home, is critical to balancing supply and demand for SPLs. Future research may consider the dynamic planning of SPLs to adjust the location to satisfy both the profit of logistics businesses and the satisfaction of residents in order to improve resilience to future uncertainties.

2. Literature Review

To date, research on the location issues of SPLs appears to be scarce. The research field of SPLs is not highly established, with only 42 research documents in the Web of Science Core Collection Database (as of August 2023, when this manuscript was written). The first study related to SPLs was published in 2014 [21]. Lemke et al. (2016) [12] and Iwan et al. (2016) [14] were among the first scholars to introduce the concept of SPLs and present descriptive evidence of benefits from both the operators' and consumers' perspectives. To the best of our knowledge, Deutsch and Golany (2018) [22] developed the first quantitative approach to determine the optimal locker location. Just as an SPL is a newly implemented delivery method, the retrieval results indicate that the research topic of SPLs is also a newly emerging field. Based on the search results, we discuss related work along two aspects: (1) SPL preference under COVID-19 and (2) the location issues of SPLs.

2.1. SPL Preference under COVID-19

The rise of online shopping has led to a significant increase in demand for dedicated delivery services to the end consumer. However, traditional home delivery has several disadvantages, including a high number of missed deliveries, which can lead to more kilometers traveled, higher pollution emissions, and additional costs [21,23]. As a self-service tool, an SPL is considered an efficient way to solve the last-mile problem in delivery, enhance the delivery network system quality, and increase customer satisfaction [24]. As a result, SPLs are growing quickly, especially in metropolitan areas [21].

The COVID-19 pandemic has led to a surge in e-commerce sales and has also changed customer behavior [4,25]. During the pandemic, the e-commerce sector in Pakistan increased by 10% in daily records [26]; the volume of parcels increased by 37% in the UK from April to May 2020 [27]; and retail e-commerce sales in Canada experienced a 99.3% increase from February to May 2020 [28]. The surge in e-commerce has further strained the last-mile delivery process, which has long been recognized in the literature as the most costly, inefficient, and environmentally burdensome component of the supply chain [25]. In the shock of COVID-19, among the delivery alternatives, those characterized by a contactless or unattended nature stand out as the favored options by consumers under the pandemic context [29]. To this end, the pandemic represents a unique disruptive factor that drives a much faster diffusion of alternative delivery methods and shapes consumers' preferences towards contactless or self-service-based options [30]. Extensive research has been conducted to investigate customer preferences, revealing a strong inclination towards simplified and self-service options [31,32]. Customers who have grown accustomed to online services may feel uneasy when faced with in-person interactions and may avoid social situations [31,33]. Consequently, the adoption of SPLs becomes not only a matter of convenience but also a health-related choice [34].

2.2. Location Issues of SPLs

The worldwide application of SPLs has exhibited remarkable exponential growth in the short term after the breakout of the pandemic. For example, in Poland, the number of SPLs surged from 7000 to 13,000 during the pandemic [10]. While the supply and demand for SPLs are both expanding, complaints from consumers and negative profits of enterprises continue. This has focused the attention of the academy on SPL location issues [22].

SPLs can be located in places where consumers can access them easily. Even though the literature shows that the spatial distribution of SPLs is concentrated in central cities, SPL location preferences vary in different countries [16]. For example, in Sweden and France, commercial areas and public transport stops were preferred locations [21,35], while in Poland, these places were identified as the worst locations [12,14]. In South Korea and Australia, high-population-density areas with public transport stops were the most preferred locations for SPLs [17,36]. In Tianjin, China, residential communities are the most preferred location for SPLs. In 2019, 92.2% of SPLs were installed in residential communities, and this percentage increased to 97.7% in 2022 [8]. Many researchers believe that SPL networks can help solve the last-mile problem, and they emphasize the importance of determining the locations of SPLs [18,22,37].

Previous studies have used a variety of methods to optimize the location of SPLs, including the Spherical Fuzzy Analytic Hierarchy Process [38], bilevel programming approach [39], cluster analysis [40], multi-criteria methods [41], and integrated algorithms [42]. However, most of these studies have focused on optimizing the algorithm or model, rather than considering the demand for SPLs [43]. This has led to a fundamental mismatch between supply and demand. The SPL system is complex and involves a variety of stakeholders, including public authorities, urban planners, users, and private companies [44]. This makes it difficult to optimize SPL locations using a single approach. Therefore, this study will fill the major gap in the literature by developing a framework to address SPL location issues in consideration of both supply and demand aspects. The framework is based on a combination of the spatial analysis of SPL accessibility and the resident purchase power of residential communities. The results of the study could provide valuable insights for urban planners and policymakers who are responsible for the distribution of SPLs.

3. Materials and Methods

3.1. Study Area

The central Tianjin region was selected for this study (Figure 1). Tianjin has 16 districts, a population of more than 13.73 million people, and a land area of 11,916.85 km^2 [45]. The study area includes six districts: Heping District, Hedong District, Hexi District, Hebei District, Nankai District, and Hongqiao District. It is the city's central urban area for political, cultural, and economic activities, with a land area of 576.160 km^2 (13.29% of Tianjin) and a population of 4.06 million people (29.26% of Tianjin, according to the People's Republic of China's Seventh Census) [45]. And the urbanization rate of the study area is 100% [30]. The central Tianjin region accounts for the majority of online shopping and express delivery services of Tianjin city due to its high population and road network density. As a result, a considerable number of SPLs have been placed in the area [8]. This makes it an ideal location for studying the supply and demand of SPLs.

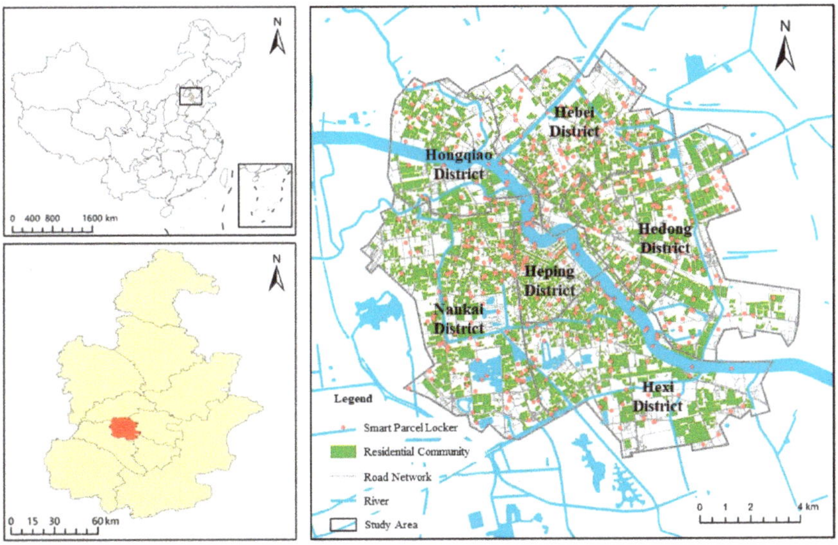

Figure 1. Study area.

3.2. Data

Three datasets were used in this study:

- Area of interest (AOI) data on 2693 residential communities in the study area. Compared with point of interest (POI) data, which have been commonly used in prior studies on residential communities and service facilities [46], AOI data provide information about the shape of a region and can more accurately present the boundaries of residential communities [47]. The AOI data were obtained from Lianjia.com (accessed on 10 May 2023), one of China's major real estate transaction service platforms.
- The geographical data for SPLs. The data were obtained via open access electronic map using Python (accessed on 19 December 2022) and includes the name, address,

longitude, and latitude of each SPL. By the end of 2022, there were 479 SPLs in the study area, with the vast majority (97.7%) installed within residential communities [8].

- The attribute information of the residential communities, including the area of the community, the number of households of the community, and the average house price of the community. The method of data collection for this dataset was the same as for the AOI data.

3.3. Pre-Experiment

We conducted a pre-experiment questionnaire survey in Tianjin in May 2022 to gain a better understanding of residents' opinions and usage habits regarding SPLs. The questionnaire survey was administered online, and the survey link is available at https://www.wenjuan.com/s/UZBZJvi7EWm/ (accessed on 31 May 2022). A total of 350 participants completed the survey, with a male-to-female participant ratio of 1:1.5. The preliminary results are summarized as follows (Table 1) [48]:

- SPLs are the most popular choice for residents among various types of SSFs.
- Residents expressed dissatisfaction with the current spatial layout of SPLs. They would like SPLs to be located within a 5 min walking distance of their homes.
- Residents rely heavily on SPLs for their parcel collection and delivery needs. This suggests that it may be challenging to replace SPLs with other types of facilities in the short term.

Table 1. Pre-experiment questionnaire survey questions and answers.

Questions	Answers
"The most frequently used sort of SSFs"	SPLs > ATMs > vending machines > shared bikes > self-checkout
"Attitude towards location of SPLs"	Satisfied (56%), Dissatisfied (44%)
"Frequency of using SPLs"	More than 3 times/week (49.5%), once a week (31.9%)
"Daily travel distance and transportation"	15 min living circle (34%), walking (34.8%)
"Expected distance from SPLs to home"	Within 5 min walk (61.0%)

The findings from the pre-experiment survey serve as a foundation for the next step, which involves developing the framework for supply and demand assessment.

3.4. Methods

To assess the supply and demand for SPLs in residential communities, an integrated framework was designed. The framework consists of three main modules:

- SPL supply assessment: This module evaluates the accessibility of SPLs on a residential community level.
- SPL demand estimation: This module estimates the demand for SPLs based on the purchase capacity of each residential community.
- Supply–demand matching analysis of SPLs: This module evaluates the supply and demand for SPLs and identifies where and how they become mismatched. The evaluation is carried out at both the macro-regional level and the micro-community level, with the former focusing on the overall supply and demand for SPLs and the latter focusing on the supply and demand for SPLs in individual residential communities.

The framework is illustrated in Figure 2.

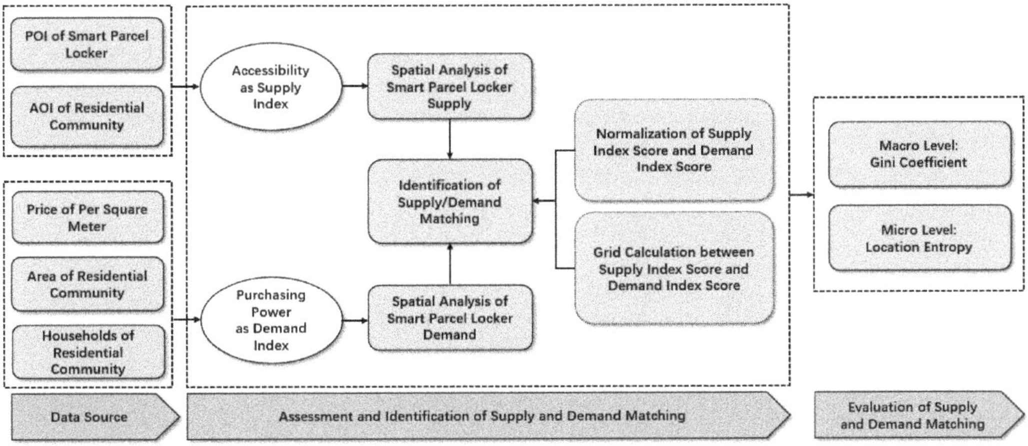

Figure 2. The framework for the assessment of supply and demand of SPLs.

3.4.1. Supply Assessment

The supply of smart parcel lockers (SPLs) includes the number, location, and size of SPLs [22]. However, the accessibility of SPLs is also a key consideration when optimizing their locations and layouts [49,50]. Therefore, this study used the walking accessibility of residential communities to SPLs as the indicator of supply. Walking accessibility was measured using walking time and distance [51]. Previous research has shown that people are less willing to walk as the distance increases, a phenomenon known as the distance attenuation law [52]. Based on these findings, this study categorized the straight-line distance from home to SPL into six grades: within 300 m, 300–600 m, 600–900 m, 900–1200 m, 1200–1500 m, and over 1500 m. The corresponding path distance for each walking time was calculated assuming a walking speed of 80 m/min [53]. Accessibility is presented as a supply index. In order to correspond to the demand index in the next step, an item of the demand index score is added (Table 2). The pre-experiment results indicated that residents generally do not prefer SPLs that are more than 1500 m away from their residences. Therefore, the category of over 1500 m was excluded from the assessment of SPL supply.

Table 2. Supply index score measuring standard.

Straight-Line Distance	Path Distance	Walking Time	Supply Index	Supply Index Score
<300 m	<400 m	<5 min	1.0	1
300–600 m	400–800 m	5–10 min	0.8	2
600–900 m	800–1200 m	10–15 min	0.6	3
900–1200 m	1200–1600 m	15–20 min	0.4	4
1200–1500 m	1600–2000 m	20–30 min	0.2	5
>1500 m	>2000 m	>30 min	0	-

The specific operational steps for assessing the supply of SPLs are as follows: First, take the demand point of each residential community in the study area as the center and the path distance as the search threshold range. Next, find the distance interval where the nearest SPL is located using ArcGIS. Finally, refer to the indicators provided in Table 2 to determine the supply index and its corresponding supply index score.

3.4.2. Demand Estimation

As previously stated, SPLs are market-driven commodities that are closely related to e-commerce. According to the e-commerce literature, household income and internet usage are important factors influencing online purchases [40]. During the pandemic, high rates of last-mile deliveries were connected with higher income, higher education levels, more internet access, and larger households [17,54]. In this case, this study considers the residential community's purchasing capacity to be an indicator of demand for SPLs. The demand index for residential community i is calculated as follows:

$$D_i = \frac{T_i \times P_i}{A_i} \quad (1)$$

where D_i represents the demand index of residential community i; T_i represents the total population of residential community i; P_i represents the average property price in the residential community i; and A_i represents the area of residential community i. We have assumed that all residents of the study area have access to the internet, so the item of internet access has been deleted from the formula. In addition, since the dataset we obtained from the real estate website provides the total number of households in the residential community instead of the entire population, the value of T_i needs to be determined indirectly. The average household size in Tianjin, based on the 2022 census data, is 2.70 people per household [55]. Therefore, the formula for calculating T_i is as follows:

$$T_i = 2.70 \times H_i \quad (2)$$

where H_i represents the number of households in the residential community i.

Furthermore, the supply index falls within the range of [0, 1], as mentioned in the preceding subsection. To ensure consistency between the supply index and demand index results, the demand index value is linearly and proportionally scaled to the [0, 1] interval. The formula for this scaling process is as follows:

$$D'_i = \frac{D_i - D_{min}}{D_{max} - D_{min}} \quad (3)$$

where D'_i is the demand index value after being scaled to the [0, 1] interval; D_{min} is the minimum value of the demand index; and D_{max} is the maximum value of the demand index. The community with the highest purchasing capability is assigned a value of 1, while the community with the lowest purchasing capability is assigned a value of 0. To be consistent with the supply index for the next step of matching assessment, the demand index score and supply index score are each set to five values from 1 to 5 (Table 3).

Table 3. Demand index and demand index score.

Demand Index	Demand Index Score
[0.8–1.0]	1
[0.6–0.8]	2
[0.4–0.6]	3
[0.2–0.4]	4
[0–0.2]	5

3.4.3. Spatial Matching Identification and Evaluation

To examine the spatial distribution of SPLs at the residential community level, we assessed the correlations between the supply index score and the demand index score. The supply index score and the demand index score both have five grades and are assigned from high to low. As a result, there are 25 possible matching outcomes, which are further classified into three groups (Table 4). This matching process was performed using ArcGIS.

Table 4. Matching outcomes of supply–demand.

Matching Type of Supply–Demand	Combination of Supply–Demand Index Scores
Supply > Demand (S > D)	12, 13, 14, 15, 23, 24, 25, 34, 35, 45
Supply = Demand (S = D)	11, 22, 33, 44, 55
Supply < Demand (S < D)	21, 31, 41, 51, 32, 42, 52, 43, 53, 54

The matching evaluation was conducted at both macro and micro levels. At the macro level, the entire study area was used as the assessment area, and the degree of supply and demand matching was assessed using the Gini coefficient and Lorenz Curve methods. The Gini coefficient is a widely used indicator in economics to measure the inequality in the distribution of social wealth within an economy [56]. It can also be used to analyze broader distributional equity aspects, including those in urban planning [57]. Given that the matching of supply and demand for SPLs is fundamentally rooted in the concept of the equitable allocation of social resources, this analysis approach is applicable. The formula for computing the Gini coefficient is as follows [58]:

$$G = 1 - \sum_{k=1}^{n} \left(D'_k - D'_{k-1}\right)\left(S_k + S_{k-1}\right) \tag{4}$$

Above, the cumulative value of the demand index score is denoted as D'_k, where $k = 1, 2, \ldots, n$, $D'_0 = 0$, $D'_n = 1$; the cumulative value of the supply index score is denoted as S_k, where $k = 1, 2, \ldots, n$, $S_0 = 0$, $S_n = 1$. In this case, the Gini coefficient is the metric that measures the balance between the supply and demand for SPLs. The closer the Gini coefficient is to 0, the higher the matching degree between the supply and demand of SPLs and the more balanced the supply and demand relationship; conversely, the closer the Gini coefficient is to 1, the more imbalanced the supply and demand of SPLs are.

At the micro level, this study uses the location quotient (LQ) to express the ratio of supply and demand in each residential community. The LQ is a ratio that can be easily used to estimate economic impact multipliers from the regional level to the local level [59]. We use the LQ to calculate how each residential community's supply–demand ratio compares to that of the overall study area. The LQ is calculated as follows [60]:

$$LQ_i = \frac{S_i}{S'} \cdot \frac{D'}{D'_i} \tag{5}$$

where LQ_i is the location quotient of the residential community i, S_i is the supply index score of the residential community i, D'_i is the demand index score of the residential community i, S' is the average supply index score of all residential communities in the study area, and D' refers to the average demand index score of all residential communities in the study area. If the LQ value of a residential community is greater than 1, it implies that the supply–demand ratio of SPLs in this residential community is higher than the overall level in the study area.

4. Results

4.1. Supply Assessment of SPLs

The spatial patterns of the supply for SPLs are illustrated in Figure 3. The figure shows a gradual increase in SPL supply index scores from the center towards the periphery. Residential communities in proximity to the city center tend to have lower supply index scores, indicating a higher availability of SPLs. Conversely, suburban residential communities are predominantly characterized by higher supply index scores, indicating a relatively lower supply of SPLs in those areas.

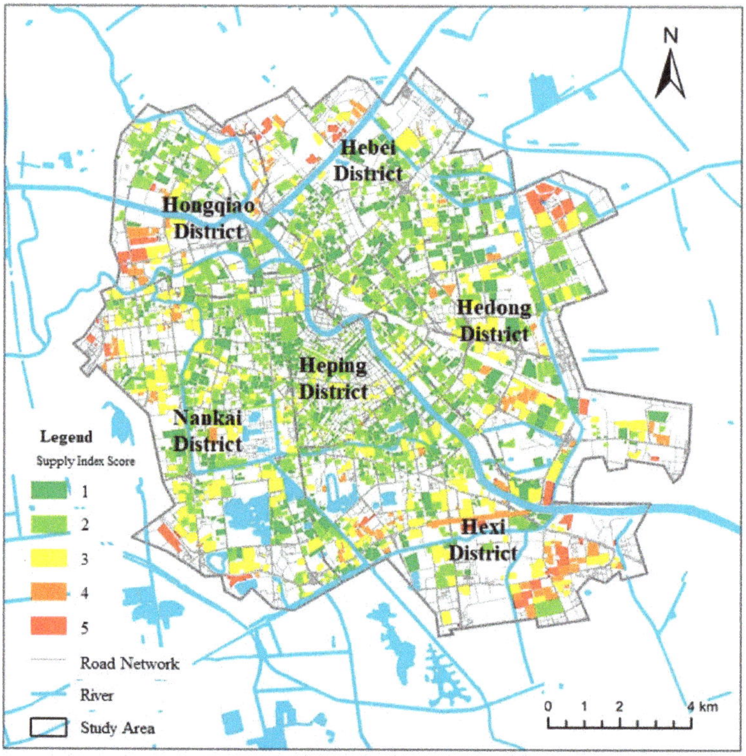

Figure 3. The spatial characteristics of supply index score.

High-supply communities (with supply index scores of 1 and 2) accounted for 72.30% of the total and were concentrated in the districts of Heping, Nankai, and Hexi. These districts each contributed 15.74%, 13.59%, and 13.18%, respectively. Medium-supply residential communities (with a supply index score of 3) were concentrated around urban centers and suburban areas, accounting for 20.01% of all residential communities. These communities were primarily located in east Hexi (24.95%), south Nankai (23.65%), and east Hedong (17.88%). Low-supply communities (with supply index scores of 4 and 5) were primarily located in further-flung areas, such as east Hexi (32.86%), north Hebei (19.05%), southwest Nankai (19.05%), and northeast Hedong (18.10%). These communities accounted for 7.69% of the total.

4.2. Demand Assessment of SPLs

Multiple high-demand areas have been identified in the study area. As shown in Figure 4, the demand index score for SPLs in residential communities generally increases from the center towards the periphery, indicating a concentration of high-demand communities in the central region. However, the high-demand areas are also expanding towards the western and southern parts of the study area.

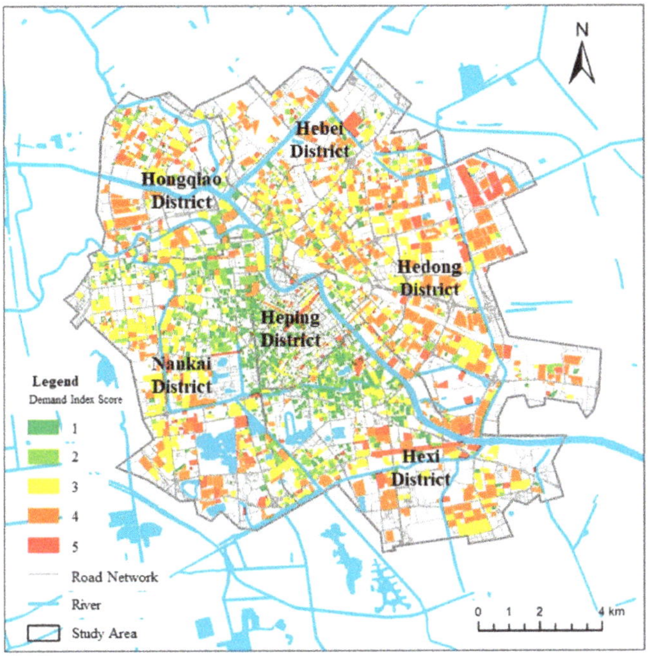

Figure 4. The spatial characteristics of demand index score.

The demand for SPLs in the study area is unevenly distributed. High-demand communities (with demand index scores of 1 or 2) account for 49.57% of the total and are primarily located in the Heping, west Hexi, and east Nankai districts. These communities are concentrated in densely populated areas with a well-connected road network. Low-demand communities (with demand index scores of 4 or 5) account for 25.99% of the total and are widely dispersed in the outskirts characterized by relatively lower population density and less developed road networks. These communities are found in 44.25% of Hedong's residential communities, 40.09% of Hongqiao's residential communities, and 35.97% of Hebei's residential communities. Medium-demand communities (with a demand index score of 3) account for 24.43% of the total and are primarily located on the outskirts of high-demand areas.

4.3. Identification of Supply and Demand Matching

The spatial distribution of supply–demand matching for SPLs is illustrated in Figure 5. Balanced communities (S = D) (25.21%) are primarily located on the densely populated left bank of the Haihe River, which also hosts the city center and a well-connected road network. Oversupplied communities (S > D) (44.78%) are mainly concentrated on the right bank of the Haihe River and in the southwest suburbs. Short-supplied communities (S < D) (30.01%) are concentrated in the west Hexi and east Nankai areas, as well as some residential communities in the suburbs of each district.

The evaluation of supply–demand matching based on straight-line distance reveals an interesting pattern (Figure 6). Only the supply within the 300–600 m range significantly surpasses the demand, while the sections below 300 m, 600–900 m, 900–1200 m, and 1200–1500 m all experience a shortage. This finding helps explain why residents participating in the pre-experiment perceive SPLs to be too far from their homes and express a desire for closer proximity. It indicates that residents generally find SPLs acceptable if they are within a distance of 300 m.

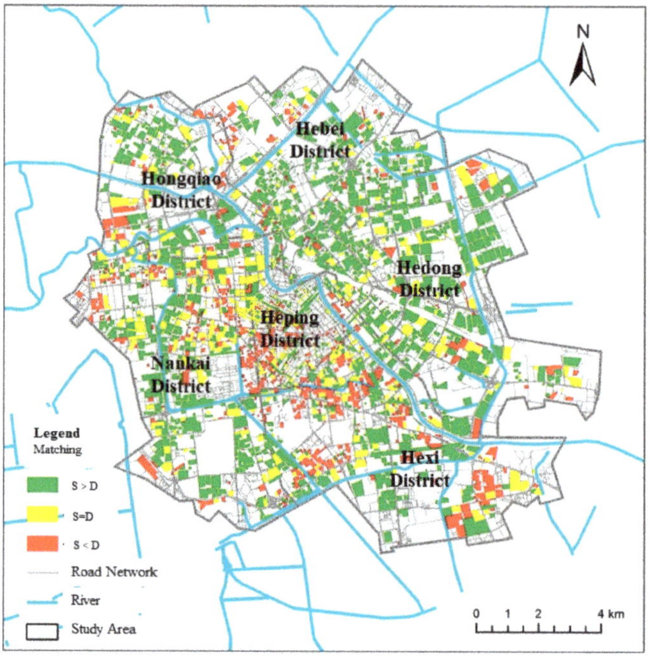

Figure 5. The spatial characteristic of supply–demand matching.

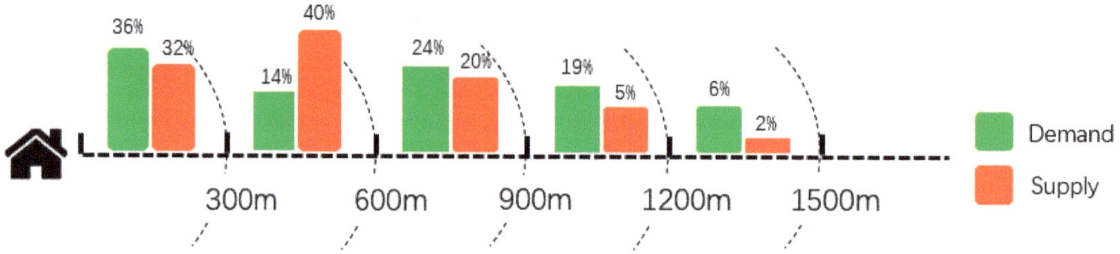

Figure 6. Supply and demand matching analysis for different distance sections.

4.4. Evaluation of Supply and Demand Matching

4.4.1. Evaluation at the Macro Level

According to the United Nations Gini System for Income Distribution Segmentation, a Gini coefficient of 0.5 or greater implies a significant difference between the wealthy and the poor [61]. Based on the evaluation results, the Gini coefficient is 0.61, indicating a severe mismatch between the supply and demand for SPLs.

According to the graphical representation of the Lorenz curve (Figure 7), the top 10% of residents' demand for SPLs accounts for 68% of the supply, while the top 20% of residents' demand account for more than 80% of the supply (Figure 7). These findings clearly indicate an irrational and unfair spatial distribution of SPLs within residential communities.

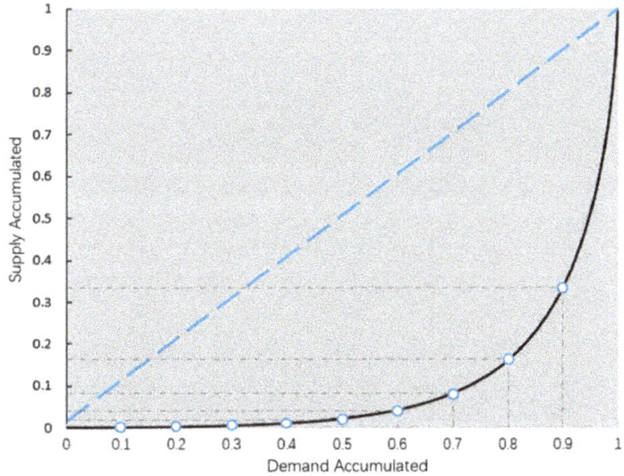

Figure 7. Lorenz Curve of the Gini coefficient.

4.4.2. Evaluation at the Micro Level

To assist the analysis of the supply–demand ratio at the community level, the value of LQ was categorized into five groups (Table 5). From the perspective of spatial distribution, the LQ of SPL supply–demand matching generally presents a circular spatial pattern (Figure 8). Extremely high and relatively high supply–demand ratios are predominantly concentrated in the central part of the study area, encompassing Heping, east Nankai, and west Hexi. Medium, relatively low, and extremely low supply–demand ratios are dispersed across each district, particularly towards the outskirts nearing the suburban boundaries.

Table 5. Classification standard of location entropy value.

Grade	Location Entropy	Remarks
Extremely Low	<0.80	The supply–demand ratio is lower than the average level of the study area
Relatively Low	0.80–3.00	
Medium	3.00–7.00	
Relatively High	7.00–15.00	
Extremely High	15.00–3329.90	The supply–demand ratio is more than fifteen times higher than the average level of the study area

The LQ values of residential communities in the study area are unevenly distributed. High-LQ communities (with LQ values greater than 1) account for 77.23% of the total and are predominantly located in Nankai (21.96%), Hexi (20.71%), and Hedong (19.91%). These communities are concentrated in the central part of the study area. Low-LQ communities (with LQ values less than 1) account for 25.77% of the total and are primarily located in Heping (39.63%), Hexi (20.32%), and Nankai (13.55%). These communities are dispersed across the study area, particularly towards the outskirts. The highest LQ value recorded was 3329.9, which is significantly higher than the average level in the study area. This suggests a possibility of significant waste in public resources.

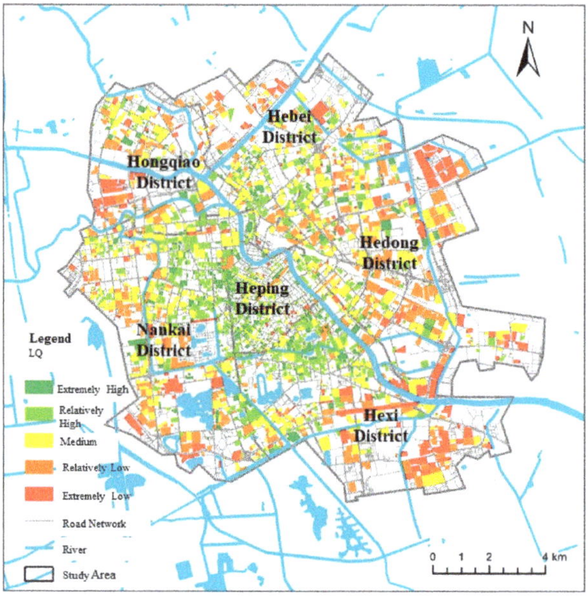

Figure 8. Location quotient of supply–demand matching.

5. Discussion

5.1. Supply and Demand Assessment of SPLs

SPLs are a newly emerging type of self-service facility that provide convenient contactless services and are associated with the digital life of smart cities [13,33,62]. However, assessing and planning SPLs remains challenging. Previous studies have mainly focused on either the supply or demand of SPLs [12,63], and this study is the first to explore the connection between SPL supply and demand at the residential community level from a city planning perspective.

We developed an analytical framework for assessing the supply and demand of SPLs, with the supply situation measured by accessibility and the demand situation measured by purchasing capacity of the residential community. In addition, we employed the Gini coefficient and location quotient to further evaluate the supply–demand matching status on the macro and micro level. This approach provides comprehensive coverage of all communities and is highly efficient, representing a novel contribution to SPL research. However, the framework only includes a few measurement indicators, and there is a lack of subjective indicators. This may cause the results to differ from the actual to some extent. Future research could expand the measurement indicator system and incorporate subjective indicators.

The assessment results of supply show that SPLs are highly accessible, especially in urban centers, where population density and road networks are more developed. Meanwhile, the assessment results of demand show that SPLs are highly needed in residential communities with relatively high incomes, which are also generally located in urban centers. These findings are supported by previous literature [17,64], such as a study of Amazon lockers in Portland, which found that they tend to be located in areas with high population density, employment density, and proximity to arterial roads [40]. Even though both the supply and demand for SPLs are high in Tianjin, there is a considerable imbalance between the two. This imbalance is due to the fact that the locations of SPLs are different from what consumers expected and what SPL providers provided. This imbalance could lead to the inefficient utilization of SPLs, waste of public spaces in residential communities, and even inequity in the city [40,65,66]. This case emphasizes the significance of taking into account the spatial distribution and layout of SPLs in city planning. A substantial quantity of low-demand SPLs and public areas could be saved for resilient utilization by properly planning the location of SPLs.

5.2. Residential Community Level Policy Initiatives

In Tianjin, the government set ambitious goals for the installation of SPLs in 2015 and 2016. However, the first five pilot SPLs were installed in 2016, and there were only 479 SPLs until the end of 2022 [8,20]. This is a significant discrepancy between the government's goals and the actual number of SPLs installed. One of the main reasons for this discrepancy is the lack of planning and guidance on the installation location and management organization of SPLs. On the one hand, SPLs are essentially profit-driven commodities for businesses, and during the initial stages of promoting and implementing SPLs, the primary focus of SPL providers is often on large-scale market penetration and the establishment of consumer habits [40]. On the other hand, SPLs have also become an important aspect of national and local policies as effective measures for contactless delivery during the pandemic [63,67]. As a result, there is a need for clear policies on the planning and management of SPLs, including where to install them and where not to and which department of the government is responsible for the planning and management of SPLs.

Without clear planning and guidance, SPLs could occupy a significant amount of public space. According to the Hive Box official website, the size of one standard SPL is 250 cm (H) × 450 cm (L) × 55 cm (W) [68]. Assume that there is a one-meter distance in front of the SPLs for the courier to drop off and the consumer to pick up. To achieve the government's goal of 5000 SPLs, 3786 m^2 of residential communities' public space will be occupied by SPLs. To avoid future unmanageable circumstances, it is crucial to set clear policies on the planning and management of SPLs during their early stage of growth. This will help to ensure that SPLs are installed in a way that is beneficial to both businesses and residents and that SPL development does not become a social problem due to their encroachment on public spaces.

5.3. Incentives for Dynamic Approach

It is important to note that both the supply and demand for SPLs are dynamic. The supply and demand are interdependent and are influenced by technological innovation, public emergencies such as COVID-19, and urbanization. As the external environment changes and becomes more uncertain, the internal supply–demand relationship also changes [16,34]. This suggests that there may not be a single optimal solution for planning SPLs, and a more flexible and dynamic approach to the planning of SPLs could be the future direction of research [43,50,68].

Policymakers should implement full life-cycle policies that guide and govern the layout of SPLs. In this study, the majority (97.7%) of SPLs were in residential communities. The government of Tianjin should focus on coming up with effective ways to guide the sustainable development of SPLs in residential communities. Residential communities with an oversupply of SPLs should consider removing them and exploring ways to repurpose the public space left vacant. Conversely, in residential communities where the demand for SPLs exceeds the supply, careful consideration should be given to determining the precise quantity, size, and location of SPLs to be installed [22,25].

In addition, SPL providers and researchers should pay more attention to dynamic changes in consumer demand. As highlighted earlier, the pandemic has caused a surge in both the supply of and demand for SPLs worldwide. Even in the absence of a pandemic, the supply of and consumer demand for SPLs will change dynamically. In other words, the environment, consumer demand, and the supply of SPLs are all uncertain. Therefore, tracking dynamic changes in customer demand and incorporating dynamic layouts of SPLs becomes crucial. For example, dynamic planning is an efficient strategy for addressing uncertain conditions [69]. It differs from traditional static planning by dividing the process into short-term stages, estimating demand for each stage, and making necessary adjustments accordingly [70]. If all stages are adequately planned, dynamic planning can mitigate encroachment on public areas and reduce resource waste caused by SPLs, thus serving as an effective approach.

6. Conclusions

In this section, the main conclusions of the study, its contributions, and recommendations for future research studies will be provided.

This paper examines the spatial pattern of the supply–demand matching of SPLs at the residential community level in central Tianjin, China. The analysis reveals spatial disparities in SPL supply and demand, as well as SPL location issues. The main conclusions can be summarized into four aspects: (1) The supply index of SPLs is determined by accessibility. In general, the residential communities in the study area are easily accessible to SPLs. The coverage rate within a 5 min walk is 31.82%, and the coverage rate within a 10 min walk is 72.30%. The areas with low accessibility are mainly distributed on the outskirts of the study area. (2) Residential communities with high demand for SPLs (49.57%) are concentrated in the urban center. This is compatible with high populations, high per capita income, and well-developed road networks. Medium- and low-demand areas are distributed in the outskirts. (3) There is a significant disparity observed between the supply and demand matching of SPLs. Among all the residential communities in the study area, only 25.21% achieve a balance between supply and demand for SPLs. Communities experiencing oversupply are located on the right bank of the Haihe River and in the southwestern suburbs, accounting for 44.78% of the total. Furthermore, communities facing insufficient supply are situated in west Hexi, east Nankai, and the outskirts of each district, accounting for 30.01%. (4) From a macro perspective, the Gini coefficient of SPL distribution in the study area's residential communities is 0.61, and the top 20% of residents have access to 80% of the SPLs. On a micro level, 77.23% of residential communities have a supply–demand ratio for SPLs that exceeds the average level in the study area. Notably, the residential community with the highest supply–demand ratio reaches an astonishing 3330 times the average level in the study area, highlighting the extreme imbalance in SPL spatial distribution.

This study proposes the following implications and recommendations for SPL providers, urban planners, and policymakers. First, SPL providers should focus on meeting the needs of consumers, rather than simply providing a high supply. The findings of this study show that even when the supply of and demand for SPLs are both high, residents may still be dissatisfied with SPL locations if there is a mismatch between supply and demand. Therefore, it is important for SPL providers to accurately assess demand and to tailor their offerings accordingly. Second, urban planners should play a more active role in SPL location issues. This is because SPLs are closely related to public spaces, facility layout, and consumer behavior. The absence of urban planning in SPL location issues is one of the main reasons for the confusion of the layout of SPLs and the ambiguity of related policies. Urban planners should work with SPL providers to ensure that SPLs are located in places that are convenient for consumers and that they do not unduly encroach on public space. Lastly, policymakers should consider both supply and demand when formulating policies for SPLs. In the case study of Tianjin, the city government set too-ambitious goals for the number of SPLs, without considering the actual supply capacity and demand situation. This led to a mismatch between supply and demand, which in turn created confusion and dissatisfaction among residents. Policymakers should work with urban planners and SPL providers to develop more realistic and sustainable policies for SPLs. Moreover, it is important to note that customer demand and the supply of SPLs are dynamic. The location planning of SPLs based on previous data is likely to lag behind the actual situation. Therefore, it is necessary for stakeholders to monitor supply and demand on a regular basis. In addition, we propose to implement dynamic planning for SSFs such as SPLs. This would allow for more flexibility in the location of SPLs and would help to ensure that they are always located in places where they are needed most.

The main contributions of this study are as follows: First, it provides the first attempt to provide an analytical framework and assess the supply–demand situation of SPLs. This is a significant contribution because the existing literature has largely focused on either supply of or demand for SPLs, ignoring the interdependence between the two. This study's new perspective of investigating the supply–demand matching of SPLs enriches the literature and provides a more comprehensive understanding of this issue. Second, this study advances the existing research by representing supply and demand status in a spatially explicit way from an urban planning perspective. This is important because it allows for a more nuanced understanding of the spatial distribution of SPLs and the factors that influence supply and demand. Third, the current study provides the breakthrough discovery that SPL location issues and consumer dissatisfaction are not due to short supply but due to oversupply and a mismatch between supply and demand. This finding has important implications for urban planning and the management of SPLs.

Despite the contributions of this study, there are some limitations that require future development. First, the case study was conducted in Tianjin, which is an e-commerce and express delivery hub in northern China. This means that the population may have a higher supply and demand for alternative delivery modes than in other cities. Future research should cross-validate the results by surveying other SSFs in other countries and cities. Second, the assessment of SPL supply and demand relies on various indicators. The assessment of supply was based solely on accessibility, and the assessment of demand was simplified to purchasing capability. Future research could consider developing a comprehensive indicator system to enhance the accuracy of assessing SPL supply and demand.

Author Contributions: Conceptualization, M.D., Y.S. and Y.H.; methodology, M.D. and Y.S.; software, M.D.; validation, M.D., Y.S. and Y.H.; formal analysis, M.D.; investigation, M.D.; resources, M.D.; data curation, M.D.; writing—original draft preparation, M.D.; writing—review and editing, M.D., Y.S. and Y.H.; visualization, M.D.; supervision, Y.S. and Y.H.; project administration, Y.S. and Y.H.; funding acquisition, Y.H. All authors have read and agreed to the published version of the manuscript.

Funding: This research was funded by Reconstructing the Architecture System based on the Coherence Mechanism of "Architecture-human-environment" in the Chinese Context, Key Project of National Natural Science Foundation of China, grant number 52038007.

Data Availability Statement: The data used to support the findings of this study are available from the authors upon request.

Acknowledgments: The authors wish to express their appreciation and gratitude to the anonymous reviewers and editors for their insightful comments and suggestions to improve the paper's quality.

Conflicts of Interest: The authors declare no conflict of interest.

References

1. de Palma, A.; Vosough, S.; Liao, F. An overview of effects of COVID-19 on mobility and lifestyle: 18 months since the outbreak. *Transp. Res. Part A Policy Pract.* **2022**, *159*, 372–397. [CrossRef] [PubMed]
2. Wang, X.; Wong, Y.D.; Sun, S.; Yuen, K.F. An investigation of self-service technology usage during the COVID-19 pandemic: The changing perceptions of 'self' and technologies. *Technol. Soc.* **2022**, *70*, 102032. [CrossRef]
3. Wang, X.; Wong, Y.D.; Chen, T.; Yuen, K.F. Consumer logistics in contemporary shopping: A synthesised review. *Transp. Rev.* **2022**, *43*, 502–532. [CrossRef]
4. Wang, X.; Wong, Y.D.; Yuen, K.F. Does COVID-19 promote self-service usage among modern shoppers? An exploration of pandemic-driven behavioural changes in self-collection users. *Int. J. Environ. Res. Public Health* **2021**, *18*, 8574. [CrossRef] [PubMed]
5. Narayanan, S.; Chaniotakis, E.; Antoniou, C. Shared autonomous vehicle services: A comprehensive review. *Transp. Res. Part C Emerg. Technol.* **2020**, *111*, 255–293. [CrossRef]
6. Tsai, Y.-T.; Tiwasing, P. Customers' intention to adopt smart lockers in last-mile delivery service: A multi-theory perspective. *J. Retail. Consum. Serv.* **2021**, *61*, 102514. [CrossRef]
7. Yuen, K.F.; Wang, X.; Ma, F.; Wong, Y.D. The determinants of customers' intention to use smart lockers for last-mile deliveries. *J. Retail. Consum. Serv.* **2019**, *49*, 316–326. [CrossRef]

8. Ding, M.; Ullah, N.; Grigoryan, S.; Hu, Y.; Song, Y. Variations in the Spatial Distribution of Smart Parcel Lockers in the Central Metropolitan Region of Tianjin, China: A Comparative Analysis before and after COVID-19. *ISPRS Int. J. Geo-Inf.* **2023**, *12*, 203. [CrossRef]
9. Carotenuto, P.; Gastaldi, M.; Giordani, S.; Rossi, R.; Rabachin, A.; Salvatore, A. Comparison of various urban distribution systems supporting e-commerce. Point-to-point vs. collection-point-based deliveries. *Transp. Res. Rec.* **2018**, *30*, 188–196. [CrossRef]
10. Milewski, D.; Milewska, B. The energy efficiency of the last mile in the e-commerce distribution in the context the COVID-19 pandemic. *Energies* **2021**, *14*, 7863.
11. Seghezzi, A.; Siragusa, C.; Mangiaracina, R. Parcel lockers vs. home delivery: A model to compare last-mile delivery cost in urban and rural areas. *Int. J. Phys. Distrib.* **2022**, *52*, 213–237. [CrossRef]
12. Lemke, J.; Iwan, S.; Korczak, J. Usability of the parcel lockers from the customer perspective–the research in Polish Cities. *Transp. Res. Procedia* **2016**, *16*, 272–287. [CrossRef]
13. Hasija, S.; Shen, Z.-J.M.; Teo, C.-P. Smart city operations: Modeling challenges and opportunities. *Manuf. Serv. Oper. Manag.* **2020**, *22*, 203–213. [CrossRef]
14. Iwan, S.; Kijewska, K.; Lemke, J. Analysis of parcel lockers' efficiency as the last mile delivery solution–the results of the research in Poland. *Transp. Res. Rec.* **2016**, *12*, 644–655. [CrossRef]
15. Lagorio, A.; Pinto, R. The parcel locker location issues: An overview of factors affecting their location. In Proceedings of the 8th International Conference on Information Systems, Logistics and Supply Chain: Interconnected Supply Chains in an Era of Innovation, Austin, TX, USA, 22–24 April 2020.
16. Leung, A.; Lachapelle, U.; Burke, M. Spatio-temporal analysis of Australia Post parcel locker use during the initial system growth phase in Queensland (2013–2017). *J. Transp. Geogr.* **2023**, *110*, 103634. [CrossRef]
17. Lachapelle, U.; Burke, M.; Brotherton, A.; Leung, A. Parcel locker systems in a car dominant city: Location, characterisation and potential impacts on city planning and consumer travel access. *J. Transp. Geogr.* **2018**, *71*, 1–14. [CrossRef]
18. Moslem, S.; Stević, Ž.; Tanackov, I.; Pilla, F. Sustainable development solutions of public transportation: An integrated IMF SWARA and Fuzzy Bonferroni operator. *Sustain. Cities Soc.* **2023**, *93*, 104530. [CrossRef]
19. Breitung, W. Enclave urbanism in China: Attitudes towards gated communities in Guangzhou. *Urban Geogr.* **2012**, *33*, 278–294. [CrossRef]
20. Tianjin Municipal People's Government. Available online: https://www.tj.gov.cn/zwgk/szfwj/tjsrmzfbgt/202005/t20200519_2370240.html (accessed on 25 April 2023).
21. Morganti, E.; Seidel, S.; Blanquart, C.; Dablanc, L.; Lenz, B. The impact of e-commerce on final deliveries: Alternative parcel delivery services in France and Germany. *Transp. Res. Procedia* **2014**, *4*, 178–190. [CrossRef]
22. Deutsch, Y.; Golany, B. A parcel locker network as a solution to the logistics last mile problem. *Int. J. Prod. Res.* **2018**, *56*, 251–261. [CrossRef]
23. Van Duin, J.H.R.; de Goffau, W.; Wiegmans, B.; Tavasszy, L.A.; Saes, M. Improving home delivery efficiency by using principles of address intelligence for B2C deliveries. *Transp. Res. Procedia* **2016**, *12*, 14–25. [CrossRef]
24. Viu-Roig, M.; Alvarez-Palau, E.J. The impact of E-Commerce-related last-mile logistics on cities: A systematic literature review. *Sustainability* **2020**, *12*, 6492. [CrossRef]
25. Olsson, J.; Hellström, D.; Pålsson, H. Framework of last mile logistics research: A systematic review of the literature. *Sustainability* **2019**, *11*, 7131. [CrossRef]
26. Bhatti, A.; Akram, H.; Basit, H.M.; Khan, A.U.; Raza, S.M.; Naqvi, M.B. E-commerce trends during COVID-19 Pandemic. *Int. J. Futur. Gener. Commun.* **2020**, *13*, 1449–1452.
27. Meiksin, A. Dynamics of COVID-19 transmission including indirect transmission mechanisms: A mathematical analysis. *Epidemiol. Infect.* **2020**, *148*, E257. [CrossRef]
28. Aston, J.; Vipond, O.; Virgin, K.; Youssouf, O. Retail E-Commerce and COVID-19: How Online Shopping Opened Doors While Many Were Closing. Available online: https://www150.statcan.gc.ca/n1/pub/45-28-0001/2020001/article/00064-eng.htm (accessed on 3 June 2023).
29. Yaprak, Ü.; Kılıç, F.; Okumuş, A. Is the COVID-19 pandemic strong enough to change the online order delivery methods? Changes in the relationship between attitude and behavior towards order delivery by drone. *Technol. Forecast. Social Chang.* **2021**, *169*, 120829. [CrossRef]
30. Wang, X.; Wong, Y.D.; Qi, G.; Yuen, K.F. Contactless channel for shopping and delivery in the context of social distancing in response to COVID-19 pandemic. *Electron. Commer. Res. Appl.* **2021**, *48*, 101075. [CrossRef]
31. Wang, X.; Wong, Y.D.; Yuen, K.F. Rise of 'lonely'consumers in the post-COVID-19 era: A synthesised review on psychological, commercial and social implications. *Int. J. Environ. Res. Public Health* **2021**, *18*, 404.
32. Kim, A.; Sung, Y. My privacy and control matter: Understanding motivations for using untact services. *Cyberpsychol. Behav. Soc. Netw.* **2021**, *24*, 426–431. [CrossRef]
33. Lee, S.M.; Lee, D. "Untact": A new customer service strategy in the digital age. *Serv. Bus.* **2020**, *14*, 1–22. [CrossRef]
34. Wang, X.; Wong, Y.D.; Kim, T.Y.; Yuen, K.F. Does COVID-19 change consumers' involvement in e-commerce last-mile delivery? An investigation on behavioural change, maintenance and habit formation. *Electron. Commer. Res. Appl.* **2023**, *60*, 101273. [CrossRef]
35. Bengtsson, C.; Vikingson, A. Exploring and Evaluating the Parcel Locker—A Swedish Consumer Perspective. *Mech. Eng.* **2015**, *10*, I10.

36. Lee, H.; Chen, M.; Pham, H.T.; Choo, S. Development of a decision making system for installing unmanned parcel lockers: Focusing on residential complexes in Korea. *KSCE J. Civ. Eng.* **2019**, *23*, 2713–2722. [CrossRef]
37. de Oliveira, L.K.; Morganti, E.; Dablanc, L.; de Oliveira, R.L.M. Analysis of the potential demand of automated delivery stations for e-commerce deliveries in Belo Horizonte, Brazil. *Res. Transp. Econ.* **2017**, *65*, 34–43. [CrossRef]
38. Yang, G.; Huang, Y.; Fu, Y.; Huang, B.; Sheng, S.; Mao, L.; Huang, S.; Xu, Y.; Le, J.; Ouyang, Y.; et al. Parcel locker location based on a bilevel programming model. *Math. Probl. Eng.* **2020**, *2020*, 5263689. [CrossRef]
39. Schaefer, J.S.; Figliozzi, M.A. Spatial accessibility and equity analysis of Amazon parcel lockers facilities. *J. Transp. Geogr.* **2021**, *97*, 103212. [CrossRef] [PubMed]
40. Keeling, K.L.; Schaefer, J.S.; Figliozzi, M.A. Accessibility and equity analysis of transit facility sites for common carrier parcel lockers. *Transp. Res. Rec.* **2021**, *2675*, 1075–1087. [CrossRef]
41. Ghaderi, H.; Zhang, L.; Tsai, P.W.; Woo, J. Crowdsourced last-mile delivery with parcel lockers. *Int. J. Prod. Econ.* **2022**, *251*, 108549. [CrossRef]
42. Che, Z.-H.; Chiang, T.-A.; Luo, Y.-J. Multiobjective optimization for planning the service areas of smart parcel locker facilities in logistics last mile delivery. *Mathematics* **2022**, *10*, 422. [CrossRef]
43. Molin, E.; Kosicki, M.; van Duin, R. Consumer preferences for parcel delivery methods: The potential of parcel locker use in the Netherlands. *Eur. J. Transp. Infrastruct. Res.* **2022**, *22*, 183–200.
44. Tianjin Municipal Bureau of Statistics & Survey Office of the Nation–Bureau of Statistics in Tianjin. Available online: https://stats.tj.gov.cn/nianjian/2022nj/zk/indexch.htm (accessed on 22 February 2023).
45. Zhou, X.; Ding, Y.; Wu, C.; Huang, J.; Hu, C. Measuring the spatial allocation rationality of service facilities of residential areas based on internet map and location-based service data. *Sustainability* **2019**, *11*, 1337. [CrossRef]
46. Hu, Y.; Gao, S.; Janowicz, K.; Yu, B.; Li, W.; Prasad, S. Extracting and understanding urban areas of interest using geotagged photos. *Comput. Environ. Urban Syst.* **2015**, *54*, 240–254. [CrossRef]
47. Ding, M.; Hu, Y.; Du, Y. Characteristics of Supply-demand and Planning of Self-service Facilities in Urban Residential Areas: An Empirical Analysis Based on the Investigation in Tianjin. *Urban Dev. Stud.* **2023**, *30*, 3–8.
48. Luo, R.; Ji, S.; Ji, Y. An active-learning Pareto evolutionary algorithm for parcel locker network design considering accessibility of customers. *Comput. Oper. Res.* **2022**, *141*, 105677.
49. Kahr, M. Determining locations and layouts for parcel lockers to support supply chain viability at the last mile. *Omega* **2022**, *113*, 102721. [CrossRef]
50. Páez, A.; Scott, D.M.; Morency, C. Measuring accessibility: Positive and normative implementations of various accessibility indicators. *J. Transp. Geogr.* **2012**, *25*, 141–153. [CrossRef]
51. Zhang, Y.; Zou, Y.; Zhu, Z.; Guo, X.; Feng, X. Evaluating Pedestrian Environment Using DeepLab Models Based on Street Walkability in Small and Medium-Sized Cities: Case Study in Gaoping, China. *Sustainability* **2022**, *14*, 15472. [CrossRef]
52. Bohannon, R.W.; Andrews, A.W. Normal walking speed: A descriptive meta-analysis. *Physiotherapy* **2011**, *97*, 182–189.
53. Figliozzi, M.; Unnikrishnan, A. Home-deliveries before-during COVID-19 lockdown: Accessibility, environmental justice, equity, and policy implications. *Transp. Res. Part D Transp. Environ.* **2021**, *93*, 102760. [CrossRef]
54. Tianjin Municipal People's Government. Statistical Bulletin of National Economic and Social Development of Tianjin in 2022. Available online: https://www.tj.gov.cn/sq/tjgb/202203/t20220314_5828933.html (accessed on 6 June 2023).
55. Shu, H.; Xiong, P. The Gini coefficient structure and its application for the evaluation of regional balance development in China. *J. Clean. Prod.* **2018**, *199*, 668–686.
56. Wang, D.; Liang, Y.; Peng, S.; Yin, Z.; Huang, J. Integrated assessment of the supply–demand relationship of ecosystem services in the Loess Plateau during 1992–2015. *Ecosyst. Health Sustain.* **2022**, *8*, 2130093. [CrossRef]
57. Teng, F.; He, J.; Pan, X.; Zhang, C. Metric of carbon equity: Carbon Gini index based on historical cumulative emission per capita. *Adv. Clim. Chang. Res.* **2011**, *2*, 134–140. [CrossRef]
58. Miller, M.M.; Gibson, L.J.; Wright, N.G. Location quotient: A basic tool for economic development analysis. *Econ. Dev. Rev.* **1991**, *9*, 65.
59. Isserman, A.M. The location quotient approach to estimating regional economic impacts. *J. Am. Inst. Plann.* **1977**, *43*, 33–41. [CrossRef]
60. Daioglou, V.; Doelman, J.C.; Wicke, B.; Faaij, A.; van Vuuren, D.P. Integrated assessment of biomass supply and demand in climate change mitigation scenarios. *Glob. Environ. Chang.* **2019**, *54*, 88–101. [CrossRef]
61. Mokhsin, M.; Ludin, M.Z.M.; Suhaimi, A.I.H.; Zainol, A.S.; Som, M.H.M.; Halim, H.A. ParcelRestBox: IoT-Based Parcel Receiving Box System Design for Smart City in Malaysia. In Proceedings of the 2021 IEEE International Conference on Computing (ICOCO), Kota Kinabalu, Malaysia, 14–16 November 2022; pp. 180–185.
62. Tang, Y.M.; Chau, K.Y.; Xu, D.; Liu, X. Consumer perceptions to support IoT based smart parcel locker logistics in China. *J. Retail. Consum. Serv.* **2021**, *62*, 102659. [CrossRef]
63. Morganti, E.; Dablanc, L.; Fortin, F. Final deliveries for online shopping: The deployment of pickup point networks in urban and suburban areas. *Res. Transp. Bus. Manag.* **2014**, *11*, 23–31. [CrossRef]
64. Yalcin Kavus, B.; Ayyildiz, E.; Gulum Tas, P.; Taskin, A. A hybrid Bayesian BWM and Pythagorean fuzzy WASPAS-based decision-making framework for parcel locker location selection problem. *Environ. Sci. Pollut. Res.* **2022**, 1–18. [CrossRef]

65. Ambrosini, X.; Bermont-Vialatte, A.; Lefort, M. Do We Need Regulation for Parcel Lockers? In *Postal Strategies: Logistics, Access, and the Environment*; Springer Nature Switzerland AG: Cham, Switzerland, 2023; pp. 101–113.
66. Liu, W.; Liang, Y.; Bao, X.; Qin, J.; Lim, M.K. China's logistics development trends in the post COVID-19 era. *Int. J. Logist. Res. Appl.* **2022**, *25*, 965–976. [CrossRef]
67. Hive Box. Available online: https://fcbox.com/index.html (accessed on 20 February 2023).
68. Schwerdfeger, S.; Boysen, N. Optimizing the changing locations of mobile parcel lockers in last-mile distribution. *Eur. J. Oper. Res.* **2020**, *285*, 1077–1094. [CrossRef]
69. Ssa, B.; Nb, A. Who moves the locker? A benchmark study of alternative mobile parcel locker concepts. *Transp. Res. Part C Emerg. Technol.* **2022**, *142*, 103780.
70. Bocewicz, G.; Nielsen, P.; Jasiulewicz-Kaczmarek, M.; Banaszak, Z. Dynamic planning of mobile service teams' mission subject to orders uncertainty constraints. *Appl. Sci.* **2020**, *10*, 8872. [CrossRef]

buildings

MDPI

Article

Towards a Healthy Architecture: A New Paradigm in the Design and Construction of Buildings

Santiago Quesada-García [1,*], Pablo Valero-Flores [2] and María Lozano-Gómez [1]

[1] Department of Architectural Design, University of Seville, Av. Reina Mercedes 2, 41012 Sevilla, Spain; mlozano@us.es
[2] Faculty of Medicine, Campus Teatinos, University of Malaga, Blvr. Louis Pasteur 32, 29010 Málaga, Spain; pvalero@uma.es
* Correspondence: sqg@us.es

Abstract: The recent COVID-19 pandemic has catalysed a new awareness of what living and working spaces should look like from a different perspective, and healthier cities and architecture have arisen because of inescapable public demand. Society has become clearly aware that there are still unhealthy concentrations within its environment. Spaces in cities are still being built that can favour the spread of diseases, in addition to using harmful construction materials. Living spaces must not only be sustainable, functional, and aesthetically beautiful but also comfortable, safe, and accessible, and, above all, they must be healthy. Healthy architecture has emerged as a new paradigm. This is the subject area of this work. This paper describes and develops the nature of this concept and proposes a novel definition of healthy architecture, aiming to compile state-of-the-art knowledge with a qualitative empirical and multi-method process, using case studies. This article provides a global perspective on new approaches and proposes a Decalogue with the basic principles that an environment or building must comply with in order to be healthy. The main contribution is to establish the basis for the creation of a new healthy architecture epistemology, focussing on cognitive, emotional, and physiological stimuli. This paper can help health professionals, designers, and architects, as well as companies and public administrations, to follow an innovative path in the planning of healthier cities and buildings.

Keywords: healthy architecture; wellbeing; architecture and health; healthy environments; neuroarchitecture; salutogenesis; physiological architecture; buildings for Alzheimer's disease; public health

check for updates

Citation: Quesada-García, S.; Valero-Flores, P.; Lozano-Gómez, M. Towards a Healthy Architecture: A New Paradigm in the Design and Construction of Buildings. *Buildings* **2023**, *13*, 2001. https://doi.org/10.3390/buildings13082001

Academic Editors: Carol Monticelli, Anna Stefańska, Daniel Mateus, Carlos C. Duarte and Nuno D. Cortiços

Received: 7 July 2023
Revised: 29 July 2023
Accepted: 2 August 2023
Published: 5 August 2023

1. Introduction

Architecture's problem is marked by its period circumstances. Today, these difficulties are caused by, among other things, climate change, environmental decline, energy inefficiency, and the social transformation brought about by the silent, digital revolution, all affecting people's health and wellbeing. An imperatively social demand for healthier cities and architecture sprang up a century ago, after a different concept of mobility arose, based on motorised transport such as motorbikes, cars, and buses. This was in addition to the ecological awareness emerging after the first oil crisis in 1973 and the technological and digital revolution of the last four decades.

Cities are places shaped by different strata that show the successive stages of their history. Some authors, composing an analogy that likens disease to destruction, maintain that a city's architecture and layers are remains of social responses to the epidemics humanity has suffered throughout time [1]. However, architecture has not been the discipline that has fought against disease; this has been medicine. Architecture's contribution to health has been to design capable environments that promote and strengthen people's physical and emotional health and build spaces that eliminate noxious and insalubrious conditions in order to prevent illnesses. Architecture can be understood as an extension of nature, like coral reefs or beaver dams [2].

People spend most of their lives inside buildings; hence, homes, offices, residences, schools, and public facilities have to be not only environmentally sustainable but also designed and understood as places where people can live better, with greater wellbeing. The spatial and residential experiences of humans are stimulated by shapes, textures, sounds, lighting, and other signals, perceived physiologically by the senses, that shape the contours of memories and behaviours. Wellbeing and health are closely related to the way the human body interacts with the environment and how it influences the body and the brain [3]. Nevertheless, the effect of architecture in improving healthcare and disease prevention has gone unnoticed. This is, perhaps, because the absence of health has been confused with disease or because the evidence of a relationship between a healthy environment and the absence of disease has not been clinically established.

Over the last century, architecture has generated progressive wellbeing and an improvement in people's quality of life, promoting the construction of buildings with elements that provide better defences against pathogenic factors present in the environment [4]. From this perspective, architecture has been shown to be one of the main agents of human health. In recent years, a new paradigm has been consolidated, aimed not only at combating pathogenic problems that arise in unhealthy environments and buildings but also at stimulating the presence of elements that favour and benefit human health in buildings and cities. It is a model whose attention is especially focused on the physiological, cognitive, and emotional influence that spaces have on people.

The World Health Organization (WHO) recognises that the design, performance, and maintenance of buildings have a significant impact on the health of their occupants and can generate or worsen diseases [5]. Exposure to inadequate architectural parameters and indicators that affect indoor air quality (IAQ), thermal comfort, noise, lighting, atmospheric and ionization, among others, condition people's quality of life in the short, medium, and long term. In the wake of the recent global public health crisis, the generation of theoretical, practical, and interdisciplinary knowledge of the characteristics of buildings that provide positive sensory experiences, as well as physical and mental health outcomes, is timely and responsible [6]. The hypothesis of this paper is that architecture is no longer defined solely by visual or geometrical parameters but also by other dimensions involving environmental, cognitive, psychological, and physiological aspects, affecting comfort, wellbeing, and the physical and mental health of people. This article aims to demonstrate this proposition, and it is the main goal of this work.

To address this objective, the methodology is mixed, first developing a theoretical character and afterwards defining an analytical framework, with some examples demonstrating this new concept of healthy architecture. On the one hand, this research's state-of-art will be defined after an examination of the main related literature following the critical review foundations: search, appraisal, analysis, and synthesis [7]. On the other hand, it uses the empirical experiences of several projects, buildings, and installations that will help to contextualise the problem and obtain certain patterns, serving as a basis for the results obtained.

These contemporary architectural examples, built over the last thirty years for groups with specific needs, have been chosen for being initiatives in which architecture is considered not only essential to the functional aspects of health, but also for the incorporated symbolic, cognitive, and emotional elements. Furthermore, a series of experimental installations were evaluated, in which the understanding of architectural space was approached from a phenomenological dimension, sensitive to cognitive, sensory, neurological, or even chemical stimuli from human beings, just as biology and neurosciences have revealed. In this article, these buildings and environments are presented and analysed as case studies, and from these, the main results are obtained and presented below.

This work has been structured in several sections. After this introduction, the previous and present theoretical background of how the architectural discipline has developed to build increasingly healthier environments is presented. The period from the hygienist movements of the 19th century up to the beginning of the 21st century will be described and contextualised. Section 3 presents the hypothesis, three contemporary approaches to healthy architecture, and a series of experimental installations. These installations demonstrate how an understanding of the physical–chemical mechanisms that govern organisms represents an important change in our way of thinking and understanding space, and not just from a visual and/or aesthetic or compositional point of view. This implies new ways of planning or designing buildings, environments, and spaces to make them healthier. Section 4 presents the results, accompanied by a Decalogue of points or criteria that an environment or building should meet to be considered healthy. Sections 5 and 6 present the discussion and conclusions, respectively, and the definition of healthy architecture.

2. Background and Current Context

2.1. Hygiene Movements: The Genesis of Building Wellness

Mid-nineteenth-century hygiene movements have been described as the starting point, arising from social demands to address the unsanitary, urban agglomerations that emerged after the Industrial Revolution. From these demands, new concepts and urban models arose, such as the Garden City or Linear City [8]. Another important step was the programmatic claims of the European architectural avant-gardes of the early 20th century. A paradigmatic example of what these meant for the improvement of human health was the introduction of individual rooms into homes, intended to be bathrooms or kitchens.

One of the catalysts in identifying hygienic solutions in buildings and dwellings was the decades-long lack of effective remedies against cholera and tuberculosis. The only thing that seemed to combat the latter disease was sunlight, cleanliness, and rest, so buildings began to be designed to promote those factors (Figure 1). Hospitals were built with large windows and sanatoriums with enormous terraces; dwellings were raised above the ground to distance them from dampness; and aerodynamic furniture was designed so that dust would not settle and host germs (Figure 2).

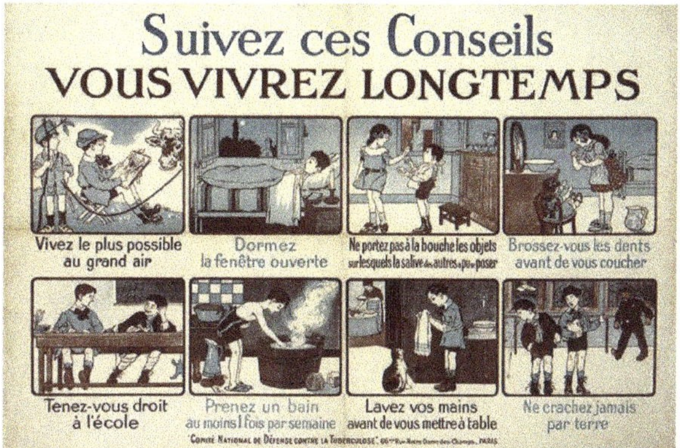

Figure 1. "Follow this advice, you will live longer". This is the translation of the title of this French poster, with eight pictures showing children taking precautions recommended to help avoid Tuberculosis. Recommendations made by the National Tuberculosis Defence Committee of France in collaboration with the Rockefeller Foundation, Paris, 1920 (Source: National Library of Medicine, Digital Collections [9]).

Those unsanitary environmental conditions motivated the pioneers of the modern movement, although dealing with these was not its objective. Its priorities were far from a linear response to the functional problems derived from a health situation. Its main concern was aesthetics, and its purpose was to create a new style. Based on the premise that all buildings should meet minimum functional standards, they proposed a new architecture that should be a faithful expression of the use for which it was intended. For the avant-gardes, the redeeming myth for humanity was the machine. Buildings, products, and objects targeted a single prototype of an individual, the mass man, whose desires and needs were seen as being common throughout the world.

Figure 2. Healthcare architecture with terraces and sunbathing areas for patients, Paimio Sanatorium, Paimio (Finland). Author: Alvar Aalto, 1932. (Source: Prepared by the authors and adapted from a photo by Leon Liao and a photo by Gustaf Welin/AAM).

With these foundations, architects constructed buildings which, like machines, resolved the required functional problems effectively, but they did not easily fulfil people's need to express emotions, recognise themselves, or self-actualise. The concepts of life and death, with their respective emotions of joy and sadness, were excluded from care and health institutions. The consequences were impersonal spaces and oppressively alienating buildings. Faithful to their programmatic principles, healthcare buildings incorporated such technical features as sanitation, accessibility, and safety. These were built on the basis of the symptoms of a disease, whereby a medical cure was considered the only useful factor in treatments. This utilitarian approach gave rise to sterile, hard surfaces; shiny, colourless spaces; long internal corridors, isolated from the outside and devoid of natural light; and artificial ventilation systems, with batteries of minimalist rooms. These features made healthcare buildings effective healing factories, but with no identity, meaningless and soulless. Rarely was any thought given to the occupant's emotional needs, whether healthy or sick, when they experienced these spaces.

In reaction to the machines that had created so much destruction during the Second World War, through the reconstruction of Europe, the person was placed at the centre of architectural thinking. At the CIAM Congresses (*Congrès Internationaux d'Architecture Moderne*, or International Congresses of Modern Architecture), which took place between 1947 and 1953, the conclusion was reached finding that it was necessary to consider the person as an individual, with their own identity and various peculiarities, causing a multiplicity of situations and requiring various environments. In 1945, Henry Sigerist, a historian and health professional, was the first to refer to the promotion of health and the environment as one of the four fundamental actions of medical care, followed by the prevention of disease, treatment, and rehabilitation [10].

Richard Neutra is also considered one of the pioneers of environmental design, guided towards the physiological needs of human beings and their natural behaviour in space. In 1946, he designed and built the Lovell Health House on Dundee Drive (Los Angeles, California) for the naturopathic doctor Phillip Lovell. The architect explored the consideration of physiology and psychology in architectural design by publishing a series of writings and collected in his book Survival Through Design (Figure 3). In it, he maintained that "we orient ourselves by physiological coordinates and we exist thanks to the sensory forms that surround and stimulate us" [11].

Figure 3. Photograph of the Lovell Health House, Los Angeles (California, USA) (author: Richard Neutra, 1929), and scanned cover of the book Survival Through Design, Richard Neutra, 1954. (Source: Prepared by the authors and adapted from Neutra, 1969 [11]; photo by Paul Narvaez).

These new, person-centred perspectives represented a decisive paradigm shift in architecture. Its assimilation was slow but progressive over time. The way in which cities inhabited their environments and how these were perceived and understood began to be modified. To understand the legibility of urban space and know more about what a person perceives, when travelling through a city, the urban planner Kevin Lynch developed the concept of a cognitive map as a hypothetical construct, created on the basis of humans' introspective stories [12]. For the first time, he made environmental sketches and information systems, obtained by experiencing his route in order to improve our understanding of the urban space. Lynch's research was a milestone in beginning to understand, in depth, the way a physical environment can have positive or negative emotional consequences for a person.

At the end of the sixties, Ian L. McHarg's book Design with Nature was published [13]. According to the American historian Lewis Mumford, in reaction to a polluted, dehumanised, and machine-dominated world, this publication established scientific, technical, and philosophical foundations for the development of a new human civilization. This would replace the current one, which, according to Mumford, was in the process of accelerated disintegration. McHarg, a town planner, landscape architect, and inspired ecologist, laid the groundwork for current environmental planning conforming to a certain ecological determinism, which was the origin of what would later be referred to with the pleonasm "sustainable architecture".

With an insightful vision of the future, McHarg finished his book with a chapter called The City: Health and Disease. In it, he asked whether health is only the absence of disease. He argued that health is a symptom of creativity and adaptation and that illness is an expression of a capacity for destruction and a lack of adaptation. He wondered in what places the physical, mental, and social health environment and illness's environment were located. He maintained that, if the healthy and unhealthy areas in cities could be identified, the environmental agents that promote health and the risk factors that cause the disease could be associated with them. From this intelligent approach, which was proposed more than fifty years ago, arose a fundamental action that all cities, counties, and countries should undertake—mapping healthy and unhealthy spaces [13].

2.2. The Emergence of Salutogenesis and Neuroarchitecture

In parallel with these urban and environmental reflections, from a more sensory, poetic, aesthetic, and phenomenological position, the developer of the polio vaccine, the doctor and researcher Jonas E. Salk, warned of the enormous importance of spaces in which the creative process, ideas, inspiration, and knowledge can flow. With that conviction, in the mid-nineteen-fifties, he commissioned Louis I. Kahn to design and build the Salk Institute for Biological Studies (Figure 4). Under the premises of that commission, in 1965, the prestigious architect built one of the masterpieces of contemporary architecture, located in San Diego, California. However, above all, he built the first example of the relationship between neuroscience and architecture. The building was designed specifically to promote the most comfortable intellectual and physical conditions, based on what was then believed about the functioning of the human brain.

Figure 4. Salk Institute for Biological Studies, San Diego (California, USA) (author: Louis I. Kahn, 1965). (Source: Adapted by the authors, photo by Brent Hellickson).

In 1988, a group of neurobiologists demonstrated that neurons are produced in the hippocampus throughout a human's lifetime [14]. Ten years later, Russell Epstein and Nancy Kanwisher discovered that a part of a person's brain is activated upon perceiving places or awareness of new spaces. The neurobiologist Fred Gage presented a key idea at a conference of the American Institute of Architecture in 2003: changes in the environment change the human brain and, therefore, also modify a person's behaviour. Thus, a novel interdisciplinary relationship between neuroscience and architecture began, which would eventually bear fruit in the new field of neuroarchitecture [15–18].

The relationship between architecture and neuroscience systematises the knowledge acquired with respect to the influence of spaces on people. Above all, this is useful in establishing a scientific method that considers the relationship between them objectively, on the one hand, the built form and the space it generates and, on the other, the person's cognitive capacities and motivation. Neuroarchitecture studies the functional requirements buildings must meet to promote the development of daily activities, the way people behave in different spaces, and the way different aspects of an architectural environmental influence brain functions, such as stress, emotion, memory, and learning. Its challenge is discovering the way the brain works given certain spatial requirements in order to understand why there are places that promote or inhibit particular emotional states and comprehend the way a person's habitat affects their mental health, emotional state, and behaviour [19,20].

Interest in understanding the influence of space on a person and the reasons why a human feels well in a certain place transcended architecture and transferred to other disciplines, such as sociology and psychology. In his book Health, Stress and Coping, the doctor and sociologist Aaron Antonovsky [21] proposed a new field of study referred to as "salutogenesis". This field focuses on the origin of health and the so-called assets of health, which were understood to be a complement to the pathogenic approach. From this perspective, it encompasses the origins of illness and its risk factors. Antonovsky maintained that an environment or space must meet three basic criteria to facilitate a person's wellbeing. Firstly, it has to be manageable, in that the space must be able to facilitate or manage resources that support the body's resistance to diseases. Secondly, it must be comprehensible, in that a person can understand the environment adequately. Finally, and most importantly, it must be meaningful; the space must have a meaning or sense recognisable to the people who inhabit and use it [22].

Illness absence and, from a salutogenic view, health assets have an influence on physical conditions, mental health, and comfort states. From this perspective, the environment is a significant factor, as it can act as a negative stressor, as well as a positive motivator [23,24]. Environmental psychology has various points in common with salutogenic theories, given that it holds that there are two factors that guide human preferences with respect to places: understanding and exploration. Understanding is the ability to appreciate the environment through cognitive frameworks. For this, the space must be coherent and legible, in that it must allow its forms to be understood within a certain cultural, historical, or aesthetic context, and it must be possible to perceive its structure and orient oneself within it properly. On the other hand, the exploration factor consists of the ability to expand one's own capacity to understand the environment and foresee new situations. To explore a place, a variety of unknown components must appear within the environment, as well as an element of mystery, which is the promise of new information following the exploration [23].

2.3. *From US Medical Programmes to the WHO Healthy Cities Network*

A space's influence on humans, in emotional and cognitive terms, is an extremely abstract problem. Therefore, to comprehend them, the research on and projects for people with cognitive deficits are particularly important. The architectural solutions developed for these groups can be extrapolated to all other buildings because of the advantages these also offer society as a whole [25]. Interest in studying the special needs of people with dementia appeared in the US in the mid-nineteen-sixties, when new care models were developed for patients with cognitive deficits who were confined to psychiatric institutions, until that point in time. The medical programmes "Medicare" and "Medicaid" provided the financial support necessary so those people could reside in centres focused on their cognitive and social needs, not simply on the symptoms of the disease. These places provided specialised care to meet the specific needs of these groups and created personalised surroundings, improving the physical environment in which the residents lived. It was a model of care that progressed very rapidly and that, in the 1980s, stimulated the development of buildings and residences, the design of which considered the emotional requirements of the users for whom they were intended.

The demand for the construction of this new type of care centre led to the publication of various architectural practice guides, with technical and compositional solutions applied to those environments. These publications prescribed measures and criteria for spatial organisation based on the project experience of the architects themselves, as well as on the empirical experiences of the caregivers and workers in the care units. Occasionally, the solutions were based on results obtained in clinical tests and research [26]. These manuals proposed that an environment intended for people with dementia had to have design guidelines that met safety, orientation, functionality, integration, and personalisation criteria [27,28].

Some of these criteria were already being considered in the construction of buildings intended for groups with physical or sensory disorders. However, the novel contribution at that time was the introduction of concepts based on a subjective perception of the space, such as integration or personalisation. To promote and support the autonomy and independence of people with cognitive deficits, Cohen and Weisman [29] introduced new concepts; support for people to help them perform the instrumental activities of daily life; optimal sensory stimulation within an environment; and the provision of spaces, making it possible to maintain patients' social bonds with their families for as long as possible. Calkins [30] argues that person-centred care provides a more cohesive basis for the designer, as it combines the various recommendations and regulations in a more meaningful way. Therapeutic purposes are still inherent in building practice recommendations but are subordinate to higher-level objectives focused on the person and, therefore, entail adaptation and a different hierarchical understanding of the environment.

All of these ideas confirmed the fact that the design of the environment had a direct effect on people with dementia, and they showed no reduced ability to engage in their daily behaviours, in contrast to what is normally associated with the deterioration and progression of the disease. Since then, there has been growing interest in understanding and knowing how and why mental health requires responsible and sustainable environments, providing wellbeing and allowing people to adopt and maintain healthy lifestyles [31].

The Ottawa Charter, established in 1986, indicated that promoting health consists of providing the population with the means necessary to improve and control it [32]. One of those means is the environment in which the person lives and works. Beginning with that declaration, various plans emerged promoted by the WHO and were intended to create healthy environments. In the same year, the WHO Healthy Cities Network was created to enhance and protect citizens' health and wellbeing by interrelating aspects that influence health and political, economic, cultural, social, and environmental sectors. The initiative had laudable intentions, and when its ambitious programmes and recommendations had the necessary funding, its visible effects were seen in cities, and the meanings of abstract concepts such as "friendly", "resilient", or "healthy cities" (a description in which a human attribute is applied to an artificial product) were specified and concrete. Finally, it also required particular measures being proposed and conducted by experts in disciplines such as public healthcare, town planning, and architecture.

3. A Hypothesis and Three Contemporary Approaches to Healthy Architecture

The context and background described above allow us to propose the hypothesis that architecture is no longer defined solely by visual or geometrical parameters but also by other materials involving environmental, cognitive, psychological, and physiological aspects. Different environmental indicators and emotional factors define a new type of non-representational space that, despite not being seen, is perceived by the human body. In other words, the expression or representation of space ceases to be strictly formal, compositional, or visual but is able to define, work, or manipulate buildings with other types of quantifiable and measurable healthy parameters or indicators that must be incorporated into the architectural project process.

Although there has never been a specific theory or scientific systematisation related to them, the health-related functional principles that the architectural avant-gardes of the 20th century incorporated into buildings (healthiness, safety, and accessibility) are inescapable today and are, to a great extent, set out in all basic construction regulations. However, in the last thirty years, there has been empirical work in architectural research that has planted a new seed that has borne fruit in a series of buildings. There are examples of architecture focusing on emotional and cognitive aspects, contributing to wellbeing and human health and supplying one more component in this new paradigm of healthy architecture. Centres for individuals with cancer or Alzheimer's disease and palliative care units have been built in various places around the world that are not focused solely on technical or regulatory aspects. Their attention is centred, in a special way, on the physiological, cognitive, and emotional influences that spaces have on people. These are buildings that have been designed for specific groups of people with illnesses, deficiencies, or particular conditions. However, the findings applied to respond to these physical, mental, and emotional needs provide solutions that, because of their comfort, effectiveness, and usefulness, can be extrapolated to the rest of the population and society. These, therefore, point to a path to follow for the implementation of healthier cities and architecture for the benefit of all [33]. These empirical experiences and the results that are extracted from them are those that will be used to demonstrate the proposed hypothesis and draw a series of useful conclusions in understanding what healthy architecture means today.

Next, a series of these case studies will be presented that correspond to a classification according to three contemporary approaches to healthy architecture: emotional, psychological, and physiological. On the one hand, for the emotional influence of architecture, Palliative Care Units in France and Maggie's Centres will be analysed. On the other hand, keeping in mind the environmental cognitive impact, an appraisal will be made of residences intended for people with Alzheimer's disease. Finally, in regard to their physiological effect on people, a selection of installations is presented.

3.1. Emotional Influence of Architectural Space

One of the main examples is that of the *Unités de Soins Palliatifs* (USPs) or Palliative Care Units in France. The first USP, named Paul Brousse, was designed by Avant-Travaux architects and was built in the city of Villejuif in 1988. The challenge the architects who built the USP faced was to design a place that did not remind users constantly of their imminent demise. To do so, they created a material and psychological environment in order to allow patients and their families to enjoy the greatest possible wellbeing while they were in the healthcare institution. Therefore, they included elements that enabled them to express their individuality or sense of belonging by personalising the spaces. Another fundamental feature of USPs is that their architectural language has strong symbolism, so these spaces elicit emotion in the people who visit and reside in them [34]. This healthcare facility model, intended for a specific group of people, reached maturity in 2006, when the Japanese architect Toyo Ito, winner of the Pritzker Architecture Prize, built a USP at the Cognacq-Jay Hospital in Paris (Figure 5).

Figure 5. Hôpital Cognacq-Jay USP, Paris, France (author: Toyo Ito & Associates, 2006). (Source: Prepared by the authors and adapted from Hôpital Cognacq-Jay USP website [35]).

Sometime after the case above, the Maggie Keswick Jencks Cancer Caring Trust network emerged in the United Kingdom. This was an initiative that began with the landscaper Maggie Keswick Jencks based on her own spatial and environmental experiences when she was diagnosed with cancer [36]. Maggie's Centres are part of an association that distinguishes itself by providing practical, emotional, and social support beyond medical treatment. These are not only for people with cancer but also for their families and friends in places conceived and designed specifically to meet these individuals' emotional needs. These centres are positioned as annexes to hospitals and are places where no direct medical treatment is provided.

These buildings are designed by qualified contemporary architects who apply their personal architectural language to the construction. In all of them, one can recognise the power of the idea that produced them and the meaning that architecture brings to a specific place. The architecture of Maggie's Centres is expressive, artistic, and high quality and creates a sense of space adapted to a certain group's specific needs. These characteristics lead patients to identify with and have a sense of belonging to a group that enjoys the privilege of using these spaces, as the Centres' buildings are places worth going to. The first Centre, designed by architect Richard Murphy, was built in 1996 on the grounds of the Western General Hospital in Edinburgh (Figure 6). Currently, there are twenty-six centres in the United Kingdom, two in Asia, and only one on the European continent, which is adjacent to Sant Pau Hospital in Barcelona, Spain (Figure 7) [37].

Figure 6. Maggie's Cancer Caring Centre (1996) with a perspective on a new extension (2001), Edinburg, United Kingdom. Author: Richard Murphy Architects. (Source: Prepared by the authors and adapted from Richard Murphy Architects [38]).

Figure 7. Maggie's network: Kálida Sant Pau Centre, Barcelona, Spain (2019). Architect: Miralles Tagliabue EMBT. (Source: Prepared by the authors and adapted from Miralles Tagliabue EMBT [39]; photo by Lluc Miralles).

3.2. Cognitive Impact of the Environment on People

Another case study is the residences intended for those who suffer from Alzheimer's disease. As with tuberculosis at the beginning of the 20th century, architecture is also now used to alleviate the symptoms of an as-yet incurable disease. Based on American care programmes intended for patients with cognitive deficits, some residences for people with Alzheimer's disease began to incorporate places for social interactions with the family and other people to arouse recollections of home and stimulate the residents' memories. The first institution built according to these criteria was the Corinne Dolan Alzheimer Center in Cleveland, Ohio, designed by Taliesin Associated Architects in 1985 [40].

At the end of the nineteen-eighties, the architectural studio Perkins Eastman developed the Woodside Place residential complex in Oakmont, Pennsylvania (Figure 8). This building was the beginning of the development of a residence typology for people with Alzheimer's that includes innovative design guidelines. It was a new type of architecture that had the peculiarity of adapting and personalising spaces for the users for whom it was intended. The small-scaled, small-sized buildings were for a limited number of inhabitants and had well-planned itineraries and routes in a simple arrangement. Consisting of a nucleus of small houses, it recreates the atmosphere of a home and opens to green areas outside. In the buildings' designs, forms, symbols, and elements are introduced that lead to reminiscence and reference cognitive archetypes, such as gabled roofs, particular types of windows and doors, chimneys, etc. These residences also include spaces intended for caregivers and care services, such as day units, medical centres, or research areas [41]. This is a new type of building with interesting examples, such as Boswijk, which EGM architects built in Vught, Holland, in 2010 (Figure 9).

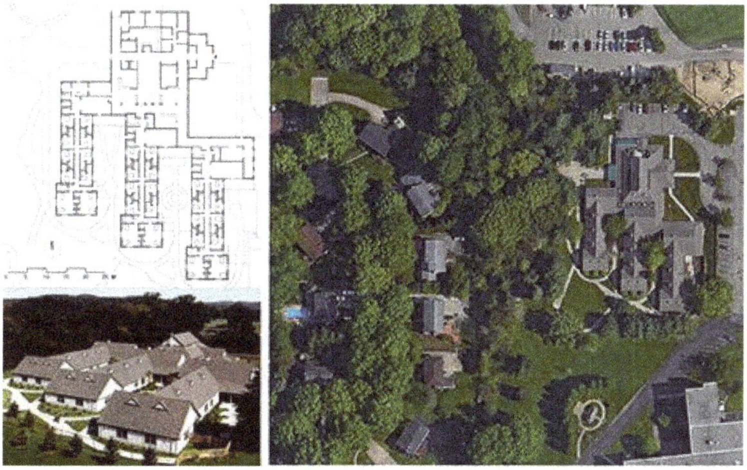

Figure 8. Woodside Place, Oakmont, Pennsylvania (1991). Author: Perkins Eastman Architects. (Sources: Prepared by the authors, Google Earth orthophoto base; map data © 2023 Airbus).

Figure 9. Boswilk Nursing Home for residents with dementia, Vught, Holland (2010). Author: EGM Architecten. (Source: Prepared by the authors, Google Earth orthophoto base; map data © 2019 Google).

All these experiences—French USPs, Maggie's Centres, and the new buildings for users with Alzheimer's disease—include spacious areas with natural light that convey a balanced sensation using scale, proportion, materials, textures, sound, colours, and odours. The arrangement of these spaces focuses on a comfortable place to be in and relax, to have a cup of tea together, or to have an informal conversation, a concept that is far removed from that of traditional hospital settings. Generally, these open to the outside, for example, outdoor spaces that are inserted in the interior of the building, producing the well-known beneficial psychological effect of natural green vegetation. Another important aspect is that all buildings "care for" the caregivers as well. The spaces these workers use are studied in detail so that they may perform their work with maximum efficiency but, at the same time, relax adequately after the innumerable moments of tension they must experience. This generates a very favourable psychological atmosphere by reducing the emotional distress caused by difficult situations [42].

3.3. *Physiological Architecture: A Novel Approach*

From 2001 onwards, at the same time as the development of the examples described in the previous section, a series of experiments was carried out by various teams of architects. These experiments and tests were presented in installations assembled at exhibitions, biennials, and shows. The aim of these projects was to investigate how space actively stimulates people's chemical, organic, and emotional mechanisms, effectively influencing their wellbeing.

This line of research was developed in North America by architects such as Elisabeth Diller and Ricardo Scofidio with Renfro (DS+R). On the occasion of the 2002 Swiss National Exhibition, as part of the "Blur" project, this team explored how the construction of an artificial atmosphere can encourage the use of the senses to generate collective experiences to improve the perception of the environment [43]. Later, in an installation presented in 2008 at the Sandreto de Rebaudengo Foundation in Venice 2008, they created an experimental montage based on virtual reality, entitled "Does the Punishment Fit the Crime?". In it, they analysed the sensations produced by a person after being locked up in a cell, designed according to the crime they had committed [44]. Their investigations continued with the installation 'Unspoken'. This installation, presented in 2016 at the Third Istanbul Biennial, studied the process of people blushing when they pass through an enclosure purposely designed to produce this emotion [45] (Figure 10).

Figure 10. Diller Scofidio + Renfro exhibitions. Above, the 2002 Swiss National Exhibition. Below left, the 2008 Sandreto de Rebaudengo Foundation in Venice installation. Below right, the 2016 Unspoken installation. (Source: Prepared by the authors and adapted from Diller Scofidio + Renfro (DS+R) [43–45]).

However, it was the Swiss team of architects Décosterd and Rahm who proposed the novel concept of physiological architecture. This is a new paradigm based on the principle that the phenomena sustaining life are constantly determined by physicochemical conditions. Depending on their presence, absence, or intensity, these constitute some of the main causes that influence a human being's dwelling and wellbeing [45]. According to Philippe Rahm, the way in which human beings deal with the sensations of inhabiting can be translated into five different actions: 1. Atmospheric, by changing the temperature through natural or artificial conditioning systems. 2. Physiological, by drinking hot or cold drinks or food. 3. Social, by changing clothing. 4. Physical, by resting to adapt the body to the environment. 5. Neurological, by stimulating a feeling of freshness within the mind through taste and smell activators.

This team of Swiss architects has developed its experimental work based on this intangible or invisible dimension of architecture with research that they carried out at installations and assemblies created in art galleries, first at the Arteplage of Expo 01 in Switzerland, on the beaches of Lake Neuchâtel, and later at the MoMA in San Francisco. In their installations, which they significantly termed Melatonin Room, the *Hormonarium*, or *Paysage électromagnétiques*, among others, they investigated the influence of different stimuli generated in specific spaces on human beings. Taim directed architectural research towards the characterisation of space and its physiological impact on human metabolism via the design and implementation of certain parameters.

The Melatonin Room installation consisted of a space for hormone stimulation, distributed in two consecutive rooms connected to each other, which could be accessed independently. This produced two alternating climates. The first was defined by the emission of electromagnetic rays at 509 nm with an intensity of 5000 lux, which suppressed the production of melatonin in the pineal gland. The space thus became a physically stimulating, motivating, and chemically exciting place. The second climate, with green light mimicking the diffusion of ultraviolet rays in a natural environment, stimulated the production of melatonin and thereby became a relaxing place. The Melatonin Room was a non-representational space that acted on the chemical mechanisms of the human organism [46].

The *Hormonorium* was another proposal for the design of a new public space. It was a space climatically defined by light, temperature, and air quality, all parameters that involve the body. The *Hormonorium* was an assembly of physiological devices that acted on the endocrine and neurovegetative systems of the human body. It was built with a luminous floor made of plexiglass to allow ultraviolet light to pass through using 528 fluorescent tubes; these emitted a white light that reproduced the solar spectrum, with UV-A and UV-B. This inverted radiation, emitted from the ground, meant the light radiation was not blocked by eyelids, eyelashes, or the natural inclination of the head. This very bright light, between 5000 and 10,000 lux, stimulated the retina and transmitted information to the pineal gland, which caused a decrease in the secretion of melatonin. In lowering the level of this hormone in the body, this environment made it possible to experience a decrease in fatigue and regulation of mood [46] (Figure 11).

Philippe Rahm continued this line of work with a study of the alterations that hormonal balance exerts on the quality and ways of life. He designed the installations Noctambulisme, Diurnisme (2007), and Digestible Gulf Stream (2008). In the latter, he set out to imitate the physical principles of the Gulf Stream to build a habitable space based on a natural climate with changing atmospheric conditions, thus freeing it from the sophisticated and expensive technical solutions of artificial thermal conditioning.

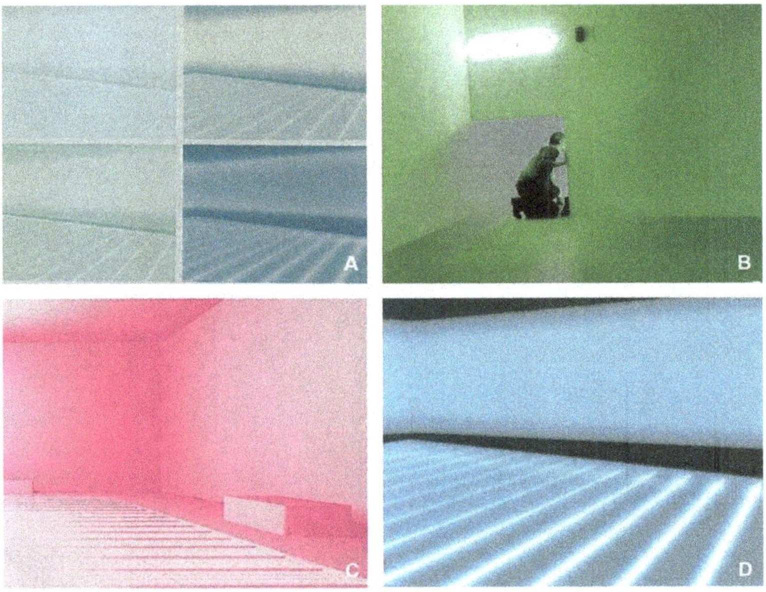

Figure 11. Images (**A**,**D**): Spaces designed to measure air quality, mounted at the Neuchâtel Arteplage of the Swiss National Exposition 2001. Image (**B**): Melatonin Room. Image (**C**): *Hormonarium*. (Source: Prepared by the authors and adapted from Décosterd & Rahm. Physiological architecture [46]).

4. Results: A Decalogue That Leads towards Healthy Architecture

The common denominator in all of the previous case studies is the creation of healthy environments. However, their principal contribution has been the construction of spaces with a comprehensible and recognisable meaning, helping to improve the emotional balance of the people who use them. In those experiences, all of the buildings provide spaces with a strong identity so that those who use them, whether healthy or sick, recognise and, moreover, feel welcome in their relaxed and non-institutional environment (Figure 12).

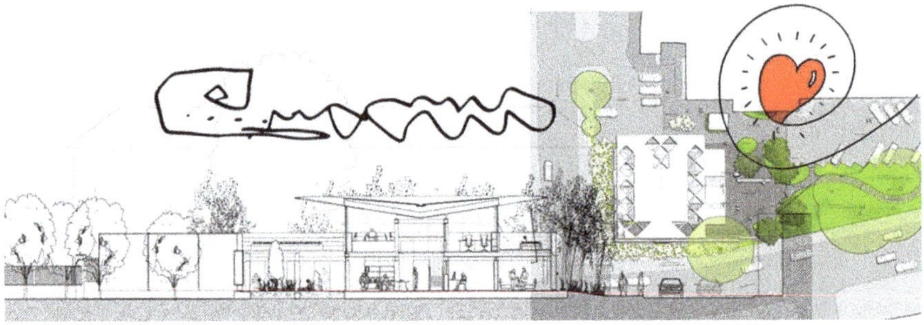

Figure 12. Maggie's West London Centre, London, UK (2008). Architects: Rogers Stirk Harbour + Partners. (Source: prepared by the authors and adapted from Rogers Stirk Harbour + Partners [47]).

In addition to the visual and tangible spatial qualities, the physiological architecture approach, offered by the exhibitions depicted above, takes into consideration the influence of electric, sound, magnetic, and thermal flows existing in today's environments. This establishes a new geography that, like any location determined by a given amount of energy, can be measured and mapped through the corresponding electric, magnetic, or climatic energy emissions. This means a contemporary space is the result of capturing or mapping the environment through the data produced by the radiation of these flows. A new spatial dimension emerges, an ambience defined by parameters with various non-metric magnitudes that nevertheless delimit environments and irremediably alter and affect human metabolism. These environments are physical and cognitive stimulators for people, as well as natural, sociological, and cultural conditions that significantly interfere with the basic and instrumental activities of daily life.

From here, architecture enters the field of physical action, leaving its strict formal function to confront the limits of space. Architecture becomes physiological, acting on the corporeality of the air and the human body without intermediaries. It is like a forcefield, the fire of a bonfire, which, when released, unfolds the conquest of space and sets in motion various sources of energy, such as heat, light, or wavelengths, necessary and essential for balance in people's metabolism. The habitable place thus becomes a modified and modifiable environment, a field without precise limits into which the human body enters and where its organs establish a physiological relationship.

Research by the European Network for Brain Evolution Research and the University of Bath found that, in addition to promoting wellbeing, well-planned environments have a positive effect on decisions and people's personalities [48]. That research highlights the fact that, depending on the experience the space produces, the environment may be understood differently, interfering with aspects such as familiarity, the relationship with the location, or even social relationships. A certain space can affect the quality of a person's spatial and social cognition, which implies that inhabiting certain environments may have either harmful or beneficial effects. For instance, some places and spaces, such as underground parking, airports, and malls, cause particular cognitive symptoms like stress, spatial and temporal disorientation, anxiety, fear, etc. As seen above, these are reactions to which people with cognitive deficits, such as Alzheimer's, autism, etc., are more prone.

There are also physical reactions to specific surroundings that are better known than cognitive and psychological reactions. These are symptoms seen in some people when they remain in certain buildings continually, the so-called Sick Building Syndrome (SBS). Symptoms consist of a suite of ailments that include headaches; eye, nasal, and buccopharyngeal discomfort; lethargy; allergies; etc. This syndrome was first noticed in the mid-1970s in offices and schools. It has varied aetiologies, from the building's formal design (e.g., some of them are hermetically sealed and have large glass surfaces) to artificial climates or construction with unhealthy materials, such as lead, asbestos, or fibrillary insulation. SBS is also caused by products and installations that release carbon monoxide, sulphur dioxide, ozone, or even carbon dioxide, which people themselves exhale in an enclosed environment. In general, it is the result of a method of building that generates and continues to emit environmental, electrical, magnetic, and chemical pollution. To combat the symptoms of SBS, in 2017, the Harvard T.H. Chan School of Public Health proposed a series of points, or criteria, that must be controlled to achieve a healthy level in a building [49]. Its document omits some important factors, as it does not mention the impact that spaces and environments have on the cognitive component of people.

Based on the previous background, analyses of international architectural experiences and examinations of experiments centred on physiological architecture concepts, the Healthy Architecture & City Research Group at the University of Seville proposed a comprehensive list of ten control indicators for use in the construction of healthy buildings. There are nine environmental factors that influence sensory, physical, and cognitive aspects simultaneously, as well as a final, more abstract and holistic factor related to the meaning, orientation, organisation, and distribution of space (Figure 13).

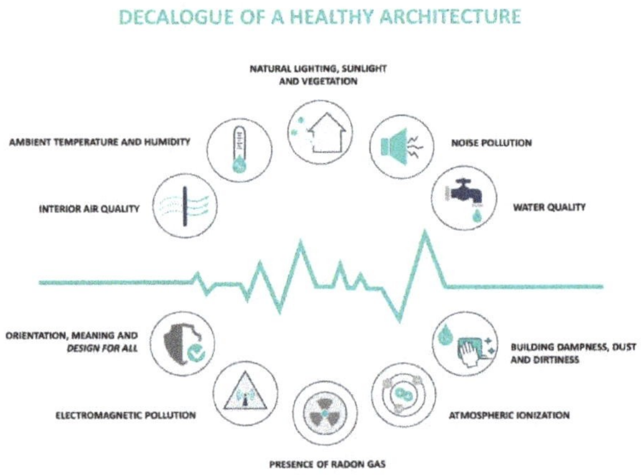

Figure 13. A Decalogue of healthy architecture (Source: Prepared by the authors).

These issues are important to human health and are set out below:

1. Interior air quality (IAQ). The most effective way to control air quality is through adequate and essential natural ventilation in all spaces. Air quality is perceived largely by the nose upon breathing and by the eyes and skin. The body's combined reaction determines whether the air is perceived to be fresh and pleasant or foul and irritating. To guarantee air quality, both natural ventilation and the choice of construction materials must be borne in mind. These should be innocuous and emit few volatile organic compounds, and the absence of pollutants such as lead, polychlorinated biphenyls, and asbestos, among others, should be verified.
2. Ambient temperature and humidity. The relationship between these parameters is very important for wellbeing, as the homeostasis of the human body depends upon these, among other factors. It is necessary to take measures to ensure indoor habitats and workplaces have a combination of temperature and humidity that is constant throughout the year. There are no constant, comparative parameters that can be applied around the world, as the sensation of comfort depends upon the climatic and geographic conditions of each location. It is, therefore, essential to establish a local range of optimum levels.
3. Natural lighting, sunlight, and green spaces. Indoor spaces should have as much natural light as possible to maintain visual comfort, but not dazzle. Moreover, spaces are needed that provide a certain number of hours of sunlight per day. An effort must be made to provide visual lines from the interior to the exterior and to introduce a view of areas with vegetation or open green spaces. This factor has a special effect on people's mental and emotional wellbeing.
4. Noise pollution. It is important to control unwanted noise pollution by protecting and insulating indoor rooms from outside noise. It is also necessary to monitor internal noise sources that may be irritating, such as mechanical equipment, electrical appliances, air-conditioning, or even a neighbour rehearsing for a forthcoming concert.
5. Water quality. Contaminants must be removed from drinking water as much as possible. To do so, in addition to the actions that suppliers are considering already, it is desirable to install domestic purification systems. Water stagnation in drains, buildings, and outdoor spaces must be prevented, as well as stagnant water in wells, puddles, etc., as these serve as breeding grounds for pests. As it is necessary to limit the use of pesticides and chemical products in pest control, it is better to avoid attracting them by eliminating where they live whenever possible.
6. Dampness and dirt in buildings. It is essential to prevent the build-up of dampness attributable to capillarity, condensation, or infiltration, as its existence promotes the

presence of fungi and bacteria that are harmful to health. Materials that, because of their characteristics and distribution, can cause adverse reactions and allergies must also be eliminated.

7. Atmospheric ionization. Clean air is usually negatively ionized and an adequate percentage of negative ions contributes to a sensation of wellbeing. Air in cities is normally charged with positive ions, similar to air in the interior of buildings. The negative ion concentration in air is reduced by air flowing through metal ducts, tobacco smoke, static electricity produced by synthetic fibres, and human activities. This condition has been related to discomfort, lassitude, stress, and the loss of mental and physical capacity. Hence, maintaining an adequate number of negative ions in the environment is another factor to consider.
8. The presence of radon gas. Radon (Rn) is a natural gas produced by the decay of radium in the uranium-238 radioactive decay chain. It is present in the Earth's crust and is water soluble, so radon can be found anywhere, although it is most commonly found in granite soils and in those containing uranium ore. Construction materials such as phosphogypsum or blocks manufactured with pieces of granite may also produce radon. The gas penetrates up to a maximum height of one meter via natural diffusion through joints between materials, cracks, or the passage of pipes, and hence, it is necessary to prevent it from concentrating. It is also essential to comply with the safety standards set out by the regulations for fire safety, the detection of carbon monoxide, etc.
9. Electromagnetic pollution. Although numerous studies have been performed in relation to electromagnetic fields (EMFs), it has not yet been possible to prove, with scientific evidence, that the consequences of prolonged exposure to the fields of high-voltage lines, telephone antennae, domestic electrical appliances, and/or workplace equipment are the cause of the fatigue, stress, or depression related to SBS. In all events, as a preventive measure, it seems important to plan and project this essential contemporary infrastructure at the right distances from residences and workplaces and also to have the necessary insulation.
10. Meaningful space with a holistic approach, clear orientation, rational distribution, and maximum safety. These are the compositional, symbolic and formal resources, supplying architectural spaces with meaning, in order to transcend their primary function. This generates comprehensible places where a narrative value, as well as a sense of belonging, can be found by the inhabitant. The connections with the symbolic and highly codified culture are reconstructed by architecture; thus, the habitat, in addition to being a shelter and a relief, is a cognitive appendix for the human being [50]. Also, the standards of design for all or universal design must be met, bearing in mind full accessibility for people with functional diversity: physical, sensory, and cognitive.

5. Discussion of a New Paradigm in the Design and Construction of Buildings

As has been shown previously, interaction with the environment defines what people are capable of and the way they characterise themselves. The physical environment is identified as a definite place, with a specific organisational structure used for precise functions within the social field, where humans live and interact with each other. It is composed of a complex, artificial epidermis that surrounds a person's habitat. This third skin, which completes that of the body itself and its textile covering of clothes, is designed and configured by architecture.

A certain spatial configuration can affect aspects of both physical health and human behaviour, positively or negatively. Perception is not solely limited to visual elements but is also formed with the other senses and cognition. The surroundings' physical and environmental features also influence spatial perceptions. The effect of a space depends on the degree of control, understanding, and sense of coherence the person can experience within it, aspects that memory, culture, training, beliefs, and individual preferences can determine.

Nevertheless, designers do not always pay attention to the potential effect a space can have on people's health, and the solutions are not usually applied to construction today. One of the reasons, as well as the main limitation of this framework, is that there is still a lack of literature and scholars who focus on this topic; its impact has been uneven until very recently. Moreover, another barrier to the implementation of this paradigm is that the few solutions proposed and the scant regulations that exist to promote healthy architecture are focused on physical aspects, and all of them are from a pathogenic point of view. The need for contributions responding to emotional and cognitive environmental influences is inescapable. It is important to establish a systematisation of key patterns from a salutogenic point of view.

As we have seen, the stimuli affecting human sensations function both internally and externally. They accumulate gradually within the body and mind and offer an information feedback system that leads people's basic needs either to be satisfied or not [51]. There are external stimuli, such as sound, smell, light, flavour, and temperature. These are detected by the five classic senses, which have their specific dispositions, and there are also multisensory stimuli, such as air quality, chemical agents, electromagnetic fields, noise, solar radiation, etc. These are detected by more than one sense and can have an effect on the body as a whole. All of these greatly influence a person's degree of comfort, quality of life, and wellbeing and can promote human health. However, they may also cause a loss of physical or cognitive abilities and illnesses. Physical and cognitive stimuli are present in all different forms of environments. If they are positive, it may be desirable to integrate them, and if they are negative, they must be eliminated because they can produce pathogenic agents.

The main contribution of this theoretical work is to establish the basis for the creation of a novel and new healthy architecture epistemology, focussing on cognitive, emotional, and physiological stimuli. Therefore, the final point of the Decalogue is the one to spotlight, as it offers a salutogenic perspective on planning, design, and construction, with a holistic approach taking into account the positive aspects of the environment. Additionally, this paper provides a series of issues, definitions, and data with which to ask new questions, opening a potential path to continue research and filling in the gaps until this healthy architecture doctrine is developed.

6. Conclusions: A Definition of Healthy Architecture

Architecture is the art of creating the best living conditions for humans by building spaces that elicit emotions. "Healthy" is an attribute that expresses the quality of the architecture. Healthy architecture builds environments that improve wellbeing and increase people's physical and cognitive capacities, thus generating assets that reduce risk factors and facilitate, enhance, and promote human health. It is a new paradigm based on five principles: 1. Building with harmless materials, zero emissions, and no environmental footprint. 2. Integrating emerging communication and information technologies safely and ethically. 3. Generating and having clean, efficient, and smart environments. 4. Designing environments that are adaptable and compatible with the development of a diversity of lifestyles. 5. Eliciting emotion with deeply meaningful architecture.

The way to address these matters, which are particularly important for people's health, does not seem to be through expensive technological solutions designed to create or perpetuate artificial environments and climates. It is a question of meeting the challenge with another way of thinking, thus applying a new model. Architecture has effective tools and resources to do this. It is able to use materials that can be recycled, control energy saving in buildings, conserve energy sources, and monitor products' ecological footprints. It is also able to build by adapting to the environment's climatic conditions and exploiting the sun's energy by capturing, accumulating, and controlling radiation processes to achieve natural heating and ventilation.

Building healthy environments with the materials available to architecture requires delving into the stimuli that their spaces generate and designing them in an intelligent and integrated manner. Thus, the third skin that envelops the human habitat is able to generate physical and cognitive stimuli in people. This is both in their memory and in the way they experience and live in a space that enhances their wellbeing, comfort, and quality of life. Delimiting, configuring, organising, and designing an environment and, above all, generating positive sensory experiences are projective actions that determine a space's value and significance [52]. This approach provides the basis of an alternative model to resolve problems and advance the knowledge necessary to have unequivocally healthy cities and architecture.

Healthy architecture involves an attitude in which, during the design and construction processes, the harmful elements that may appear during use, ageing, or demolition are eliminated. The materials and techniques to be used are chosen because of their positive effects on the environment, climate, or ecology. The solutions are adapted to the place, such that the minimum possible power supply is required to ventilate, heat, cool, or light the environment. However, healthy architecture's primary differentiating feature is that it considers the parameters that influence people physically and cognitively. A design oriented towards people's health produces social, economic, and environmental benefits. At the same time, it generates added value in planning cities and produces sustainable contexts and environments that conform better with the current demands of society. Everywhere people live and work—homes; residents' associations; factories; offices; and the city itself—must be a healthy environment.

The inhabitants of a hyper-connected society know perfectly well what is beneficial for them: what foodstuffs they must or must not consume; what activities are best for their physical fitness; and the importance of both their physical and psychological conditions. The same is true of materials, spaces, environments, and cities. Humans have always surrounded themselves with what they consider most beneficial for their health. Looking after the environment and oneself offers significant, long-term savings for users, companies, authorities, and the state, as this prevents certain problems from reaching the hospital setting.

It is more cost-effective to invest in an architectural building design that considers health indicators and parameters, both physical and mental, than to retroactively resolve the harm their absence causes. One must understand, plan, and build the spaces where everyday life takes place to produce a balanced set of stimuli with assets that reduce risk factors and promote people's health. After efficient, green, and sustainable architecture, a new paradigm has emerged that contemporary society demands should be implemented: healthy architecture.

Author Contributions: Conceptualization, S.Q.-G.; methodology, S.Q.-G. and P.V.-F.; validation, S.Q.-G.; formal analysis, M.L.-G.; investigation, S.Q.-G. and P.V.-F.; data curation, P.V.-F.; writing—original draft preparation, S.Q.-G.; writing—review and editing, M.L.-G.; visualization, M.L.-G.; supervision, S.Q.-G.; funding acquisition, S.Q.-G. All authors have read and agreed to the published version of the manuscript.

Funding: Grant PID2020-115790RB-I00 funded by MCIN/AEI/10.13039/501100011033.

Data Availability Statement: Not applicable.

Acknowledgments: This paper was possible thanks to funding obtained competitively from the PID2020-115790RB-I00 grant, funded by MCIN/AEI/10.13039/501100011033.

Conflicts of Interest: The authors declare no conflict of interest.

References

1. Colomina, B. *X-ray Architecture*; Lars Müller Publishers: Zurich, Switzerland, 2019.
2. Quesada-García, S. *Imitatio Naturae. el Paisaje como Referente en la Arquitectura Contemporánea*; University of Seville: Sevilla, Spain, 2006; p. 11; ISBN 978-84-695-7024-1.
3. Gifford, R.; McCunn, L.J. Appraising and Designing Built Environments that Promote Well-Being and Healthy Behaviour. In *Environmental Psychology*, 2nd ed.; Steg, L., Groot, J.L.M., Eds.; Wiley: Hoboken, NJ, USA, 2018; pp. 104–112. [CrossRef]

4. Setola, N.; Naldi, E.; Arnetoli, M.V.; Marzi, L.; Bologna, R. Hospital responses to COVID-19: Evidence from case studies to support future healthcare design research. *Facilities* **2022**, *40*, 131–145. [CrossRef]
5. Von Schirnding, Y. *Health in Sustainable Development Planning: The Role of Indicators*; World Health Organization: Geneva, Switzerland, 2002. Available online: https://apps.who.int/iris/bitstream/handle/10665/67391/WHO_HDE_HID_02.11.pdf (accessed on 17 July 2023).
6. Folkman, S. Stress, Health, and Coping: An Overview. In *The Oxford Handbook of Stress, Health and Coping*; Folksman, S., Ed.; Oxford University Press: Oxford, UK, 2012. [CrossRef]
7. Grant, M.J.; Booth, A. A typology of reviews: An analysis of 14 review types and associated methodologies. *Health Inf. Libr. J.* **2009**, *26*, 91–108. [CrossRef] [PubMed]
8. Benévolo, L. *Storia Dell'Architettura Moderna*; Laterza: Bari, Italy, 1960.
9. National Library of Medicine, Digital Collections. Suivez Ces Conseils, Vous Vivrez Longtemps. Tuberculosis Defense Committee with Rockefeller Fundation. Paris, France. 1920. Available online: https://collections.nlm.nih.gov/catalog/nlm:nlmuid-101452788-img (accessed on 26 June 2023).
10. Sigerist, H. *Primitive and Archaic Medicine*; Oxford University Press: New York, NY, USA, 1951.
11. Neutra, R. *Survival Through Design*; Oxford University Press: Oxford, UK, 1969.
12. Lynch, K. *The Image of the City*; The MIT Press: Cambridge, MA, USA, 1960.
13. McHarg, I.L. *Design with Nature*; John Wiley & Sons: New York, NY, USA, 1969.
14. Eriksson, P.S.; Perfilieva, E.; Björk-Eriksson, T.; Alborn, A.-M.; Nordborg, C.; Peterson, D.A.; Gage, F.H. Neurogenesis in the adult human hippocampus. *Nat. Med.* **1998**, *4*, 1313–1317. [CrossRef] [PubMed]
15. Eberhard, J. *Brain Landscape: The Coexistence of Neuroscience and Architecture*; Oxford University Press: Oxford, UK, 2008.
16. Mallgrave, H.F. *The Architect's Brain: Neuroscience, Creativity, and Architecture*; Wiley-Blackwell: Malden, MA, USA, 2010.
17. Robinson, S.; Pallasmaa, J. *Mind in Architecture: Neuroscience, Embodiment, and the Future of Design*; The MIT Press: Cambridge, UK, 2015.
18. Dance, A. Science and Culture: The brain within buildings. *Proc. Natl. Acad. Sci. USA* **2017**, *114*, 785–787. [CrossRef] [PubMed]
19. Edelstein, E.A. Building health. *Health Environ. Res. Des. J.* **2008**, *1*, 54–59. [CrossRef] [PubMed]
20. Coburn, A.; Vartanian, O.; Chatterjee, A. Buildings, Beauty, and the Brain: A Neuroscience of Architectural Experience. *J. Cogn. Neurosci.* **2017**, *29*, 1521–1531. [CrossRef] [PubMed]
21. Antonovsky, A. *Health, Stress, and Coping*; Jossey-Bass: San Francisco, CA, USA, 1979.
22. Mittelmark, M.B.; Sagy, S.; Eriksson, M.; Bauer, G.F.; Pelikan, J.M.; Lindström, B.; Espnes, G.A. (Eds.) *The Handbook of Salutogenesis*, 1st ed.; Springer: Cham, Switzerland, 2017; pp. 3–40. [CrossRef]
23. Kaplan, R.; Kaplan, S. *The Experience of Nature: A Psychological Perspective*; Cambridge University Press: Cambridge, UK, 1989.
24. Clements-Croome, D. *Designing Buildings for People: Sustainable Liveable Architecture*; The Crowood Press: Marlborough, UK, 2020.
25. Golembiewski, J.A. Salutogenic Architecture in Healthcare Settings. In *The Handbook of Salutogenesis*, 1st ed.; Mittelmark, M.B., Sagy, S., Eriksson, M., Bauer, G.F., Pelikan, J.M., Lindström, B., Espnes, G.A., Eds.; Springer: Cham, Switzerland, 2017; pp. 267–276. [CrossRef]
26. Weisman, G.D.; Calkins, M.; Sloane, P. The environmental context of special care units. *Alzheimer's Dis. Assoc. Disord.* **1994**, *8*, 308–320.
27. Lawton, M.P. *Environment and Aging*; Center for the Study of Aging: New York, NY, USA, 1986.
28. Calkins, M. *Design for Dementia: Planning Environments for the Elderly and the Confused*; National Health Publishing: Owing Mills, MD, USA, 1988.
29. Cohen, U.; Weisman, G.D. *Holding on to Home: Designing Environments for People with Dementia*; Johns Hopkins University Press: Baltimore, MD, USA, 1991.
30. Calkins, M.P. From Research to Application: Supportive and Therapeutic Environments for People Living with Dementia. *Gerontologist* **2018**, *58*, S114–S128. [CrossRef]
31. Golembiewski, J.A. The Designed Environment and How it Affects Brain Morphology and Mental Health. *Health Environ. Res. Des. J.* **2016**, *9*, 161–171. [CrossRef]
32. World Health Organization. Ottawa Charter for Health Promotion. 1986. Available online: https://www.who.int/teams/health-promotion/enhanced-wellbeing/first-global-conference (accessed on 8 June 2020).
33. Quesada-García, S.; Valero-Flores, P. Proyectar Espacios para Habitantes con Alzhéimer, una Visión desde la Arquitectura. *Arte. Individ. y Soc.* **2017**, *29*, 89–108. [CrossRef]
34. Degremont, N. Soins palliatifs et architecture: De l'hôpital vers l'homme. *Eur. J. Palliat. Care* **1998**, *5*, 127–129.
35. Hôpital Cognacq-Jay. Available online: https://www.hopital.cognacq-jay.fr/presentation/votre-hopital (accessed on 4 July 2023).
36. Dominiczak, M.H. Illness and culture: Maggie's Centres. *Clin. Chem.* **2013**, *59*, 333–334. [CrossRef] [PubMed]
37. Maggie's. Available online: https://www.maggies.org/our-centres/search-results/?location=uk (accessed on 3 July 2023).
38. Maggie's Centre Extension Submitted for Planning Permission. Winter 2015 Newsletter of Richard Murphy Architects. Available online: https://www.richardmurphyarchitects.com/Winter-2015 (accessed on 3 July 2023).
39. Centro Kálida Sant Pau/Miralles Tagliabue EMBT. Available online: https://www.archdaily.cl/cl/916841/centro-kalida-sant-pau-miralles-tagliabue-embt?ad_medium=gallery (accessed on 26 June 2023).

40. Quesada-García, S.; Valero-Flores, P. *Arquitectura y Alzhéimer: 12 Nuevos Edificios Para Necesidades Emergentes*; Romero-Vergara, G., Ed.; Universidad de Sevilla: Sevilla, Spain, 2018. Available online: https://idus.us.es/handle/11441/70199 (accessed on 12 September 2020).
41. Quesada-García, S.; Valero-Flores, P.; D'Ospina, A. *Architettura E Alzheimer (II): 16 Nuovi Edifici Per Necessità Emergente*; HAC University books: Sevilla, Spain, 2019. Available online: https://idus.us.es/handle/11441/88994 (accessed on 25 August 2020).
42. Gharib, M.A.; Golembiewski, J.A.; Moustafa, A.A. Mental health and urban design—Zoning in on PTSD. *Curr. Psychol.* **2020**, *39*, 167–173. [CrossRef]
43. Diller Scofidio + Renfro (DS+R). Blur Building, Swiss Expo 2020, Yverdon-Les-Bains, Switzerland. Available online: https://dsrny.com/project/blur-building (accessed on 26 June 2023).
44. Diller Scofidio + Renfro (DS+R). Does the Punishment Fit the Crime? *Fondazione Sanderetto de Rebaudengo, Venide, Italy.* Available online: https://dsrny.com/project/does-the-punishment (accessed on 26 June 2023).
45. Diller Scofidio + Renfro (DS+R). Unspoken. *3rd Istanbul Design Biennial.* Available online: https://dsrny.com/project/unspoken-1 (accessed on 26 June 2023).
46. Décosterd, J.; Rahm, P. *Physiological Architecture*; Birkhäuser: Basel, Switzerland, 2002.
47. Rogers-Stirk-Harbour+Partners. Available online: https://rshp.com/projects/health-and-science/maggies-west-london-centre/ (accessed on 26 June 2023).
48. Proulx, M.J.; Todorov, O.S.; Taylor Aiken, A.; de Sousa, A.A. Where am I? Who am I? The Relation between Spatial Cognition, Social Cognition and Individual Differences in the Built Environment. *Front. Psychol.* **2016**, *7*, 64. [CrossRef] [PubMed]
49. Allen, G.; Bernstein, A.; Cao, X.; Eitland, E.; Flanigan, S.; Gokhale, M.; Goodman, J.; Klager, S.; Klingensmith, L.; Laurent, J. The 9 Foundations of A Healthy Building. Harvard T.H. Chan School of Public Health. 2017. Available online: https://forhealth.org/9_Foundations_of_a_Healthy_Building.February_2017.pdf (accessed on 15 June 2020).
50. Quesada-García, S.; Valero-Flores, P. Architecture. In *The Routledge Companion to Health Humanities*, 1st ed.; Crawford, P., Brown, B., Charise, A., Eds.; Taylor & Francis Group: London, UK, 2020; pp. 335–340.
51. Erskine, R.G. The Therapeutic Relationship: Integrating Motivation and Personality Theories. *Trans. Anal. J.* **1998**, *28*, 132–141. [CrossRef]
52. Quesada-García, S.; Valero-Flores, P.; Lozano-Gómez, M. Active and Assisted Living, a Practice for the Ageing Population and People with Cognitive Disabilities: An Architectural Perspective. *Int. J. Environ. Res. Public Health* **2023**, *20*, 5886. [CrossRef] [PubMed]

buildings

MDPI

Article

The Evolution of Multi-Family Housing Development Standards in the Climate Crisis: A Comparative Analysis of Selected Issues

Agnieszka Starzyk [1], Mikołaj Donderewicz [1], Kinga Rybak-Niedziółka [1], Janusz Marchwiński [2,*], Magdalena Grochulska-Salak [3], Przemysław Łacek [1], Łukasz Mazur [4], Ivanna Voronkova [1] and Polina Vietrova [1]

[1] Institute of Civil Engineering, Warsaw University of Life Sciences—SGGW, Nowoursynowska 159, 02-776 Warsaw, Poland; agnieszka_starzyk@sggw.edu.pl (A.S.); m.donderewicz@gmail.com (M.D.); kinga_rybak@sggw.edu.pl (K.R.-N.); przemyslaw_lacek@sggw.edu.pl (P.Ł.); ivanna_voronkova@sggw.edu.pl (I.V.); polina_vietrova@sggw.edu.pl (P.V.)
[2] Faculty of Architecture, University of Ecology and Management in Warsaw, Olszewska 12, 00-792 Warsaw, Poland
[3] Faculty of Architecture, Warsaw University of Technology, Koszykowa 55, 00-659 Warsaw, Poland; magdalena.salak@pw.edu.pl
[4] Faculty of Engineering, Helena Chodkowska University of Technology and Economics, Jagiellońska 82f, 03-301 Warsaw, Poland; arch.lukaszmazur@gmail.com
* Correspondence: j.marchwinski@wseiz.pl

check for updates

Citation: Starzyk, A.; Donderewicz, M.; Rybak-Niedziółka, K.; Marchwiński, J.; Grochulska-Salak, M.; Łacek, P.; Mazur, Ł.; Voronkova, I.; Vietrova, P. The Evolution of Multi-Family Housing Development Standards in the Climate Crisis: A Comparative Analysis of Selected Issues. *Buildings* **2023**, *13*, 1985. https://doi.org/10.3390/buildings13081985

Academic Editor: Morten Gjerde

Received: 30 June 2023
Revised: 27 July 2023
Accepted: 30 July 2023
Published: 3 August 2023

Abstract: Contemporary problems related to the consequences of climate change and exposure to changing investment and implementation conditions are prompting the development of programmes adapting to climate change. Issues of adaptation and actions in relation to climate change are being discussed in the architectural, urban planning, and governmental communities. Models are being developed for shaping the functional and spatial structure, buildings and infrastructure in the city in relation to the projected climate change. Multi-criteria and interdisciplinary research is being carried out and solutions are being implemented for retaining water, minimising the heat island effect, reducing emissions and environmental impact by analysing the carbon footprint and introducing circular economy principles. The research is focused on the analysis of design and implementation conditions for multi-family housing projects in Poland, and the development of design guidelines enabling adaptation and mitigation of the negative effects of climate change, including heat island effects, smog, overheating, drought, and flooding in housing. Conclusions from the overview of the indicated documents and legal provisions for the implementation of sustainable development principles and adaptation to climate change in the investments under preparation (urban and architectural projects) enable the forecasting of development directions and ideological assumptions for shaping urbanised areas, providing the basis for shaping the resilience of the functional and spatial structure and the natural system in urban areas subject to transformation. Issues of implementing pro-environmental technologies and developing new urban planning standards disseminate the solutions of compact cities in which the development of multifunctional building complexes with public spaces equipped with greenery linked to the buildings are realised.

Keywords: housing standards; climate crisis; humanitarian crisis

1. Introduction

The dream of creating an ideal space for life has persisted throughout the history of social development. This dream changed under the influence of many events. Wars, epidemics, the development of technologies, and changes in economic conditions contributed to the changes in the perception of human needs and patterns of housing, work, and recreation [1]. One of the main events affecting the quality of life and its standards today is the climate crisis.

The ideological basis of modern sustainable urban design dates back to the 1970s. In 1975, Richard Register founded Urban Ecology, whose mission was, and still remains, the reconstruction of modern cities towards a natural balance. In 1987, Eco City was defined as "an urban environmental system in which inputs/resources and production/waste are organized". In 1990, Urban Ecology organised the first international Eco City conference in Berkeley. A set of criteria which a model Eco City should correspond to was developed at the end of the 20th century. It should operate on the basis of a self-sufficient economy, use local resources, and have an entirely carbon-neutral energy production from renewable sources. Therefore, the Eco City is a modern sustainable city that takes up the challenge of regeneratively creating urban space while respecting environmental, social, spatial, and economic conditions.

The shaping of urbanised areas in Europe has been influenced by the findings of the urban planning principles adopted by the European Council of Town Planners [2] and the 'Leipzig Charter' on Sustainable European Cities [3]. Directions for the development of European cities were also indicated in the EU Strategy "Europe 2020—A strategy for smart, sustainable and inclusive growth" (COM(2010) 758) and in the EU Climate Change Adaptation Strategy of 16 April 2013 (COM(2013) 216 final). In conjunction with the indicated development strategies and activities of the circles responsible for shaping urbanised areas and natural systems, further guidelines are being introduced for the shaping of cities in Poland.

The Polish Ministry of Infrastructure and Development has published a study entitled National Urban Policy 2023 [4] (the document was adopted by the resolution of the Council of Ministers on 20 October 2015). Problem areas and guidelines are indicated for the interdisciplinary and multi-criteria transformation of cities and urban infrastructure. The issue of shaping guidelines for investment areas in the city is directly linked to the transformation of degraded areas and revitalisation issues. The Law on Revitalisation [5] has been passed.

Sustainable development is one of the three priorities of the EU's Europe 2020 strategy. In this document, the Member States have committed themselves to implementing and achieving sustainable development at all levels of governance—national, regional, and local, declaring actions in three development zones—social, economic, and environmental, for a lasting improvement in the quality of life of the local communities. According to the Climate Neutrality Implementation Strategy, the building sector is scheduled to be completely decarbonised by 2050.

In order to achieve the decarbonising goal in the construction industry, it is necessary not only to accelerate the rate of retrofitting existing stock from 1% to 3% per year, but also ensure that the buildings constructed today are nearly zero energy. The EU's guidelines in Developing the Strategy: Climate Neutrality 2050, a long-term strategy—a vision for a prosperous, modern, competitive and climate-neutral EU economy provide indications for action on economic activities.

In implementing climate neutrality assumptions, it is advisable to prepare investments in the model of sustainable construction, defined by the Polish Green Building Council (PLGBC) as a sustainable building which is economical, comfortable and created with respect for the natural environment. By designing, constructing and using green buildings, we are simultaneously meeting our current needs and ensuring that future generations will also be able to meet their needs. Methods of conserving natural resources and caring for the environment are considered at all life stages of such a building.

CO_2 emissions from the building sector are not just due to the use of buildings, but also include an embedded carbon footprint, i.e., emissions associated with the entire life cycle of a building and its components—from production, through transport, construction, use, demolition and finally, to disposal or recycling. An action to slow climate change, which the building sector has significant influence upon, is the creation of sustainable buildings that meet energy efficiency standards, are comfortable and healthy for their residents, and respect the environment. Methods of preserving natural resources and caring for the environment are considered at all stages of a building's life. Sustainable design is characterised by high energy efficiency, responsible water management, and high quality of the indoor environment measured by acoustic and thermal comfort levels, access to daylight, ventilation and low levels of air pollutants.

Climate itself plays a minor role in how we design our modern homes, as any space can be artificially altered to achieve indoor comfort [6]. However, this is the biggest problem for ecology. The average temperature on Earth has risen sharply, which leads to global temperature anomalies. Carbon emissions, deforestation, and changes in land use contribute to this issue. The construction industry is considered one of the largest consumers of electricity, as well as a producer of waste that has a harmful effect on the environment [6–14].

Today, the European and global situation in terms of climate and environmental challenges requires special attention with regard to reducing energy consumption and increasing the energy efficiency of buildings. Carbon emissions can be reduced through the widespread use of more energy measures and clean technologies, such as highly insulating building envelopes, heat pumps, photovoltaics, centralised energy, etc. [15,16]. It is possible to increase energy efficiency in the residential sector by using renewable sources instead of traditional energy, as well as by carrying out energy modernization of old buildings. This should not only reduce energy consumption and greenhouse gas emissions, but also allow these buildings to adapt to new regulatory requirements and standards [6].

Architecture in relation to the applicable norms and standards is the art of creating order in the environment. It was assumed that it is a discipline that organises and shapes space in real forms necessary to satisfy human material and spiritual needs. It is, therefore, a reflection of social and economic reality, a specific stage of its forms of development [17]. Contemporary architecture should meet the tasks resulting from both complex forms of life organization, and rapid social and economic changes; it must not only correspond to the conditions of the present times, but also, taking into account social, economic, technical and scientific changes, anticipate as far as possible the satisfaction of different needs in the future [18,19].

Housing is one of the most desirable forms of ownership and one of the most common forms of investment. Therefore, in the climate crisis era, it is necessary to focus considerable attention on improving the quality of housing operation with minimal damage to the environment. For example, green construction is being developed in China to ease the burden of energy consumption of traditional buildings for environmental resources and social development. A green building can save resources (energy, land, water and material saving) throughout its life cycle, protect the environment, reduce pollution, provide people with health, fitness and efficient living space, and coexist in harmony with nature [8]. In Saudi Arabia, alternative energy sources are being actively implemented in the household sector, i.e., photovoltaic systems that work on solar energy [10]. A large amount of energy can be saved with a correct and adequate design of the living space. It is possible to minimise dependence on means for heating and cooling premises if heat and air in buildings are directed naturally [20]. An important role in the construction of buildings is played by the details of architectural structures, especially those that support the continuity of thermal insulation. This is why passive house construction technology is gaining popularity today, not only providing a high level of comfort with very low energy consumption, but being also ecological and safe for the environment [21–23].

Currently, the understanding of architecture includes a number of pro-environmental and pro-climate factors [24], which are taken into account in the design of structures and equipping buildings with technical infrastructure, which is emphasised by global and EU directives. In 2020, the European Commission, as part of the European Green Deal, launched the "Renewal Wave for Europe", the goal of which was to double the annual energy renovation rates of residential and non-residential buildings by 2030, and with deep energy renovation [25]. In December 2021, the European Commission proposed a revision of the Energy Performance of Buildings Directive (EPBD) as part of the "Fit for 55" package to achieve a minimum 55% reduction in greenhouse gas (GHG) emissions in the EU by 2030 [26].

In Poland, housing policy is conditioned by many local regulations, largely taking into account global and European guidelines [27]. In terms of shaping the housing estates based on the cooperation of architects and developers, there are many factors and trends indicating the positive aspects of such cooperation, expressed in successful and socially recognised construction projects [28,29].

The development industry, or rather construction developers are investors in the private or public sector who invest in the construction of real estate, including residential houses. The whole process, the origin of which oscillates primarily in financial and investment initiatives, aims at the creation of construction utility that, after a wide assessment of possibilities and prospects, will finally reach its recipient. The range of interests and investment opportunities is quite large and the market offers many investment opportunities and variants [30,31].

The aim of this study was to analyse the issues shaping residential architecture based on various standards and demands in the context of the climate crisis. Good features occurring in multi-family housing architecture are emphasised here, and positive examples of functional and spatial solutions are indicated. The study seeks to define various contemporary ways of designing residential architecture, as well as: (1) specify the social, locational, economic, spatial, environmental, cultural and technological conditions reflected in investments and the demand in the context of climate crisis; and (2) characterise the developers' standards, taking into account factors affecting the attractiveness of cultural, environmental, infrastructural, organizational and functional aspects.

2. Material and Methods

During the study, the following investments and investment considerations taking into account the typology of climate change were analysed: Accumulation of heat energy heating and overheating—heat island effect; accumulation of precipitation water through localised heavy rain and snowmelt; drought and periodic limited access to groundwater and lack of rainfall; significant daily temperature amplitudes; accumulation of violent and intense wind flows and changes in the atmospheric pressure system; accumulation of air, water, and soil pollution; accumulation of pathogens, organic pollutants and bacteria; accumulation of stress factors influencing changes in the resilience of urban ecosystems [32,33].

The analysis was based, e.g., on documents providing guidelines for sustainable design and model methodologies described as the Polish Green Building Council's (PLGBC) Green House (pl. *Zielony Dom*) Certification Criteria and the Criteria for Assessing Architectural Executions for Climate Responsible Solutions. Within the Green House certification framework, the analysis and recommended solutions concerned the areas according to the following criteria: Management of the building project; site and location; materials and resources; water management; user health and comfort; and energy optimisation. Documents relevant to the design of contemporary multi-family housing were analysed, taking into account the principles of sustainable development and climate change adaptation, including but not limited to: UN Sustainable Development Goals; European Green Deal Strategy; Communication from the Commission to the European Parliament; Polish legislation; Sustainable Energy Action Plan for Warsaw [34]; Green House Certification Criteria of the Polish Green Building Council (PLGBC); and programmes and standards of the City of Warsaw, including: Environmental Protection Programme and Warsaw Housing Standard.

The presented analysis used two research methods: (1) critical literature overview, and (2) the authors' professional experience of design and implementation work between architects and developers in the multi-family building sector. The specified aspects are based on the issues that arise during the entire process of creating developer architecture objects [35]. The literature overview considered scientific publications that relate to housing standards and the climate crisis at the same time. The extraction of these publications allowed the results to be formulated with a breakdown of factors: (i) economic, which directly affect design solutions [36,37], (ii) environmental, and (iii) educational issues related to the impact of architecture on the climate crisis [12,38]. As the research was based on examples of good communication between architects and developers, the most positive elements of this cooperation were identified, and their effects used in the designed buildings were specified and described [39,40]. The summary of the results focuses on detailing exemplary aspects related to the social and environmental attractiveness of multi-family architecture in the context of architect-developer collaboration in times of global climate crisis [41].

Specifying the methodological subdivision into construction developers and architects (Figure 1) is justified primarily by a different way of treating individual problems and solutions. By definition, investors will be interested in reaching customers, meeting their needs and requirements from an economic perspective. Architects, on the other hand, will be interested in the general well-being and balance of the designed solutions, which should meet a number of formal and legal requirements, in accordance with the art of construction, with respect for the environment and the satisfaction of future recipients. The different views of these two groups will have conflicts on many levels, but will also be convergent on many issues. The aim of the article is, among others, to highlight only the positive aspects of this cooperation, as well as the separate and individual thoughts, ideas, proposals and practices. The synergy between developers and architects turns into creativity saturated with a multidimensional point of view, and the fruits of this creativity are presented in the form of selected examples that were analysed and drawn on this basis, as well as a summary and list of the best proposals for the contemporary way of designing multi-family buildings. Not only was criticism of the literature review used, but also personal experience resulting from the cooperation of the authors (architects) with development centres. In addition, the professional experience of the authors can be subdivided into scientific and design practice in this area. The scientific work of the authors, oscillating, e.g., in the subject of multi-family buildings, focuses primarily and briefly on: aspects of pro-environmental design, energy-efficient design, sociology and the relationship between architecture and environmental, eco-tech trends. The design work of the authors in the context of this study concerns, e.g., finished conceptual, construction and executive designs of various multi-family buildings: newly designed, modernised, and historic, including historic tenement houses. The considerations and observations resulting from this were taken into account when creating the results and summaries, and based on them, the final conclusions.

2.1. Social Factors

Currently, the most active target group on the market is the 30–40 year-olds (Figure 2a). This requires the creation of new housing standards and new forms of marketing that are able to reach customers. The above statistics indicate that the main target group includes relatively young people, for whom, for worldview reasons, very important factors related to a limited impact on the environment will be of crucial significance when buying a home. Energy-efficient and environmentally friendly solutions will be both a marketing attraction and something very positive in the understanding of the modern philosophy of life, with which many people are increasingly identifying themselves. A factor influencing the change in trends among projects is a new generation that requires the use of innovation. The dominant generation is made up of people (not only in Poland) who have a high level of knowledge about real estate and environmental protection. They are characterised by a high awareness of their needs and insight into the market. In addition, the most important thing for buyers will be the type of the investment, its main theme, style of architecture, additional benefits, landscape formation, as well as greenery solutions [42–46].

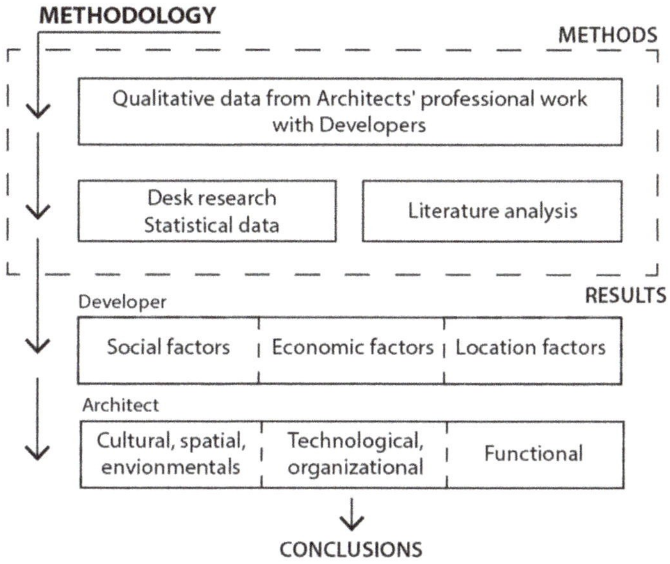

Figure 1. Diagram of the methodology of study.

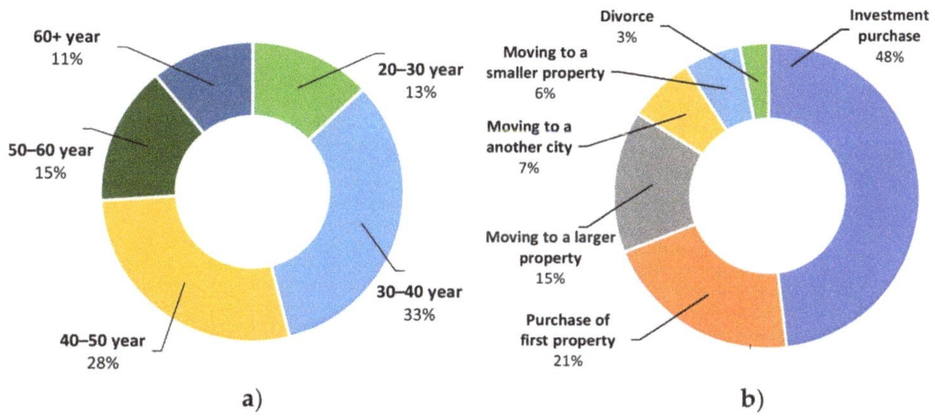

Figure 2. Statistics on the real estate market in Poland in Q4 2022: (**a**) age of housing buyers, (**b**) reasons for buying a property (Profil Kupujących).

The cited report presents information on the age of the purchasers and their derivatives regarding profession, as well as the reasons for concluding the transactions, the preferred area and the value of purchases. Among other things, the study determines the so-called buyer profile, which is illustrated in the graphs. On the basis of these graphs and the information contained in them (Figure 2a,b), it is possible to distinguish the age characteristics of potential recipients of developer apartments. According to the report, the largest group of homebuyers are between 30 to 40 years old, accounting for 33% of the total market. This is followed by the age group between 40 to 50, with 28% of the total, followed by the 50 to 60 age group, with an estimated 15% of the total. Finally, with an equal share of 13%, the age groups from 20 to 30 and those over 60 were defined with a share of 11%. Therefore, purchasers of apartments are primarily people who are on the verge of life changes by choosing new places of residence, enlarging their families with their first or next child, which increases the need to buy a completely new or larger apartment [45].

2.2. Economic Factors

The economic aspects of buying a home are one of the main factors considered by potential buyers (Table 1). The conducted analyses, in addition to indicating the fact that the main participants on the market are young people aged 30+, show that the dominant profession is a full-time employee, with a 44% share in apartment purchases. Entrepreneurs are classified with a much lower share—25%, followed by managers—15%, pensioners—10%, top managers—3%, and finally students, with a share of 3%. The profession can undoubtedly affect the reasons for investing in housing. Further, 48% of people declare with a strictly investment purpose, and 21% with the purchase of their first property. Moving to a new, larger apartment concerns 15% of the respondents, change of urban residence with a 7% share, while moving to a smaller apartment with 6%. At the very end, there is also a 3% share because of divorces. Last but not least, data on the price and area of flats were indicated, which reflect the most frequently chosen properties among particular age groups, professional groups, and by virtue of a purchase decision. The average area of apartments oscillates around 51 square metres, but the apartment area for each of the professional groups separately is also worth quoting in this case. Pensioners choose an average area of 59 square metres, managers—55 square metres, full-time employees—50, and entrepreneurs—51. The largest apartments are chosen by top managers—61 square metres, while the smallest by students—on average 45 square metres [45].

Table 1. Average price and area of property in Poland in Q4 2022 (Profil Kupujących).

Age	Price	Area [m^2]	Price per m^2
20–30	PLN 352,000	46	PLN 7652
30–40	PLN 376,000	54	PLN 6963
40–50	PLN 408,000	53	PLN 7698
50–60	PLN 399,000	46	PLN 8674
60+	PLN 496,000	58	PLN 8552
Profession	**Price**	**Area [m^2]**	**Price per m^2**
Full-time employee	PLN 366,000	50	PLN 7320
Entrepreneur	PLN 374,000	51	PLN 7333
Manager	PLN 450,000	55	PLN 8182
Pensioner	PLN 498,000	59	PLN 8441
Top Manager	PLN 525,000	61	PLN 8607
Student	PLN 370,000	45	PLN 8222
Reason of buying the property	**Price**	**Area [m^2]**	**Price per m^2**
Investment purchase	PLN 376,000	45	PLN 8356
Purchase of first property	PLN 380,000	49	PLN 7755
Moving to another city	PLN 397,000	52	PLN 7635
Moving to a larger property	PLN 470,000	69	PLN 6812
Moving to a smaller property	PLN 473,000	68	PLN 6956
Divorce	PLN 378,000	56	PLN 6750

2.3. Location Factors

One of the indicators helping to assess the potential of a given area is the Migration Balance Index. It is a determinant of places to which the flow of population is at the highest level, which may mean that investing in these locations is worthwhile. Such a study has been prepared by the GUS (Central Statistical Office in Poland), in which data on the migration of residents at the level of the whole country, voivodships, communes, and even at the level of larger cities can be acquired. The data indicate where people move out and move in [47].

The administrative subdivision of Poland according to the county measure is presented (Figure 3a). There are 308 counties and 65 cities with county laws. Indications of net migration should be understood by the unit of a migrant person in relation to the thousandth population of a given county. (Figure 3b) shows the administrative subdivision of the Mazowieckie (Masovian) voivodship according to the measure of communes. In this voivodship, there are 42 counties and 5 cities with county laws, which are divided into 314 communes. Indications illustrating the net migration should be understood as the unit of a migrant person in relation to the thousandth population of a given commune.

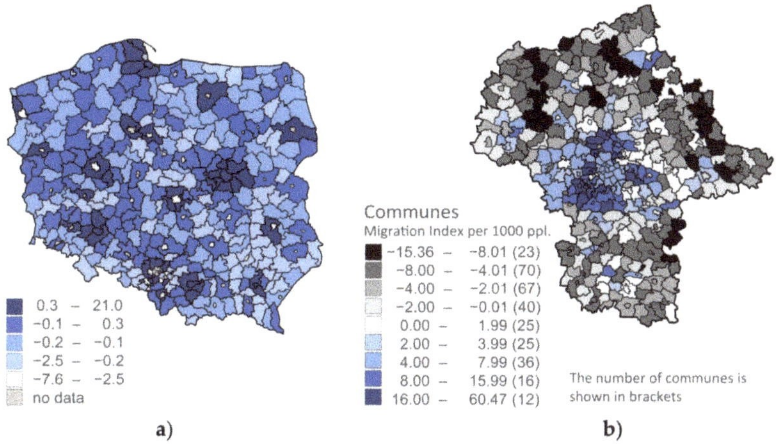

Figure 3. (**a**) Migration Balance Index in Poland for 2021, based on data available from the GUS; (**b**) Map of migration for permanent residence in the Mazowieckie (Masovian) voivodship for 2019, based on data available from the GUS [47].

3. Results

At this point, the current state of the construction market in Poland is worth presenting, by showing what problems it is struggling with and how this issue translates into the sale and construction of apartments (Figure 4). The data show that there is a certain crisis in the real estate development industry, which may affect other sectors of the economy. However, despite this, the developing Polish construction sector is still causing an oversupply of housing. Based on the GUS data, the number of households in Poland in 2021 was 14.811 million, while the number of completed apartments was approximately 15.616 million, which resulted in approximately 0.806 million excess apartments than needed for households. However, these figures do not include refugees, tourists, students, and immigrants. An attempt to estimate the exact sum of these groups is difficult, but may lead to the conclusion that the oversupply is not that large. Expert forecasts from the EKF (European Financial Congress) state that in 2023, Poland's economic growth may slow down to about 0.5–0.8%, compared to about 4.6–5.1% in 2022. In their opinion, the factor contributing to the downward trend is the housing market, which has been in poor condition since the pandemic in 2020. Developers are significantly limiting the number of new investments, which means that the number of flats available for sale this year will decrease. The reasons for the lower number of construction sites include: poor availability of bank loans; collapse in the housing demand; rising costs of materials and services; negative economic balance; and division of the investment into smaller stages. In 2022, developers built nearly 30% less apartments than in the same period of 2021. This will result in a decrease in the number of finished units available for sale in 2023 and 2024. The development market is a very important segment of the economy, related to many other industries. A decline in this sector will affect the entire economy and the labour market. According to a report by PKO BP analysts, the average transaction prices of apartments in the first half of 2023 will be lower than in the first half of 2022, and the reasons for these downward trends will be primarily: a decrease in credit demand for apartments due to high interest rates; factual decline in earnings of the population; oversupply of flats in a situation of a large number of projects launched before the pandemic period; and an increase in the expected rates of return by the investors. According to experts, a certain brake on the decline in housing prices will be: the still high rental rates; slow recovery of lending; and gradual demand among the refugees from Ukraine [47–50].

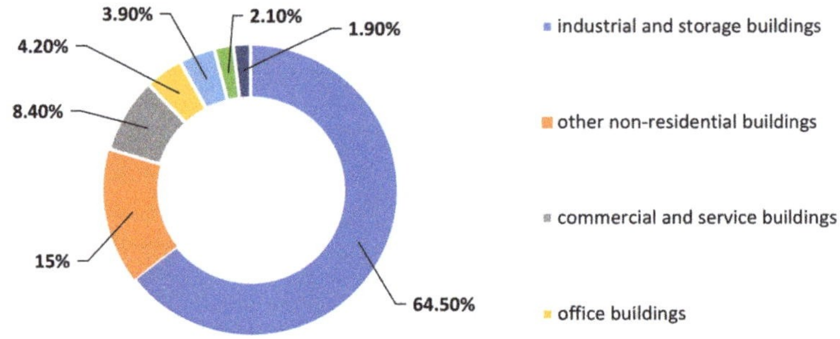

Figure 4. Structure of usable area of non-residential buildings commissioned in Q1 2023. Data available from GUS.

In the first quarter of 2023, 5.4 thousand new non-residential buildings were commissioned and 595 were extended (by 4.5% and 1.8% less than in the previous year, respectively). The total usable area of new and extended non-residential buildings amounted to 4.5 million square metres, 6.1% more than in the corresponding period of 2022. Industrial and storage buildings (64.5%) prevailed in the structure of completed space. Other non-residential buildings (15.0%), and commercial and service buildings (8.4%) also had significant shares. An increase (year/year) in the area put into use was recorded for: industrial and storage buildings (by 23.4%), hotels and tourist accommodation buildings (by 8.9%), and transport and communication buildings (by 0.2%) [51].

In the first three months of 2023, the construction of 38.6 thousand apartments began, i.e., 14.7 thousand apartments (27.6%) less than in the previous year. Dwellings built for sale or rent accounted for 60.7% of the total, and dwellings in individual construction—38.0%. The remaining dwellings were recorded in the cooperative, communal, social rental, and company forms of construction. In the discussed period, building permits were issued or notifications were made with a construction project for the construction of 52.1 thousand flats, i.e., 25.9 thousand flats (33.2%) less than in the previous year. The average projected area of a flat was 87.9 square metres, which meant a decrease by 2.2 square metres compared to the first quarter of 2022. The planned average area of flats in single-family buildings was 130.0 square metres, and in multi-family buildings—52.9 square metres (Table 2). Taking into account the structure of the number of dwellings for which building permits have been issued or a construction project has been submitted, by type of construction, the largest shares were recorded for constructions for sale or rent (66.6%), and individual constructions (30.0%). The remaining flats will be implemented in the cooperative, communal, social rental, and company form of construction.

Table 2. New residential buildings with building permits issued or with a construction project submitted in Q1 2023. Data available from GUS.

Itemization	Building Permits and Notifications of Buildings	Buildings	Flats	Usable Area in Total [m^2]	Usable Area of Flat [m^2]
Total	16,831	21,294	50,820	4,465,847	87.9
Single-family buildings	16,432	20,604	23,044	2,995,450	130.0
Multi-family buildings	399	690	27,776	1,470,397	52.9

3.1. Ways of Influencing the Design of Multi-Family Buildings in Poland by Developers

3.1.1. Social Factors

The social context may be included, e.g., in the so-called lifestyle, which determines that the attractiveness of the apartment is also created by the chosen philosophy of life and emotions, which become an easy identifier. It is possible to distinguish among developers the tendency to categorise a given investment thematically so as to reach a specific target group, taking into account additional design solutions [46,52].

An example of such a project is the Centrum 50+ housing estate in Gliwice, Poland, dedicated mainly to seniors. The complex of facilities is designed for the elderly and for people with disabilities, with buildings adapted to their unique needs, equipped with new technologies, services, catering, and administrative premises, as well as an outdoor gym [53]. Another alternative to thematic marketing is the concept of a housing estate that gathers mainly a religious community. The project called Housing Estate Only for Catholics was supposed to be the first such project in Poland. It includes the creation of a "Family estate" intended for a specific social group, together with the graphic identification and PR strategies [54].

The graphic identification of the building, its interiors and semi-public spaces is another distinguishing feature of modern housing estates. Developers decide to meet the requirements of potential buyers, which is why they try to make their investments stand out with individuality and ideas. On the example of the modernization of a modernist tenement house at Lubomira 6 Str. in Warsaw, the developer decided to give the investment a cinematic theme. This is reflected in the unique designs of the corridor floors consisting of a monochromatic film-strip mosaic. In addition, the corridors were decorated with graphics and floor indicators in the form of "film claps". The dominant feature and the most characteristic identifier is a large-scale mural depicting a film camera with a reel, painted on one of the blind walls. In the contemporary development point of view, there are also various types of identifying motifs, repeating colour palettes, logotype designs, and unique solutions identifying a given place, which may be easier for future residents to assimilate with.

3.1.2. Economic Factors

The area of flats purchased will change for those who need a flat for strictly personal or investment purposes, depending on the budget they have, the period during which they will use the property, and the profession or social status to which they qualify. The issues of development and design economics should harmonise with the initial assumptions, the possible theme of the investment, its status, or target social groups. These findings will also result in such information as the level of the required finish standard, and a multitude of alternative technological, material, functional, and spatial applications [55].

It can be assumed that the subject of potential housing development is also able to be determined around another completely arbitrary category of the purpose of people—only young residents or only with children, multi-person families, or around those who are looking for their investments in the real estate market for the purpose of their rental (short or long-term), taking as examples the target groups of students, tourists, or workers. It is more often obligatory to give a clear direction to the shape of a given project thematically in terms of marketing and planning, even in terms of economics and the finishing standard of a given property [56].

According to Kuba Karliński, an experienced investor and expert in multiplying money on real estate, various potential investment categories are distinguished, and the main and most common are flats for rent, quick sale of flats, commercial premises, tenement houses, development projects, and land investments [57]. One of the most frequently chosen forms of investment is residential property, which is an attractive form of investing capital for as many as 75% of Poles [57–59].

3.1.3. Location Factors

The location of the property is one of the key factors distinguishing features that allow matching the apartment to a specific group. Employees of large companies will be looking for apartments in guarded housing estates located in well-connected areas that allow quick access to work or business meetings. Social groups such as students will be interested in the proximity of the apartment to university centres or libraries. For tourists or commuters from outside the city, as well as refugees (in the context of sudden geopolitical change), locations with well-planned and fast communication, railway stations or public transport hubs will be desirable. An important location factor is also the relationship of a given place with the tourism industry, which may determine the investment value in the context of seasonal rentals [57,60].

The phenomenon of urban sprawl, which is associated with the economic and social development of countries or cities [61], affects the changes in the popularity of housing estates near workplaces or factories in the suburbs. It is explained that with the development of public transport, people no longer have to live right next to factories, and they can get to work at an acceptable time from a distance of several dozen kilometres [57]. According to migration data in Poland compiled by GUS, since the early 1990s, most people move towards large cities. A phenomenon on the migration map of Poland is the vicinity of Warsaw, which is the only metropolis in Poland that still has an increasing population, which is continuously spreading out; and according to demographic forecasts, this trend should continue for a long time [47]. Location aspects that will affect investment decisions and design solutions include:

- Transport infrastructure and distance to workplaces or the city centre;
- Landscape values, environmental cleanliness, noise level, climate security;
- Presence and type of neighbouring buildings;
- Availability of recreational, entertainment, commercial, service, educational, or health facilities.

3.2. *Ways of Creating Attractive Residential Architecture*

Cultural, Spatial, and Environmental Factors

A more diversified program offer with solutions that in reality will not only be a form of marketing is worth proposing (Table 3). In the cultural sphere, it seems a good idea to organise social workshops, educational activities, all kinds of creative meetings, or other forms integrating the local community. Integration of the local community is a phenomenon that is lacking nowadays and its growth is worth investing in by generating creative thought. The answer will be planning solutions for the designed common space, which the residents will be happy to use, relax, and integrate in [62].

Table 3. Cultural, spatial, and environmental factors—a list of issues in the context of ways of creating attractive residential architecture. Based on a critical literature review, knowledge and design practice of the authors.

Space
▪ Organization of shared private or semi-public space in a cultural context; ▪ Proposals for recreation, sport, learning, integration, and rest; ▪ Adaptations of design solutions for specific social groups;
Greenery
▪ Prevention against excessive violation of the environment in the investment process; ▪ Using the potential of existing nature for design purposes; ▪ Planting trees and shrubs, and supplementing the natural environment; ▪ The use of system solutions for green roofs; ▪ Covering vertical elements of buildings with vines, ivy, moss, or algae; ▪ General intensification of the biologically active surface.
Water
▪ Creating rain gardens in the ground and in containers; ▪ Organizing basins and retention ponds; ▪ Unsealing the surface that does not allow for natural vegetation; ▪ Use of hardened structural and permeable substrates; ▪ Creation of absorbent wells; ▪ Use of open rainwater drainage systems; ▪ Use of rainwater retention systems for irrigation purposes.
Sun
▪ Using passive solutions for sun protection; ▪ Utilizing the energy potential of solar energy.

An example of an ideological development solution is the "Art City" housing estate in Kraków, Poland, where the leading theme of the investment is culture, and the motto is "to live with culture". The investor decided to implement the idea of a city of art by organizing a series of artistic workshops for children, exhibitions, culture clubs, and film screenings. These activities focus on the integration and contact of various social groups. As a result, information about the investment reaches people who are interested in buying an apartment, but who are also sensitive to art and want to communicate with it every day [63].

Another valuable aspect in organizing space is the environment. An example of a single-family housing estate is in Nowa Wola, Poland, where during the autumn season colourful bird booths were hung on the wall surrounding the construction site, through which the visualisation of the object could be seen by looking inside the hollow. This composition was a preview of the educational campaign "Novisa Garden—Patron of Nature", under which over half a thousand students from local primary schools and kindergartens took part in free lectures about nature, ornithology, and measures of protecting local fauna and flora. The choice of this type of campaign initiated a dialogue about local nature [46].

The basic issue in the context of nature protection is to prevent the destruction of the existing greenery. An area rich in natural vegetation should be interfered with as little as possible. It is advisable to revitalise it, and to create the main theme of the estate project from the existing landscape value. It is important to create a so-called ecological minimum plan. The assumptions of such a plan may include the use of horizontal and vertical surfaces of buildings, to cover them with vegetation. A good idea to protect buildings from overheating is to use all kinds of vertical and horizontal green surfaces, such as façades covered with ivy or roofs with a biologically active surface, also performing a water retention function. This is a particularly attractive and desirable proposal for the top floors of the building due to the prevention of the negative effects of droughts [64].

Currently, solutions using existing greenery or terrain conditions are preferred to organise rainwater retention systems. These include the process of planting trees and shrubs, the creation of green areas, the use of green walls and roofs, the creation of rain gardens in the ground or in a container, the creation of basins and retention ponds, the unsealing of surfaces and ground, the use of permeable paved surfaces, and the use of absorbent manholes and open rainwater sewage systems [65].

3.3. Technological and Organizational Factors

Technology, in the sense of solutions used in constructions, is an extremely extensive and dynamically changing issue. However, it is possible to distinguish individual fields or technological issues that have a strong relationship with multi-family housing (Table 4). These solutions can be attractive for customers looking for new alternatives available on the housing market. A certain technological aspect is the possibility of using the potential of virtual reality for a better, more intuitive presentation of the offered architecture, which may contribute to the development of the possibility of personalizing architecture for its users [66].

In the future, thanks to virtual reality technology, it will be possible to change the appearance of the apartment interior according to a personalised idea. This is a definite step towards a completely new quality of modern architecture presentation, as well as streamlining the stage of planning, execution, renovation, and thus, perhaps even reducing material expenditures, which may have a positive impact on the environment [67]. VR is a technology worth investing in for reasons including: faster return on investment by decision-making time reduction by up to 30%; faster verification; better communication with the client; replacement of 2D drawings (paper savings) with an interactive spatial model; increasing competitiveness through visually impressive product presentations; greater efficiency; ease and speed of introducing possible changes in the project; possibility of using mobile devices such as tablets, smartphones, or even special goggles. Architects are also increasingly choosing a BIM (Building Information Modeling) software system, which has a positive effect on understanding and shortening the time of introducing certain arrangements, especially using platforms based on the Internet Cloud System [68]. By working on a single model, companies involved in the design, construction, and management of a building can significantly increase their efficiency and reduce errors throughout the documentation process. Digital design data combined with innovative parametric information modeling technology gives significant advantages over traditional design and building methods [69].

The willingness to use certification in the construction industry is being demonstrated more and more often. Ecological solutions, including buildings with environmental certificates, are just entering the market and usually concern investments carried out at a higher standard, in which apartments are offered at a higher price. In such projects, emphasis is placed on the consumption of thermal energy. Gravitational ventilation in apartments is considered ineffective, so it should be replaced with a mechanical system with pressure diffusers. Thanks to this, air exchange is more suited to the real needs of interior ventilation, and reduces heat loss. The use of energy-saving LED lighting in common areas is also distinguished. Other solutions used include: solar collectors, photovoltaic panels, ground heat exchangers, filters and anti-smog vegetation, rainwater recovery and irrigation systems, environmentally friendly materials, hygienic certificates, heat recuperators, highly efficient heaters, or quality window frames. The certificates used in constructions include, e.g., BREEAM, "Wymogi Narodowego Funduszu Ochrony Środowiska i Gospodarki Wodnej" (Requirements of the National Fund for Environmental Protection and Water Management), ISO 14001 [70–73].

Table 4. Technological and organizational factors—a list of issues in the context of ways of creating attractive residential architecture. Based on a critical literature review, knowledge and design practice of the authors.

Computer Technology	
▪ Use of VR technology for better presentation, creation of concepts, personalization of projects; ▪ Use of BIM technology to multiply organizational possibilities, reduce costs, time, and problems in the investment process.	
Pro-environmental certification	
▪ BREEAM; LEED, GREEN BUILDING (PLGBC); ▪ Requirements of the "NFOŚiGW" (National Fund for Environmental Protection and Water Management); ▪ ISO 14001.	
System technologies	
▪ High-quality window and door elements with a low heat transfer coefficient and a high level of acoustic insulation; ▪ High-quality building thermal insulation materials; ▪ Filters and anti-smog plantings; ▪ Environmentally friendly building materials with hygienic certificates; ▪ Organization of small fauna and flora ecosystems; ▪ Infiltration systems; ▪ Rainwater retention systems for land irrigation; ▪ Charging stations for electric cars; ▪ High-quality heating systems; ▪ Mechanical ventilation with a recuperation system; ▪ Pressure diffusers; ▪ Automatised LED lighting systems; ▪ Photovoltaic solutions; ▪ Installations of ground heat exchangers; ▪ Gas installations—as a still attractive and efficient variant.	
Passive architectural sun protection solutions	
Location of the building	Appropriate to the existing buildings, urban tissue, cardinal points, shape and surface of the plot.
Architectural form	Location of the building as appropriate in relation to the existing buildings, urban and law factors, orientation of the cardinal directions, shape and surface of the plot.
Interior spaces	Making the best use of the conditions for lighting interiors with daylight, using a number of illuminating elements and spaces such as atriums, passages, galleries, curtain walls with large glazed surfaces.
Functional arrangement	Appropriate for recognizing the thermal needs of the functional division of the building, in accordance with their intended use, specifying day or night zones.
Façades	Taking into account the energy demand so as to minimise losses and maximise the acquisition of thermal energy from the sun.
Passive material sun protection solutions	
Glazing elements	Tinted glass, glass with reflective solar control coatings, double-function glass, printed glass, glazing with light diffusing systems, and solar control glazing with variable properties.
Shading elements	Shading elements such as roller blinds, marquises and marquiselets, external and internal lamellar blinds, louvres sunscreens, shading and light sun-shading shelves, shutters and automatic control systems for mobile elements.
Finishing elements	Finishing materials as reflectors and light absorbers including specific ones whose surfaces are characterised by high values of the solar radiation reflectance coefficient and those with photoabsorption abilities.
Building materials	Building materials as a thermal mass including those that form building elements that do not constitute thermally insulated building partitions, and are so massive that they perfectly play the role of heat accumulation.
Greenery	Greenery as an excellent option when used in the form of shading elements or in the case of the so-called biotic roofs or walls—they naturally reduce the temperature and the heating of building elements.

Gas producers and suppliers seem to be convinced that gas installation is still attractive, this fuel being cheap and safe. It is stated that there is no problem with building a gas connection where the appropriate infrastructure is present, as evidenced by the annual increase in the number of new connections (Table 5). Natural gas can be replaced by many other energy carriers: centralised heat, electricity, fuel oil, or solid fuels such as coal, coke, briquettes, but in the long run, gas is one of the cleanest ecological energy carriers, the use of which is very profitable [74]. The interest in having a gas installation among dwelling users is confirmed by the fact that the share of dwellings with a municipal gas installation connected to the market has remained relatively constant over the years; research carried out again in 2021 showed that this indicator increased to 57.2%. This may indicate that the potential of using gas for heating and utility purposes is still worth considering.

Table 5. Changes in the number and share of houses and premises connected to the gas network in Poland in the years 1999–2016, and 2021, based on GUS.

Year of Research	Number of Houses and Premises with a Gas Connection (in Millions)	Number of Premises and Houses in Poland (in Millions)	Percentage Share of Objects with a Gas Connection to the Number of Houses and Premises in Poland
1999	6.61	11.76	56.2% -
2000	6.72	11.84	56.8% ↑
2001	6.83	11.95	57.2% ↑
2002	6.87	12.44	55.2% ↓
2003	6.98	12.60	55.4% ↑
2004	7.03	12.68	55.4% -
2005	7.09	12.78	55.5% ↑
2006	7.15	12.88	55.5% -
2007	7.24	12.99	55.7% ↑
2008	7.31	13.15	55.6% ↓
2009	7.51	13.30	56.5% ↓
2010	7.60	13.47	56.4% ↓
2011	7.65	13.59	56.3% ↓
2012	7.71	13.72	56.2% ↓
2013	7.77	13.85	56.1% ↓
2014	7.80	13.98	55.8% ↑
2015	7.86	14.12	55.7% ↓
2016	7.92	14.27	55.5% ↓
-	-	-	-
2021	8.77	15.34	57.2% ↑

Another characteristic technological trend in Poland is the turn towards renewable energy sources. Undoubtedly, this is the result of EU policy. In the Renewable Energy Directive rules for the EU, the European Commission set out to achieve a 32% renewables target by 2030 [75]. As a result, in February 2021, the "Energy Policy of Poland until 2040" was adopted ("Polityka energetyczna Polski do 2040"; PEP 2040) [76]. On this basis, a slow departure from coal as the main energy fuel is assumed in favour of renewable energy sources. According to the data of the Energy Market Agency, between January and November 2022, renewable energy sources generated over 31,000 GWh of electricity in Poland—this means a 125% increase compared to the result from 2021. In November 2022, the capacity of all energy sources in Poland amounted to 60 GW, as much as 22 GW (36%) of which came from RES [77].

The goals set in the Renewable Energy Directive are to be attained primarily through the development of PV systems and off-shore wind farms. Photovoltaics has the largest share in the energy transformation. As much as 53.4% of the electricity generated from renewable sources in Poland relates to electricity from solar radiation (by comparison, the share of wind energy amounts to 36.4%) [77]. In PEP 2040, it is assumed that there will be a significant increase to 5–7 GW in 2030, and 10–16 GW in 2020 in the capacity of PV systems [76]. Poland is now one of the Top 5 solar PV investment markets in Europe. In 2021 alone, the country added around 3.2 GW of solar PV installations (Figure 5). With a cumulative installed solar PV capacity of 7.1 GW at the end of 2021, Poland is now a major European solar energy market, with many investors developing large-scale projects far exceeding the 100 MW project scale [78].

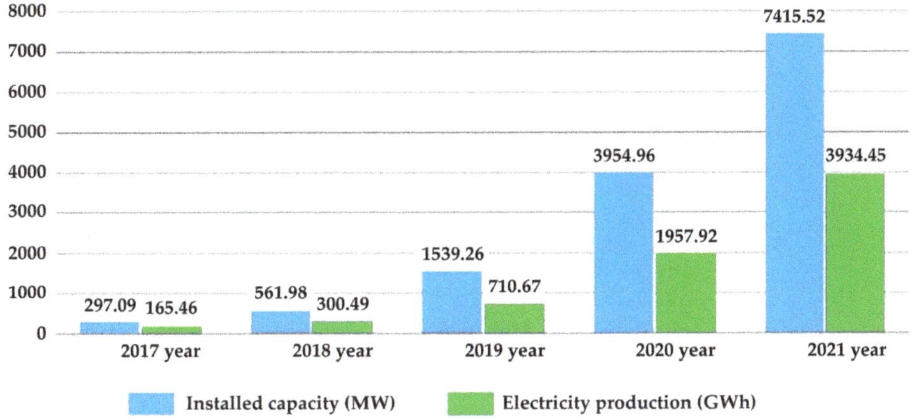

Figure 5. Installed capacity and electricity from photovoltaic cells in Poland between 2017 and 2021 [79].

Despite the barriers resulting from economical aspects, i.e., high initial investment costs, the dynamic development of photovoltaics in Poland is undoubtedly largely due to governmental and local programs that encourage the use of PV in buildings through various forms of financial support. The government program "My Electricity" ("Mój Prąd") launched in 2020 and addressed to individual recipients, i.e., owners of single-family buildings, turned out to be a great success. During the first two editions of this program, almost 3 of 70 MWp PV capacity was installed [80]. This is more than twice as much as expected in the EU study of 2017 residential solar PV capacity by 2030 for Poland (151 MWp was assumed) [81]. The success of this program resulted in extending the forms of support to multi-family buildings—cooperatives and housing communities. In 2020, an equivalent of the "My Electricity" program was launched, a program dedicated to multi-family buildings—housing communities and cooperatives under the name "Słoneczne Dachy" ("Solar Roofs"). The program, supported by the Voivodship Fund for Environmental Protection (WFOŚ) in Poznań, was dedicated to multi-family buildings located in the Wielkopolskie Voivodship. Although it did not repeat the success of the "My Electricity" program, it became a valuable experience, highlighting the problems arising from the differences between individual and community prosumer recipients. The analyses showed that the reason for the relatively low interest was primarily the limitation of the possibility of using the electricity produced only for the needs of common parts of buildings, excluding the individual needs of owners of residential premises, which had a significant impact on the extended payback period [82]. In the current year (2023), a modified program addressed to multi-family buildings was launched—it is already a national government program called "Tenant Prosumer" ("Prosument Lokatorski"). Cooperatives, housing associations, as well as local governments with multi-family buildings will be able to apply for subsidies for photovoltaics. The budget for subsidies to investments in RES is over PLN 448 million (~EUR 100 million) [83].

The immense interest in photovoltaics in Poland, including residential housing, results mainly from the interest to save money in the conditions of increasing prices for energy from conventional sources. However, it is also a manifestation of ecological awareness, aimed at reducing the carbon footprint in the entire process of using buildings, as well as striving to stimulate the economy not only on a national but also local scale [84]. Another important factor related with the application of RES technologies is the passive building design. The shape of the building and the choice of architectural and material solutions are of great importance for the energy processes taking place between the building and its surroundings. In addition, one of the most important issues (in this regard) is the penetration of solar energy through external partitions into interiors; with regard to the Polish climate zone, solutions should be sought that allow the maximum use of heat and sunlight in the cold and transitional periods, and provide protection against their excess in the warm periods. Overheating of buildings can be prevented, e.g., by passive solutions. In multi-family buildings, the greatest threat is excessive solar gain. Passive solar solutions are building elements the shape of which, as well as the appropriate location in relation to the sun, creates a system aimed at the optimal use of solar energy for natural air conditioning and lighting of the internal space of the building or its fragment. In the thermal aspect, these solutions draw solar energy using its natural physical phenomena, without the need for power from an external source. This means that their role is not only the natural use of heat from insulation, but also the protection against its excess. There are two main groups of solutions that passively contribute to reducing the risk of room overheating. These include spatial solutions, as well as material and construction solutions. The passivity of spatial solutions can be defined by the location of the building, its form, the arrangement of internal space, the layout of functions, and the façade cladding. Passive building protection against overheating includes material solutions that can be briefly divided into: glazed elements, spatial shading elements, finishing materials, insulation materials, and greenery and landscaping [85].

Functional Factors

One of the current trends in broadly understood housing utility is co-living, the intention of which is to create targeted support for integration at the social level. In addition, it is important to provide residents or employees with a program offer that is based on the economy of sharing space, combining the residential function with recreation, work, and basic needs. This leads to the promotion of interpersonal activity and integration in leisure time [18].

The idea of shared workspaces has been very well received in Poland, especially if the presence of co-living is combined with co-working. This is a very common trend in the context of global change or sudden geopolitical changes, potential cataclysms or social threats. Thanks to the possibility of organizing work in the place of residence or its vicinity, certain threats or nuisances may be eliminated. It may also have a positive impact on the reduction of car traffic and less frequent use of public transport, especially during rush hours. This may also have benefits on a purely social level, because due to remote work, the private time saved can be used for healthy activities [86].

Buildings should have spaces for meetings, celebrations, or networking events. Finding oneself in an environment of similar people creates ideal living conditions, and for apartment users it will also be a cheaper alternative to classic rental in the city centre. Research and practice show that Poles are looking for slightly smaller apartments, and at the same time, they are ready to pay for additional functions. Individual projects in the co-living trend include intelligent control systems, parcel lockers at receptions, assigned places for bicycles, or chargers for electric cars in garages. JLL's research shows that every fifth respondent is more likely to buy an apartment in a housing estate with services such as a self-service laundry with a drying room, a scooter for minutes, or a parking space for a rented car. The presence of these paid services is expected to be more relevant to buyers over the age of 45 [87].

Factors that influence the way flats are shaped (Table 6) include such phenomena as new diverse attitudes and views, e-work, and the crisis of the traditional family. There are three main, general types of multi-family housing that follow the spirit of the times and adapt to the dynamically developing society. These include: a dependent apartment, a flexible apartment, and a comfortable apartment. In addition, the types of multi-family architecture are defined thematically. Global, green, spectacular, defensive, mobile, dynamic, vernacular, and the so-called architecture of deconcentrated concentration are specified [88]. Not all people are interested in buying premises in buildings characterised by the so-called higher standard of finishing, technology, or functional and spatial solutions. This group is dedicated to houses described as "accessible". It is a vernacular architecture, restrained and with a traditional form of development. These houses are cheap to build and operate [89].

Table 6. Functional factors—a list of issues in the context of ways of creating attractive residential architecture. Based on a critical literature review, knowledge and design practice of the authors.

The Idea of Social Integration/Co-living/Co-working	
Attractive solutions for employees, young and middle-aged people, students and graduates. These groups will be interested in alternative proposals which may additionally be a cheaper form of rental. Very important trends in modern times, in the face of threats and geopolitical changes. Integrated forms of amenities and accompanying services will be offered near the buildings.	
Characteristics of contemporary apartments [88]	
Dependent apartment	Defined by the new philosophy of life of people whose everyday life is often limited by traditional forms of spending time, eating, or cleaning. These needs can be met outside the place of residence, using the local environment.
Flexible apartment	Prone to making spatial changes in them during their inhabitation, which will result in the minimization of structural elements, walls, columns, and shafts. The functional layout of these flats may be easily changed.
Comfortable apartment	Striving to reproduce the comfort of everyday life characteristic of single-family housing, which means that the standard and size of individual premises will be high. Large gardens are common, and apartments are sometimes multi-level, with access to elements such as mezzanines, spacious terraces, balconies, and verandas.
Types of contemporary multi-family architecture [88]	
Global	Fits in with the topic of unification of world architecture projects through the traditional, local, regional, or national context.
Green	Concerns the idea of sustainable development by subordinating multi-family housing to the natural environment and exposing pro-ecological solutions.
Spectacular	Characterised by unconventional, effective architectural forms, focused on publicity and image success.
Defensive	Tends to create defensive and protective forms against all external factors; shapes partitions and building elements with a specific, solid barrier character.
Mobile	Departs from the fixed and traditional spatial context, characterises buildings with easy disassembly and the possibility of assembly in another place.
Dynamic	Negates the static forms of residential development and introduces new, dynamic ones, which sometimes reflect the current technical possibilities.
Deconcentrated concentration	Opposes the proposals of concentric forms and creates loose, deliberately broken structures. Single flats or their fragments will have a shape and a definitely multi-element character.
Vernacular	Being modest, simple, restrained, resulting strictly from the demand and universality of housing needs. Corresponding to the requirements of the majority of society.

Table 6. *Cont.*

The idea of accessibility [89]
The essence of this idea is mainly the lower costs of renting apartments, which in turn depends on the costs of construction and maintenance, which result from design solutions. This will force the use of staircase-free house solutions, the limitation of common areas, staircases, the concentration of installation risers, and the flexibility of functional layouts. The blocks of the buildings will be characterised by compactness, which is important from the point of view of energy saving and care for the environment. The concentrated nature of the development on the plot may be necessary to implement the accessible house model.

4. Discussion

The presented principles related to designing and investing within residential architecture refer to the situation and expectations in Poland. It is worth noting that the described strategy of conduct, taking into account close cooperation between the architect and the developer, is in line with the principles of conduct expressed in EU law [90–92] and is reflected in the policies of other European countries [93–96]. The described factors and requirements related to the quality of architecture are part of a broader policy of searching for appropriate aspects, the presence of which would guarantee the appropriate level of architecture, while ensuring its economic profitability [97,98]. In addition, the presented content refers to the challenges and opportunities related to shaping contemporary architecture [99]. Aspects related to market expectations presented herein concern local specificity, but also reflect a global trend related to specialist housing needs [100]. The proposed division into types of contemporary architecture reflects the current social conditions and expectations of different age groups related to living. The idea of co-living is part of a similar trend related to sharing, which is currently an important element of everyday life and lifestyle, visible both in everyday functioning, for example, in the social media system, and the organization of life integrated around them, and consequently, in housing systems [101–103]. Ideas related to reaching individual social or religious groups are implemented all over the world and are nothing particularly original on the European market, but they draw attention to a certain opposite trend, where despite widespread appeals regarding the inclusiveness of space, areas that are exclusive in their own way are still being sought to meet the expectations of specific customer groups [104,105].

In order to find the most optimal unified model of a modern, comfortable, compact and environmentally friendly model of a residential unit, it is necessary to consider the residential cell on three scales:

- private space (apartment or single-family house);
- communal spaces (staircases, elevator units, corridors, basements, and utility rooms);
- public space (objects located in the nearest surroundings that represent added value, such as public service facilities, green areas, parking lots, etc.).

The more services and amenities are available around a private space or a separate residential unit, the less is the need for the volume and multifunctionality of the internal private space of the apartment itself. The validity of this hypothesis can be traced to the example of Warsaw—in the latest project of the Warsaw Studio [106], urban solutions are mainly focused on polycentricity (creation of independent, fully functional local centres within 15 min of walking (or cycling)), multi-family housing development, and sustainable solutions that help to decrease transport emissions, and reduce the need for residents to use individual and public transport.

Recently, the trend of Poles purchasing homes with larger living spaces is increasing, which is not only caused by the desire for more comfortable living space, but also reflects the effects of the pandemic when residents were able to work remotely, as the need for office space has fallen sharply. In itself, this trend can be seen as positive, as staying at home saves a lot of time on the daily commute and reduces the carbon footprint, which is still one of the largest environmental problems in modern society. A striking example of this trend is the office complex in Warsaw's Mokotów district (the so-called "Mordor"), where a decision was made to demolish several office complexes and build residential buildings.

The pandemic has forced us to rethink the value of private territory for full-fledged isolated living and the need to create an optimal, not a minimum, level of comfort in the living unit.

Therefore, along with the need to minimise the heated space of residential apartments, attention should be paid to the following factors that minimise the need to use private space:

1. Availability of services and amenities around the residential property that fulfil the basic household needs [107];
2. Availability of full-fledged common spaces for residents of apartment buildings, which ensure socialization of residents and contribute to increasing the ability to build neighbourly cooperation and optimise the use of resources (courtyards, internal common spaces) [108];
3. Versatility of the interior spaces of apartments with the possibility of internal reconstruction to meet the different needs of the family at different periods of its existence (including the possibility of dividing and combining apartments, multifunctionality of utility rooms); the provision of an adequate level of sunlight when dividing rooms, and the possibility of effective shading in the case of excessive sunlight to optimise cooling and heating systems [109].

Based on the conducted research, the following guidelines for programming housing estates were developed:

- Location of the development in relation to the proposed amenities and access to public transport, reducing the need to use private cars, and the proximity of available services and transport stops in the vicinity of the housing development within a radius of 700 m from the main entrance to the housing development site (distance calculated along a safe short walk line);
- Inventory of the housing development site and the need to undertake site adaptation measures; analysis of the feasibility of using existing buildings that will be redeveloped; determination of the percentage of the new development site that was previously developed land;
- Possible remediation of the housing development site, in case of contamination of the development site;
- Solutions for reduction of the heat island effect (the phenomenon of the urban heat island consists in a significant increase in temperature in the city in relation to peripheral areas, which is responsible, e.g., for the accumulation of heat by the urban tissue, including strong heating of the surfaces of walls, roofs, and hardened ground); exemplary solutions: use of surfaces made of light-coloured materials with a solar reflectance SRI of at least 75, or permeable surfaces (min. 50% perforation) with vegetation in the openings. Realisation of flat roofs using a final layer in the form of a light-coloured membrane, covering with light-coloured gravel or other roof finishing material with a solar reflectance SRI of at least 75, or as a green roof. Construction of façades, including the surfaces of loggias and balconies, in light-coloured materials with a solar reflectance (SRI) of at least 0.50, and a thermal emissivity of at least 0.85;
- Using existing vegetation (primarily trees) or designing new planting to provide shade within the project site;
- Shaping the landscape on the basis of natural inventories defining the condition and potential of the local biodiversity, measures to protect the local biodiversity, solutions taking care of the safety of birds, habitat-appropriate plants, drought and flooding-resistant plants;
- Environmental monitoring before and during implementation;
- Introducing drought-tolerant plant species—not requiring intensive watering and care;
- Provision for the development of electromobility, infrastructure for car charging: creating the possibility for the owner of an individual car charging infrastructure/installation of chargers for electric cars in a dedicated parking space;
- Availability of bicycle infrastructure, including bicycle paths at a distance of a maximum 300 m, to which there is safe access along existing pavements, or those designed

and constructed within the framework of the residential development under construction; bicycle storage solutions in the residential development and carsharing (private vehicles made available on the basis of a platform associating users using this service—creation of a platform promoting carsharing for the residents of the residential development or by a business entity offering a rental service for an uninterrupted period);
- When developing the plot, the natural context should be taken into account and a system of biologically active areas should be developed with the addition of tree planting on the plot;
- Implementation of water retention systems within the plot boundaries;
- Providing recreational areas and internal common use spaces on the plot;
- Application of solutions such as green roofs and green walls;
- Analysis of the building life cycle and its potential to function change;
- Use of shading façade elements;
- Energy-efficient solutions, energy-efficient internal transport systems, and air-tightness of the building;
- Reduction of atmospheric emissions and impact on smog formation;
- Indoor air quality control systems, use of natural ventilation;
- Ensuring thermal and acoustic comfort;
- Ensuring daylight access;
- Use of vegetation in the building and biophilic solutions in the design of common space, e.g., exteriors with bamboo planting or shade-loving climbers and, in selected locations in the lobby, walls/elements and illuminated surfaces made of salt, improving conditions and air quality (decontaminating);
- Shaping biodiversity, i.e., species and habitat richness and proportions;
- Shaping an ecospot, i.e., a functioning natural area left without external interference, allowing shelter for small animals, and free vegetation of plants;
- Efficient water management in the project, i.e., re-use of grey water, retention of rainwater and its use, for e.g., garden irrigation and/or toilet flushing and/or infiltration into the ground, use of safe natural methods for water purification (grey water, rainwater), use of technologies reducing water consumption;
- Use of RES, renewable, non-fossil energy sources (wind, solar, aerothermal, geothermal, hydrothermal, biomass, biogas, agricultural biogas and bioliquids);
- Introduction of principles of universal and inclusive design—providing access and use by as many people as possible without the need for additional adaptations or specialised solutions, regardless of age, gender, or degree of disability of the users;
- Use of recycled materials and elements—in line with the principles of a closed loop economy, in order to minimise the use of natural resources and reduce greenhouse gas emissions associated with the manufacture of new building elements; it is advisable to use those which are already in circulation;
- Limiting light pollution, external lighting installation, including the lighting of internal circulation routes within the residential development and lighting elements located on the façade, which should comply with the following conditions:
    - All façade luminaires, including sconces having a flat glass and shielded light source, should be directed downwards;
    - Luminaires with a luminaire emitting light only downwards—the colour of the light should be between 2600 K and 3700 K;
    - Bottom-up illumination of trees and buildings should be abandoned;
    - External lighting installed outside the façade may not exceed the height of 4 m—control of external lighting by an astronomical clock or twilight detectors with the possibility of using night breaks—in the case of external lighting of balconies and loggias, use of lighting with time switches;
- Implementation of green and blue infrastructure solutions on the development site, linked to the urban greening and water management system of the city.

In the discussion, attention should also be drawn to the contemporary humanitarian problem related to the war in Ukraine. In 2022–2023, over a million Ukrainians affected by the effects of the war settled in Poland. The challenge does not only concern Poland, but also Europe and countries outside Europe. Will this have an impact on the shaping of housing development in Poland? This topic is very important. Due to the definition of the research in question, but also the extensiveness of the issue of the humanitarian crisis related to the war, it has not been discussed in this study. This research question remains open as a contribution to further research.

5. Conclusions

The assumption made at the beginning was to collect various aspects present in the process of creating multi-family residential buildings. A clear binder in this was the development context and additional conditions determining a modern approach to design solutions, taking into account factors that are not always obvious from the point of view of both the designer and future users. As the issues presented in the study show, the number of such factors is extremely high, but so are the number of solutions that can be implemented to better reach customers and meet their housing needs. There are trends from the investment point of view to search for a set of solutions, the implementation of which will be the most appropriate for individual social groups for financial, location, or ideological reasons, and then to develop these guidelines within a full-fledged architectural and construction project. The differentiation of these factors, due to location, leading theme, and finally target groups, together with the definition of investment and economic intentions and type of finish standard offered leads to naming the criteria on the basis of which design proposals are differentiated. These criteria, depending on the intentions, can be divided into solid, spatial, material, technological, design, or purely marketing solutions. Depending on the lower or higher investment standard, other characteristic elements are present.

Author Contributions: A.S., M.D., K.R.-N., J.M., M.G.-S., P.Ł., Ł.M., I.V. and P.V.; methodology, A.S., K.R.-N., P.Ł., Ł.M. and M.G.-S.; software, M.D.; validation, A.S., M.D., K.R.-N. and M.G.-S.; formal analysis, A.S., M.D., K.R.-N., J.M., M.G.-S., P.Ł., Ł.M., I.V. and P.V.; resources, A.S., M.D., K.R.-N., J.M., M.G.-S., P.Ł., Ł.M., I.V. and P.V.; data curation, A.S., M.D., K.R.-N., J.M., M.G.-S., P.Ł., Ł.M., I.V. and P.V.; writing—original draft preparation, A.S., M.D., K.R.-N., J.M., M.G.-S., P.Ł., Ł.M., I.V. and P.V.; writing—review and editing, A.S., P.Ł. and Ł.M.; visualization, M.D.; supervision, A.S.; project administration, A.S. and J.M.; funding acquisition, A.S. All authors have read and agreed to the published version of the manuscript.

Funding: This research received no external funding.

Data Availability Statement: No data availability.

Conflicts of Interest: The authors declare no conflict of interest.

References

1. Kocur-Bera, K. Impact of the COVID-19 Pandemic Era on Residential Property Features: Pilot Studies in Poland. *Int. J. Environ. Res. Public Health* **2022**, *19*, 5665. [CrossRef]
2. The New Charter of Athens 2003. Available online: http://www.itc.cnr.it/ba/re/Documenti/The%20New%20Charter%20of%20Athens%202003.htm (accessed on 24 July 2023).
3. Adopted at the Meeting of Ministers for Urban Development and Territorial Cohesion of the EU in Leipzig 2007. Available online: https://ec.europa.eu/regional_policy/sources/policy/what/territorial-cohesion/territorial_agenda_leipzig2007.pdf (accessed on 24 July 2023).
4. Krajowa Polityka Miejska 2023. Available online: https://archiwum.pte.pl/pliki/2/21/KrajowaPolitykaMiejska.pdf (accessed on 24 July 2023).
5. Ustawa z Dnia 9 Października 2015 r. o Rewitalizacji. Available online: https://isap.sejm.gov.pl/isap.nsf/DocDetails.xsp?id=wdu20150001777 (accessed on 24 July 2023).
6. Rahmani, K.; Ahriz, A.; Bouaziz, N. Development of a New Residential Energy Management Approach for Retrofit and Transition, Based on Hybrid Energy Sources. *Sustainability* **2022**, *14*, 4069. [CrossRef]

7. Taylor, C.; Roy, K.; Dani, A.A.; Lim, J.B.P.; De Silva, K.; Jones, M. Delivering Sustainable Housing through Material Choice. *Sustainability* **2023**, *15*, 3331. [CrossRef]
8. Ma, W.; Ren, Z.; Ke, H. Green Housing Subsidy Strategies Considering Consumers' Green Preference. *Sustainability* **2022**, *14*, 2748. [CrossRef]
9. Kim, K.H.; Jeon, S.-S.; Irakoze, A.; Son, K. A Study of the Green Building Benefits in Apartment Buildings According to Real Estate Prices: Case of Non-Capital Areas in South Korea. *Sustainability* **2020**, *12*, 2206. [CrossRef]
10. Imam, A.A.; Al-Turki, Y.A.; R., S.K. Techno-Economic Feasibility Assessment of Grid-Connected PV Systems for Residential Buildings in Saudi Arabia—A Case Study. *Sustainability* **2019**, *12*, 262. [CrossRef]
11. Mejia, A.S.; Moya, M.; Martínez-Gómez, J. Evaluation of the Design of Ecological Buildings Using the Matrix of Eco-Efficiency in Residential Sectors of Ecuador. *Sustainability* **2023**, *15*, 7585. [CrossRef]
12. Mazur, Ł.; Bać, A.; Vaverková, M.D.; Winkler, J.; Nowysz, A.; Koda, E. Evaluation of the Quality of the Housing Environment Using Multi-Criteria Analysis That Includes Energy Efficiency: A Review. *Energies* **2022**, *15*, 7750. [CrossRef]
13. Yang, Y.; Javanroodi, K.; Nik, V.M. Climate Change and Energy Performance of European Residential Building Stocks—A Comprehensive Impact Assessment Using Climate Big Data from the Coordinated Regional Climate Downscaling Experiment. *Appl. Energy* **2021**, *298*, 117246. [CrossRef]
14. Rybak-Niedziółka, K.; Starzyk, A.; Łacek, P.; Mazur, Ł.; Myszka, I.; Stefańska, A.; Kurcjusz, M.; Nowysz, A.; Langie, K. Use of Waste Building Materials in Architecture and Urban Planning—A Review of Selected Examples. *Sustainability* **2023**, *15*, 5047. [CrossRef]
15. Palladino, D.; Di Turi, S. Energy and Economic Savings Assessment of Energy Refurbishment Actions in Italian Residential Buildings: Comparison between Asset and Tailored Calculation. *Sustainability* **2023**, *15*, 3647. [CrossRef]
16. Marchwiński, J.; Starzyk, A.; Kopyłow, O.; Kurtz-Orecka, K. Impact of Atrium Glazing with and without BIPV on Energy Performance of the Low-Rise Building: A Central European Case Study. *Energies* **2023**, *16*, 4683. [CrossRef]
17. Fang, Y.; Cho, S. Design Optimization of Building Geometry and Fenestration for Daylighting and Energy Performance. *Sol. Energy* **2019**, *191*, 7–18. [CrossRef]
18. Balakina, A.; Pavlyuk, A. Colivings are new type of hostel in the structure of a modern campus. *Constr. Archit.* **2022**, *10*, 61–65. [CrossRef]
19. European Commission. *Directorate-General for Regional Policy Cities of Tomorrow: Challenges, Visions, Ways Forward*; European Commission: Luxembourg, 2011.
20. Qureshi, R.A.; Shah, S.J.; Akhtar, M.; Abbass, W.; Mohamed, A. Investigating Sustainability of the Traditional Courtyard Houses Using Deep Beauty Framework. *Sustainability* **2022**, *14*, 6894. [CrossRef]
21. Godlewski, T.; Mazur, Ł.; Szlachetka, O.; Witowski, M.; Łukasik, S.; Koda, E. Design of Passive Building Foundations in the Polish Climatic Conditions. *Energies* **2021**, *14*, 7855. [CrossRef]
22. Janusz, M.; Agnieszka, S. Problematyka Projektowania Budynków Przedszkoli Ze Szczególnym Uwzględnieniem Aspektów Ekologiczno-Energetycznych.Projekt Energoefektywnego Przedszkolaw Michałowicach Cz. 2. *Builder* **2021**, *286*, 44–49. [CrossRef]
23. Winkler, J.; Jeznach, J.; Koda, E.; Sas, W.; Mazur, Ł.; Vaverková, M. Promoting Biodiversity: Vegetation in a Model Small Park Located in the Research and Educational Centre. *J. Ecol. Eng.* **2022**, *23*, 146–157. [CrossRef]
24. Ryńska, E. *Architekt W Procesie Tworzenia Harmonijnego Środowiska*; Oficyna Wydawnicza Politechniki Warszawskiej: Warsaw, Poland, 2004.
25. Communication from the Commission to the European Parliament, the Council, the European Economic and Social Committee and the Committee of the Regions. A Renovation Wave for Europe—Greening Our Buildings, Creating Jobs, Improving Lives. Available online: https://eur-lex.europa.eu/legal-content/EN/TXT/?qid=1603122220757&uri=CELEX:52020DC0662 (accessed on 26 June 2020).
26. Wilson, A. Energy Performance of Buildings Directive. Available online: https://www.europarl.europa.eu/RegData/etudes/ATAG/2023/739377/EPRS_ATA(2023)739377_EN.pdf (accessed on 26 June 2023).
27. Lis, P.; Rataj, Z.; Suszyńska, K. *Zmiany Sytuacji Mieszkaniowej w Polsce w Latach 2015–2019*; Polskie Towarzystwo Ekonomiczne Oddział w Poznaniu: Poznań, Poland, 2020.
28. Housing at the Center of the New Urban Agenda. Nairobi: UN Nations Human Settlements Programme. Available online: https://unhabitat.org/sites/default/files/download-manager-files/Housing_at_the_centre.pdf (accessed on 26 June 2023).
29. Weck, S.; Madanipour, A.; Schmitt, P. Place-Based Development and Spatial Justice. *Eur. Plan. Stud.* **2022**, *30*, 791–806. [CrossRef]
30. Dekker, K.; Van Kempen, R. Large housing estates in europe: Current situation and developments. *Tijd. Voor Econ. Soc. Geog.* **2004**, *95*, 570–577. [CrossRef]
31. Gaspareniene, L.; Venclauskiene, D.; Remeikiene, R. Critical Review of Selected Housing Market Models Concerning the Factors That Make Influence on Housing Price Level Formation in the Countries with Transition Economy. *Procedia—Soc. Behav. Sci.* **2014**, *110*, 419–427. [CrossRef]
32. Hurlimann, A.C.; Moosavi, S.; Browne, G.R. Climate Change Transformation: A Definition and Typology to Guide Decision Making in Urban Environments. *Sustain. Cities Soc.* **2021**, *70*, 102890. [CrossRef]
33. Kalbarczyk, E.; Kalbarczyk, R. Typology of Climate Change Adaptation Measures in Polish Cities up to 2030. *Land* **2020**, *9*, 351. [CrossRef]

34. Sustainable Energy Action Plan for Warsaw [SEAP]. Available online: http://adaptcity.pl/wp-content/uploads/2016/09/Analiza_dokumentow_wg_metodologii.pdf (accessed on 26 June 2023).
35. Pedersen, D.B. Integrating Social Sciences and Humanities in Interdisciplinary Research. *Palgrave Commun.* **2016**, *2*, 16036. [CrossRef]
36. Hełdak, M.; Raszka, M. Evaluation of the Local Spatial Policy in Poland with Regard to Sustainable Development. *Pol. J. Environ. Stud.* **2013**, *22*, 395–402.
37. Chen, H.; Zhang, Y.; Zhang, N.; Zhou, M.; Ding, H. Analysis on the Spatial Effect of Infrastructure Development on the Real Estate Price in the Yangtze River Delta. *Sustainability* **2022**, *14*, 7569. [CrossRef]
38. Nowysz, A.; Mazur, Ł.; Vaverková, M.D.; Koda, E.; Winkler, J. Urban Agriculture as an Alternative Source of Food and Water Security in Today's Sustainable Cities. *Int. J. Environ. Res. Public Health* **2022**, *19*, 15597. [CrossRef]
39. Gilbertson, J.; Grimsley, M.; Green, G. Psychosocial Routes from Housing Investment to Health: Evidence from England's Home Energy Efficiency Scheme. *Energy Policy* **2012**, *49*, 122–133. [CrossRef]
40. Mazur, Ł. Wybrane Czynniki Przyrodnicze Kształtujące Budownictwo Mieszkaniowewe Współczesnych Miastach. *Acta Sci. Pol.—Archit. Bud.* **2020**, *19*, 83–91. [CrossRef]
41. Renigier-Biłozor, M.; Źróbek, S.; Walacik, M. Modern Technologies in the Real Estate Market—Opponents vs. Proponents of Their Use: Does New Category of Value Solve the Problem? *Sustainability* **2022**, *14*, 13403. [CrossRef]
42. Tunali, H.; Yerdelen, F. The Analysis of Factors Affecting Investment Choices of Households in Turkey with Multinomial Logit Model. *Int. Res. J. Financ. Econ.* **2010**, *40*, 186–202.
43. Ge, J.; Hokao, K. Research on Residential Lifestyles in Japanese Cities from the Viewpoints of Residential Preference, Residential Choice and Residential Satisfaction. *Landsc. Urban Plan.* **2006**, *78*, 165–178. [CrossRef]
44. Tsou, W.-L.; Sun, C.-Y. Consumers' Choice between Real Estate Investment and Consumption: A Case Study in Taiwan. *Sustainability* **2021**, *13*, 11607. [CrossRef]
45. Profil Kupujących Wg Barometr Metrohouse i Gold Finance—Najnowsze Dane z Rynku Mieszkaniowego i Kredytów Hipotecznych. 2019. Available online: https://metrohouse.pl/ckfinder/userfiles/files/Barometr/Barometr_Metrohouse_Gold_Finance_03Q_2019.pdf (accessed on 26 June 2023).
46. Bałaj, M. Deweloperskie Niestandardy. Available online: https://www.nieruchomosci-online.pl/porady/deweloperskie-niestandardy-3657.html?utm_source=pprn&utm_medium=artykul&utm_campaign=estate-01-2017 (accessed on 29 July 2020).
47. Gospodarka Mieszkaniowa w 2020 Roku. Available online: https://stat.gov.pl/obszary-tematyczne/infrastruktura-komunalna-nieruchomosci/nieruchomosci-budynki-infrastruktura-komunalna/gospodarka-mieszkaniowa-w-2020-roku,14,4.html (accessed on 26 June 2023).
48. Polska Gospodarka 2023—Prognozy. Kryzys w Branży Deweloperskiej, Oznacza Problemy w Kolejnych Sektorach Gospodarki. Available online: https://www.muratorplus.pl/inwestycje/inwestycje-mieszkaniowe/polska-gospodarka-2023-prognozy-kryzys-w-branzy-deweloperskiej-oznacza-problemy-w-kolejnych-sektorach-gospodarki-aa-4M9n-9wV1-eAoN.html (accessed on 24 July 2023).
49. Frączyk, J. Luka Mieszkaniowa. Oto Ile Już Mamy Mieszkań i Czy Naprawdę Wciąż Ich Brakuje. Available online: https://businessinsider.com.pl/gospodarka/luka-mieszkaniowa-oto-ile-mamy-mieszkan-i-czy-naprawde-wciaz-ich-brakuje/lbx1679 (accessed on 24 July 2023).
50. Ceny Mieszkań Pójdą w Dół. Oto Dlaczego Tak Się Stanie. Available online: https://www.money.pl/gospodarka/ceny-mieszkan-pojda-w-dol-oto-dlaczego-tak-sie-stanie-6879217054337824a.html (accessed on 24 July 2023).
51. GUS 2023. Available online: https://stat.gov.pl/ (accessed on 24 July 2023).
52. Halik, Ł. Information and Communication Systems Used for Keeping the Register of Real Estate Prices and Values (Rrepv) in Poland. *Real Estate Manag. Valuat.* **2018**, *26*, 45–53. [CrossRef]
53. CENTRUM 50+ Nowe Mieszkania w Mieście. Available online: https://gliwice.eu/aktualnosci/miasto/centrum-50-nowe-mieszkania-w-miescie (accessed on 30 July 2020).
54. Pod Warszawą Powstaje Osiedle Dla Katolików. "Jesteśmy Otwarci Na Wszystkich, Którzy Chcą Żyć w Zgodzie". Available online: https://www.national-geographic.pl/artykul/pod-warszawa-powstaje-osiedle-dla-katolikow-jestesmy-otwarci-na-wszystkich-ktorzy-chca-zyc-w-zgodzie (accessed on 19 June 2023).
55. Tversky, A.; Kahneman, D. The Framing of Decisions and the Psychology of Choice. *Science* **1981**, *211*, 453–458. [CrossRef]
56. Baker, H.K.; Kumar, S.; Goyal, N. Personality Traits and Investor Sentiment. *Rev. Behav. Financ.* **2021**, *13*, 354–369. [CrossRef]
57. Karliński, K. *Zarabiajmy Na Nieruchomościach*; Impera Kuba Karliński: Warsaw, Poland, 2018; ISBN 978-83-952376-0-7.
58. Raport-Stan Mieszkalnictwa w Polsce. Available online: https://www.gov.pl/web/rozwoj-technologia/raport-o-stanie-mieszkalnictwa (accessed on 26 June 2023).
59. Kang, Y.; Zhang, F.; Peng, W.; Gao, S.; Rao, J.; Duarte, F.; Ratti, C. Understanding House Price Appreciation Using Multi-Source Big Geo-Data and Machine Learning. *Land Use Policy* **2021**, *111*, 104919. [CrossRef]
60. Gromadzki, J.; Lewandowski, P. Refugees from Ukraine on the Polish Labour Market. *Soc. Insur. Theory Pract.* **2023**, *155*, 29–40. [CrossRef]
61. Nowak, P. Eksurbanizacja Współczesnych Miast/Urban Sprawl of Contemporary Cities. *Stud. Miej.* **2015**, *20*, 133–140.
62. Carmona, M.; Carmona, S.; Clarke, W. *A Bibliography of Design Value, for the Commission for Architecture and the Built Environment*; The Bartlett School of Planning University College London: London, UK, 2001.

63. Ekopark—Art City. Available online: https://www.ekopark.pl/content/uploads/2018/08/katalog-ArtCity.pdf (accessed on 30 July 2020).
64. Krzyżanowska, A. Przed Nami Era Deweloperów Myślących o Przyrodzie. Available online: https://www.rp.pl/nieruchomosci/art909881-przed-nami-era-deweloperow-myslacych-o-przyrodzie (accessed on 31 July 2020).
65. Zielono-Błękitne Rozwiązania Dla Osiedli Mieszkaniowych. Broszura Dla Wspólnot Warszawskich Zrealizowana Przez Fundację Sendzimira Na Zlecenie Zarządu Zieleni m. St. Warszawy, 2017. Available online: https://sendzimir.org.pl/publikacje/zielono-blekitne_rozwiazania_dla_osiedli/ (accessed on 29 July 2020).
66. Belaroussi, R.; Dai, H.; González, E.D.; Gutiérrez, J.M. Designing a Large-Scale Immersive Visit in Architecture, Engineering, and Construction. *Appl. Sci.* **2023**, *13*, 3044. [CrossRef]
67. Gómez-Tone, H.C.; Martin-Gutierrez, J.; Bustamante-Escapa, J.; Bustamante-Escapa, P. Spatial Skills and Perceptions of Space: Representing 2D Drawings as 3D Drawings inside Immersive Virtual Reality. *Appl. Sci.* **2021**, *11*, 1475. [CrossRef]
68. Łuciów, D. Wirtualna Rzeczywistość w Budownictwie i Architekturze. Available online: https://magazynprzemyslowy.pl/artykuly/wirtualna-rzeczywistosc-w-budownictwie-i-architekturze (accessed on 30 July 2020).
69. Chrząstowska, M. Wirtualna Rzeczywistość w Budownictwie i Architekturze. Available online: https://www.muratorplus.pl/technika/programy/technologia-bim-projekt-budowa-zarzadzanie-budynkiem-aa-nQFt-dMH6-MdL3.html (accessed on 30 July 2020).
70. BREEAM. Available online: https://bregroup.com/products/breeam/ (accessed on 31 July 2020).
71. NFOSIGW. Available online: https://www.gov.pl/web/nfosigw/narodowy-fundusz-ochrony-srodowiska-i-gospodarki-wodnej (accessed on 31 July 2020).
72. ISO. Available online: https://www.iso.org.pl/uslugi-zarzadzania/wdrazanie-systemow/zarzadzanie-srodowiskowe/iso-14001/ (accessed on 31 July 2020).
73. Jakie Rozwiązania Ekologiczne Są Stosowane w Nowych Mieszkaniach? Available online: https://bank.pl/jakie-rozwiazania-ekologiczne-sa-stosowane-w-nowych-mieszkaniach/ (accessed on 31 July 2020).
74. Kalinowska, B. Deweloperzy Nie Lubią Gazu. Available online: https://www.rp.pl/artykul/148842-Deweloperzy-nie-lubia-gazu.html (accessed on 31 July 2020).
75. Wicki, L.; Pietrzykowski, R.; Kusz, D. Factors Determining the Development of Prosumer Photovoltaic Installations in Poland. *Energies* **2022**, *15*, 5897. [CrossRef]
76. Zdonek, I.; Tokarski, S.; Mularczyk, A.; Turek, M. Evaluation of the Program Subsidizing Prosumer Photovoltaic Sources in Poland. *Energies* **2022**, *15*, 846. [CrossRef]
77. Odnawialne Źródła Energii w 2022 Roku. Sprawdź Najnowsze Statystyki i Prognozy Na 2023 (MediaRoom). Available online: https://biznes.pap.pl/pl/news/all/info/3392411,odnawialne-zrodla-energii-w-2022-roku{-}{-}sprawdz-najnowsze-statystyki-i-prognozy-na-2023-(mediaroom) (accessed on 20 June 2023).
78. Schnell, C. *Overview of Solar Investments in Poland*; Solar Plaza: Rotterdam, The Netherlands, 2022.
79. Berent-Kowalska, G.; Jurgaś, A.; Kacprowska, J.; Moskal, I.; Kapica, K. Energia Ze Źródeł Odnawialnych w 2021 Roku. Available online: https://stat.gov.pl/obszary-tematyczne/srodowisko-energia/energia/energia-ze-zrodel-odnawialnych-w-2021-roku,3,16.html (accessed on 26 June 2023).
80. Olczak, P.; Kryzia, D.; Matuszewska, D.; Kuta, M. "My Electricity" Program Effectiveness Supporting the Development of PV Installation in Poland. *Energies* **2021**, *14*, 231. [CrossRef]
81. Study on Residential Prosumers in the European Energy Union JUST/2015/CONS/FW/C006/0127 Framework Contract EAHC/2013/CP/04. Available online: https://commission.europa.eu/system/files/2017-11/study-residential-prosumers-energy-union_en.pdf (accessed on 26 June 2023).
82. Nie Będzie Kontynuacji Programu Słoneczne Dachy. W Planie Nowe Pilotaże. Available online: https://zielona-energia.cire.pl/artykuly/serwis-informacyjny-cire-24/179866-nie-bedzie-kontynuacji-programu-sloneczne-dachy-w-planie-nowe-pilotaze (accessed on 19 June 2023).
83. Prosument Lokatorski—Startuje Nabór Wniosków o Dopłaty. Available online: https://www.gov.pl/web/rozwoj-technologia/prosument-lokatorski{-}{-}startuje-nabor-wnioskow-o-doplaty (accessed on 26 June 2023).
84. Rakowska, J.; Ozimek, I. Renewable Energy Attitudes and Behaviour of Local Governments in Poland. *Energies* **2021**, *14*, 2765. [CrossRef]
85. Marchwiński, J.; Zielonko-Jung, K. *Ochrona Przeciwsłoneczna w Budynkach Wielorodzinnych. Pasywne Rozwiązania Architektoniczno-Materiałowe*; Oficyna Wydawnicza WSEiZ w Warszawie: Warsaw, Poland, 2013.
86. Kacprowska, K. Remote Work during the COVID-19 Pandemic. *CetL* **2021**, *32*, 83–94. [CrossRef]
87. Roguski, A. Filozofia Mieszkania Zmienia Także Polski Rynek. Available online: https://www.rp.pl/Mieszkaniowe/306069849-Nowa-filozofia-mieszkania-zmienia-takze-polski-rynek.html (accessed on 31 July 2020).
88. Pallado, J. *Zabudowa Wielorodzinna Podstawy Projektowania*; Wydawnictwo Politechniki Śląskiej: Gliwice, Poland, 2016.
89. Pallado, J. *Architektura Wielorodzinnych Domów Dostępnych*; "Śląsk" Sp. z o.o. Wydawnictwo Naukowe: Katowice, Poland, 2007.
90. Directive (EU) 2018/844 of the European Parliament. Available online: http://data.europa.eu/eli/dir/2018/844/oj (accessed on 26 June 2023).
91. Council of the European Union. Council Conclusions on the World Plan for Culture 2019-2022. *Off. J. Eur. Union* **2018**, *460*, 12–25.

92. Council Conclusions on the Culture, High-Quality Architecture and Build Environment as Key Elements of the New European Bauhaus Initiative. Available online: https://eur-lex.europa.eu/legal-content/EN/TXT/?uri=CELEX:52021XG1213(03) (accessed on 26 June 2023).
93. The National Affordable Homes Agency. *721 Housing Quality Indicators (HQI) Form*, 4th ed.; The National Affordable Homes Agency: London, UK, 2008.
94. Andrew, M.; Meen, G. House Price Appreciation, Transactions and Structural Change in the British Housing Market: A Macroeconomic Perspective. *Real. Estate Econ.* **2003**, *31*, 99–116. [CrossRef]
95. Otto, P.; Schmid, W. Spatiotemporal Analysis of German Real-Estate Prices. *Ann. Reg. Sci.* **2018**, *60*, 41–72. [CrossRef]
96. Curto, R.; Fregonara, E.; Semeraro, P. Listing Behaviour in the Italian Real Estate Market. *Int. J. Hous. Mark. Anal.* **2015**, *8*, 97–117. [CrossRef]
97. Gann, D.; Salter, A.; Whyte, J. Design Quality Indicator as a Tool for Thinking. *Build. Res. Inf.* **2003**, *31*, 318–333. [CrossRef]
98. Goodchild, M.F. Citizens as Sensors: The World of Volunteered Geography. *GeoJournal* **2007**, *69*, 211–221. [CrossRef]
99. Friedman, A. Challenges and Opportunities of Contemporary Architecture. *Architecture* **2021**, *1*, 3–4. [CrossRef]
100. Pardo-Bosch, F.; Aguado, A.; Pino, M. Holistic Model to Analyze and Prioritize Urban Sustainable Buildings for Public Services. *Sustain. Cities Soc.* **2019**, *44*, 227–236. [CrossRef]
101. Bresson, S.; Denèfle, S. Diversity of Self-Managed Co-Housing Initiatives in France. *Urban Res. Pract.* **2015**, *8*, 5–16. [CrossRef]
102. Coudroy De Lille, L. Housing Cooperatives in Poland. The Origins of a Deadlock. *Urban Res. Pract.* **2015**, *8*, 17–31. [CrossRef]
103. Put, B.; Pasteels, I. Motivational Barriers to Shared Housing: The Importance of Meanings of "Home" in the Diffusion of Housing Innovations. *Hous. Theory Soc.* **2022**, *39*, 257–274. [CrossRef]
104. Jarvis, H. Towards a Deeper Understanding of the Social Architecture of Co-Housing: Evidence from the UK, USA and Australia. *Urban Res. Pract.* **2015**, *8*, 93–105. [CrossRef]
105. Lang, R.; Carriou, C.; Czischke, D. Collaborative Housing Research (1990–2017): A Systematic Review and Thematic Analysis of the Field. *Hous. Theory Soc.* **2020**, *37*, 10–39. [CrossRef]
106. Studies of Conditions and Directions of Spatial Development of the City of Warsaw. Available online: https://architektura.um.warszawa.pl/-/dokument-projektu-studium-materialy1 (accessed on 18 June 2023).
107. Gehl, J. *Life between Buildings: Using Public Space*; Island Press: Washington, DC, USA, 2011; ISBN 978-1-59726-827-1.
108. Vietrova, P.; Vasyliev, P.; Maksymiv, L. Participatory Design Method for Improvement of Urban Public Spaces—Case Study. *Acta Sci. Pol. Archit.* **2023**, *21*, 15–23. [CrossRef]
109. Khechiba, A.; Djaghrouri, D.; Benabbas, M.; Leccese, F.; Rocca, M.; Salvadori, G. Balancing Thermal Comfort and Energy Consumption in Residential Buildings of Desert Areas: Impact of Passive Strategies. *Sustainability* **2023**, *15*, 8383. [CrossRef]

buildings

MDPI

Article

The Influence of Residential Block Form on Summer Thermal Comfort of Street Canyons in the Warm Temperate Zone of China

Xiaoyi Song, Guangbin Wang, Qingtan Deng *, Siyu Wang and Chenxia Jiao

School of Architecture and Urban Planning, Shandong Jianzhu University, Jinan 250101, China; xiaoyi11song@163.com (X.S.); 2020055258@stu.sdjzu.edu.cn (G.W.); wangsiyu_chn@163.com (S.W.); 2021055213@stu.sdjzu.edu.cn (C.J.)
* Correspondence: dengqingtan@126.com

Abstract: The residential block is the cognitive unit for residents to constitute urban imagery. As one of the most frequently used urban outdoor public spaces by residents, the thermal comfort of street canyons is an essential indicator for assessing sustainable and livable cities. The problem of the uncomfortable summer climate in the warm temperate zone of China has not been adequately studied. The study aims to analyze the influence of the building layout form of residential block units and block configuration on the outdoor summer thermal comfort of street canyons. Outdoor air temperature (T_a), mean radiant temperature (T_{mrt}), wind speed (V_a), and physiological equivalent temperature (PET) were simulated using ENVI-met. A new index, PET_{ws}, was introduced based on a statistical analysis of the PET index to assess the overall street canyon thermal comfort of the block. The results indicate that the number of rows of buildings in the building row layout has a more significant effect on the summer thermal comfort PET of street canyons than the number of columns in the warm temperate zone, especially on N–S-oriented streets. Reducing the number of rows can increase the overall thermal comfort PET_{ws} of street canyons by a maximum of 2.2%. The best choice for the number of building columns is two columns. Adopting different block configurations can increase the thermal comfort PET_{ws} of street canyons by up to 2.5%. An optimal block form has been created to improve the overall street canyon summer thermal comfort of the block.

Keywords: block form; street canyon; thermal comfort; PET; ENVI-met

Citation: Song, X.; Wang, G.; Deng, Q.; Wang, S.; Jiao, C. The Influence of Residential Block Form on Summer Thermal Comfort of Street Canyons in the Warm Temperate Zone of China. *Buildings* **2023**, *13*, 1627. https://doi.org/10.3390/buildings13071627

Academic Editors: Carol Monticelli, Anna Stefańska, Daniel Mateus, Carlos C. Duarte and Nuno D. Cortiços

Received: 31 May 2023
Revised: 24 June 2023
Accepted: 25 June 2023
Published: 27 June 2023

1. Introduction

With rapid urbanization and continuous human intervention in the natural environment, the climatic environment of urban areas has changed significantly. The urban heat island (UHI) effect, deterioration of habitat, and public health crisis have led to the realization that sustainable practices are needed to avoid human thermal comfort problems caused by extreme microclimatic environments [1–3]. A good microclimate environment can effectively extend the time of outdoor activities of residents and improve the sustainability and livability of urban space and environment [4]. Early involvement in urban microclimate issues before the design decision stage and the adoption of appropriate urban design strategies is a necessary and effective way to address climate change and improve outdoor thermal comfort [5,6].

Blocks are the basic building units of cities. Blocks with different morphologies and interfaces are bound to have different heat transfer characteristics in their internal spaces and thus exhibit different microclimatic characteristics. As one of the representative urban activity places and behavioral spaces in the built environment, street canyons are units with specific microclimate and play an important role in improving urban microclimate. In previous studies, how street canyon geometry such as building height to street width (H/W) [7–9], orientation [7,9,10], sky view factor (SVF) [8], width and length [11] affects the microclimate and thermal comfort of street canyons in urban blocks has been investigated. It was found that H/W and orientation have a greater effect on the thermal comfort of street canyons [12]. Under hot and dry climatic conditions, the intensity of UHI increases as the H/W ratio decreases [13]. The high aspect ratio can increase wind speed and shade in street canyons, thus improving thermal comfort at the pedestrian level, especially in summer. However, as the ratio between canyon length and the height of buildings (L/H) increases, there is no obvious change in thermal comfort [11]. Chatzidimitriou et al. [14] discussed that the most comfortable in summer are deep canyons, while the most comfortable in winter are canyons of medium width. By evaluating wind flow and temperature variation, it has been demonstrated that the ratio of H/W = 1 and the ratio of the street length to width (L/W) = 2 are optimal values for controlling UHI [15]. Almost all studies have shown that east–west (E–W)-oriented street canyons have the worst thermal comfort relative to other orientations [16,17]. For hot and humid environments, north–south (N–S)-oriented street canyons can obtain the most comfortable time, followed by northwest-southeast (NW–SE)-oriented and northeast–southwest (NE–SW)-oriented [18]. In a study conducted in Tabriz, analysis of T_a, T_{mrt}, and PET showed that 135°, 145°, and 155° are the best street orientations for thermal comfort in the relatively hot summer, and 135° is the most comfortable choice in the cold winter [19]. The research on street canyons conducted in a cold city has indicated significant quadratic correlations between SVF and thermal environmental elements of the street canyon [20].

The inappropriate arrangement of buildings in urban blocks causes less efficient air exchange in the canyon. It is difficult for outside winds to enter and heat to escape from the canyon, resulting in heat buildup in the canyon in summer. A study in Nanjing, China, showed a correlation between building layout patterns and microclimate elements. The building layout had a more significant impact on T_{mrt} and wind speed than on T_a [17]. Taleghani et al. [21] conducted a comparative study of the thermal comfort of five different building forms and layouts at pedestrian height at the hottest time of summer. The results indicated that the courtyard type can provide the best thermal comfort for the human body in summer compared to other spatial forms of building complexes. Shareef et al. [22] demonstrated that in hot climate regions, a meandering building layout reduced the block air temperature by 1.9 °C compared to the baseline case of a grid arrangement. However, a grid arrangement of pavilions with straight canyons was more beneficial to increase the wind speed within the canyons than a block with staggered canyons. Ma et al. [23] found that when the ratio of surrounding building height to street building height (SH/h) $\geq$ 1.8 and building coverage ratio (BCR) $\geq$ 47%, street pedestrian thermal comfort can be achieved for at least three comfort hours regardless of street orientation or layout form. Among the studies that have been conducted so far, some through numerical models [24,25] and some based on field measurements [26,27].

As a result of climate change, the combined effect of the buildings and environment around the street on the microclimate and thermal comfort of the street canyon is becoming more and more significant. Therefore, the thermal comfort performance of street canyons should be considered for analysis at the block scale rather than in an isolated street space [28]. Referring to the above literature review, most of the urban morphological parameters involved in the current studies are related to street geometry, including street aspect ratio, orientation, SVF, etc. Among them, studies on orientation generally focus on the variation of the thermal environment in a single orientation or the comparison between different orientations. However, fewer studies have focused on the impact of street block unit building layout and block configuration around streets on the thermal comfort performance of street canyons. There is a lack of an index that considers the combined effect of the thermal environment of canyons with different street orientations at the block scale to assess the thermal comfort performance of canyons. At present, research related to thermal comfort in street canyons tends to target specific climatic regions or areas, with more research on humid and hot climates [29,30]. The conclusions are not applicable to other climatic conditions. Studies in China in recent years have also been conducted mainly in Guangzhou [31,32], Nanjing [17], and Harbin [33]. Existing studies have not paid attention to the frequent hot summer weather, the significant reduction of outdoor thermal environment quality in urban residential blocks, and the decline of thermal comfort of residents in the warm temperate zone of China in recent years. In the design of residential blocks, there is a need to provide a spatial form that creates a comfortable outdoor thermal environment.

This study takes residential blocks in Jinan, China, as the object of study to investigate the effects of block unit building layout and block configuration on the summer thermal comfort of street canyons under specific climatic conditions in the warm temperate zone. Based on a quantitative simulation study, this paper provides an optimal block configuration that can improve the summer thermal comfort of street canyons. It helps to reasonably control the spatial form of the block at the beginning of block planning and design to create an excellent outdoor thermal environment at the lowest cost and provides a reference for the design of climate adaptability residential blocks.

2. Methods

2.1. Study Area

The warm temperate zone is one of the climate zones in China classified by the standard " Names and codes for climate regionalization in China-Climatic zones and climatic regions" [34]. The warm temperate zone is hot and rainy in summer and cold and dry in winter. Little attention has been paid to the thermal comfort of warm temperate summer in China. In recent years, the high summer temperatures in China's warm temperate zone have been frequent, and the quality of the thermal environment in residential blocks has declined significantly. Jinan (36.40° N, 117.00° E) is one of the representative cities in the warm temperate zone of China. The summer in Jinan is hot and rainy. The topography of Jinan is high in the south and low in the north, surrounded by mountains on three sides, thus forming a unique semi-basin terrain in Jinan, which makes it difficult to obtain convection in the stratosphere. It is one of the reasons for the high temperature in the summer. The frequency of extreme temperatures shows an overall upward trend in Jinan. In 2022, the extremely highest temperature exceeded 40 °C. According to the data of the National Meteorological Information Center, the monthly temperature and relative humidity of Jinan from 2010 to 2020 are shown in Figure 1. Average monthly temperature variation throughout the year ranges from −0.2 °C to 27.8 °C. The annual average number of high-temperature (daily maximum temperature ≥ 35 °C) days is 12 d. In addition, July has the highest temperature, with a monthly average temperature of 27.8 °C. In summer (June–August), the relative humidity is between 44% and 79%. The summer thermal environment in Jinan shows an overheating trend.

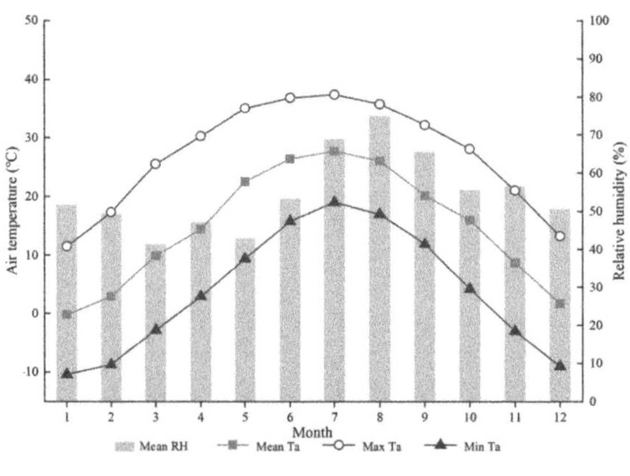

Figure 1. Monthly temperature and relative humidity in Jinan (2010–2020) (National Meteorological Information Center (http://data.cma.cn/), accessed on 27 October 2022).

Jinan has a diverse urban morphology due to its long history, and the block form is diverse as it has gone through different stages of urban development. Influenced by the ancient traditional city layout and foreign planning ideas brought by the self-opening of commercial ports in the Republic of China, the road layout of residential blocks in the old city of Jinan mostly adopted the planning model of small blocks. In the process of expanding and spreading the urban space form of the main city of Jinan in all directions, due to the restriction of the Yellow River in the north and the mountainous area in the south, the urban space shows the form of developing to the east and west flank axes, forming the new urban area in the east and west. The " Standard for urban residential area planning and design" [35] proposes that "residential areas should adopt the traffic organization mode of 'small block, dense road network', and the road network density should not be less than 8 km/km^2; the spacing between urban roads should not exceed 300 m, and it is appropriate to be 150–250 m and should be combined with the layout of residential blocks". The scale of residential blocks in new urban areas is also gradually decreasing under the influence of the theory of New Urbanism and the practice of New Urbanism in Jinan. In this study, the main urban area within the Jinan City bypass expressway was selected as the study area (Figure 2), and small-scale residential blocks were used as the study object.

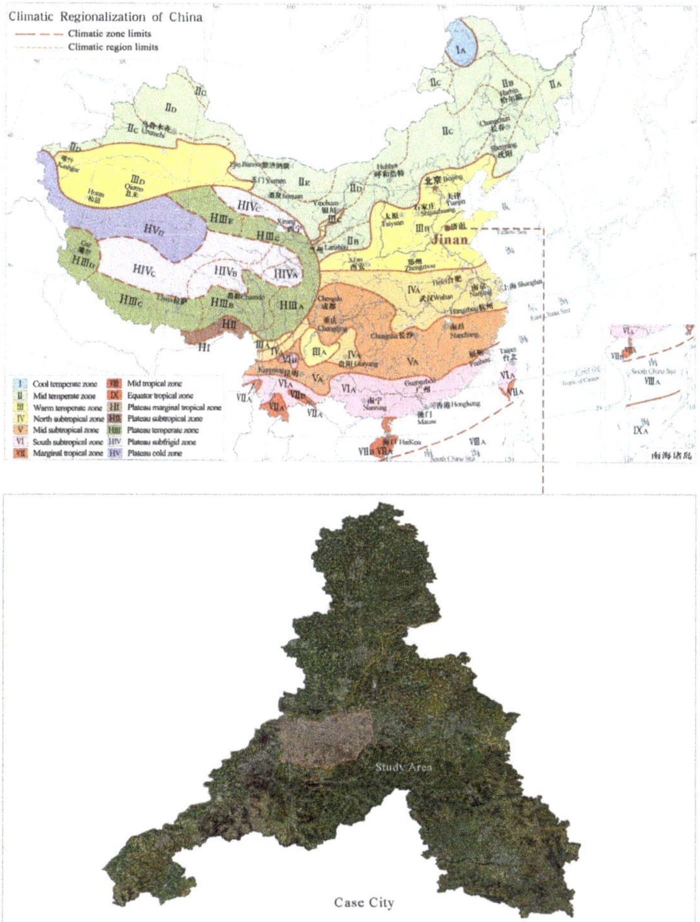

Figure 2. Location of the study area (Geographic Data Sharing Infrastructure, College of Urban and Environmental Science, Peking University (http://geodata.pku.edu.cn),accessed on 18 June 2023).

2.2. Research Framework

The study analyzes the effect of small-scale residential block forms on the thermal comfort of street canyons in Jinan. The methodological framework of this study is shown in Figure 3.

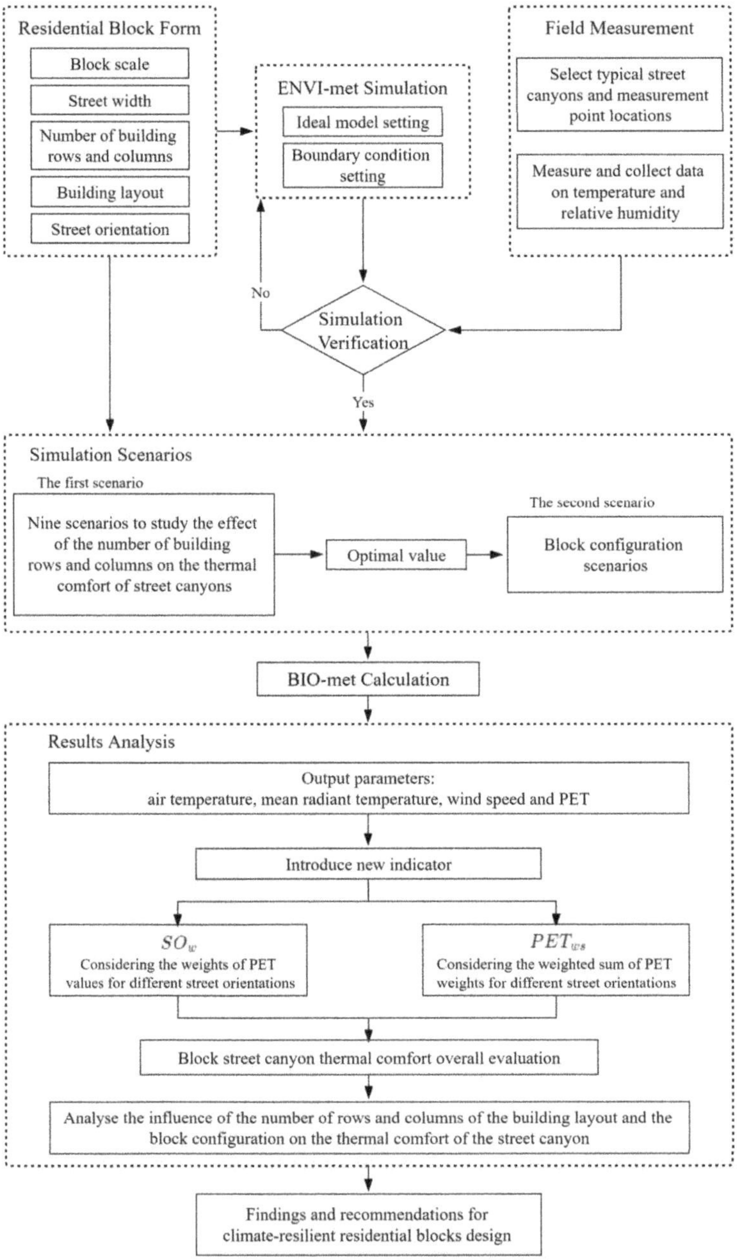

Figure 3. Research framework.

2.3. Building Ideal Models of Residential Blocks

2.3.1. Block Scale Setting

This paper is based on the analysis and study of the actual road network scale of residential blocks in the study area, and according to "the Technical Regulations for the Management of Urban and Rural Planning in Jinan" and other relevant specifications on road width, building setback distance and building spacing, 100 m × 100 m is selected as the building line range of the block unit. It is offset 8 m outward as the boundary line of the road, and then offset 10 m outward as the boundary line of the block unit, forming a block unit with a scale of 136 m × 136 m. The block unit is gridded in 2 × 2 to generate an ideal residential block of 272 m × 272 m (Figure 4).

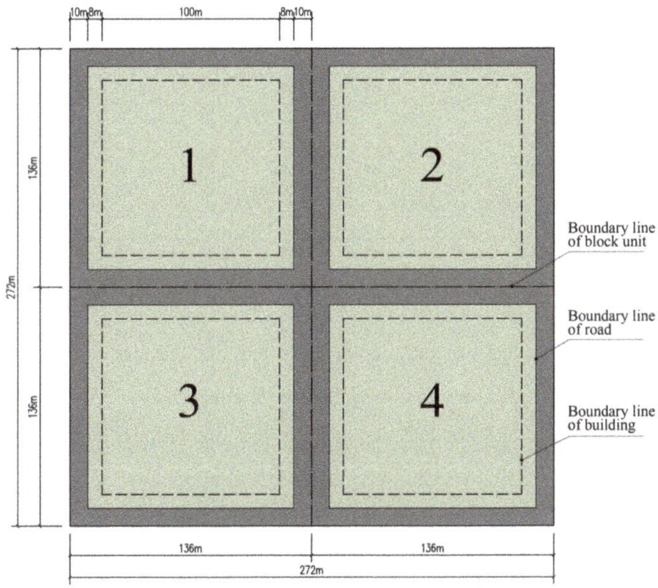

Figure 4. Ideal residential block scale model.

2.3.2. Basic Block Morphology Abstraction

In this paper, through the research of small-scale residential blocks in the study area, the row layout is selected as the building plan combination form. Row layout is conducive to light and ventilation and is a more commonly used building group layout mode. Nine building layout types were extracted using the number of rows and columns of buildings as the parameter variables (as shown in Figure 5, the number of rows refers to the number of buildings in north–south orientation, and the number of columns refers to the number of buildings on east–west orientation) in the row layout (Figure 6). Considering the scale of the street block and building size, the number of rows and columns is set to 1–3, respectively. To keep the building floor area ratio in the block constant, the building height decreases with the increase in the number of rows. According to the regulation that the height of residential buildings in Jinan shall not be greater than 80 m, the building height is set at 72 m (24 floors), 36 m (12 floors), and 24 m (8 floors). Combining the requirements for daylight spacing in the above relevant specifications and taking into account the differences in the design of daylight spacing between multi-story and high-rise buildings, the N–S and E–W spacing of buildings is set. In order to facilitate the calculation of floor area and data processing analysis, it is assumed that the building density is the same in the model with the same number of rows. Only N–S and E–W orientations are selected for street orientation in this study, which is the main orientation of streets in residential blocks in

Jinan. Nine different base block unit models were formed by combining the number of rows and columns (Table 1) and labeling them based on rows and columns.

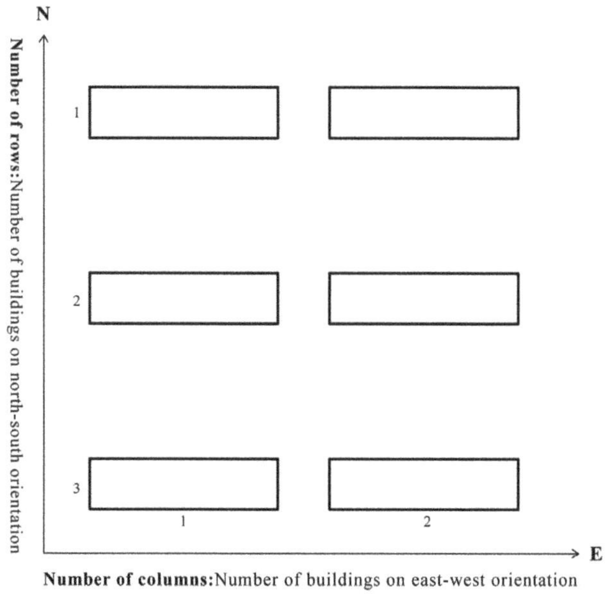

Figure 5. Definition of row and column numbers.

Figure 6. Building layout forms extracted in the study area.

Table 1. Basic block unit abstraction and classification.

Prototypes	R1C1	R2C1	R3C1
Schematic diagram	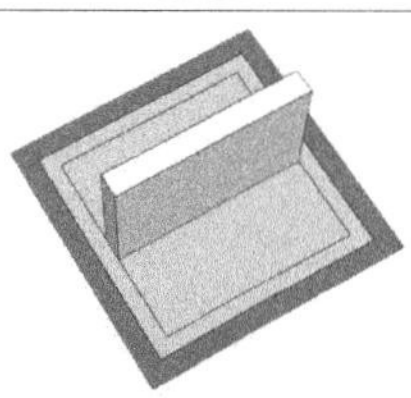	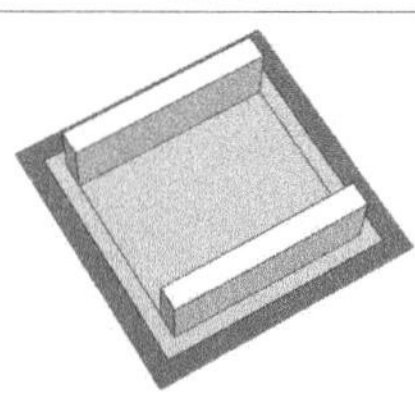	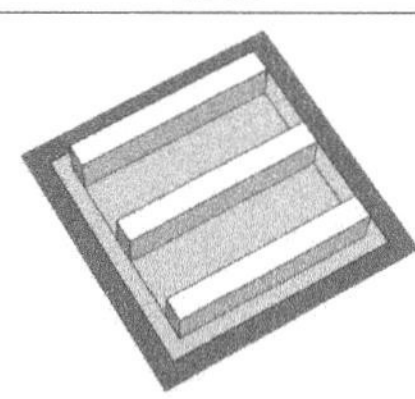

Table 1. *Cont.*

Prototypes	**R1C1**	**R2C1**	**R3C1**
Plane diagram	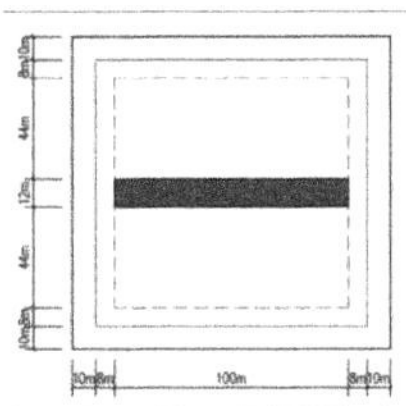	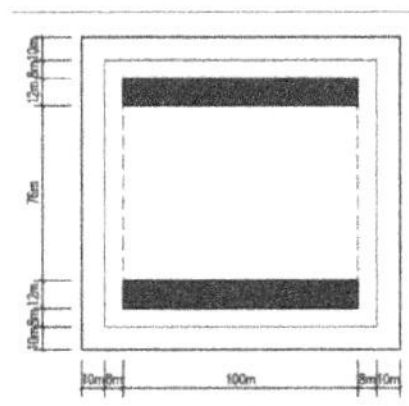	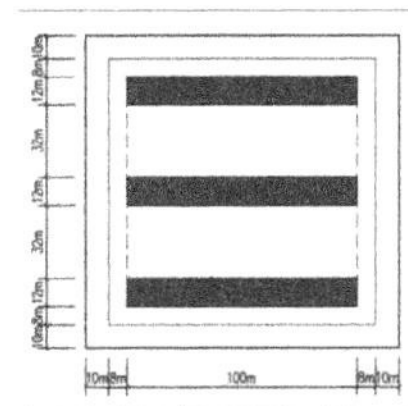
Building height Row Column	72 m 1 1	36 m 2 1	24 m 3 1
Prototypes	**R1C2**	**R2C2**	**R3C2**
Schematic diagram	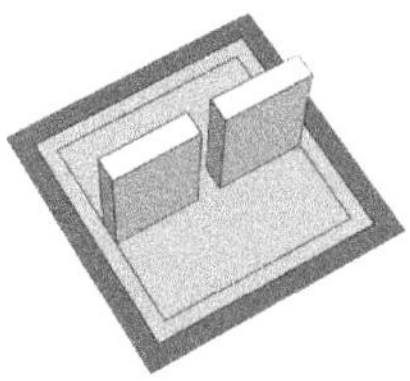	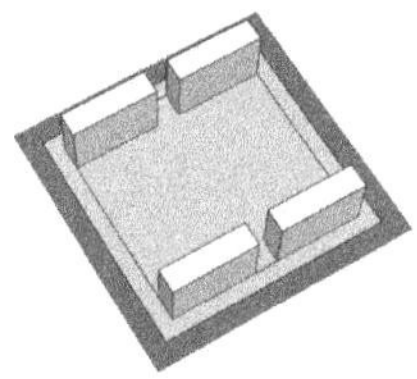	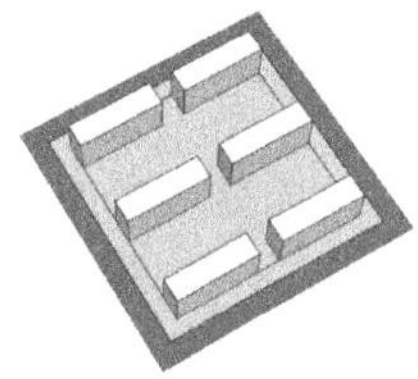
Plane diagram	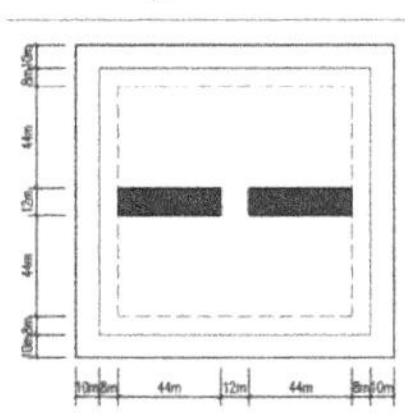	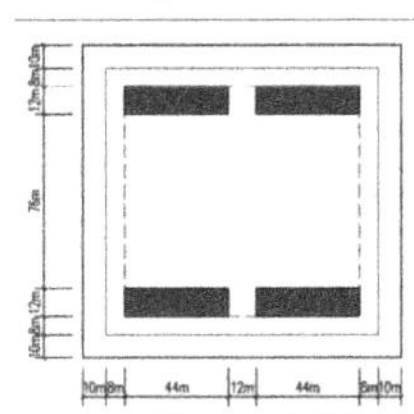	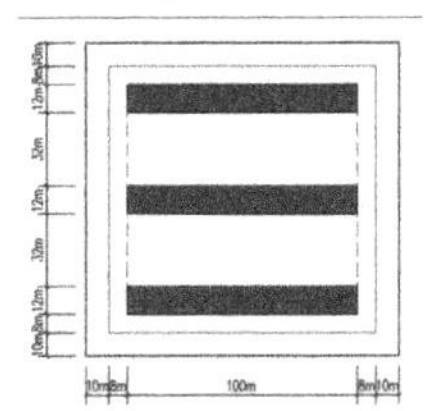
Building height Row Column	72 m 1 2	36 m 2 2	24 m 3 2
Prototypes	**R1C2**	**R2C2**	**R3C2**
Schematic diagram	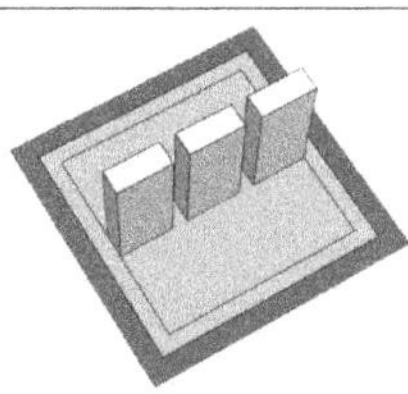	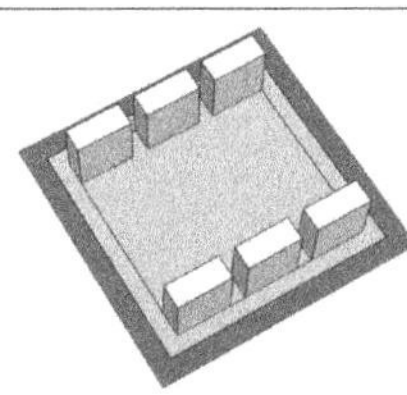	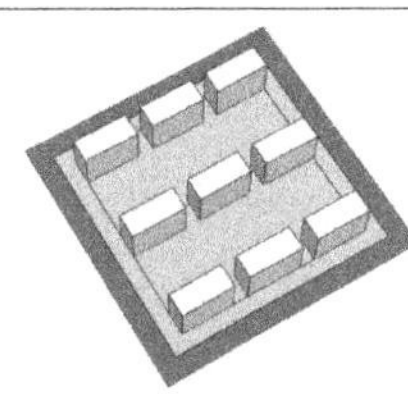
Plane diagram	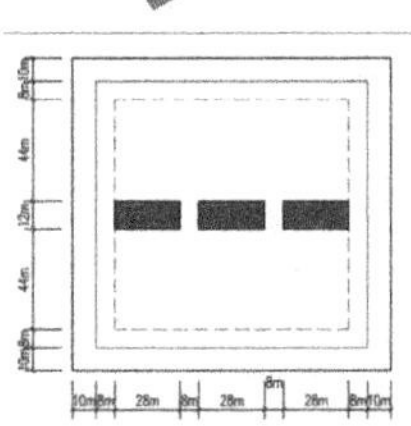	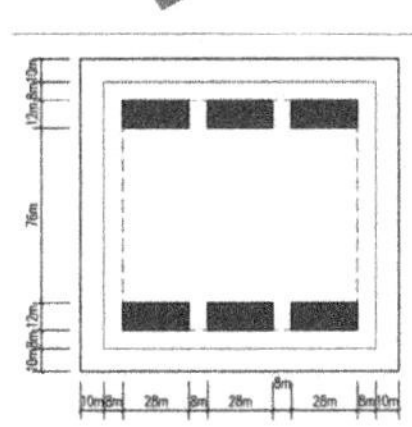	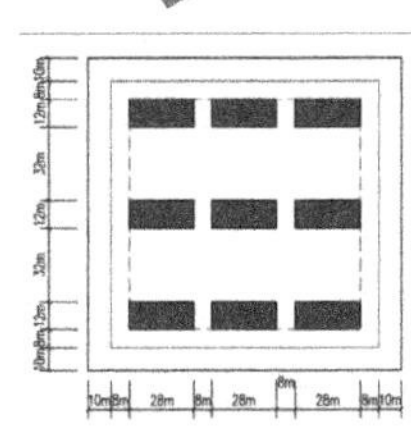
Building height Row Column	72 m 1 3	36 m 2 3	24 m 3 3

2.3.3. Urban Residential Blocks Design Scenario

In this study, two urban residential block design scenarios are set up. First, the building layout of the four block units is the same form, which can form nine block forms, as shown in Figure 7.

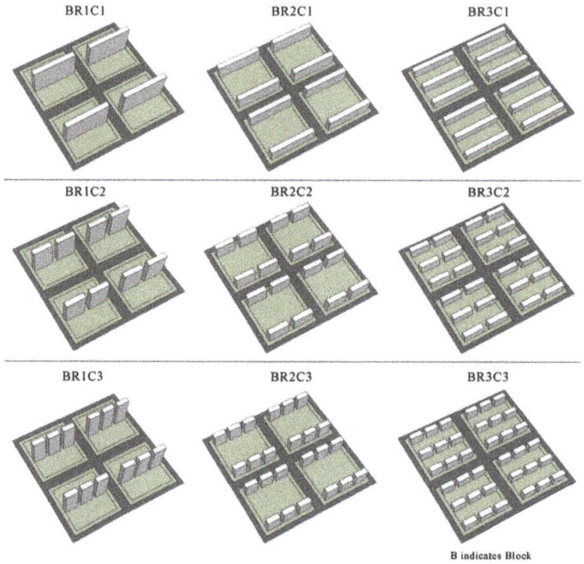

Figure 7. Basic block form abstraction.

By simulating and analyzing the thermal comfort of N–S- and E–W-oriented street canyons, the number of building columns has no significant effect on the thermal comfort of N–S-oriented street canyons. Therefore, in the second scenario, the number of building columns is considered to be constant. The number of building columns with optimal thermal comfort in E–W-oriented street canyon was selected, and the block configuration was carried out with three basic block unit models, R1C2, R2C2, and R3C2, depending on the number of rows. Considering the issues of daylighting, shading, and ventilation, the building heights on both sides of the E–W-oriented street are set to the same height or north high and south low in block configuration scenarios, without considering the south high and north low. Building height restrictions are not set on both sides of the N–S-oriented street. This generates 36 different block configuration scenarios (Figure 8), including the three block configuration scenarios BR1C2, BR2C2, and BR3C2 in the first scenario.

2.4. Simulation Study

In recent years, the ENVI-met V5.0 3D urban microclimate simulation software [36] developed and continuously improved by Bruse et al. has become increasingly mature for outdoor microclimate simulation studies due to its strong relevance and applicability [37,38], especially in residential environments [39]. Researchers have applied it to a variety of scenarios including streets, blocks, botanical gardens, industrial areas, and campuses. Its reliability has been verified in studies in several climatic zones [40,41]. In this study, outdoor thermal comfort was calculated using the BIO-met V5.0 in ENVI-met. BIO-met can simulate the universal thermal index, including the standard effective temperature, physiologically equivalent temperature, predicted mean vote, and universal thermal climate index.

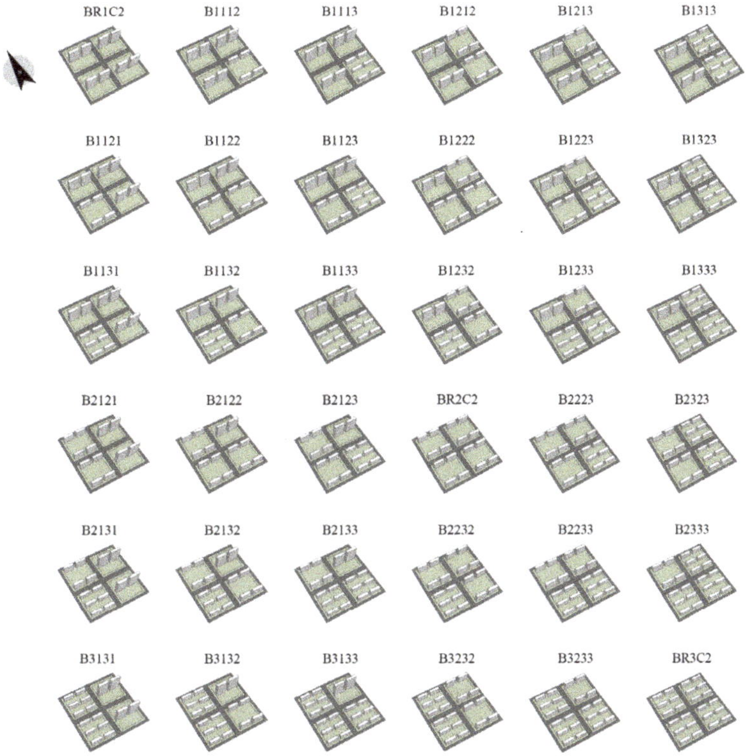

Figure 8. Block configuration scenarios.

2.4.1. Initialization Model and Boundary Conditions

The initial model of typical residential blocks in Jinan was established by using ENVI-met software. The computational domain was finally set to 320 m × 320 m × 152 m (x × y × z) based on a series of pre-simulations. The grid resolution of all models was 4 m × 4 m × 4 m (x × y × z).

As mentioned earlier, July is the hottest month of the year in Jinan. In this paper, 8 July 2022 is selected as the simulation date, which is the hottest day in July, and there is no rain in the two days before and after that day. In addition, the meteorological data of that day were obtained from the Jinan meteorological station (area station number: 54823, 36.36° N, 117.00° E, altitude of the observation site: 170.3 m), which was used as the initial meteorological conditions for the simulation. The specific meteorological data are shown in Table 2. In addition, considering the daily temperature variation, the total simulation duration was set to 18 h in order to avoid the effect of initial conditions. The main input parameters of the simulation are set as shown in Table 3, and the rest of the parameters are default values.

Table 2. Meteorological data from Jinan Meteorological Station on 8 July 2022.

Time	0:00	1:00	2:00	3:00	4:00	5:00	6:00	7:00	8:00	9:00	10:00	11:00
T_a	27.1	27.4	26.8	26.6	26.9	27.3	28	29.1	30.2	31.3	31.9	32.7
RH	67	64	69	69	64	57	53	53	52	50	47	46
Time	**12:00**	**13:00**	**14:00**	**15:00**	**16:00**	**17:00**	**18:00**	**19:00**	**20:00**	**21:00**	**22:00**	**23:00**
T_a	33.8	34	34.4	34.1	34.2	33.7	33.1	32	30.3	29.9	30	29.5
RH	45	44	46	43	45	46	47	50	54	52	52	52

Table 3. Input parameters for simulations with ENVI-met.

Parameters	Input Data
Date	8 July 2022
Domain size (dx, dy, dz)	320 × 320 × 152 m
Size of grid cells	4 × 4 × 4 m
Duration	2:00–20:00
Initial T_a and RH	Hourly data from weather station
Wind speed and direction at inflow border	2.7 m/s, 180°
Albedo of walls/roofs	0.3
Thermal conductivity of walls/roofs	1.74 W/(m·K)
Albedo of pavement material	Asphalt: 0.2, cement brick: 0.3
Body parameters	A 35-year-old male, 1.75 m, 75 kg
Static clothing insulation	0.4 clo
Walking speed	1.21 m/s
Total metabolic rate	86.21 W/m^2

A total of 18 points on each block model N–S- and W–E-oriented street canyon were selected as the output data, and the average of the 18 receptors for each street was the data for that street (Figure 9). The same street cross-section has three output points: the center point of the carriageway and a point on the sidewalk on both sides (one point on each side), with each output point located at a height of 1.2 m from the ground (Figure 10).

2.4.2. Simulation Validation

A typical residential block (positioned at 36.43° N, 117.06° E) was selected as the measured site. The field measurement was conducted on 8 July 2022 from 6:00 to 20:00. As shown in Figure 11a, the red outline indicates the location of the measured street. The TRM-GPS1 handheld network weather station instrument used for the field measurements is shown in Figure 11b to record the hourly T_a and RH at 1.2 m height. To calibrate the simulation model, Figure 11c provides a photograph of the street canyon taken with a fisheye lens, which was used to calculate the SVF.

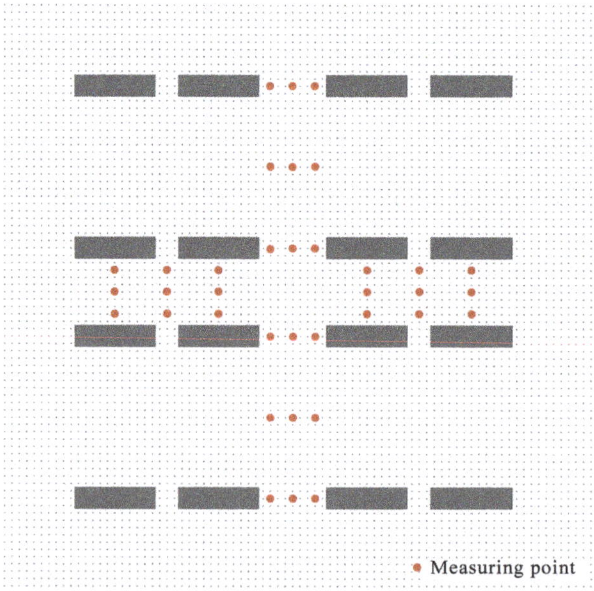

Figure 9. Output measuring points of the simulation data.

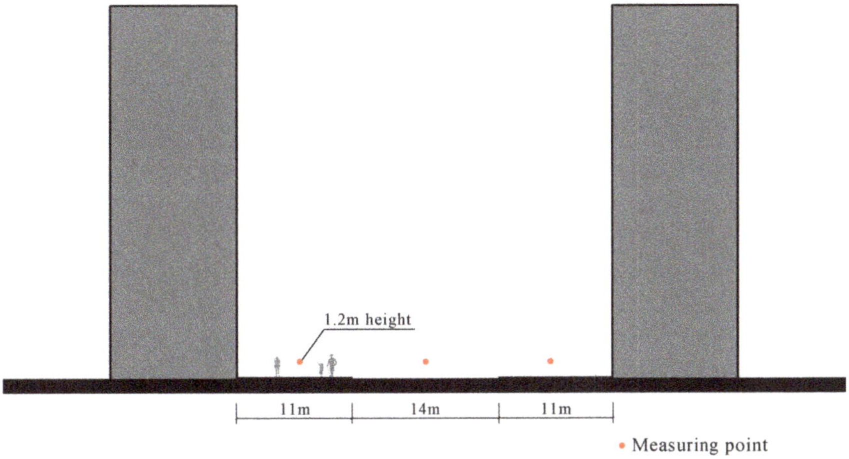

Figure 10. Description of the research model.

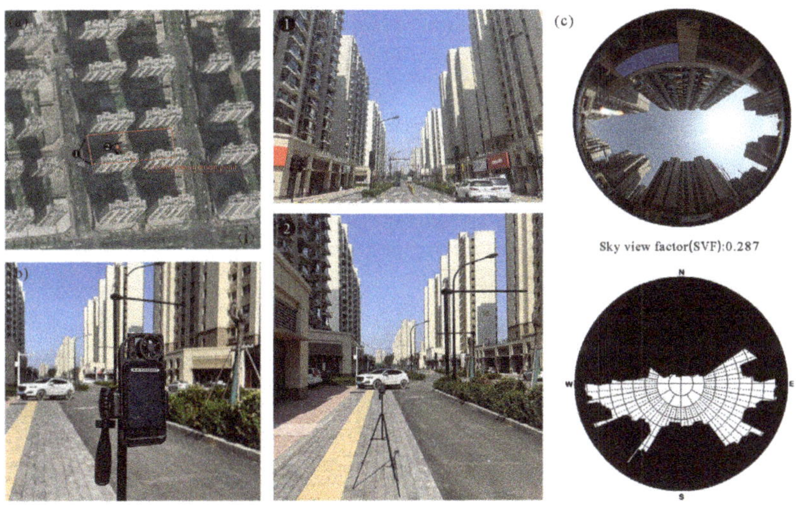

Figure 11. On-site measurement scenario; (**a**) the street canyon (red outline) and the on-site measurement point; (**b**) a setup of the measuring instrument; (**c**) fisheye lens image and SVF of the measurement location.

As shown in Figure 12, the simulated results of the environmental elements of the street canyon have a strong correlation with the measured results. It can also be found that the simulated results are lower than the measured values. In a natural environment, T_a and RH are affected by various complex factors, such as changes in cloudiness, wind fields, reflections from the surrounding environment, etc., and the measured values fluctuate. However, during the simulation, ENVI-met cannot take into account environmental changes and human interference. Considering the error range of the instruments used (temperature $\pm$ 0.4 °C, relative humidity $\pm$ 3%) and the measurement range (temperature: -30~70 °C, relative humidity: 5~100%), the relevant simulations using ENVI-met are reliable.

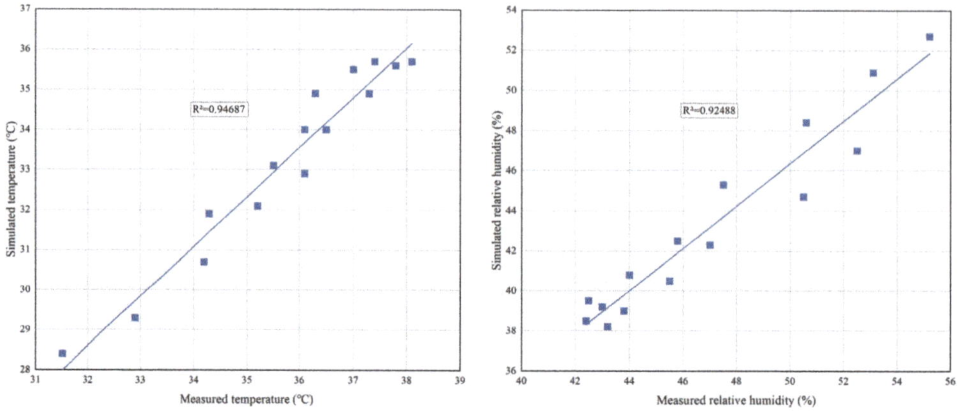

Figure 12. Simulated and measured values of (**a**) T_a, (**b**) RH.

2.4.3. Thermal Comfort Evaluation

The Physiological Equivalent Temperature (PET) Index is currently one of the most used indexes for outdoor thermal comfort evaluation. Höppe et al. [42] proposed PET based on the Munich energy balance model for individuals (MEMI). The PET index takes into account various factors such as the main meteorological parameters, activity intensity, clothing, and individual parameters, and is therefore relatively more objective. This index is now used in different climatic regions [43]. PET has been shown to be applicable in cold regions of China [44–46]. Residents in different climate zones have different temperature tolerance, and the range of heat index may also vary. This paper selects the PET range defined in Table 4 to evaluate pedestrian thermal comfort [47].

Table 4. PET value and thermal perception.

Thermal Perception	PET (°C)
Very cold	<4
Cold	4~8
Cool	8~13
Slightly cool	13~18
Neutral	18~23
Slightly warm	23~29
Warm	29~35
Hot	35~41
Very hot	>41

2.5. Introducing PET_{ws} and SO_w

In this study, considering that the thermal comfort of street canyons with different orientations in the block has different trends with the change in block morphological parameters, PET_{WS} and SO_W were introduced in order to accumulate different orientation PET values and thus evaluate the overall street canyon thermal comfort of the block. PET_{WS} is a weighted sum taking into account the weight of PET values for different street canyons orientations (SO_W).

The objective weighting method is based directly on the raw information of the indicator and obtains the weights after processing by statistical methods. Entropy method is a common method in objective assignment methods. The greater the variance of an evaluation indicator, the lower the entropy value, the more information the indicator contains and transmits, and the greater the corresponding weight.

In an assessment problem with m indicators and n evaluated objects, the entropy H_i of the *i*th indicator is defined by Equation (1):

$$H_i = -k\sum_{j=1}^{n} p_{ij}lnp_{ij}, i = 1, 2, \ldots, m \tag{1}$$

where $k = -ln(n)^{-1}, p_{ij} = x_{ij}/\sum_{j=1}^{n} x_{ij}$, when $p_{ij} = 0$, $p_{ij}lnp_{ij} = 0$. x_{ij} is the *j*th evaluated object of the *i*th indicator.

The entropy weight w_i of the *i*th indicator is defined as in Equation (2):

$$w_i = \frac{1 - H_i}{m - \sum_{i=1}^{m} H_i} \tag{2}$$

where $0 \leq w_i \leq 1$, $\sum_{i=1}^{m} w_i = 1$.

Calculate the PET_{ws} for each configuration block according to Equation (3):

$$PET_{ws} = \sum_{i}^{m} PET_i \times SO_{wi} \tag{3}$$

where PET_i is the PET value in *i* orientation and SO_{wi} is the weight in *i* orientation.

3. Results

This paper focuses on T_a, T_{mrt}, V_a, and PET, which are environmental factors affecting outdoor thermal comfort, as the basic factors for studying the thermal comfort of street canyons. The average of T_a, T_{mrt}, V_a, and PET in street canyons of each block from 6:00 to 20:00 were counted for statistical analysis. This time period is the main time for residents' outdoor activities.

3.1. Impact of Building Row Number on T_a, T_{mrt}, V_a, PET

Figure 13 illustrates that in summer, the average of T_a, T_{mrt}, V_a, and PET change significantly with the increase in the number of rows and columns of the building layout, and the change trend is different. There are significant differences in the influence of the same factors by the number of building rows and columns on streets with different orientations.

	E-W			N-S		
	BR1C1	BR2C1	BR3C1	BR1C1	BR2C1	BR3C1
T_a	32.87	33.16	33.15	33.33	33.51	33.64
T_{mrt}	53.93	51.74	52.94	50.87	47.97	46.38
V_a	0.44	0.16	0.13	2.08	2.02	1.89
PET	43.73	45.16	46.54	40.23	39.03	38.39
	BR1C2	BR2C2	BR3C2	BR1C2	BR2C2	BR3C2
T_a	32.99	33.25	33.24	33.38	33.51	33.67
T_{mrt}	53.93	51.82	52.86	50.85	47.96	46.32
V_a	0.76	0.47	0.35	2.10	1.97	1.86
PET	42.76	43.99	45.48	40.31	39.04	38.40
	BR1C3	BR2C3	BR3C3	BR1C3	BR2C3	BR3C3
T_a	33.06	33.22	33.18	33.47	33.52	33.64
T_{mrt}	53.95	51.73	52.71	50.87	47.87	46.16
V_a	0.78	0.09	0.07	1.99	1.96	1.85
PET	42.77	46.89	48.38	40.38	39.01	38.32

Figure 13. Average of indicators for nine block scenarios.

As the number of building rows increases, in E–W-oriented street canyons, the average of T_a first increases and then decreases, but the decreasing trend is not significant. The average value is minimized when the number of rows is 1. The T_a of BR2C1 is 0.29 °C higher than that of BR1C1. The average of T_{mrt} decreases and then increases. The average value is maximum when the number of rows is 1. The T_{mrt} of BR2C3 decreased by 2.22 °C compared to BR1C3, and BR3C3 increased by 0.98 °C compared to BR2C3. The wind speed gradually decreases with the rows. The average of PET increases gradually with the increase in the number of rows, and the maximum variation trend between BR1C3, BR2C3, and BR3C3 increased by 4.12 °C and 1.49 °C, respectively. In N–S-oriented street canyons, the average of T_a increases with the increase in the number of rows. The T_a of BR3C1 is 0.31 °C higher than that of BR1C1. The average of T_{mrt}, V_a, and PET all decrease with the increase in the number of rows. The T_{mrt}, V_a, and PET of BR3C3 are reduced by 0.14 m/s, 4.71 °C, and 2.06 °C, respectively, compared with BR1C3. The change trends of T_a, T_{mrt}, V_a, and PET average values are similar in significance with increasing number of rows between different columns, but the trends of all four factors are different between streets with different orientations. The change trends of PET average values in E–W- and N–S-oriented street canyons are opposite, resulting in a gradual increase in the mean difference between differently oriented street canyons as the number of building rows increases. The mean difference in PET between different orientations in BR1C3 is 2.39 °C, while in BR3C3 the difference reaches 10.06 °C.

Among the block models with different numbers of building columns, the average T_a of the E–W-oriented street canyon is the lowest when the number of columns is 1, and BR1C1 is the model with the lowest T_a at 32.87 °C in 1 column. The average of T_{mrt} is the smallest when the number of columns is 3. The average T_{mrt} of BR1C1, BR1C2, and BR1C3 models are almost unaffected by the number of columns. The reason for this result may be related to its building height and the location of the building in the block units. The average of PET decreases and then increases with the increasing number of columns, and it is the smallest when the number of columns is 2. Figure 14a presents the curves of PET in the E–W street canyon at different time points for each block model. BR2C1, BR2C2, and BR2C3 mutate at 11:00am and show a decreasing trend, different from the trend of the other block models, and then gradually increase. The average T_a of the N–S-oriented street canyon is the lowest when the number of columns is 1. The average of T_{mrt} decreases with the increasing number of columns, but the decreasing trend is not obvious. The maximum value of the change is 0.16 °C. The effect of building column number on the average V_a and PET of N–S-oriented street canyon is not significant (Figure 14b), with variations ranging from 0.01 to 0.11 m/s, and 0.01 to 0.08 °C, respectively.

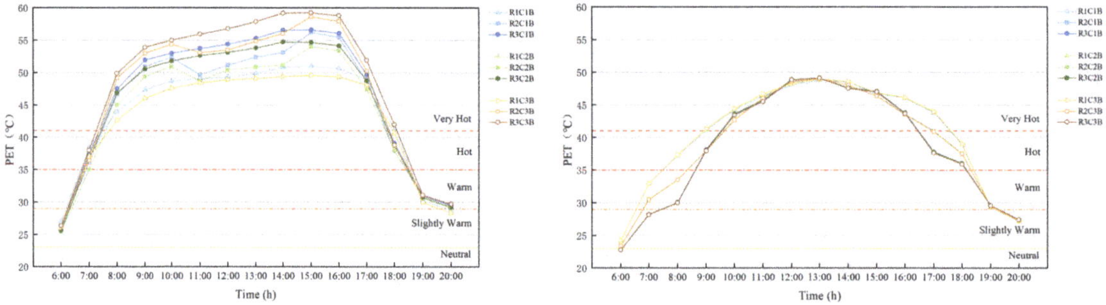

(a) (b)

Figure 14. PET changes in BR1C1–BR3C3 block models on (**a**) E–W- and (**b**) N–S-oriented streets.

The results indicate that the influence of the number of building rows on the four indicators is more significant than the number of building columns. For example, in E–W-oriented street, the maximum variation values of average T_a, T_{mrt}, V_a, and PET with the number of building rows are 0.29 °C, 2.22 °C, 0.71 m/s, and 4.12 °C, respectively, while the maximum values with the number of building columns are 0.12 °C, 0.15 °C, 0.11 m/s, and 2.9 °C, respectively. The influence of the number of building columns on the N–S orientation of the street canyon PET is not significant compared to the E–W orientation. In the same block model, the thermal comfort of the N–S-oriented street canyon is significantly better than the E–W orientation. Therefore, considering four factors comprehensively, the number of building columns with the lowest PET value of E–W-oriented street canyon 2 columns, is selected for block configuration with different numbers of building rows. The second scenario is configured based on the results of this section and the optimal choice of the number of columns.

3.2. Impact of Block Configuration on T_a, T_{mrt}, V_a, PET

According to the results in Figure 15, when the other blocks have the same building layout, the average of T_a in E–W-oriented street canyons gradually increases as the number of rows in block unit 1 in Figure 4 increases. The average of T_a increases from B1233 to B3233 by 0.06 °C and 0.07 °C, respectively. Both T_{mrt} and PET mean values decrease and then increase. The mean value of T_{mrt} decreases by 0.6~0.8 °C when the number of rows changes from one to two and increases by 0.07~0.17 °C when it changes from two to three. The change trend of PET between B1131, B2131, and B3131 is the largest, decreasing by 0.71 °C and increasing by 0.51 °C. The change in wind speed is not obvious. When the number of building rows in unit 1 changes from one to two, the average T_a of N–S-oriented street canyon increases. When the number of rows changes from two to three, there is no significant change in T_a. The average of N–S orientation T_{mrt} and PET decreases with the increase in the number of building rows in block unit 1. When other block building layouts are identical, as the number of building rows in block unit 2 increases in Figure 4, the trends of the E–W-oriented and N–S-oriented street canyons T_a, T_{mrt}, V_a, and PET averages are the same as in block unit 1. The maximum values of E–W orientation T_a, T_{mrt}, V_a, and PET variations are 0.09 °C, 0.75 °C, 0.03 m/s, 0.61 °C, and N–S orientation are 0.07 °C, 0.87 °C, 0.05 m/s, 0.47 °C, respectively. When the building layout of other blocks is the same, in the E–W-oriented street canyons, the mean value of T_a first increases and then decreases as the number of rows in block unit 3 in Figure 4 increases. The average of T_{mrt} decreases and then increases, and the value is maximum when the number of rows is three. The wind speed decreases gradually. The mean value of PET gradually increases. The average of T_a and PET gradually increases in the N–S-oriented street canyon. The average of V_a and T_{mrt} decreases. When other block building layouts are identical, as the number of building rows in block unit 4 increases in Figure 4, the trends of the E–W-oriented and N–S-oriented street canyons T_a, T_{mrt}, V_a, and PET averages are the same as in block unit 3.

The mean difference in T_a, T_{mrt}, V_a, and PET in E–W-oriented street canyons between B1313 and B3131 is 0.01 °C, 0.04 °C, 0.02 m/s, 0.03 °C, respectively, and the difference in N–S orientation is 0.05 °C, 0.62 °C, 0.04 m/s, 0.3 °C, respectively. The difference in E–W orientation between B1133 and B1313 is 0.11 °C, 0.52 °C, 0.20 m/s, 1.00 °C, respectively, and the difference in N–S orientation is 0.00 °C, 0.48 °C, 0.04 m/s, 0.23 °C, respectively. The average of T_a, T_{mrt}, V_a, and PET for two street block units of the same building layout configured on the north and south sides of the street differ significantly from those on the east and west sides, especially in E–W-oriented streets.

3.3. Impact of Block Configuration on PET_{ws}

Using the entropy method to calculate the mean PET values of the 36 block configuration scenarios, the PET weight of the E–W-oriented street canyon is 0.61 and the N–S orientation weight is 0.39. The street canyon PET_{ws} values for each block model are shown in Figure 16.

E-W

	BR1C2	B1112	B1113	B1212	B1213	B1313
T_a	32.99	33.09	33.02	33.15	33.06	33.14
T_{mrt}	53.93	53.32	53.98	52.97	53.44	53.55
V_a	0.76	0.63	0.59	0.67	0.62	0.59
PET	42.76	43.63	43.97	43.29	43.51	44.04
	B1121	B1122	B1123	B1222	B1223	B1323
T_a	33.09	33.11	33.07	33.15	33.10	33.18
T_{mrt}	53.52	53.16	53.60	52.36	53.08	53.18
V_a	0.61	0.43	0.41	0.46	0.42	0.40
PET	43.65	44.49	44.73	44.06	44.47	44.92
	B1131	B1132	B1133	B1232	B1233	B1333
T_a	33.03	33.08	33.03	33.15	33.07	33.16
T_{mrt}	53.98	53.59	54.03	53.04	53.49	53.60
V_a	0.57	0.41	0.39	0.44	0.39	0.38
PET	44.21	44.93	45.04	44.53	45.01	45.31
	B2121	B2122	B2123	BR2C2	B2223	B2323
T_a	33.13	33.16	33.13	33.25	33.19	33.27
T_{mrt}	52.90	52.54	52.98	51.82	52.27	52.36
V_a	0.66	0.46	0.45	0.47	0.44	0.42
PET	43.25	44.18	44.50	43.99	44.07	44.64
	B2131	B2132	B2133	B2232	B2233	B2333
T_a	33.05	33.10	33.07	33.20	33.13	33.21
T_{mrt}	53.37	52.99	53.41	52.27	52.71	52.79
V_a	0.60	0.41	0.39	0.43	0.40	0.38
PET	43.50	44.51	44.94	44.18	44.42	44.89
	B3131	B3132	B3133	B3232	B3233	BR3C2
T_a	33.13	33.18	33.14	33.26	33.20	33.28
T_{mrt}	53.51	53.12	53.55	52.36	52.80	52.86
V_a	0.57	0.39	0.38	0.41	0.38	0.35
PET	44.01	44.92	45.20	44.70	44.87	45.48

N-S

	BR1C2	B1112	B1113	B1212	B1213	B1313
T_a	33.38	33.48	33.49	33.51	33.53	33.54
T_{mrt}	50.85	50.06	49.62	49.19	48.77	48.32
V_a	2.10	2.03	1.99	2.06	2.02	1.97
PET	40.31	39.91	39.76	39.43	39.30	39.16
	B1121	B1122	B1123	B1222	B1223	B1323
T_a	33.46	33.46	33.50	33.47	33.53	33.54
T_{mrt}	50.22	49.46	49.01	48.48	48.15	47.70
V_a	2.01	1.99	1.96	2.01	1.98	1.94
PET	40.00	39.70	39.51	39.21	39.05	38.91
	B1131	B1132	B1133	B1232	B1233	B1333
T_a	33.46	33.49	33.54	33.51	33.57	33.59
T_{mrt}	49.90	49.11	48.66	48.24	47.79	47.35
V_a	1.96	1.96	1.93	1.99	1.96	1.92
PET	39.90	39.55	39.39	39.07	38.92	38.78
	B2121	B2122	B2123	BR2C2	B2223	B2323
T_a	33.48	33.46	33.51	33.51	33.56	33.59
T_{mrt}	49.56	48.81	48.34	47.96	47.52	47.05
V_a	2.02	1.99	1.98	1.97	1.94	1.92
PET	39.65	39.34	39.14	39.04	38.86	38.65
	B2131	B2132	B2133	B2232	B2233	B2333
T_a	33.49	33.51	33.57	33.56	33.62	33.65
T_{mrt}	49.27	48.48	48.01	47.64	47.19	46.71
V_a	1.97	1.96	1.94	1.94	1.90	1.89
PET	39.56	39.22	39.04	38.91	38.76	38.55
	B3131	B3132	B3133	B3232	B3233	BR3C2
T_a	33.49	33.52	33.57	33.57	33.64	33.67
T_{mrt}	48.94	48.15	47.68	47.27	46.81	46.32
V_a	1.93	1.92	1.90	1.91	1.89	1.86
PET	39.46	39.12	38.93	38.74	38.59	38.40

Figure 15. Average of indicators for each block scenario.

Prototypes	BR1C2	B1112	B1113	B1212	B1213	B1313
PET_{ws}	41.80	42.18	42.33	41.78	41.87	42.14
Prototypes	B1121	B1122	B1123	B1222	B1223	B1323
PET_{ws}	42.23	42.62	42.69	42.17	42.36	42.58
Prototypes	B1131	B1132	B1133	B1232	B1233	B1333
PET_{ws}	42.53	42.83	42.84	42.4	42.63	42.76
Prototypes	B2121	B2122	B2123	BR2C2	B2223	B2323
PET_{ws}	41.85	42.29	42.41	42.06	42.04	42.3
Prototypes	B2131	B2132	B2133	B2232	B2233	B2333
PET_{ws}	41.96	42.45	42.64	42.12	42.21	42.42
Prototypes	B3131	B3132	B3133	B3232	B3233	BR3C2
PET_{ws}	42.24	42.66	42.75	42.38	42.42	42.72

The lowest PET_{ws}
The highest PET_{ws}

Figure 16. PET_{ws} values of block street canyons.

B1212 and B1133 are the block configuration models with the lowest and highest PET_{WS} of block street canyons, respectively. The PET_{WS} of B1212 decreases by 2.5% compared to B1133. The value of PET_{WS} increases as the number of building rows increases when the block is evenly configured. The PET_{WS} of BR3C2 increases by 2.2% compared to BR1C2, reaching 42.72.

When configuring mixed blocks, by comparing the PET_{WS} values of B1131, B2131, B3131, and B1113, B1213, and B1313, respectively, it can be observed that as the number of building rows in the block unit on the north side of the street increases, the PET_{WS} value first decreases and then increases. The value is maximum when the number of rows is one. B2131 compared to B1131 and B1213 compared to B1113 thermal comfort improves by about 1.3% and 1%, respectively. By comparing the PET_{WS} values of B1112, B1122, B1132, and B2131, B2132, and B2133, respectively, as the number of building rows in the block unit on the south side of the street increases, the PET_{WS} value gradually increases. B1112 compared to B1132 and B2131 compared to B2133 thermal comfort improves by about 1.5% and 1.6%, respectively. Comparing the PET_{WS} values of B1212, B2121, B1122 and B1313, B3131, B1133, respectively, it can be found that the change in PET_{WS} is not significant when the two street block units of the same building layout are configured on the east or west side of the street. However, there is a significant difference in the PET_{WS} when they are configured on the north and south sides of the street compared to the east and west sides. The PET_{WS} of B1212 decreases by about 2% compared to B1122. In the B2131, B1213, B1123, and B1132 block models, the block units adopt the same building layout but are configured in different locations on the block. The PET_{WS} of B1213 decreases by about 2.2% compared to B1132.

4. Discussion

The extensive urbanization development in China has led to urban characteristics such as excessive block scale and wide roads. This urban morphology of wide roads, sparse road network, and large block outline changes the characteristics of the urban underlying surface, forming the UHI. The urban microclimate environment is deteriorating year by year, leading to a decrease in the quality of the living environment. Jinan adopts the planning layout of small blocks and dense road networks in community life circle planning, which has inherent advantages in regulating microclimate and energy saving and carbon reduction. Analyzing the specific impact mechanism between the morphology of small-scale residential blocks and microclimate provides a reliable basis for residential block planning. Jinan, as a typical city with a warm temperate climate in China, has hot summer and poor ventilation potential. Adjusting the building layout form of street block units and block configuration is an effective way to improve the thermal comfort of residential blocks in summer.

In this paper, the effect of residential block morphology on the microclimate and thermal comfort of street canyons in summer is simulated and analyzed using ENVI-met. The study first explored the effects of different building rows and columns on the microclimate and thermal comfort of street canyons in residential blocks based on the same floor area ratio. The results show that the number of rows of buildings in the row layout has a higher significance than the number of columns on the summer thermal comfort PET in the block street canyon. On this basis, the optimal number of columns is determined as two columns. Then, with the same number of columns, the four block units were combined according to the different number of building rows in the block units to generate 36 different block configuration models. Based on the analysis of the above simulation results:

The thermal comfort of street canyons is mainly influenced by T_{mrt}, T_a, V_a, and other factors. The prevailing summer winds in Jinan are in a southerly direction. N–S-oriented streets are significantly better ventilated than E–W-oriented streets. In addition, the solar altitude angle in the study area is larger in summer, and the shadow area shaded by buildings is smaller for N–S-oriented streets and E–W-oriented streets, and the variability of the influence of solar radiation is not significant. Therefore, in the above simulation results, it can be seen that the thermal comfort of N–S-oriented street canyons in the same block is significantly better than E–W oriented. In summer, the PET of the E–W-oriented street canyon gradually increases with the number of building rows and first decreases and then increases with the number of building columns. The PET of N–S orientation decreases with the number of building rows, while the number of building columns has no significant effect on the PET of N–S orientation. With the increase in the number of rows and columns of buildings in the building layout, the trend of the average value of PET varies among different trends. Therefore, a new index, PET_{ws}, is introduced to evaluate the thermal comfort of residential blocks, and the thermal comfort of streets with different orientations is evaluated comprehensively, so as to evaluate the outdoor thermal comfort of blocks as a whole.

When the four block units have the same building layout form, the number of building rows increases from one to three rows and the PET_{ws} value increases by 2.2%. When the four block units were combined according to the different forms of building layout in the block units, the PET_{ws} of B1212, which has the best thermal comfort, were reduced by 2.5% compared to the PET_{ws} of B1133, which has the worst thermal comfort. As the number of building rows in the block units on the north side of the street increases, the PET_{ws} values first decrease and then increase. As the number of building rows in the block units on the south side of the street increases, the PET_{ws} values gradually increase. PET_{ws} can be reduced by about 2% for 2 block units of the same building layout configured on the east and west sides of the street compared to the north and south sides. Block units with the same building layout but configured in different locations on the block can reduce PET_{ws} by about 2.2%.

In a block design, the form of the building layout within block units can simultaneously influence the spatial form of the canyon with different street orientations. Therefore, when assessing the thermal comfort of block street canyons, attention should be paid to the block morphology around the streets, instead of considering isolated, single-oriented street spaces. The thermal comfort variation of street canyons with different orientations should be evaluated comprehensively to form an appropriate block morphology. PET_{ws} can be used as an indicator to assess the overall outdoor thermal comfort of a block. In the block planning with row layout for warm temperate cities, when the blocks are uniformly configured, the building layout can be in one row and two columns in order to make the street canyon have better thermal comfort in summer, as in BR1C2. When the blocks are mixed configuration, the block units on the north side of the street can be laid out with two building rows, and the south side can be laid out with one building row. Block units of the same building layout are configured on the east or west side of the street.

This study only analyzes the influence of the building row layout pattern on the summer thermal comfort of street canyons in the warm temperate city of Jinan, China, which can provide a research framework for the study of urban microclimate and thermal comfort belonging to other climate zones. In addition, the study was conducted based on the field measurement and the software simulation. The response of people may be included in future studies in order to better understand the human body's perception of thermal comfort in outdoor environments. The field measurement was conducted on a typical day and at specific hours of the summer. In a realistic environment, vehicles, pedestrians, and surrounding buildings all have an impact on the thermal comfort of street canyons in summer, and a better model is needed to evaluate the impact of block building layout forms on the thermal comfort of street canyons.

5. Conclusions

The objective of this research paper is to explore the effects of building layout form and block configuration on the summer thermal comfort of street canyons in residential blocks. The present study takes Jinan, a typical city in the warm temperate zone of China, as the study area, extracts the typical forms of residential blocks, and establishes ideal block models. Environmental factors such as Ta, T_{mrt}, and V_a in residential blocks are simulated using ENVI-met, and PET is calculated by Bio-met. The data of street canyon space of each block model are extracted. The impact of the number of building rows and columns on outdoor thermal comfort is analyzed by comparing the thermal comfort of nine uniformly configured block models for block configuration. Using the entropy method, the PET of street canyons in 36 block configuration models was calculated statistically. PET_{WS} considering PET weights of different oriented streets are used to evaluate the thermal comfort of the whole street canyon in the block. The block morphology model with the best summer thermal comfort is selected, in other words, the block with the lowest PET_{WS} value.

The study is based on the same building floor area ratio and the number of building rows and columns as variables for block configuration to explore its effect on the microclimate and thermal comfort of residential summer block street canyons. In addition, the results can provide references and guidance for urban residential block planning and street design. The results of the study are summarized as follows:

The number of rows of buildings in the building row layout has a higher significance than the number of columns on the summer thermal comfort PET in the block street canyon, especially on the N–S-oriented street canyons. The best choice for the number of building columns is two columns. The number of building rows had opposite trends in the effect of PET on the summer thermal comfort of E–W-oriented and N–S-oriented street canyons in the block. The N–S-oriented street canyon thermal comfort PET is significantly better than the E–W-oriented in the same block.

When four block units have the same form of building layout, by adjusting the number of building rows and columns, the overall summer thermal comfort of the block street canyon PET_{WS} can be improved by a maximum of 2.2%. When the four block units are combined according to the different forms of building layout in the block units, the maximum increase in thermal comfort PET_{WS} of the block street canyon can be 2.5%. The thermal comfort of each block model street canyon was evaluated by statistical calculation of PET_{WS}, and B1212 was the best block form to improve the overall summer thermal comfort of the street canyon in the warm temperate zone.

Author Contributions: Conceptualization, X.S. and Q.D.; methodology, X.S.; software, X.S.; validation, X.S., Q.D., G.W. and S.W.; formal analysis, X.S., Q.D., G.W. and C.J.; investigation, X.S., Q.D., G.W. and S.W.; resources, X.S. and Q.D.; data curation, X.S. and C.J.; writing—original draft preparation, X.S.; writing—review and editing, Q.D. and G.W.; visualization, X.S.; supervision, Q.D.; project administration, X.S. and Q.D.; funding acquisition, Q.D. All authors have read and agreed to the published version of the manuscript.

Funding: This research received no external funding.

Data Availability Statement: Data are available on request due to restrictions. The data presented in this study are available on request from the corresponding author.

Conflicts of Interest: The authors declare no conflict of interest.

References

1. Kotharkar, R.; Ramesh, A.; Bagade, A. Urban heat island studies in South Asia: A critical review. *Urban CLim.* **2018**, *24*, 1011–1026. [CrossRef]
2. Ramakreshnan, L.; Aghamohammadi, N.; Fong, C.S.; Ghaffarianhoseini, A.; Ghaffarianhoseini, A.; Wong, L.P.; Hassan, N.; Sulaiman, N.M. A critical review of Urban Heat Island phenomenon in the context of Greater Kuala Lumpur, Malaysia. *Sust. Cities Soc.* **2018**, *39*, 99–113. [CrossRef]
3. Leng, H.; Liang, S.; Yuan, Q. Outdoor thermal comfort and adaptive behaviors in the residential public open spaces of winter cities during the marginal season. *Int. J. Biometeorol.* **2020**, *64*, 217–229. [CrossRef]
4. Sun, C.; Lian, W.; Liu, L.; Dong, Q.; Han, Y. The impact of street geometry on outdoor thermal comfort within three different urban forms in severe cold region of China. *Build. Environ.* **2022**, *222*, 109342. [CrossRef]
5. Wang, W.; Liu, K.; Zhang, M.; Shen, Y.; Jing, R.; Xu, X. From simulation to data-driven approach: A framework of integrating urban morphology to low-energy urban design. *Renew. Energy* **2021**, *179*, 2016–2035. [CrossRef]
6. Qiu, G.Y.; Zou, Z.; Li, X.; Li, H.; Guo, Q.; Yan, C.; Tan, S. Experimental studies on the effects of green space and evapotranspiration on urban heat island in a subtropical megacity in China. *Habitat Int.* **2017**, *68*, 30–42. [CrossRef]

7. Memon, R.A.; Leung, D.Y.; Liu, C.-H. Effects of building aspect ratio and wind speed on air temperatures in urban-like street canyons. *Build. Environ.* **2010**, *45*, 176–188. [CrossRef]
8. Qaid, A.; Ossen, D.R. Effect of asymmetrical street aspect ratios on microclimates in hot, humid regions. *Int. J. Biometeorol.* **2015**, *59*, 657–677. [CrossRef]
9. Ali-Toudert, F.; Mayer, H. Numerical study on the effects of aspect ratio and orientation of an urban street canyon on outdoor thermal comfort in hot and dry climate. *Build. Environ.* **2006**, *41*, 94–108. [CrossRef]
10. Shishegar, N. Street design and urban microclimate: Analyzing the effects of street geometryand orientation on airflow and solar access in urban canyons. *J. Clean Energy Technol.* **2013**, *1*, 52. [CrossRef]
11. Muniz-Gäal, L.P.; Pezzuto, C.C.; de Carvalho, M.F.H.; Mota, L.T.M. Urban geometry and the microclimate of street canyons in tropical climate. *Build. Environ.* **2020**, *169*, 106547. [CrossRef]
12. Mohajeri, N.; Gudmundsson, A.; Kunckler, T.; Upadhyay, G.; Assouline, D.; Kämpf, J.; Scartezzini, J. A solar-based sustainable urban design: The effects of city-scale street-canyon geometry on solar access in Geneva, Switzerland. *Appl. Energy* **2019**, *240*, 173–190. [CrossRef]
13. Bakarman, M.A.; Chang, J.D. The influence of height/width ratio on urban heat island in hot-arid climates. *Procedia Eng.* **2015**, *118*, 101–108. [CrossRef]
14. Chatzidimitriou, A.; Yannas, S. Street canyon design and improvement potential for urban open spaces; the influence of canyon aspect ratio and orientation on microclimate and outdoor comfort. *Sust. Cities Soc.* **2017**, *33*, 85–101. [CrossRef]
15. Karimimoshaver, M.; Khalvandi, R.; Khalvandi, M. The effect of urban morphology on heat accumulation in urban street canyons and mitigation approach. *Sust. Cities Soc.* **2021**, *73*, 103127. [CrossRef]
16. Andreou, E. Thermal comfort in outdoor spaces and urban canyon microclimate. *Renew. Energy* **2013**, *55*, 182–188. [CrossRef]
17. Deng, J.-Y.; Wong, N.H. Impact of urban canyon geometries on outdoor thermal comfort in central business districts. *Sust. Cities Soc.* **2020**, *53*, 101966. [CrossRef]
18. Srivanit, M.; Jareemit, D. Modeling the influences of layouts of residential townhouses and tree-planting patterns on outdoor thermal comfort in Bangkok suburb. *J. Build. Eng.* **2020**, *30*, 101262. [CrossRef]
19. Karimimoshaver, M.; Shahrak, M.S. The effect of height and orientation of buildings on thermal comfort. *Sust. Cities Soc.* **2022**, *79*, 103720. [CrossRef]
20. Li, G.; Ren, Z.; Zhan, C. Sky View Factor-based correlation of landscape morphology and the thermal environment of street canyons: A case study of Harbin, China. *Build. Environ.* **2020**, *169*, 106587. [CrossRef]
21. Taleghani, M.; Kleerekoper, L.; Tenpierik, M.; Van Den Dobbelsteen, A. Outdoor thermal comfort within five different urban forms in the Netherlands. *Build. Environ.* **2015**, *83*, 65–78. [CrossRef]
22. Shareef, S.; Altan, H. Urban block configuration and the impact on energy consumption: A case study of sinuous morphology. *Renew. Sust. Energ. Rev.* **2022**, *163*, 112507. [CrossRef]
23. Ma, X.; Leung, T.; Chau, C.; Yung, E.H. Analyzing the influence of urban morphological features on pedestrian thermal comfort. *Urban Clim.* **2022**, *44*, 101192. [CrossRef]
24. Pearlmutter, D.; Berliner, P.; Shaviv, E. Integrated modeling of pedestrian energy exchange and thermal comfort in urban street canyons. *Build. Environ.* **2007**, *42*, 2396–2409. [CrossRef]
25. Arnfield, A.J. Street design and urban canyon solar access. *Energy Build.* **1990**, *14*, 117–131. [CrossRef]
26. Bourbia, F.; Boucheriba, F. Impact of street design on urban microclimate for semi arid climate (Constantine). *Renew. Energy* **2010**, *35*, 343–347. [CrossRef]
27. Krüger, E.; Diniz, W. Relationship between indoor thermal comfort conditions and the Time Weighted Preservation Index (TWPI) in three Brazilian archives. *Appl. Energy* **2011**, *88*, 712–723. [CrossRef]
28. Futcher, J.A.; Kershaw, T.; Mills, G. Urban form and function as building performance parameters. *Build. Environ.* **2013**, *62*, 112–123. [CrossRef]
29. Emmanuel, R.; Johansson, E. Influence of urban morphology and sea breeze on hot humid microclimate: The case of Colombo, Sri Lanka. *Clim. Res.* **2006**, *30*, 189–200. [CrossRef]
30. Zhang, Y.; Du, X.; Shi, Y. Effects of street canyon design on pedestrian thermal comfort in the hot-humid area of China. *Int. J. Biometeorol.* **2017**, *61*, 1421–1432. [CrossRef]
31. Yin, S.; Lang, W.; Xiao, Y. The synergistic effect of street canyons and neighbourhood layout design on pedestrian-level thermal comfort in hot-humid area of China. *Sust. Cities Soc.* **2019**, *49*, 101571. [CrossRef]
32. Cao, A.; Li, Q.; Meng, Q. Effects of orientation of urban roads on the local thermal environment in Guangzhou city. *Procedia Eng.* **2015**, *121*, 2075–2082. [CrossRef]
33. Yin, Q.; Cao, Y.; Sun, C. Research on outdoor thermal comfort of high-density urban center in severe cold area. *Build. Environ.* **2021**, *200*, 107938. [CrossRef]
34. *GB/T 17297-1998;* Names and Codes for Climate Regionalization in China-Climatic Zones and Climatic Regions. The State Bureau of Quality and Technical Supervision: Beijing, China, 1998.
35. *GB 50180-2018;* Standard for Urban Residential Area Planning and Design. China Construction Industry Press: Beijing, China, 2018.
36. Bruse, M.; Fleer, H. Simulating surface–plant–air interactions inside urban environments with a three dimensional numerical model. *Environ. Modell. Softw.* **1998**, *13*, 373–384. [CrossRef]

37. Emmanuel, R.; Fernando, H.J. Urban heat islands in humid and arid climates: Role of urban form and thermal properties in Colombo, Sri Lanka and Phoenix, USA. *Clim. Res.* **2007**, *34*, 241–251. [CrossRef]
38. Fahmy, M.; Sharples, S. On the development of an urban passive thermal comfort system in Cairo, Egypt. *Build. Environ.* **2009**, *44*, 1907–1916. [CrossRef]
39. McRae, I.; Freedman, F.; Rivera, A.; Li, X.; Dou, J.; Cruz, I.; Ren, C.; Dronova, I.; Fraker, H.; Bornstein, R. Integration of the WUDAPT, WRF, and ENVI-met models to simulate extreme daytime temperature mitigation strategies in San Jose, California. *Build. Environ.* **2020**, *184*, 107180. [CrossRef]
40. Bande, L.; Afshari, A.; Al Masri, D.; Jha, M.; Norford, L.; Tsoupos, A.; Marpu, P.; Pasha, Y.; Armstrong, P. Validation of UWG and ENVI-met models in an Abu Dhabi District, based on site measurements. *Sustainability* **2019**, *11*, 4378. [CrossRef]
41. Tsoka, S.; Tsikaloudaki, A.; Theodosiou, T. Analyzing the ENVI-met microclimate model's performance and assessing cool materials and urban vegetation applications–A review. *Sust. Cities Soc.* **2018**, *43*, 55–76. [CrossRef]
42. Höppe, P. The physiological equivalent temperature–a universal index for the biometeorological assessment of the thermal environment. *Int. J. Biometeorol.* **1999**, *43*, 71–75. [CrossRef]
43. Lin, T.-P.; Matzarakis, A. Tourism climate and thermal comfort in Sun Moon Lake, Taiwan. *Int. J. Biometeorol.* **2008**, *52*, 281–290. [CrossRef] [PubMed]
44. Lai, D.; Guo, D.; Hou, Y.; Lin, C.; Chen, Q. Studies of outdoor thermal comfort in northern China. *Build. Environ.* **2014**, *77*, 110–118. [CrossRef]
45. Su, Y.; Zhao, Q.; Zhou, N. Improvement strategies for thermal comfort of a city block based on PET Simulation—A case study of Dalian, a cold-region city in China. *Energy Build.* **2022**, *261*, 111557. [CrossRef]
46. Yang, Y.; Zhou, D.; Gao, W.; Zhang, Z.; Chen, W.; Peng, W. Simulation on the impacts of the street tree pattern on built summer thermal comfort in cold region of China. *Sust. Cities Soc.* **2018**, *37*, 563–580. [CrossRef]
47. Zhang, W.; Gao, Z. Research progress of outdoor thermal comfort index. *J. Environ. Health* **2015**, *32*, 836–840.

buildings MDPI

Article

Energy-Based Design: Improving Modern Brazilian Buildings Performance through Their Shading Systems, the Nova Cintra Case Study

Daniel Mateus [1,*] and Gonçalo Castro Henriques [2]

1 CIAUD, Research Centre for Architecture, Urbanism and Design, Lisbon School of Architecture, Universidade de Lisboa, 1349-063 Lisboa, Portugal
2 Faculdade de Arquitetura e Urbanismo, Universidade Federal do Rio de Janeiro, LAMO-PROURB, Rio de Janeiro 21941-901, Brazil; gch@fau.ufrj.br
* Correspondence: dmateus@fa.ulisboa.pt or mateus.dani@gmail.com

Abstract: Current research applies an energy-based design model to improve performance in existing modern buildings, in Rio de Janeiro, from the 1940's, improving these buildings' shading systems. This article proposes a methodology tested through a case study, the Nova Cintra building. The methodology starts by analysing the original shading system performance, regarding insolation, illuminance and air temperature. Using these results, proposes two computacional methods to improve performance: (1) a combinatorial modelling process, recombining the existing shading systems positions in the building's north façade; and (2) a transformation process, using parametric and algorithmic–parametric modelling, to improve the existing shading systems performance. Both processes use optimization algorithms. The results of these modelling and optimization methods are compared with the results of the original system and suggests an improvement between 111.1% and 590.4% for insolation; between 360.9% and 84.4% for illuminance; and between 2.9% and 3.0% for air temperature, considering winter and summer solstices. This improvement aims at reducing the buildings' energy consumption and foresees the production of renewable energy from solar harvesting, to mitigate climate change.

Keywords: energy-based design; building performance; shading systems; carioca modern façades; insolation; illuminance; air temperature; computational methods; combinatorial, parametric and algorithmic–parametric modelling; optimization

Citation: Mateus, D.; Henriques, G.C. Energy-Based Design: Improving Modern Brazilian Buildings Performance through Their Shading Systems, the Nova Cintra Case Study. *Buildings* **2023**, *13*, 2543. https://doi.org/10.3390/buildings13102543

Academic Editor: Gerardo Maria Mauro

Received: 29 August 2023
Revised: 29 September 2023
Accepted: 3 October 2023
Published: 8 October 2023

1. Introduction

This article addresses the analysis and improvement of buildings' energy performance, through computational methods such as performance simulation, combinatorial modelling, parametric modelling and algorithmic–parametric modelling, used jointly with optimization algorithms.

It describes an ongoing research project, entitled Carioca Modern Façades, that studies the shading systems of a set of Brazilian modern buildings, from the carioca modernist school, from Rio de Janeiro. These buildings are from the 1940/50's, prior to the use of air-conditioning systems, and present original shading systems that are efficient in controlling excessive interior daylight, and also the overheating temperature in the building's interior space. This research project intends to analyse and improve the performance of the buildings by improving the existing shading systems. The present article describes the case study realized for one of these selected buildings, the Nova Cintra. It approaches in particular the Nova Cintra's north façade shading systems, the façade more exposed to the sun along the year.

The shading system design is critical in the building envelope design. The research uses the concept of building envelope proposed by Sadineni, Madala and Boehm [1], as a key factor that determines the quality and controls the indoor conditions inside a building, irrespective of transient outdoor conditions. The building envelope behaves similarly to the human skin, as a barrier between the external environment and the organism. In fact, in the literature, the words skin or envelope are often used [2]. The shading systems have the capacity to filter the entry of natural daylight inside a building, which determines the illuminance (amount of natural light) and the air temperature of the building's interior space. Having a major impact on the energy consumption of the building. Hence, they are a crucial element of the building envelope. Beyond their direct influence in the energy consumption of a building, the shading systems also have the potential for renewable solar energy harvesting and for the production of electric energy, through the conversion of solar energy by the use of photovoltaic technology [3–5]. This fact makes shading systems elements that can improve the energy performance of a building, contributing also positively to mitigate climate change, diminishing the need of energy consumption from pollutant sources and increasing the energy production from renewable sources.

1.1. Literature Review

The literature review considers previous studies that use computational methods for analysis and to improve buildings' energy performance, considering the building envelope and/or the shading systems. References include Caldas and Norford's research [6] that proposes a design tool based on genetic algorithms and a design optimization method, to search the best design solutions regarding the windows of an office building, determining their position and size. The design optimizes the building's thermal and illuminance properties. Venâncio [7] develops and tests a parametric model for shading devices in early design stages, adopting multi-criteria analysis to optimize solutions to assure shading, blocking the incident solar radiation in the interior of rooms, and providing simultaneously adequate daylight levels. Jalali, Noorzai and Heidari [8] optimize an office building façade using genetic algorithms and multi-objective optimization, namely the Strength Pareto Evolutionary Algorithm (SPEA-2), to find a parametric set of optimal solutions regarding usable space, reduced thermal load and improved building natural daylight. Several other researchers use parametric modelling, performance analysis and optimization algorithms to find optimized design solutions for buildings [9–19]. Henriques, Duarte and Leal [20] develop an alternative strategy to find design solutions for a responsive skylight system with adequate illuminance performance, in a parametric solutions space. Instead of optimization algorithms, they use heuristics (sets of rules and methods to drive the discovery, invention and resolution of problems) to save time and resources to achieve good solutions. It is also worth mentioning the research by Vazquez, Duarte and Poerschke [21] that presents a digital framework to design masonry screen walls with optimized performance, associated with vernacular construction rules. Vazquez's research uses a generative design system based on shape grammars [22] and on existing construction rules. That design system is translated into a parametric model, connecting the system to a simulation engine to calculate the daylight results and cooling energy loads. Finally, Vazquez uses genetic algorithms to find families of optimized solutions, with visual feedback, to understand the trade-offs between such solutions. Granadeiro [23] develops a similar approach, realizing a shape grammar for Frank Lloyd Wright's prairie houses, converted to a parametric design system. Granadeiro connects this parametric design system with an energy simulation software, using genetic algorithms to find optimized building envelopes design solutions for the prairie houses, regarding energy consumption, in the parametric space of solutions generated through the prairie houses-shape grammar.

The research described in this article, as the researches mentioned above, uses parametric modelling/design with optimization algorithms as well to improve the energy performance of a building and its shading systems.

Parametric design models digital objects, such as buildings and their construction elements. Oxman [24] defines parametric design as an act of design thinking based on the process of exploring and re-editing associative relationships, in a geometrical solution space. Parametric design is thus associated with the definition of a mathematical model. Accordingly, Caetano, Santos and Leitão [25] define parametric design as an approach that describes a design symbolically, using parameters. Therefore, parametric design uses a defined set of parameters to establish the shape of a model, or object, from the possible shapes that it can assume in a universe of solutions. This research defines a set of parameters for an existing shading system and develops a parametric modelling process to improve the shading system's performance. Here, we understand the term parametric modelling as a modelling process that stimulates parametric design thinking, providing a multiplicity of improved solutions.

Parametric modelling has the advantage of enabling quick exploration of a universe of shape solutions that share a set of common features, defined by a correspondent set of parameters. It presents, however, the disadvantage that the universe of shape solutions is limited, being the limitation larger or smaller depending on the number and type of parameters that define the solution universe. Thus, parametric modelling is restricted to the number of shapes that compose the universe of solutions. This restriction is a gap in the parametric modelling type.

Regarding optimization algorithms, genetic algorithms are used in the previously mentioned research. Caldas and Norford [6] highlight that a genetic algorithm is a procedure loosely based on the Darwinian notions of survival of the fittest, which uses selection and recombination operators to search, among candidate solutions, high-performance solutions. Genetic algorithms search for and evaluate design solutions, in a search guided by the results evaluation. The designer interacts with this loop process. According to Touloupaki and Theodosiou [19], genetic algorithms dominate the field of building design optimization regarding building envelopes [26], overall building form [27,28], HVAC [29] and renewable energy system [12] optimization. This research uses genetic algorithms as well, jointly with parametric modelling and algorithmic modelling, to find shading systems that have optimized energy performance.

1.2. Research Objectives

This research, regarding the identified parametric modelling gap, intends on the following:

1. Develop and apply different modelling types for the shading systems of the Nova Cintra building's north façade that complement the parametric universe of shape solutions for the shading systems and that improve the energy performance;
2. Implement an energy-based design process, proposed and developed in previous research [30,31], applying shape transformation processes on the Nova Cintra building's north façade shading systems, to improve energy performance. Shape transformations use different modelling processes, some addressed in the original energy-based design process and others developed and incorporated through this research.

The research associates combinatorial modelling with algorithmic–parametric modelling, to overcome parametric modelling limitations, increasing the universe of modelled solutions, thus increasing the possibility of improving energy performance.

A previous research stage developed combinatorial modelling [32], combining a set of different objects with a set of different positions. The objects combined are different shading systems combined in a set of positions, defined by the modern modular structure of the building, to attain the best performance regarding multi-objective goals.

Algorithmic–parametric modelling is a mixed process. It uses algorithmic modelling, which is the act of modelling objects through an algorithmic design process. Algorithmic design, according with Caetano, Santos and Leitão [25], is a design process based on algorithms. The Cambridge Dictionary [33] defines an algorithm as a "set of mathematical instructions or rules that, especially if given to a computer, will help to calculate an answer to a problem". Oxman [25] considers algorithmic design as the coding of explicit instructions to generate digital forms, while Queiroz and Vaz [34] state that algorithmic design allows the user to design directly through code manipulation. Similarly to the parametric process in this research, algorithmic modelling is a modelling process that acts as an algorithmic design process. Algorithmic modelling creates shape variations through transformation operations, defined previously in the shape grammar entitled Building Envelopes Grammar [30,31]. The Building Envelopes Grammar operations can combine different arrangements and sequences, composing different algorithms that result in different shapes with improved energy performance, specifically regarding the harvesting of solar energy, to convert it to electric energy using photovoltaic technology. Therefore, this research uses Building Envelopes Grammar as an algorithmic modelling process, to transform the shading system's shape, improving its energetic performance. This research complements algorithmic modelling with parametric modelling, resulting in a mixed algorithmic–parametric process. The parametric modelling part of the mixed process uses the same parameters previously defined for the shading system, in the aforementioned initial single parametric modelling.

Combinatorial modelling and algorithmic–parametric modelling have different advantages and disadvantages. Combinatorial modelling has the advantage of permitting a very quick exploration of a set of previously defined shape solutions, in a set of different positions, to identify which solutions have the best performance for each position. However, it has the disadvantage of using a more limited set of shape solutions for each position.

Algorithmic–parametric modelling presents the point advantages of two modelling types. Algorithmic modelling allows fast shape creation or transformation operation that improves the performance of the modelled object. Parametric modelling allows complementing the algorithmic modelling process, exploring a set of solutions defined by parameters related with the shape solution created by the algorithmic process. The parametric exploration of solutions allows for further improving the performance of the modelled object. The algorithmic–parametric modelling process presents the overall advantage of exploring a wider universe of shape solutions with improved performance, having a higher probability of achieving the highest performance. Algorithmic–parametric modelling, however, has the disadvantage of slowing down regarding the combinatorial and parametric modelling types, to improve performance.

Comparing the three modelling types, algorithmic–parametric modelling obtains the largest universe of solutions, which leads to higher performance results. Parametric modelling allows for achieving the second largest universe of solutions. Combinatorial modelling presents the most restrict universe of solutions. Regarding the speed of calculation, combinatorial is the fastest, followed by the parametric and the algorithmic–parametric modelling types.

This energy-based design research uses parametric, combinatorial and algorithmic–parametric modelling types, together with optimization algorithms. The research improves the former energy-based design model [30,31], with parametric and combinatorial modelling types and through the junction of a parametric modelling process with the previously developed algorithmic modelling process. The use of optimization algorithms is also an improvement for the energy-based design model. The performances analysed and improved concern insolation (solar energy harvesting for conversion to electric energy), illuminance (quantity of natural illumination in the interior spaces of the building) and air temperature, with the main goal of obtaining buildings that minimize energy consumption and maximize energy production through renewable sources such as the Sun, enabling energy production for their own use.

1.3. Research Stages

The present article describes an ongoing research project, entitled Carioca Modern Façades, applying the energy-based design model on the shading systems of the selected set of modern Brazilian buildings, using as case study the Nova Cintra building.

The research proceeds in two stages. The first stage starts by analysing the Nova Cintra building. General analyses consider insolation (incident solar radiation) of the overall building envelope and illuminance analysis of the interior rooms of the building's north façade. Followed the detailed insolation and illuminance analyses of the north façade shading systems, the research uses combinatorial modelling and multi-objective optimization, regarding insolation and illuminance, to generate alternative solutions for the shading systems.

The second research stage selected the upper left corner module of the north façade to generate alternative solutions improving the shading systems performance. This selection intends to shorten the analysis and generation process of optimized shading systems solutions by up to 3 months. This stage also analyses the air temperature in the room near the selected façade module. The research developed parametric modelling and algorithmic–parametric modelling, together with single-objective and multi-objective optimization, to generate alternative solutions for the shading systems in the selected façade module. The results obtained generated alternative shading solutions to improve performance.

The developments in the energy-based design model achieved with the realization of the first and second research stages intend to allow their appliance in a larger scale, to design shading systems of other new and refurbished buildings.

2. Materials and Methods

This section describes the research methodology. It starts by revealing the research framework, establishing a relation between the Carioca Modern Façades research project and the energy-based design model. This section describes the selected set of eight buildings for the Carioca Modern Façades and their façade characteristics as well.

Then, the research details the selected case study, the Nova Cintra building. After this, the methodology for the first research stage is revealed, which consisted in the 3D modelling of Nova Cintra, in the energy analysis of the north façade, considering insolation and illuminance, and in the combinatorial modelling and multi-objective optimization for the shading systems, regarding energy performance improvement through insolation and illuminance. Later, the research details the methodology for the second stage of the research, namely the development of the parametric and the algorithmic–parametric modelling processes, together with single-objective and multi-objective optimization processes, to improve energy performance for the Nova Cintra building.

The methodology of this research is replicable to the other buildings of the Carioca Modern Façades research project. Our methodology is also flexible, and might be used to analyse and improve all the building types that contain shading systems.

2.1. Research Framework: Carioca Modern Façades, an Energy-Based Design Model

This article is part of a wider research entitled Carioca Modern Façades. The main goal of the Carioca Modern Façades is to analyse and improve the energy performance of a set of eight buildings from the carioca modernist school, from Rio de Janeiro, Brazil. Following an energy-based design model, which constitutes a design model that uses shape generation and transformation processes, codified in shape grammars, to improve buildings' energy performance. Below (Figure 1) are displayed the selected eight buildings for the research.

These buildings of the carioca modern building heritage use the façade as an architectural filter, relating interior and exterior conditions through constructive systems, structural modulation and material properties [35] with improved energy performance capacity, adapting modern architecture to the climate [36]. Authors grouped the buildings, and their architectonic filter/threshold, into four categories:

1. The façade intermediated by balconies: Julio Barros Barreto building (1947), Bristol building (1950) and Nova Cintra building (1948) (1,2,3 Figure 1);
2. The façade intermediated by glass planes: Nova Cintra building (1948) and Barão Gravatá building (1952) (3,4 Figure 1);
3. The façade intermediated by coupled concrete brise-soleil: MMM Roberto building (1945) and Dona Fátima and Finúsia building (1951) (5,6 Figure 1);
4. The façade intermediated by filters, cobogós, shutters and trusses: Ramirez building (1954) and Sambaíba building (1953) (7,8 Figure 1).

Our case study Nova Cintra building belongs in two categories as it uses the north façade as a filter, intermediated by balconies, and the south façade intermediated by glass planes.

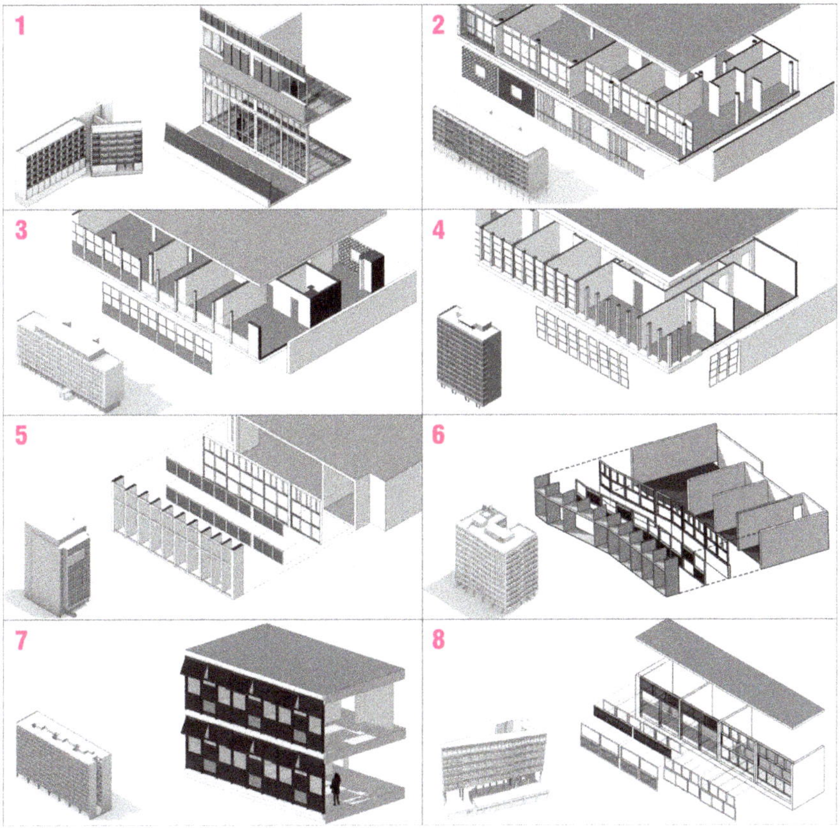

Figure 1. Carioca Modern Façades selected buildings, volume and façade system. **1**- Barreto building, by Irmãos Roberto, 1947; **2**- Bristol building, by Lucio Costa, 1950; **3**- Nova Cintra building, by Lucio Costa, 1948; **4**- Barão Gravatá building, by Sérgio Bernardes, 1952; **5**- MMM Roberto, by Irmãos Roberto, 1945; **6**- D. Fátima building, by Irmãos Roberto, 1951; **7**- Angel Ramirez building, by Irmãos Roberto, 1954; **8**- Sambaíba building, by Irmãos Roberto, 1953. Image adapted from Mara Eskinazi [32].

2.2. Case Study Description: The Nova Cintra Building

The architect Lucio Costa designed the Nova Cintra building in 1948. According to Lucas and Bastos [37], the Nova Cintra is a modern Brazilian example of passive bioclimatic adaptation. The adaptation addresses the following: (1) Climatic context—for most of the year, the city of Rio de Janeiro has a hot and humid tropical weather, with high temperature and relative humidity, with more rain in summer and dry periods in winter. This weather produces thermal discomfort and a fundamental resource to mitigate this is to have interior air renovation, combined with shading, reducing the interior thermal gain, avoiding excessive direct solar light, but taking advantage of natural daylight. (2) Building implantation—Nova Cintra main axe is longitudinal to the street (66×44 m) with a deviation of 20 degrees to north. The building position and the façade envelope minimize the thermal load: the blind façade in the north and east; the west façade only with ventilation areas for the toilets; (3) Wind permeability—the building orientation benefits of the region's prevailing south/southeast wind, but also of its position in the terrain, in steps, over pilotis, with an indented upper floor and especially a balcony in the north façade, with a shading system. (4) Façade and shading composition—the retreated structure should be enhanced in relation to the façade, with a 4 m metric in the north and south elevations, with internal walls aligned and perpendicular to the larger façade. The north façade shading uses brises and cobogós, followed by a balcony that separates the rooms with a sliding window frame. The balconies' shading and the internal walls' alignment favours natural ventilation, along the year, and controls excessive solar daylight, favouring passive comfort of the building, without air-conditioning systems.

The authors selected the north façade intermediated by balconies and by a shading system, mainly comprised of ceramic cobogós and concrete brise-soleil, as displayed in Figure 2.

Figure 2. Nova Cintra building north façade, with the façade three main shading systems: the ceramic cobogó with window, the ceramic cobogó without window and the concrete brise-soleil. There are two other shading systems, the lattices and the shutters, both from wood, applied only in the building's first floor. Photo credits, the authors.

2.3. *First Research Stage: Nova Cintra 3D Modelling, North Façade Analysis, Considering Insolation and Illuminance, Combinatorial Modelling and Multi-Objective Optimization*

2.3.1. Nova Cintra 3D Modelling and Analyses

The ongoing research project Carioca Modern Façades comprises two research developments, the Nova Cintra 3D model energy performance analysis and the combinatorial modelling with multi-objective optimization, to recombine the position of the three main shading systems in the north façade. An article presented and published at the international conference SIGraDi 2022 addresses this research [32].

The research resorted to visual and textual programming using Grasshopper and Python to automate tasks. The Nova Cintra 3D model (Figure 3) adapted previous 3D models focusing on the information organization to analyse and simulate performance. The performance analyses included the following: (1) insolation (or solar irradiation), calculating the solar irradiation per unit area, in kWh/m^2, and the total solar irradiation, in kWh, during the year in the building envelope (roof and façades, including shading systems). Figure 4 shows the insolation in the Nova Cintra building (left) and the insolation detail in Nova Cintra's north façade (right); (2) the illuminance, after being filtered by the shading systems, in the building interior space. Using this information, we can identify the interior spaces that have appropriate daylight, between 100 and 3.000 lux [38]. Below 100 lux, there is lack of daylight and above 3.000 lux, there is excess. Higher levels of illuminance increase interior temperature and might require artificial cooling.

These analyses show building surfaces with higher and lower insolation across the year. Figure 4 shows the Nova Cintra building insolation graphical analysis. The roof surface has the higher insolation, 1.241 kWh/m^2, followed by the north façade, with 587 kWh/m^2. The roof receives the yearly higher result regarding the total value of solar irradiation (TVSI), 1.698.142 kWh (42.8% of the entire envelope), obtained by multiplying the insolation by the surface area. The roof is followed by the north façade, which obtains 972.971 kWh of TVSI (24.5% of the envelope). In the north façade, the shading system cobogó with window registers the higher insolation, 696 kWh/m^2, and next the cobogó without window, with 651 kWh/m^2, and the brise-soleil, with 495 kWh/m^2. However, regarding the TVSI, is the brise-soleil that achieves the higher, 339.656 kWh (75.2% total shading), due to a higher surface area, 682 m^2, followed by cobogó with window, with 68.924 kWh (15.3%) and 101 m^2, and the cobogó without window, with 43.057 kWh (9.5%) and 66 m^2.

The research then analysed the interior rooms' performances from the first, fourth and seventh (and last) floor, near the north façade, regarding illuminance. To calculate the cases for the other floors, the research resorted in interpolation, to diminish the analysis time. Analyses comprehend winter solstice (21st June at 12 p.m.), summer solstice (21st December at 12 p.m.) and autumn equinox (20th March at 12 p.m.), with the Sun in intermediate position. For the winter solstice, the average illuminance is 4.152 lux. The value decreases for 1.528 lux in the autumn equinox, reaching 640 lux in summer solstice. As such, in the summer solstice and in the autumn equinox, the illuminance is adequate, between 100 and 3.000 lux. For the winter solstice, the average exceeds 3.000 lux, which can provoke overheating and require artificial cooling. Thus, despite the shading systems not being efficient throughout the whole year, they have a sun-blocking capacity. Figure 5 displays the illuminance for the intermediate floor (the 4th), with and without shading (image below at left and right).

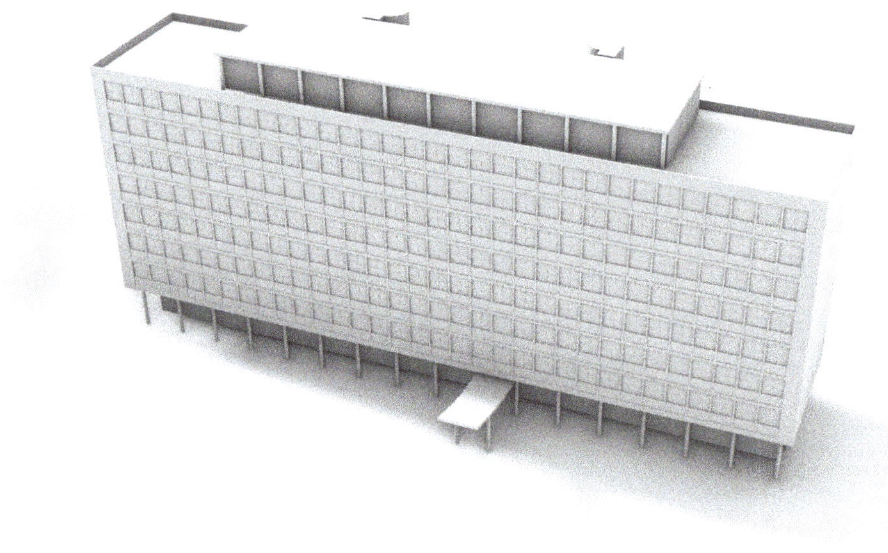

Figure 3. 3D model of Nova Cintra building.

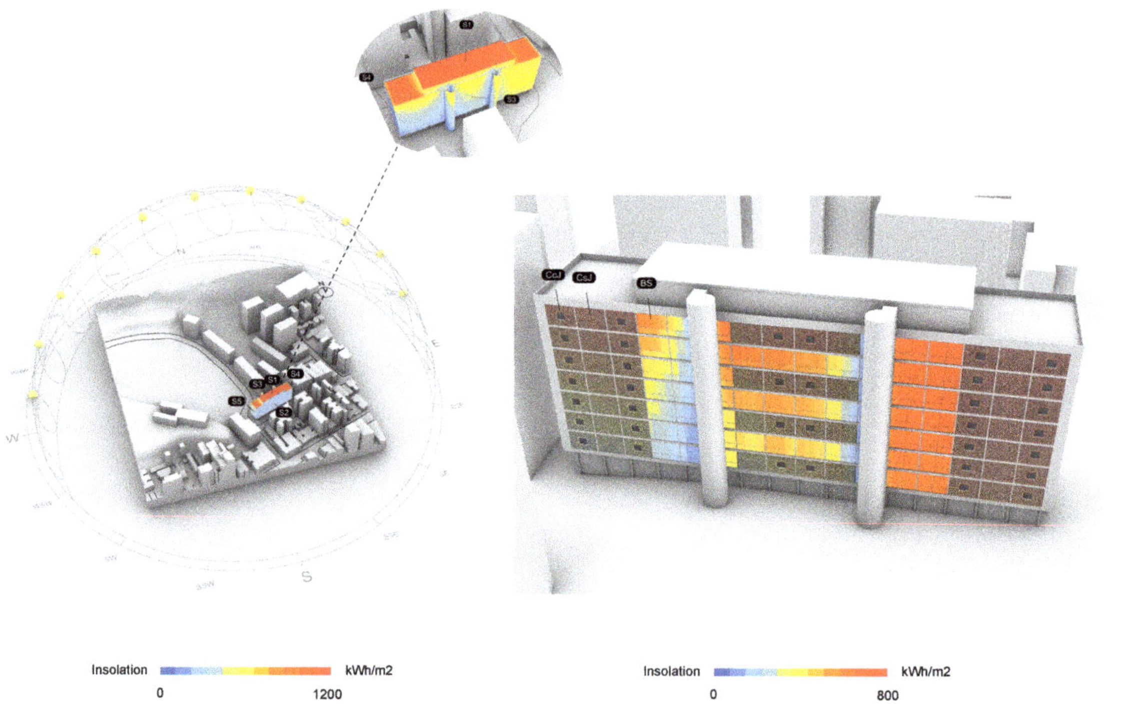

Figure 4. Nova Cintra building envelope insolation (**left**) with a north façade detailed view (**above**) and insolation in the shading systems of the north façade (**right**).

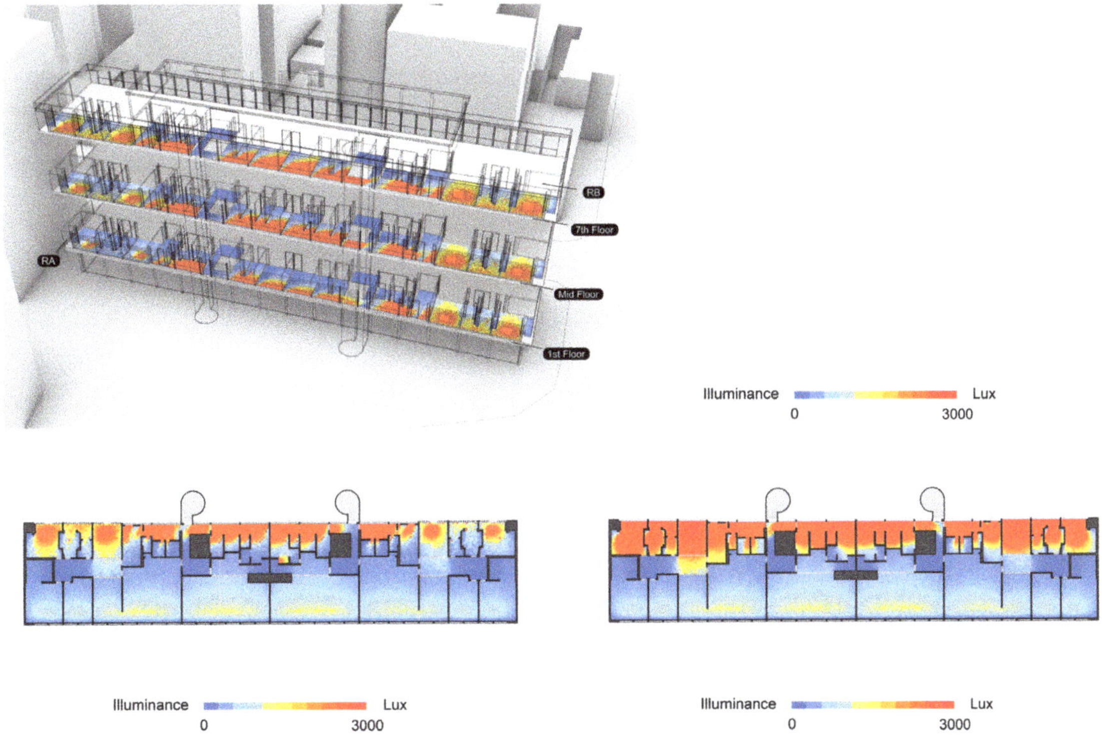

Figure 5. Nova Cintra building: illuminance in winter solstice (21st June) for the rooms of the 1st, 4th and 7th floors, on the north façade (**above left**); illuminance in the 4th floor in winter solstice (21st June), with and without the existing shading systems (**below left** and **right**).

2.3.2. Detailed Analyses, Combinatorial Modelling and Multi-Objective Optimization

After analysing the Nova Cintra building envelope for insolation and illuminance, the research performed detailed analyses of the north façade's shading systems for insolation and illuminance. For insolation and illuminance, analyses rely on the Ladybug and Honeybee plug-ins (versions 1.2.0 and 0.0.63, respectively). The initial insolation analyses comprised the whole year, while illuminance analyses considered the winter solstice (21st June at 12 p.m.), summer solstice (21st December at 12 p.m.) and autumn equinox (20th March at 12 p.m.), to decrease calculation requirements. The solstice's results are the best- and worst-case scenarios, regarding illuminance performance.

The detailed analyses evaluated the north façade's performance, composed of three main shading types: cobogó with window (identified with Letter A), cobogó without window (Letter B) and brise-soleil (Letter C). This discarded two shading systems, wood lattice and wood shutter, were used only on the first floor, for commercial use. The three shading systems's detailed analyses address the winter and summer solstices, regarding insolation and illuminance, at 10 a.m., 12 a.m. and 4 p.m.

To calculate individual performance, we performed individual façade analysis for each shading system (letters A, B and C) comprising the $14 \times 7 = 98$ north façade modules. Calculating for each shading system: (1) insolation in the exterior face of the shading elements, in Wh/m^2; (2) insolation in the plane immediately after the shading elements, in Wh/m^2; (3) insolation in the plane after the shading elements, but without shading, in Wh/m^2; (4) average illuminance in the north façade rooms, in lux; and (5) points number in the north façade rooms, with illuminance between 300 and 750 lux.

Detailed analyses feed a combinatorial modelling process to select the best shading systems, in each module, to improve the building performance by minimizing solar radiation in interior rooms (objective 1), maximizing the points number with adequate illuminance (objective 2) and maximizing average illuminance (objective 3), to assure adequate interior daylight. To automate data analyses with visual and textual programming in Python, we created an alphanumeric label to identify each simulation, such as "CintraWkwmCIE21jun10:00_LetraA" with the reference to the building name, type of performed analysis, analysis day and hour. Concerning the three objectives mentioned, we realized $1 \times 2 \times 1 \times 2 \times 3 \times 3 = 36$ simulations considering the building envelope in the urban context, climatic data and materials used in the construction. To interpret the emergent façade patterns, we developed a graphic representation, using colour gradients in meshes.

To compare the results, the research normalized results on a scale of 0 to 1. The research tested different methods to find the best performance for each scenario (21st June and 21st December, at 10 a.m., 12 p.m. and 4 p.m.) for the 98 modules of the north façade. Initially, we tested random combinations of the shading types letters (A, B, and C) using genetic algorithms to maximize the performance, through the use of the multi-objective optimization plug-in Octopus in Grasshopper. After, a quick sorting algorithm with conditionals proved more straightforward in selecting the best performance for each module. The façade solutions correspond to a list of 98 letters A, B and C. The textual algorithm in Python (inside grasshopper), compares the results using different weights for each objective, in a sensitivity analysis, to understand how each factor affects global selection and the results range.

The multi-objective optimization results are present in Section 3 Results.

2.4. Second Research Stage: Parametric and Algorithmic–parametric Modelling, Single-Objective and Multi-Objective Optimization

The Carioca Modern Façades second research stage model parametrically and algorithmically parametrically the Nova Cintra north façade shading systems. A single-objective and a multi-objective optimization, regarding insolation, illuminance and air temperature, looks for formal configurations that improve performance. Finally, we compare the performance of the modelled shading systems obtained through the combinatorial, parametric and algorithmic–parametric modelling processes used together with the optimization methods. The purpose is to identify the advantages of each modelling process.

2.4.1. Parametric Modelling and Performance Optimization

The parametric modelling process transforms an existing shading system, the cobogó with window, labelled letter A. The change affects two parameters: the central window amplitude and the circular components' thickness amplitude of the original cobogó. This simple parametric definition enables quick insolation and illuminance optimization. The optimization considers, together with insolation and illuminance, a third energy parameter: interior air temperature.

The new parametric cobogó intends on avoiding overheating in the interior compartment. Figure 6 displays cobogó with window type A parameters.

The performance of this cobogó in the north façade upper left corner module, considers: (1) insolation in the cobogó superior face, in kWh/m^2 and kWh; (2) number of room points with illuminance between 300 and 750 lux; and (3) room temperature in °C, using Honeybee plug-in (version 1.6.0).

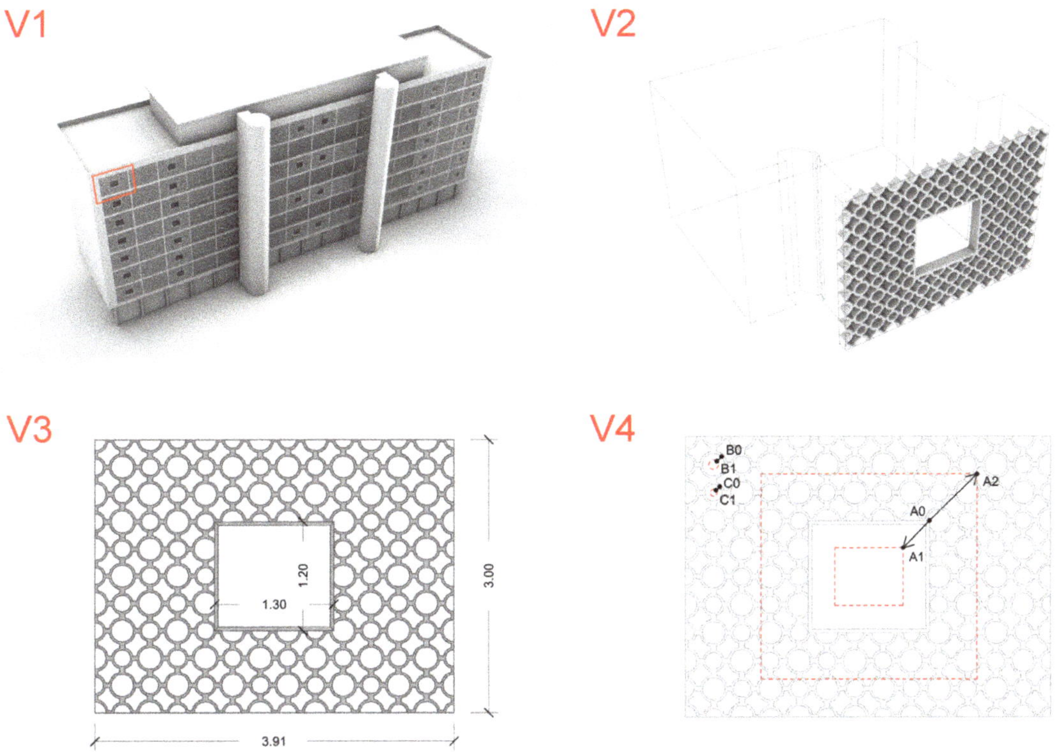

Figure 6. Shading system cobogó with window (type A) located upper left corner module (Nova Cintra north façade) (image **V1**); axonometric perspective cobogó with window (image **V2**); front view cobogó with window (image **V3**); parameterization cobogó with window. Parameter 1: central window amplitude, smaller (point A0 to A1), or larger (point A0 to A2), ranging from 0.0 to 1.8; Parameter 2: thickness amplitude of largest and smallest circles of the cobogó (points B0 to B1 and C0 to C1, respectively), ranging from 0.0 to 1.0 (image **V4**).

Single-objective optimization uses separated performance analyses for the cobogó type A, regarding insolation, illuminance and air temperature. After single-objective optimization, multi-objective optimization was conducted to improve all goals simultaneously. Single-objective optimization relies in Galapagos (version 0.2.0448) and multi-objective in Octopus (version 0.4), both plug-ins for Grasshopper. Galapagos [39] is a plug-in that uses genetic algorithms to discover the optimal combination of values for a given set of variables, applying the Darwinian theory of evolution on the design of alternatives. The result after several iterations and the elimination of unfit solutions is a pool of optimized design alternatives [17] for an objective, a single function, hence the use of Galapagos for single-objective optimization. Octopus [40] is a plug-in that utilizes a genetic algorithm for multi-objective (multiple function) optimization, named the SPEA-2 algorithm, published by Zitzler, Laumanns and Thiele [41]. There is a description of the implementation of the algorithm in the plug-in in the work of Vierlinger [42]. Galapagos and Octopus are therefore used in this research to obtain optimized results for the energy performance of the developed shading system, cobogó type A, for insolation, illuminance and air temperature. Section 3 presents the Cobogó type A results using single and multi-objective optimizations.

2.4.2. Algorithmic–Parametric Modelling and Performance Optimization

To improve cobogó type A energy performance, the original shape underwent additional transformations, using the Building Envelopes Grammar [30,31]. The Building Envelopes Grammar is an algorithmic–parametric modelling process. It contains different shape creation and transformation operations, which form the algorithmic part of the modelling process. Combinations can follow different arrangements, using different algorithms to form efficient buildings to harvest solar energy.

The cobogó type A transformation uses the normalization and division operations of the Building Envelopes Grammar. Transformations constitute the algorithmic part of the algorithmic–parametric modelling process. The normalization operation transforms a target surface into a surface perpendicular to the sun's rays, in a specific geographic location and time. Surfaces perpendicular to the sun's rays are more efficient at harvesting solar energy. The division operation divides the surface into two equal surfaces.

Next, it is necessary to divide a created plane, coincident with the exterior face of the cobogó type A, present in the upper left corner module (Nova Cintra north façade), in two equal horizontal surfaces. The created plane ranges from the floor to the ceiling and from the left room wall to the right room wall. The top horizontal surface requires normalization, but to the bottom horizontal surface does not, to allow the solar rays to reach the floors below. The top horizontal surface normalization considers the Sun's positions across the year, hourly, located more to north than the north façade. Another constraint considered for the normalized surfaces was that their four vertices should not be more than 1.5 m away from the façade's exterior plane.

Later, we adjusted the obtained set of possible normalized surfaces, moving each surface from their point located at a greater distance inside the building, to their perpendicular point, coincident with the exterior façade plane. This process assures that the selected set of normalized surfaces are totally outside the building, with only one vertex coincident with the north façade plane. Next, we moved the upper vertices of the bottom horizontal divided surface to make them coincident with the lower vertices of the normalized surfaces. This procedure assures a set of bottom horizontal surfaces, each one connected to a normalized surface. Figure 7 presents a schematic explanation of the division and normalization operations, applied on the upper left corner module.

Research then applies the parametric part of the process selecting, from the set of normalized surfaces, the normalized surface with the best performance for solar energy harvesting (insolation). Best performance relies on single-objective optimization using Galapagos.

After selecting the normalized surface with the best insolation performance, the algorithmic procedure returns, being the original cobogó type A projected in the selected normalized surface and the bottom horizontal surface associated. Current research developed the projection operation of cobogó type A, that is now part as well of the Building Envelopes Grammar. The algorithmic procedure of the algorithmic–parametric modelling finishes with the conclusion of the projection operation.

After finished the algorithmic part, the parametric part of the algorithmic–parametric process is also concluded, being realized with additional single-objective optimizations using Galapagos, considering parametric variations for the projected cobogó. The parameters are those adopted in the previous parametric modelling process (Section 2.4.1): the central window amplitude and the circular components' thickness amplitude. The optimization considers insolation, illuminance and temperature, separately.

Octopus performs additional multi-objective optimization for the normalized and parameterized cobogó, considering simultaneously insolation, illuminance and temperature. The next section presents the single- and multi-objective optimizations results.

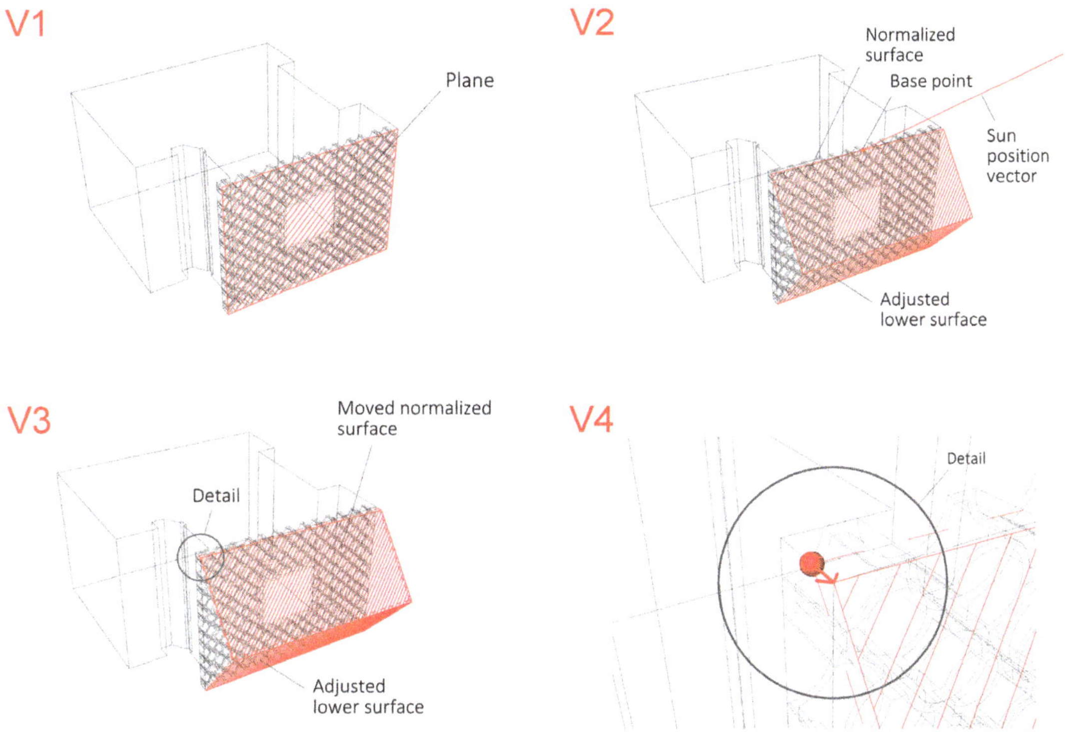

Figure 7. Plane coincident with exterior cobogó face (image **V1**). Plane division in two equal horizontal surfaces, top horizontal surface normalization, turning it perpendicular to one sun position (in this example, May 2nd at 12 p.m.), considering as base point for normalization operation the midpoint of the plane upper edge; adjustment of the bottom surface, connecting their two upper vertices to the normalized surface two lower vertices (image **V2**). Movement of normalized surface, from the point located more in the interior of the room to the perpendicular point located in the plane coincident with the cobogó exterior face (image **V3**). Detailed view of the movement of normalized top surface, showing the movement reference points (image **V4**).

3. Results

This section presents the results of the first and second stages of the Carioca Modern Façades research:

1. The results achieved in combinatorial modelling, recombining the original shading systems in the Nova Cintra north façade, used together with multi-objective optimization, considering insolation and illuminance measures. The results achieved in insolation and illuminance are also highlighted for the best performance shading system, obtained through the combinatorial modelling process, in the upper left corner module (Nova Cintra north façade). We use this module in the parametric and algorithmic–parametric modelling and optimization processes. These last results will enable us to compare the results obtained for this module in all the modelling and optimization processes used in the first and second stages of the research.
2. The original shading type results of the upper left corner module (Nova Cintra north façade) considering insolation, illuminance and air temperature. These results enable a comparison of the original shading performance with the performance of the shading developed with the modelling and optimization processes.

3. Results using parametric and algorithmic–parametric modelling processes over the original shading system cobogó with window (type A), together with single-objective and multi-objective optimization, regarding insolation, illuminance and air temperature.

3.1. Results of the Combinatorial Modelling and Multi-Objective Optimization

The Nova Cintra north façade results, using combinatorial modelling together with multi-objective optimization, are relative to three shading types: the cobogó with window (letter A), cobogó without window (letter B) and brise-soleil (letter C), which are displayed in Table 1. The multi-objective optimization considers three goals: (1) to minimize insolation after the shading, in Wh/m^2; (2) maximize points number with ideal illuminance between 300 and 750 lux in the interior rooms; and (3) maximize the interior rooms' average illuminance, in lux. Results without ponderation (the three first columns of Table 1: A, B, C) that consider equally the scores for insolation, points number with ideal illuminance and average illuminance (ins 1, illuPts 1, illuAvg 1), show that both in June (Jun) and December (Dec), the letter B prevails, at 10 a.m. and 12 p.m., while at 4pm (Jun and Dec), the best solution is letter C.

Table 1. Multi-objective analysis of the shading types (A, B, C) for insolation, points number with ideal illuminance between 300–750 lux and average illuminance.

Letters	A	B	C	A	B	C	A	B	C
MULTI-OBJECTIVE	1 ins	1 illuPts	1 illuAvg	0.2 ins	0.6 illuPts	0.2 illuAvg	0.2 ins	0.2 illuPts	0.6 illuAvg
Jun 10 a.m.	0	88	10	6	72	20	4	53	41
Jun 12 p.m.	0	97	1	0	97	1	0	12	86
Jun 4 p.m.	1	30	67	12	30	56	1	30	67
Dec 10 a.m.	0	95	3	0	95	3	0	95	3
Dec 12 p.m.	29	47	22	36	25	37	29	47	22
Dec 4 p.m.	36	20	42	58	2	38	36	20	42
SUM	66	377	145	112	321	155	68	257	257
AVERAGE	1	66	16	9	51	29	3	39	42

Taking more in consideration the objective points with ideal illuminance (ins 0.2, illuPts 0.6, illuAvg 0.2), solution B continues to prevail, with the exceptions of 4 p.m. Jun and 12 p.m. Dec where the best solution is C, and of 4 p.m. Dec, where letter A prevails.

Finally, favouring the objective average illuminance (ins 0.2, illuPts 0.2, illuAvg 0.6), there are three best solutions with the letter C (Jun 12 p.m. and 4 p.m., Dec 4 p.m.), with letter B being better in the other three situations (Jun 10 a.m., Dec 10 a.m. and 12 p.m.).

In summary, letter B, cobogó without window, is the best global solution. Letter C is the second best solution, and there are less situations in which the letter A dominates. We present a graphical representation of Table 1, using the corresponding shading types, in Figure 8. Through this graphical representation, it is possible to verify that in Dec 4 p.m., considering the different ponderations, letter A (red) is the best solution for many modules, suggesting low illuminance in this period, due to the Sun's high position. In the afternoon periods letter C (blue), brise-soleil, is also frequent, especially in Jun 4 p.m.

In the upper left corner module (Nova Cintra north façade), used in the other modelling and optimization processes of the research, letter B (cobogó without window, treated by type B in the other modelling and optimization processes) is the one that achieves the best results for the June 12 p.m. and December 12 p.m. periods. These are the periods analysed in the other modelling and optimization processes of the research.

Table 2 displays insolation, illuminance and also air temperature obtained for the shading cobogó without window (type B), in the upper left corner module (Nova Cintra north façade), for the periods of June 12 p.m. and December 12 p.m.

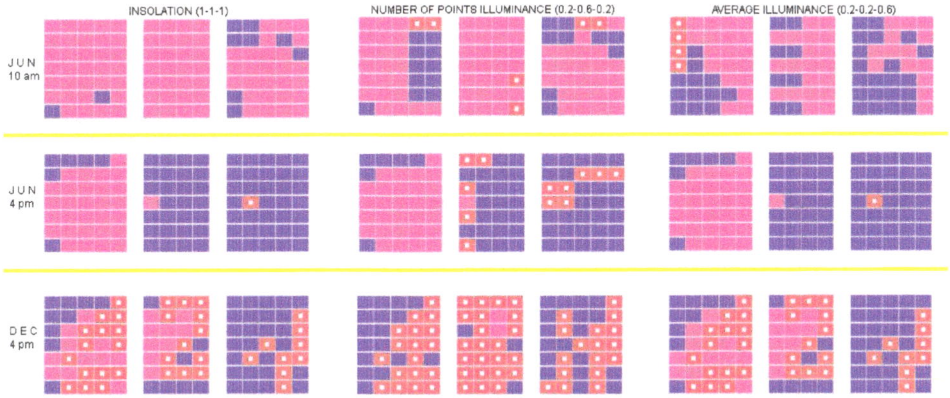

Figure 8. Graphical results combinatorial modelling and multi-objective optimization of the shading types letter A (red), B (lilac) and C (blue), resulting in a string (ACCBAB...) optimized according to objectives and weights.

Table 2. Insolation, points number with ideal illuminance and air temperature, simulated for the shading cobogó without window (type B), located at upper left corner module (Nova Cintra north façade).

Shading Analyses: Cobogó without Window (Type B)	Insolation (kWh ǀ kWh/m²)	N° of Points with Ideal Illuminance (300–750 lux) ǀ Total Number of Points	Air Temperature (°C)
Winter solstice, 21st June 12 p.m.	1.89 kWh ǀ 0.63 kWh/m²	23 ǀ 178 points	28.35 °C
Summer solstice, 21st December 12 p.m.	0.44 kWh ǀ 0.15 kWh/m²	64 ǀ 178 points	33.62 °C

Shading type B presents the following results: in June 12 p.m., 1.89 kWh and 0.63 kWh/m² for insolation, 23 points with illuminance between 300 and 750 lux in a total of 178 possible points and 28.35 °C of air temperature; in December 12 p.m., 0.44 kWh and 0.15 kWh/m² for insolation, 64 points with illuminance between 300 and 750 lux in a total of 178 possible points and 33.62 °C of air temperature.

Research discards average illuminance here, as in the other cases addressed.

Regarding insolation, the values considered for the combinatorial modelling and multi-objective optimization consider a plane behind the interior face of the shading. However, the insolation indicated in Table 2 is relative to that obtained in the exterior face of the shading, meaning that there is referent to sun energy harvesting, that can be used for the generation of renewable energy through the use of photovoltaic technology. The research considers this in the other modelling and optimization processes, hence their indication in Table 2, in order to be comparable with the insolation obtained in the other modelling and optimization processes of the research.

The insolation in the exterior face of the shading relates directly with the insolation behind the interior face of the shading: the bigger the insolation in the shading exterior face, the smaller the corresponding insolation behind the shading.

3.2. Results Original Shading Type, Cobogó with Window

We selected one module, the upper left corner module of the Nova Cintra building north façade, and their original shading type A (cobogó with window), for the realization of the other modelling and optimization processes of the research, due to the following reasons:

1. To save time in the development of the single-objective and multi-objective optimizations for insolation, illuminance and air temperature, for the parametric and algorithmic–parametric modelling processes;
2. The selected shading type A, cobogó with window, is the original of the upper left corner module of the Nova Cintra north façade and is the shading type that has the worst results in the combinatorial modelling and multi-objective optimization; so, it needs improvement;
3. The upper left corner module is the first module in the façade, if we interpret the façade as a mathematical matrix or as a sentence that starts from top to bottom, from left to right.

Therefore, we obtained the results for the original shading type A, regarding insolation, illuminance and air temperature, in the upper left corner module of the Nova Cintra north façade, in order to compare them with the results obtained for the combinatorial, parametric and algorithmic–parametric modelling processes. Table 3 shows results obtained for the original shading type A, cobogó with window: 2.26 kWh for insolation and 0.63 kWh/m^2 for insolation efficiency indicator, regarding shading harvesting by square meter, in the exterior face of the shading, 28 points with ideal illuminance (between 300 and 750 lux) out of 178 possible points and 29.20 °C of air temperature, during the winter solstice, 21st June 12 p.m.

Table 3. Results of cobogó with window (type A), for the winter and summer solstices: insolation, points number with ideal illuminance and air temperature.

Shading Analyses: Original Cobogó with Window (Type A)	Insolation (kWh I kWh/m^2)	N° of Points with Ideal Illuminance (300–750 lux) I Total Number of Points	Air Temperature (°C)
Winter solstice, 21st June 12 p.m.	2.26 kWh I 0.63 kWh/m^2	28 I 178 points	29.20 °C
Summer solstice, 21st December 12 p.m.	0.52 kWh I 0.10 kWh/m^2	112 I 178 points	34.51 °C

In the summer solstice, 21st December 12pm, the results are 0.52 kWh and 0.10 kWh/m^2 for insolation, 112 points with ideal illuminance (between 300 and 750 lux) out of 178 possible points and 34.51 °C air temperature.

Figure 9 displays a graphical representation of the results of insolation, illuminance and air temperature, for the shading type cobogó with window, at the upper left corner module (Nova Cintra north façade), for the winter and summer solstices.

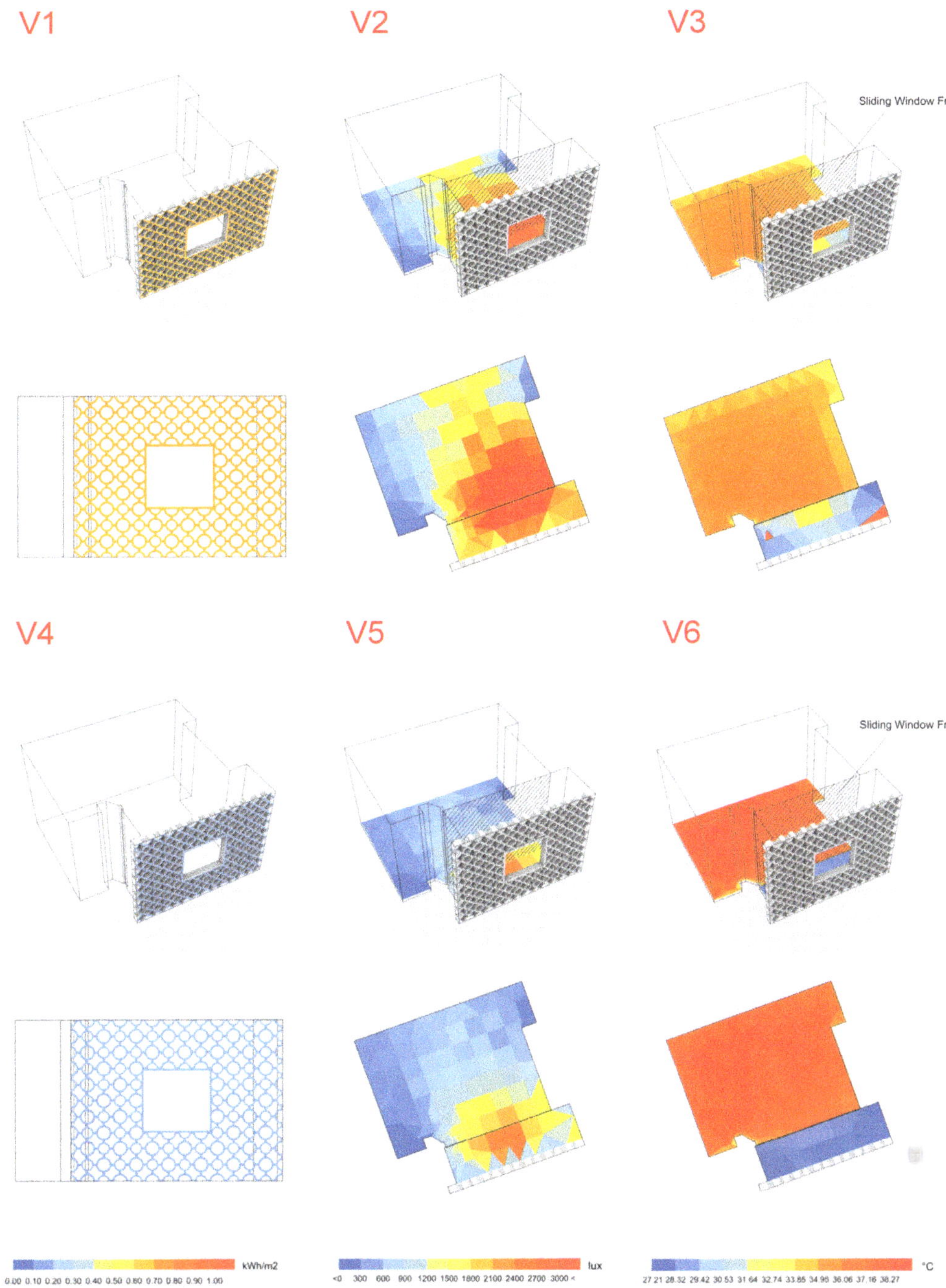

Figure 9. Results insolation (left column), points number with ideal illuminance (middle column) and air temperature (right column) original shading cobogó with window (type A), winter solstice, June 21st 12 p.m. (images **V1**, **V2** and **V3**) and summer solstice, December 21st 12 p.m. (images **V4**, **V5** and **V6**).

3.3. Results of the Parametric Modelling, with Single-Objective and Multi-Objective Optimization

The parametric modelling, as defined in Section 2.4.1, uses two parameters for the single-objective optimization of the shading type cobogó with window (type A).

The universe of solutions is defined by parameter 1, the window central void amplitude, with values that range from 0.0 to 1.8, with the value 1.0 being the original size of the cobogó with window (1.30 m length and 1.20 m height); and parameter 2, the circular modules' thickness amplitude, that range from 0.0, the maximum thickness (0.145 m in the bigger circles and 0.118 m in the smaller circles) to 1.0, the minimum and original thickness (0.04 m in the bigger circles and 0.035 m in the smaller circles).

Table 4 presents the single-objective optimization results obtained for insolation, points number with ideal illuminance (300–750 lux) and air temperature, for the winter and summer solstices. For instance, the single-objective optimization of the insolation for the winter solstice has a result of 4.77 kWh and 0.62 kWh/m^2, with 0.1 (0.13 m length and 0.12 m height) for parameter 1 and 0.0 (thickness of 0.145 m in bigger circles and of 0.118 m in smaller circles) for parameter 2. The single-objective optimization of the insolation for the summer solstice presents the result of 1.11 kWh and 0.14 kWh/m^2, with 0.0 (0 m length per 0 m height) for parameter 1 and 0.0 (thickness of 0.145 m in bigger circles and of 0.118 m in smaller circles) for parameter 2.

Figure 10 displays a graphical representation with colours of the results of the single-objective optimization processes, for the winter and summer solstices.

Table 4. Single-objective optimization results of the parametric model realized for the cobogó with window (type A), regarding insolation, points number with ideal illuminance and air temperature, for the winter and summer solstices.

Single-Objective Optimization of the Parametric Model	Parameter 1: Central Window Ampitude	Parameter 2: Circular Thickness Amplitude	Insolation (kWh I kWh/m^2)	Points number with Ideal Illuminance (300–750 lux) I Total Number of Points	Air Temperature (°C)
Winter solstice, 21st June 12 p.m.	0.1	0.0	4.77 kWh I 0.62 kWh/m^2	-	-
	0.1	0.3	-	95 I 178 points	-
	0.5	0.0	-	-	28.34 °C
Summer solstice, 21st December 12 p.m.	0.0	0.0	1.11 kWh I 0.14 kWh/m^2	-	-
	0.3	1.0	-	118 I 178 points	-
	0.5	0.0	-	-	33.62 °C

The research performed as well a multi-objective optimization for the shading cobogó with window (type A) to achieve an optimized shape simultaneously considering insolation, points number with ideal illuminance and air temperature performance.

Figure 11 shows a perspective of the multi-objective optimized shape obtained for the shading cobogó with window, illustrating with colour the insolation-optimized result for the winter solstice.

Table 5 shows the results obtained in the multi-objective optimization realized for the parametric model of the cobogó with window (type A), regarding insolation, points number with ideal illuminance and air temperature, in the winter and summer solstices. The winter solstice results are 4.67 kWh and 0.63 kWh/m^2 for insolation, 106 out of 178 possible points for ideal illuminance, 29.09 °C for air temperature, 0.1 (0.13 m length and 0.12 m height) for parameter 1 and 0.2 (thickness of 0.124 m for bigger circles and of 0.101 m for smaller circles) for parameter 2. Summer solstice results are 1.03 kWh and 0.10 kWh/m^2 for insolation, 96 out of 178 possible points for ideal illuminance, 33.63 °C for air temperature, 0.8 (1.04 m length and 0.96 m height) for parameter 1 and 0.0 (thickness of 0.145 m in bigger circles and of 0.118 m in smaller circles) for parameter 2.

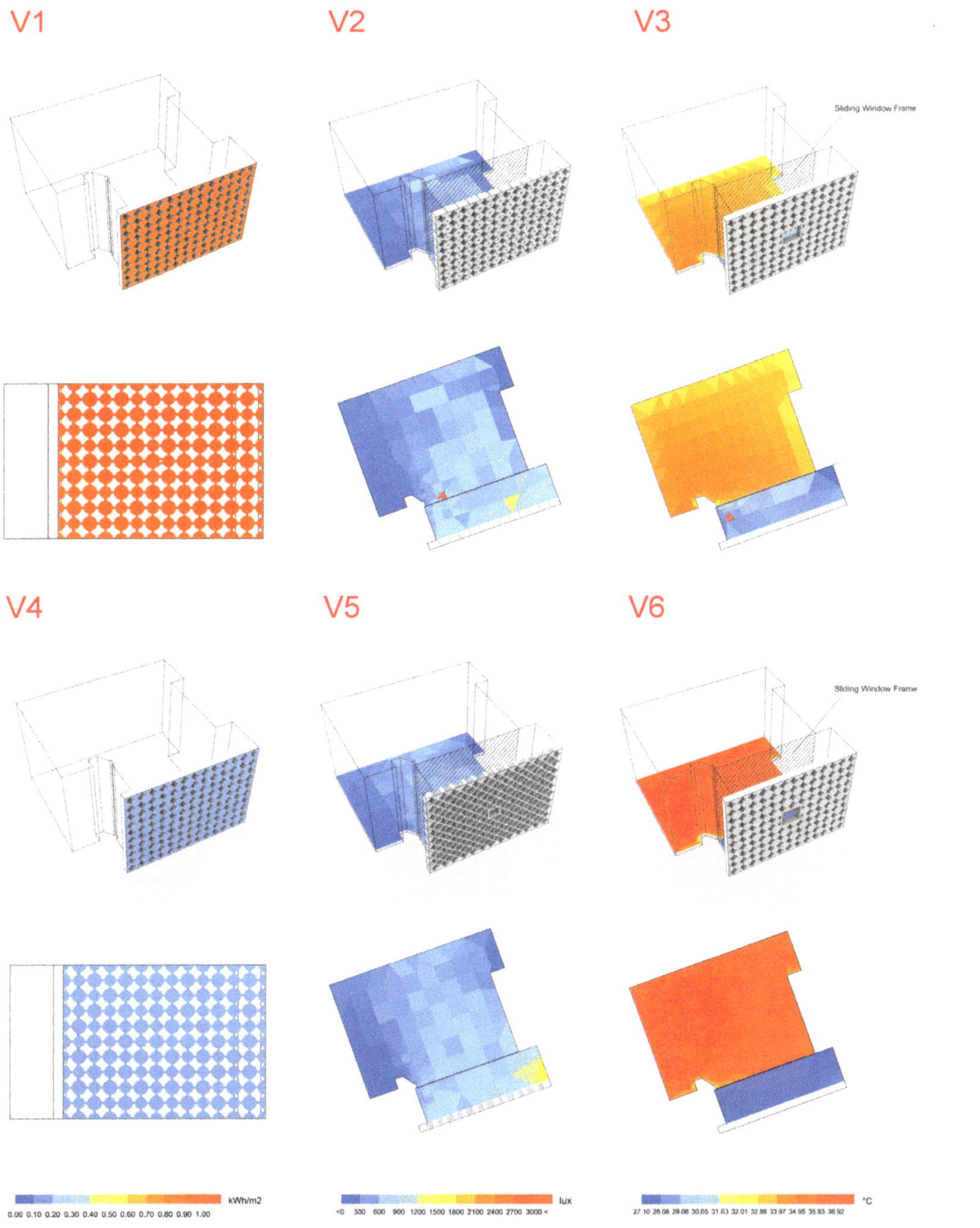

Figure 10. Results of the single-objective optimizations for winter solstice, June 21st 12 p.m. (images **V1**, **V2** and **V3**) and summer solstice, December 21st 12 p.m. (images **V4**, **V5** and **V6**). Results regarding insolation (**left** column), points number with ideal illuminance (**middle** column) and air temperature (**right** column).

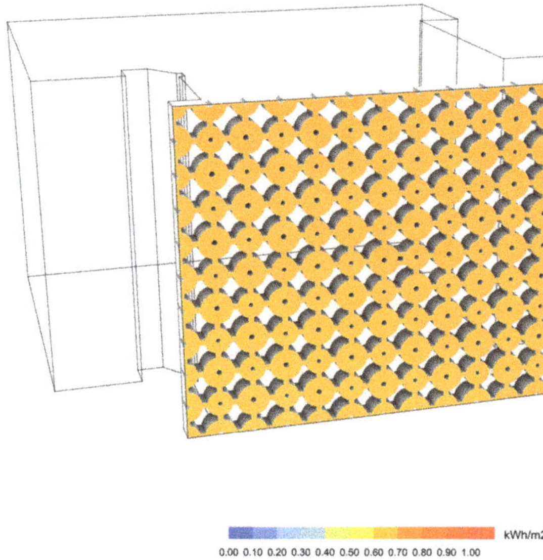

Figure 11. Multi-objective optimization result of the parametric model realized for the cobogó with window (type A), regarding insolation, points number with ideal illuminance and air temperature, in the winter solstice, 21st June 12 p.m.

Table 5. Results of the multi-objective optimization realized for the parametric model of the cobogó with window, regarding insolation, points number with ideal illuminance and air temperature.

Multi-Objective Optimization of the Parametric Model	Parameter 1: Central Window Amplitude	Parameter 2: Circular Thickness Amplitude	Insolation (kWh ǀ kWh/m^2)	Points Number with Ideal Illuminance (300–750 lux) ǀ Total Number of Points	Air Temperature (°C)
Winter solstice, 21st June 12 p.m.	0.1	0.2	4.67 kWh ǀ 0.63 kWh/m^2	106 ǀ 178 points	29.09 °C
Summer solstice, 21st December 12 p.m.	0.8	0.0	1.03 kWh ǀ 0.10 kWh/m^2	96 ǀ 178 points	33.63 °C

3.4. Results of the Algorithmic–Parametric Modelling with Single-Objective and Multi-Objective Optimization

The algorithmic–parametric modelling processes of division and normalization, described in Section 2.4.2, generate 350 possible positions to normalize the top horizontal surface, considering a planar surface from the upper left corner module (Nova Cintra north façade), coincident with the exterior surface of the cobogó with window (type A) and divided in two equal horizontal surfaces, top and bottom surfaces. The 350 possible positions for normalize the top horizontal surface, aligning this surface perpendicular to different sun positions across the year, are optimized regarding insolation, to maximize solar energy harvesting. We performed a single-objective optimization process for the winter and summer solstices. The Sun normalization's best performance is on September 1st at 12 p.m., for both solstices. Table 6 displays the best insolation performance results for a normalized top horizontal surface, obtained through the single-objective optimizations. After the selection of the normalized top horizontal surface with best performance and their associated bottom horizontal surface, the parametric model realized for the original cobogó with window (type A) is projected in these two surfaces. With this projection the algorithmic part of the algorithmic–parametric modelling process is complete.

Table 6. Best insolation performance for winter and summer solstices, using single-objective optimization for the normalized top horizontal surface, considering a divided planar surface from the upper left corner module (Nova Cintra north façade).

Normalization Cobogó with Window (Type A)	Sun Position	Insolation (kWh/m^2 I kWh)
Winter solstice, 21st June 12 p.m.	1 September at 12 p.m.	0.87 kWh/m^2 I 7.77 kWh
Summer solstice, 21st December 12 p.m.	1 September at 12 p.m.	0.65 kWh/m^2 I 5.36 kWh

The projected parametric model forms the remaining parametric part of the algorithmic–parametric modelling process, after the completion of the parametric optimization of the normalized top horizontal surface. The projected parametric model considers the same two parameters as the single parametric modelling process: the window central void amplitude and the circular modules thickness amplitude. Table 7 presents results of a single-objective optimization, considering these two parameters, of the divided and normalized cobogó with window, regarding insolation, points number with ideal illuminance (300–750 lux) and air temperature, for the solstices.

Table 7. Single-objective optimization results of the algorithmic–parametric model realized for the cobogó with window (type A), regarding insolation, points number with ideal illuminance and air temperature, for the winter and summer solstices.

Single-Objective Optimization of the Algorithmic–Parametric Model	Parameter 1: Central Window Amplitude	Parameter 2: Circular Thickness Amplitude	Insolation (kWh I kWh/m^2)	Points Number with Ideal Illuminance (300–750 lux) I Total Number of Points	Air Temperature (°C)
Winter solstice, 21st June 12 p.m.	0.1	0.1	4.77 kWh I 0.45 kWh/m^2	-	-
	0.1	0.3	-	86 I 184 points	-
	0.4	0.0	-	-	28.79 °C
Summer solstice, 21st December 12 p.m.	0.1	0.1	3.37 kWh I 0.32 kWh/m^2	-	-
	0.3	0.5	-	91 I 184 points	-
	0.1	0.0	-	-	33.49 °C

For instance, the single-objective optimization of the insolation for the winter solstice presents a result of 4.77 kWh and 0.45 kWh/m^2, with 0.1 (0.13 m length and 0.12 m height) for parameter 1 and 0.1 (thickness of 0.134 m in bigger circles and of 0.108 m in smaller circles) for parameter 2. For the summer solstice, the single-objective optimization of the insolation has a result of 3.37 kWh and 0.32 kWh/m^2, with 0.1 (0.13 m length and 0.12 m height) for parameter 1 and 0.1 (thickness of 0.134 m in bigger circles and of 0.108 m in smaller circles) for parameter 2. Figure 12 displays a graphical representation with colours of the results of single-objective optimization processes, for winter and summer solstices.

We also developed a multi-objective optimization of the algorithmic–parametric model realized for the cobogó with window (type A), considering the same parameters that were considered in the parametric modelling process. The optimized shape solution considers simultaneously the performance obtained for insolation, points number with ideal illuminance and air temperature. Figure 13 displays the shape obtained for the divided and normalized shading cobogó with window, in the multi-objective optimization realized for the winter solstice day, illustrating with colour the insolation-optimized result.

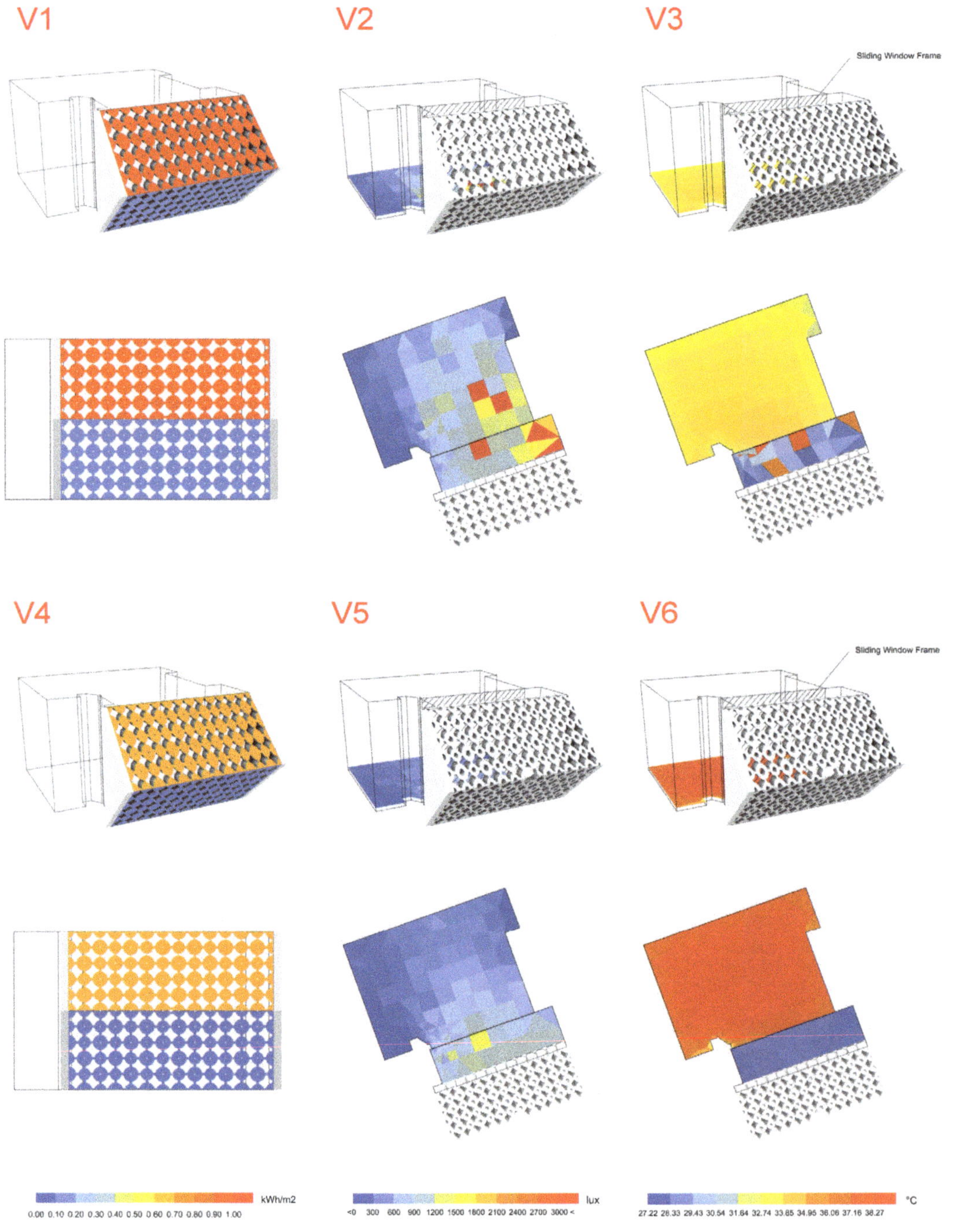

Figure 12. Single-objective optimizations results for winter solstice, June 21st 12 p.m. (images **V1**, **V2** and **V3**) and summer solstice, December 21st 12 p.m. (images **V4**, **V5** and **V6**). Results for insolation (**left** column), points number with ideal illuminance (**middle** column) and air temperature (**right** column).

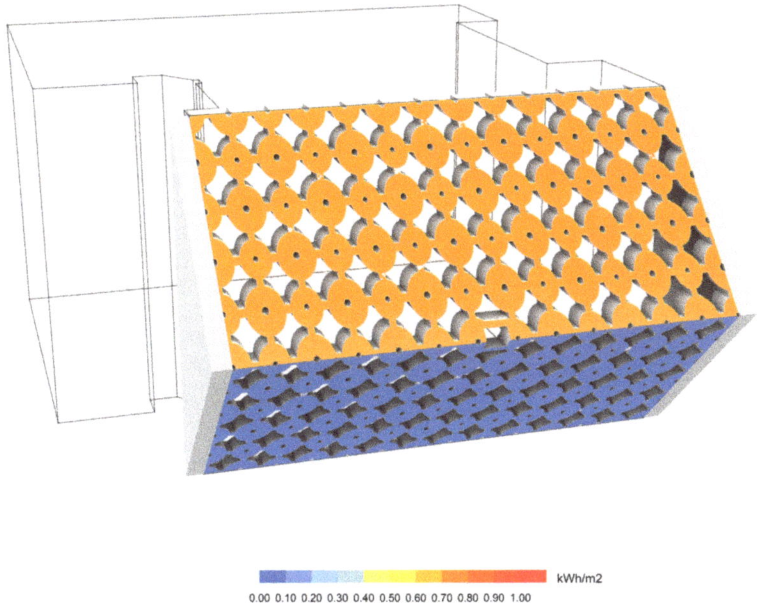

Figure 13. Multi-objective optimization results, algorithmic–parametric cobogó with window, for insolation, points number with ideal illuminance and air temperature, in the summer solstice, 21st December 12 p.m.

Table 8 presents the results obtained in the multi-objective optimization realized for the algorithmic–parametric model of the cobogó with window (type A), regarding insolation, points number with ideal illuminance and air temperature, in the winter and summer solstices.

The winter solstice results are 4.70 kWh and 0.46 kWh/m^2 for insolation, 92 out of 184 possible points for ideal illuminance, 29.37 °C for temperature, 0.2 (0.26 m length and 0.24 m height) for parameter 1 and 0.3 (thickness 0.109 m in bigger circles and 0.089 m in smaller circles) for parameter 2. Summer solstice results are 3.59 kWh and 0.34 kWh/m^2 for insolation, 86 out of 184 possible points for ideal illuminance, 34.48 °C of air temperature, 0.2 (0.26 m length and 0.24 m height) for parameter 1 and 0.2 (thickness 0.122 m in bigger circles and 0.098 m in smaller circles) for parameter 2.

Table 8. Results of the multi-objective optimization realized for the algorithmic–parametric model of the cobogó with window, regarding insolation, points number with ideal illuminance and air temperature.

Multi-Objective Optimization of the Algorithmic–Parametric Model	Parameter 1: Central Window Amplitude	Parameter 2: Circular Thickness Amplitude	Insolation (kWh ǀ kWh/m^2)	Points Number with Ideal Illuminance (300–750 lux) ǀ Total Number of Points	Air Temperature (°C)
Winter solstice, 21st June 12 p.m.	0.2	0.3	4.70 kWh ǀ 0.46 kWh/m^2	92 ǀ 184 points	29.37 °C
Summer solstice, 21st December 12 p.m.	0.2	0.2	3.59 kWh ǀ 0.34 kWh/m^2	86 ǀ 184 points	34.48 °C

4. Discussion

This research explores computational methods as combinatorial, parametric and algorithmic–parametric modelling, together with single-objective and multi-objective optimization processes, regarding insolation, points number with ideal illuminance and air temperature, to improve the performance of the existing shading systems of the Nova Cintra building north façade.

Addressed in particular the module of the upper left corner of the north façade, analysing cobogó with window (type A) performance and exploring new shading possibilities.

Tables 9–11 compare the original shading solution with the generated solutions using the modelling and optimization processes. To compare the shading performance regarding insolation, points number with ideal illuminance and air temperature, we present the percentage difference in brackets. The generated solutions improved the performance for insolation, points number with ideal illuminance and air temperature. There are, however, some exceptions, when performance diminishes:

1. Insolation performance during the winter and summer solstices, using a combinatorial modelling with multi-objective optimization: insolation is 1.9 kWh (−16.4%) and 0.4 kWh (−15.4%);
2. Insolation performance during the winter solstice, using parametric modelling and single-objective optimization, algorithmic–parametric modelling and single-objective optimization, as well as algorithmic–parametric modelling and multi-objective optimization, where energy harvesting is 0.6 kWh/m^2 (−1.6%), 0.5 kWh/m^2 (−28.6%) and 0.5 kWh/m^2 (−27.0%), respectively;
3. Air temperature during the winter solstice, through algorithmic–parametric modelling and the multi-objective optimization process, where temperature increases for 29.4 °C (+0.6%), regarding the original shading.

The generated shading solutions performance achieve the best improvements regarding insolation, points number with ideal illuminance and air temperature, in the following cases:

1. Winter solstice, for insolation, with parametric modelling and single-objective optimization and with algorithmic–parametric modelling and single-objective optimization, where total insolation is 4.8 kWh (+111.1% than original shading). There is no energy harvesting efficiency improvement in these cases, the original shading efficiency value is 0.6 kWh/m^2, the same in combinatorial modelling and multi-objective optimization, as well as in parametric modelling and multi-objective optimization. In the summer solstice, the best improvements for insolation are in the algorithmic–parametric modelling and multi-objective optimization process, with a total insolation 3.6 kWh (+590.4%) and 0.3 kWh/m^2 (+240%) energy harvesting-efficiency;
2. For the points number with ideal illuminance, in winter solstice, the parametric modelling and multi-objective optimization process, obtains 106 points (+360.9% than the original shading). During the summer solstice, the best result is with parametric modelling and single-objective optimization process, with 118 points (+84.4%);
3. In the winter solstice, for air temperature, with parametric modelling and single-objective optimization, which reaches 28.3 °C (−2.9% than the original shading). During the summer solstice, with the algorithmic–parametric modelling and single-objective optimization, 33.5 °C (−3.0%) is obtained.

Table 9. Insolation in the original shading and the generated shading, winter and summer solstices.

Insolation Analyses	Original Shading kWh I kWh/m^2	Combinatorial Modelling and Multi-Objective Optimization kWh I kWh/m^2	Parametric Modelling and Single-Objective Optimization kWh I kWh/m^2	Parametric Modelling and Multi-Objective Optimization kWh I kWh/m^2	Algorithmic-Parametric Modelling and Single-Objective Optimization kWh I kWh/m^2	Algorithmic-Parametric Modelling and Multi-Objective Optimization kWh I kWh/m^2
Winter solstice, 21st June 12 p.m.	2.3 I 0.6	1.9 I 0.6 (−16.4% I –)	4.8 I 0.6 (+111.1% I −1.6%)	4.7 I 0.6 (+106.6% I –)	4.8 I 0.5 (+111.1% I −28.6%)	4.7 I 0.5 (+108.0% I −27.0%)
Summer solstice, 21st December 12 p.m.	0.5 I 0.1	0.4 I 0.2 (−15.4% I +50%)	1.1 I 0.1 (+113.5% I +40%)	1.0 I 0.1 (+98.1% I –)	3.4 I 0.3 (+548.1% I +220%)	3.6 I 0.3 (+590.4% I +240%)

Table 10. Points number with ideal illuminance (between 300 and 750 lux) in the original shading and the generated shading, in winter and summer solstices.

Points with Ideal Illuminance (300–750 lux) Analyses	Original Shading Points I Points	Combinatorial Modelling and Multi-Objective Optimization Points I Points	Parametric Modelling and Single-Objective Optimization Points I Points	Parametric Modelling and Multi-Objective Optimization Points I Points	Algorithmic-Parametric Modelling and Single-Objective Optim Points I Points	Algorithmic-Parametric Modelling and Multi-Objective Optimization Points I Points
Winter solstice, 21st June 12 p.m.	23 I 178	28 I 178 (+21.7%)	95 I 178 (+313.1%)	106 I 178 (+360.9%)	86 I 184 (+261.7%)	92 I 184 (+287.0%)
Summer solstice, 21st December 12 p.m.	64 I 178	112 I 178 (+75%)	118 I 178 (+84.4%)	96 I 178 (+50%)	91 I 184 (+37.6%)	86 I 184 (+30.0%)

Table 11. Air temperature (°C) in the original shading and the generated shading, in winter and summer solstices.

Air Temperature Analyses	Original Shading °C	Combinatorial Modelling and Multi−Objective Optimization °C	Parametric Modelling and Single−Objective Optimization °C	Parametric Modelling and Multi−Objective Optimization °C	Algorithmic−Parametric Modelling and Single−Objective Optimization °C	Algorithmic−Parametric Modelling and Multi−Objective Optimization °C
Winter solstice, 21st June 12 p.m.	29.2	28.4 (−2.9%)	28.3 (−2.9%)	29.1 (−0.4%)	28.8 (−1.0%)	29.4 (+0.6%)
Summer solstice, 21st December 12 p.m.	34.5	33.6 (−2.6%)	33.6 (−2.6%)	33.6 (−2.6%)	33.5 (−3.0%)	34.5 (−0.1%)

As the main findings, we highlight that the generated shading improvements in insolation are higher in the algorithmic–parametric modelling and respective optimization processes, in both solstices, especially in the summer solstice, the Sun's highest position, when the top normalized surface of the generated shading is further exposed to the Sun. In winter solstice, the sun lowest position, the improvement in the shading efficiency is considerable, but not as significant as in the summer solstice. The summer solstice has a greater improvement in the efficiency of harvesting sun energy, because the normalized top horizontal shading surface exposes more to the sunrays, comparing to the original shading surface.

Regarding the points number with ideal illuminance, the parametric modelling and multi-objective optimization process is more efficient during the winter solstice and the parametric modelling and single-objective optimization process has improved efficiency during the summer solstice. A hypothesis for the improved performance of the parametric modelling process, with single-objective and multi-objective optimizations, comparatively to the algorithmic–parametric modelling process, is that the parametric modelling process occurs in a vertical planar surface, allowing a homogeneous entry of natural daylight in the rooms. The algorithmic–parametric process generates two different shading surfaces, creating a distinct and heterogeneous daylight entry through these two surfaces. Future research intends to explore different parameterization processes in the algorithmic–parametric process: one parameterization process for the normalized top surface and another different parameterization process for the bottom-shading surface. This process expects to achieve different daylight entries through the shading, for the top and bottom surfaces, to improve the efficiency in the points number with ideal illuminance inside the rooms.

For the air temperature, during the winter solstice, the parametric modelling and single-objective optimization process seems to be the most efficient, while for summer solstice the algorithmic–parametric modelling and single-objective optimization process have the best result. As well as in the case of the points number with ideal illuminance, there is the hypothesis that the number of shading surfaces and their orientation, one vertical surface in the parametric modelling process and two different surfaces, oriented in different ways, in the algorithmic–parametric process, influence the temperature performance. Therefore, future research for air temperature intends to verify also if the appliance of two different parameterization processes, for the two different surfaces of the algorithmic–parametric shading, allow improving shading performance.

Comparing the improvements obtained in this research with the improvements achieved by the researches described in the literature review, we can verify that Caldas and Norford [6] found an improvement of 15% between the worst and the best result achieved for the annual electrical energy consumption of an office building. These results are optimized with a genetic algorithm. The parametric space of solutions are the window dimensions in the façades of the building.

Jalali, Noorzai and Heidari [8], in the design of the overall form and façades windows of an office building, realized through parametric modelling, mention improvements of 8% (reduction in cooling load), 21% (reduction in heating load) and 37% (increase in Useful Daylight Illuminance). These improvements are also achieved with optimization algorithms, applied on the parametric space of solutions defined for the overall form and façades windows of the office building.

Fang and Cho [11], also using parametric modelling and optimization algorithms for the geometry definition of buildings and their façades, report improvements in energy savings of 30% in office buildings and of 1 to 3% in apartment buildings.

González and Fiorito [15], using the same processes for the optimization of external shadings in an office building, indicate improvements of 35% in energy savings and of 48% in the reduction of CO_2 emissions for the building.

Therefore, the improvements achieved in the reference researches, which are related as well with energy performance measures, are in a range between 1% and 48%.

In the present research the improvements, all of them obtained using optimization algorithms as well, are between 98.1% reached by parametric modelling and 590.4% attained by algorithmic–parametric modelling, for the insolation.

For the points number with ideal illuminance, the improvements are between 21.7%, obtained by combinatorial modelling, and 360.9%, reached by parametric modelling.

For air temperature, improvements are between 0.1%, achieved by algorithmic–parametric modelling, and 3.0%, attained also by algorithmic–parametric modelling.

Thus, the improvements obtained in this research are in general higher comparing with the improvements achieved by the researches described in the literature review, considering the values obtained for insolation and points number with ideal illuminance. In insolation, the best improvement values are achieved through algorithmic–parametric modelling, being considerably higher. In points number with ideal illuminance, the higher values are reached by parametric modelling. Regarding air temperature, the best improvement values obtained by algorithmic–parametric modelling are lower, being in the range achieved by the researches present at the literature review.

In general, algorithmic–parametric modelling demonstrated that can be more efficient than parametric modelling to obtain higher improvements in the energy performance of the studied shading system. Despite the improvements in points number with ideal illuminance are higher with parametric modelling, we believe that is possible to improve the algorithmic–parametric process, in order to achieve also for this energy measure better results than parametric modelling.

It should be referred also that combinatorial modelling has in general lower improvement results than parametric and algorithmic–parametric modelling types. This is due it having a narrow space of shape solutions. However, it has the advantage of being the quickest process of achieve improved optimized performance solutions for the shading system, of all the three tested modelling types in this research.

5. Conclusions

This research develops an energy-based design model, to improve the energy performance of a carioca modern building, the Nova Cintra. The research was conducted to improve the building performance using computational methods, such as combinatorial modelling, parametric modelling and algorithmic–parametric modelling, jointly with single-objective and multi-objective optimizations, applied over the existing shading systems on the building north façade. The proposal is an exploratory and prospective research, not intends at least in this stage to build new shading systems in a real context. The shading solutions are developments of the original shading systems, improving their performance regarding insolation, illuminance and air temperature. These improvements intend to avoid air-conditioning systems inside the building, as well as reduce the need of artificial lighting, diminishing the energy consumption. On the other hand, improvement in the performance regarding insolation explores the solar energy harvesting capacity in the building shading systems, which acquire the potential of generating renewable energy using photovoltaic technology.

The methodology for the first stage of the research modelled in 3D the Nova Cintra building using visual and textual programming, analysed in general all the building envelope and in detail their north façade, for insolation and illuminance, and explores also combinatorial modelling and multi-objective optimization processes, using the three original shading systems of the north façade. This improves the insolation and illuminance performance of the building façade.

The second research stage, for the upper left corner module of the north façade, generated solutions to improve shading, using parametric and algorithmic–parametric modelling with single-objective and multi-objective optimization processes, for insolation, illuminance, and air temperature.

Conclusions regarding the generated shading solutions performance:

1. The research use of modelling and optimization processes, namely the combinatorial, parametric and algorithmic–parametric modelling, with single and multi-objective optimizations, can improve the performance of the building's shading solutions, regarding insolation, illuminance and air temperature;
2. The algorithmic–parametric modelling, jointly with single-objective and multi-objective optimization processes, can improve better the insolation. The parametric modelling, with single-objective and multi-objective optimization, can improve better the illuminance. Both parametric (for winter solstice) and algorithmic–parametric (for summer solstice) modelling procedures, with single-objective optimization, achieved better performance for the air temperature;
3. The research developed delivers energy-based design to improve buildings' energy performance. Research confirms that shape transformation and modelling processes can improve energy performance.

This research performance improvement is in general higher than in previous researches described in the literature review. Algorithmic–parametric modelling in our research proved to be more efficient than parametric modelling and combinatorial modelling, improving the energy performance of the shading systems, especially regarding insolation.

This research has the following limitations: (1) the time expended on the optimization processes, using the Galapagos and Octopus plug-ins. Most of the optimizations took 24 to 48 h to be accomplished. (2) The period established for the realization of the research: 3 months. These aspects limited the realization of further algorithmic–parametric modelling processes in obtaining higher performance. This research also chose not to apply the algorithmic–parametric modelling process to the entire north façade of the Nova Cintra building.

Future work should thus address the development of the algorithmic–parametric modelling type, looking to make it more efficient than parametric modelling to improve the number of points with ideal illuminance. Expecting to also increase the results already achieved for air temperature, it intends as well to develop a method to apply the study to the entire Nova Cintra north façade.

This research uses an energy-based design model to transform building-shading systems, using computational methods to improve the buildings energy performance for factors such as insolation, illuminance and air temperature. Energy-based design models can apply as well to design other building elements such as roofs and walls, or design entirely buildings. In the case of this research, the use of energy-based design in building-shading systems can minimize the building's energy consumption needs in devices such as air-conditioning systems and artificial lighting. Simultaneously, the improvements achieved in insolation can stimulate the harvesting of solar energy in the shading systems and the production of renewable energy. Using photovoltaic technology in the future, research might transform buildings beyond energy consumers to energy producers. The potential gain in the overall buildings' energy performance can diminish the use of energy from pollutant sources such as coal and increase the production and use of energy from renewable sources such as the Sun. This is a significant contribution, by human action, to mitigate climate change, a fundamental quest for the environmental sustainability of our planet.

Author Contributions: Conceptualization, D.M. and G.C.H.; methodology, D.M.; software, D.M. and G.C.H.; validation, D.M. and G.C.H.; formal analysis, D.M. and G.C.H.; investigation, D.M. and G.C.H.; resources, D.M. and G.C.H.; data curation, D.M. and G.C.H.; writing—original draft preparation, D.M.; writing—review and editing, D.M. and G.C.H.; visualization, D.M.; supervision, D.M. and G.C.H.; project administration, D.M. and G.C.H.; funding acquisition, G.C.H. All authors have read and agreed to the published version of the manuscript.

Funding: This research was funded by Brazilian scientific initiation grants and artistic and cultural initiation grants (PIBIC and PIBIAC) for Taiane Nepomuceno and Ana Clara Mouro (PIBIAC IC 2022 CLA/UFRJ Edital 449/2022), and Ronaldo Lee (PIBIAC IC 2021, CLA/UFR, Edital 212/2021) that we acknowledge and thank for their research contribution.

Data Availability Statement: Data sharing is not applicable to this article.

Conflicts of Interest: The authors declare no conflict of interest.

References

1. Sadineni, S.B.; Madala, S.; Boehm, R.F. Passive building energy savings: A review of building envelope components. *Renew. Sustain. Energy Rev.* **2011**, *15*, 3617–3631. [CrossRef]
2. Sommese, F.; Badarnah, L.; Ausiello, G. A critical review of biomimetic building envelopes: Towards a bio-adaptive model from nature to architecture. *Renew. Sustain. Energy Rev.* **2022**, *169*, 112850. [CrossRef]
3. Tablada, A.; Kosoric, V.; Huang, H.; Chaplin, I.K.; Lau, S.-K.; Yuan, C.; Lau, S.-Y. Design Optimisation of Productive Façades: Integrating Photovoltaic and Farming Systems at the Tropical Technologies Laboratory. *Sustainability* **2018**, *10*, 3762. [CrossRef]
4. Svetozarevic, B.; Begle, M.; Jayathissa, P.; Caranovic, S.; Shepherd, R.F.; Nagy, Z.; Hischier, Z.; Hofer, J.; Schlueter, A. Dynamic photovoltaic building envelopes for adaptive energy and comfort management. *Nat. Energy* **2019**, *4*, 671–682. [CrossRef]
5. Nagy, Z.; Svetozarevic, B.; Jayathissa, P.; Begle, M.; Hofer, J.; Lydon, G.; Willmann, A.; Schlueter, A. The Adaptive Solar Facade: From concept to prototypes. *Front. Archit. Res.* **2016**, *5*, 143–156. [CrossRef]
6. Caldas, L.G.; Norford, L.K. A design optimization tool based on a genetic algorithm. *Autom. Constr.* **2002**, *11*, 173–184. [CrossRef]
7. Venâncio, R. Sombreamento com iluminação: Desenvolvimento e teste de modelo paramétrico para facilitar o projeto de proteções solares. *Ambiente Construído* **2020**, *20*, 59–77. [CrossRef]
8. Jalali, Z.; Noorzai, E.; Heidari, S. Design and optimization of form and façade of an office building using the genetic algorithm. *Sci. Technol. Built Environ.* **2020**, *26*, 128–140. [CrossRef]

9. Kirimtat, A.; Koyunbaba, B.K.; Chatzikonstantinou, I.; Sariyildiz, S.; Suganthan, P.N. Multi-objective optimization for shading devices in buildings by using evolutionary algorithms. In Proceedings of the 2016 IEEE Congress on Evolutionary Computation (CEC), Vancouver, BC, Canada, 24–29 July 2016. [CrossRef]
10. Lin, S.-H.E.; Gerber, D.S. Designing-in performance: A framework for evolutionary energy performance feedback in early stage design. *Autom. Constr.* **2014**, *38*, 59–73. [CrossRef]
11. Fang, Y.; Cho, S. Building Geometry Optimization with Integrated Daylighting and Energy Simulation. In Proceedings of the 15th IBPSA Conference, San Francisco, CA, USA, 7–9 August 2017. [CrossRef]
12. Evins, R. A review of computational optimisation methods applied to sustainable building designs. *Renew. Sustain. Energy Rev.* **2013**, *22*, 230–245. [CrossRef]
13. Caruso, G.; Fantozzi, F.; Leccese, F. Optimal theoretical building form to minimize direct solar irradiation. *Sol. Energy* **2013**, *97*, 128–137. [CrossRef]
14. Caruso, G.; Kämpf, J.H. Building shape optimisation to reduce air-conditioning needs using constrained evolution algorithms. *Sol. Energy* **2015**, *118*, 186–196. [CrossRef]
15. González, J.; Fiorito, F. Daylight Design of Office Buildings: Optimisation of External Solar Shadings by Using Combined Simulation Methods. *Buildings* **2015**, *5*, 560–580. [CrossRef]
16. Konin, K.; Gamas, A.; Kensek, K. Passive performance and building form: An optimization framework for early-stage design support. *Sol. Energy* **2016**, *125*, 161–179. [CrossRef]
17. Touloupaki, E.; Theodosiou, T. Optimization of Building form to Minimize Energy Consumption through Parametric Modelling. *Procedia Environ. Sci.* **2017**, *38*, 509–514. [CrossRef]
18. Zhang, A.; Bokel, R.; Dobbelsteen, A.V.D.; Sun, Y.; Huang, Q.; Zhang, Q. Optimization of thermal and daylight performance of school buildings based on a multi-objective genetic algorithm in the cold climate of China. *Energy Build.* **2017**, *139*, 371–384. [CrossRef]
19. Touloupaki, E.; Theodosiou, T. Performance Simulation Integrated in Parametric 3D modelling as a Method for Early Stage Design Optimization—A Review. *Energies* **2017**, *10*, 637. [CrossRef]
20. Henriques, G.C.; Duarte, J.P.; Leal, V. Strategies to control daylight in a responsive skylight system. *Autom. Constr.* **2012**, *28*, 91–105. [CrossRef]
21. Vazquez, E.; Duarte, J.; Poerschke, U. Masonry screen walls: A digital framework for design generation and environmental performance optimization. *Archit. Sci. Rev.* **2021**, *64*, 262–274. [CrossRef]
22. Stiny, G. Introduction to shape and shape grammars. *Environ. Plan. B* **1980**, *7*, 343–351. [CrossRef]
23. Granadeiro, V. Early Design and Optimization of the Building Envelope: Integrating Architectural Generative Design Systems and Energy Performance Calculation. Ph.D. Thesis, Instituto Superior Técnico da Universidade Técnica de Lisboa, Lisbon, Portugal, 2013.
24. Oxman, R. Thinking difference: Theories and models of parametric design thinking. *Des. Stud.* **2017**, *52*, 4–39. [CrossRef]
25. Caetano, I.; Santos, L.; Leitão, A. Computational design in architecture: Defining parametric, generative, and algorithmic design. *Front. Archit. Res.* **2020**, *9*, 287–300. [CrossRef]
26. Huang, Y.; Niu, J. Optimal building envelope design based on simulated performance: History, current status and new potentials. *Energy Build.* **2016**, *117*, 387–398. [CrossRef]
27. Jin, J.-T.; Jeong, J.-W. Optimization of a free-form building shape to minimize external thermal load using genetic algorithm. *Energy Build.* **2014**, *85*, 473–482. [CrossRef]
28. Zhang, L.; Zhang, L.; Wang, Y. Shape optimization of free-form buildings based on solar radiation gain and space efficiency using a multi-objective genetic algorithm in the severe cold zones of China. *Sol. Energy* **2016**, *132*, 38–50. [CrossRef]
29. Palonen, M.; Hasan, A.; Siren, K. A genetic algorithm for optimization of building envelope and HVAC system parameters. In Proceedings of the 11th Conference of IBPSA, Glasgow, UK, 27–30 July 2009; pp. 159–166.
30. Mateus, D. Projeto baseado em Energia: Projeto Digital de Envelopes de Edifícios. Ph.D. Thesis, Lisbon School of Architecture of the University of Lisbon, Lisbon, Portugal, 2020.
31. Mateus, D.; Duarte, J.; Romão, L. Energy-Based Design: A Digital Design System for the Design of Energy-Harvesting Building Envelopes. In Proceedings of the SIGraDi 2021 | Designing Possibilities, XXV International Conference of the Iberoamerican Society of Digital Graphics, Ubiquitous Conference, Online, 8–12 November 2021. [CrossRef]
32. Mateus, D.; Henriques, G.C.; Eskinazi, M.; Menna, R.L.; Nepomuceno, T.M. Carioca modern façades: Expanding passive shading systems through computational methods. In Proceedings of the SIGraDi 2022 | Critical Appropriations, XXVI International Conference of the Iberoamerican Society of Digital Graphics, Lima, Peru, 7–11 November 2022. [CrossRef]
33. Algorithm. Available online: https://dictionary.cambridge.org/dictionary/english-portuguese/algorithm (accessed on 22 September 2023).
34. Queiroz, N.; Dantas, N.; Nome, C.; Vaz, C. Designing a Building envelope using parametric and algorithmic processes. In Proceedings of the 19th Conference of the Iberoamerican Society of Digital Gaphics, Florianópolis, Brazil, 23–27 November 2015; pp. 797–801.
35. Eskinazi, M.O.; Penter, P.E. A Fachada como interface, de Lucio Costa a Irmãos Roberto: Repertório de projeto. In Proceedings of the 13º Seminário DOCOMOMO, Salvador, Brazil, 7–10 October 2019.

36. Barber, D.A. *Modern Architecture and Climate: Design before Air Conditioning*; Princeton University Press: Princeton, NJ, USA, 2020. [CrossRef]
37. O Valor Bioclimático na Arquitetura Moderna Tropical. Available online: https://vitruvius.com.br/revistas/read/arquitextos/21.251/8048 (accessed on 24 July 2023).
38. Mardaljevic, J.; Andersen, M.; Roy, N.; Christoffersen, J. Daylighting metrics: Is there a relation between useful daylight illuminance and daylight glare probability? In Proceedings of the First Building Simulation and Optimization Conference, Loughborough, UK, 10–11 September 2012.
39. Galapagos. Available online: https://www.grasshopper3d.com/group/galapagos (accessed on 23 September 2023).
40. Octopus. Available online: https://www.grasshopper3d.com/group/octopus?overrideMobileRedirect=1 (accessed on 23 September 2023).
41. Zitzler, E.; Laumanns, M.; Thiele, L. *SPEA2: Improving the Strength Pareto Evolutionary Algorithm*; Report; ETH Zurich, Computer Engineering and Networks Laboratory: Zürich, Switzerland, 2021. [CrossRef]
42. Vierlinger, R. Multi Objective Design Interface. Master's Thesis, University of Applied Arts, Vienna, Austria, 2013.

buildings

MDPI

Article

An Automated Prefabricated Facade Layout Definition for Residential Building Renovation

Kepa Iturralde [1,2,*], Samanti Das [1,2], Aravind Srinivasaragavan [1,2], Thomas Bock [1] and Christoph Holst [2]

[1] Chair of Building Realization and Robotics, School of Engineering and Design, Technical University of Munich, 80333 Munich, Germany; ge75guy@mytum.de (S.D.); bockrobotics@web.de (T.B.)

[2] Chair of Engineering Geodesy, School of Engineering and Design, Technical University of Munich, 80333 Munich, Germany; christoph.holst@tum.de

* Correspondence: kepa.iturralde@tum.de

Abstract: The rising global energy demand has made it essential to reduce energy consumption in the residential building stock. Adding a new insulating envelope with Renewable Energy Sources (RESs) onto the existing buildings' facade is one way to achieve zero energy consumption in residential buildings. The ENSNARE project aims to semi-automate this process by using prefabricated facade modules and developing new building data acquisition techniques. Prior to this research project, an analysis was carried out and several research gaps were identified. One of the obstacles to using prefabricated modules with RES is that the layout needs to be drawn and adjusted during different phases of the project. That is time-consuming. For this reason, this article describes two new solutions: (1) automated drafting of the optimized layout of prefabricated modules of the facade and the number of solar panels based on the existing residential building model, and (2) automated adjustment of the layout depending on the phase of the renovation project and the accuracy of the measurements in each step. The proposed semi-automated approach has the potential to significantly reduce the time used in drafting the layout of the prefabricated modules, which benefits the whole renovation process, contributing to a more sustainable future for the residential building stock.

Keywords: online; building model; prefabrication; renovation

check for updates

Citation: Iturralde, K.; Das, S.; Srinivasaragavan, A.; Bock, T.; Holst, C. An Automated Prefabricated Facade Layout Definition for Residential Building Renovation. *Buildings* **2023**, *13*, 2981. https://doi.org/10.3390/buildings13122981

Academic Editor: Asimina Athanatopoulou-Kyriakou

Received: 30 September 2023
Revised: 15 November 2023
Accepted: 20 November 2023
Published: 29 November 2023

1. Introduction

One of the key strategies for addressing the challenge of climate change is to reduce global energy consumption [1]. In recent years, efforts to achieve zero-energy consumption of residential buildings have focused on insulating existing structures and installing renewable energy sources (RES) on their rooftops [2–4]. Other approaches include optimizing building envelopes to facilitate better harvesting of solar energy [5–7]. However, manually implementing these measures carries the potential risks of user intrusion, disruptions, and hazardous activities carried out at elevated heights. To circumvent these issues, prefabricated modules are now being manufactured off-site, incorporating insulation, RES, windows, and waterproofing elements [8]. Previous studies have explored the use of automated, robotic facade renovation with prefabricated modules, which can be categorized into three sub-categories [9–11]: information or data flow, off-site manufacturing of the modules, and on-site installation of the modules (see Figure 1).

The adoption of prefabricated elements in residential building renovation has not yet gained widespread market acceptance due to its lack of competitiveness compared to traditional manual methods. The ENSNARE research project [12] aims to reduce data acquisition and processing time by 90%, while also reducing the duration of the manufacturing and installation processes. It is crucial to ensure that any reduction in working time does not result in a loss of quality and adheres to regulations and standards [13]. Errors in the data flow can lead to deviations in the manufactured modules, resulting in water and heat leaks, collisions, or impeded installation processes.

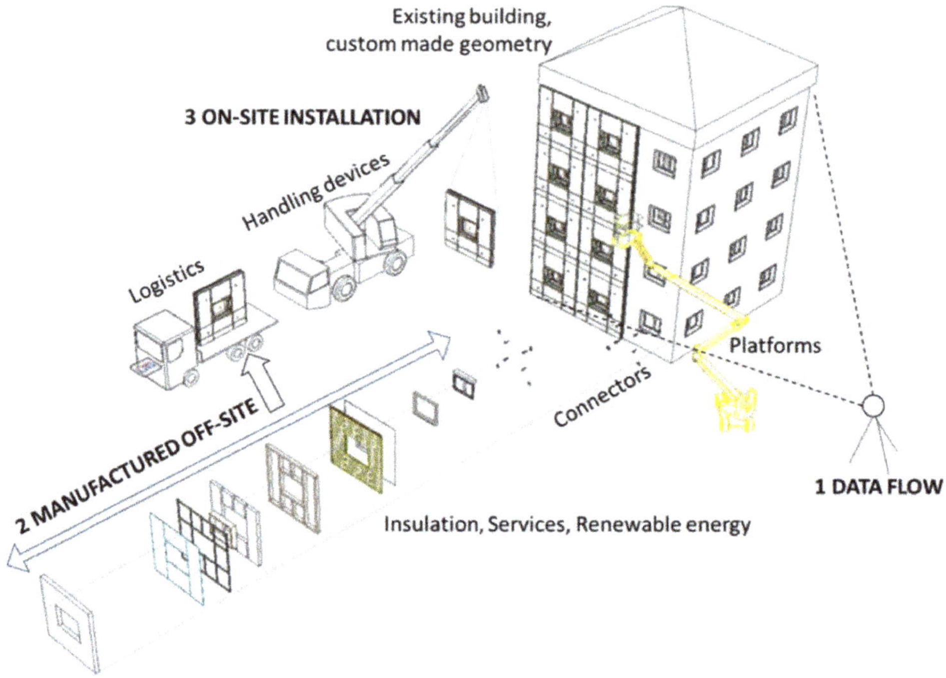

Figure 1. Building renovation with prefabricated modules.

The successful implementation of robotics and automated solutions in residential building renovation largely depends on their economic feasibility [14–17]. Furthermore, it is essential to address the issue of managing and preparing such technologies for the market [18–20]. Previous research initiatives, including BERTIM [21] and HEPHAESTUS [22], have examined this matter. Earlier studies [23] have identified up to 15 research gaps (RG) in this field.

In the field of residential building renovation, the utilization of prefabricated modules for energy-saving and generation purposes often presents less competitive alternatives compared to manual procedures due to the requirement for more comprehensive and detailed planning. When it comes to prefabricated modules for building renovation, the involved parties, including building owners, promoters, or engineers, must possess a comprehensive understanding of the building's solar energy generation capabilities, investment costs, and the necessity for insulation in the preliminary stages of the project. To achieve this, it is crucial to have a geo-located three-dimensional (3D) model of the building that can represent the building's shape, structure, and capability to accommodate prefabricated modules and solar panels. In this context, the layout of prefabricated walls and solar panels assumes a significant role, providing a clear depiction of the number of solar panels that can fit onto the building envelope, the requisite amount of insulation, and the corresponding investment costs. Concerning data acquisition and flow, and to explain the context of this article, it has been determined that the following measures are necessary:

- **RG1.1:** In the initial phases, it is necessary to generate a building model online, meaning without visiting the site. In previous stages of the research, a method to generate building models online was achieved based on facade images and OpenStreetMaps data.
- **RG1.2:** Typically, in building renovation, the initial definition of the prefabricated module layout is a manual process and can take up to 5–20 h for a low-rise building.

The aim should be to automatize defining the layout of prefabricated modules with the information of the building modeled online in RG1.1.

- **RG1.3:** On the next stages of the project, once the "client" or building owner approves the budget, the engineering team can afford to visit the building and measure accurately. Using Total Station or even a 3D Laser Scanner can be a time-consuming activity. A faster method was developed based on OpenCV Apriltags [24].
- **RG1.4:** The measurement accuracy is much higher in RG1.3 than with the technique described in RG1.1. This means that the layout needs to be re-adjusted based on more accurate measurements. Adjusting the layout by manual means is time-consuming. Therefore, we developed a technique to adjust the layout automatically.

This article focuses on describing solutions in order to mitigate research gaps **RG1.2** and **RG1.4** in the following sections. As said before, RG1.1 and RG1.3 were developed in a previous phase of the research.

2. Addressing RG1.2 by Defining the Layout of the Prefabricated Modules

In this section, we explain RG1.2, which is based on the *ENSNARE_MODULE_PLACEMENT*, which is a FreeCAD [25] plugin developed to generate the layout of prefabricated facade modules with solar panels for residential building upgrading. The tool represents a concerted effort to improve the cost-effectiveness and efficiency of residential building renovation within the ENSNARE project. The flow chart in Figure 2 explains the process of the solution.

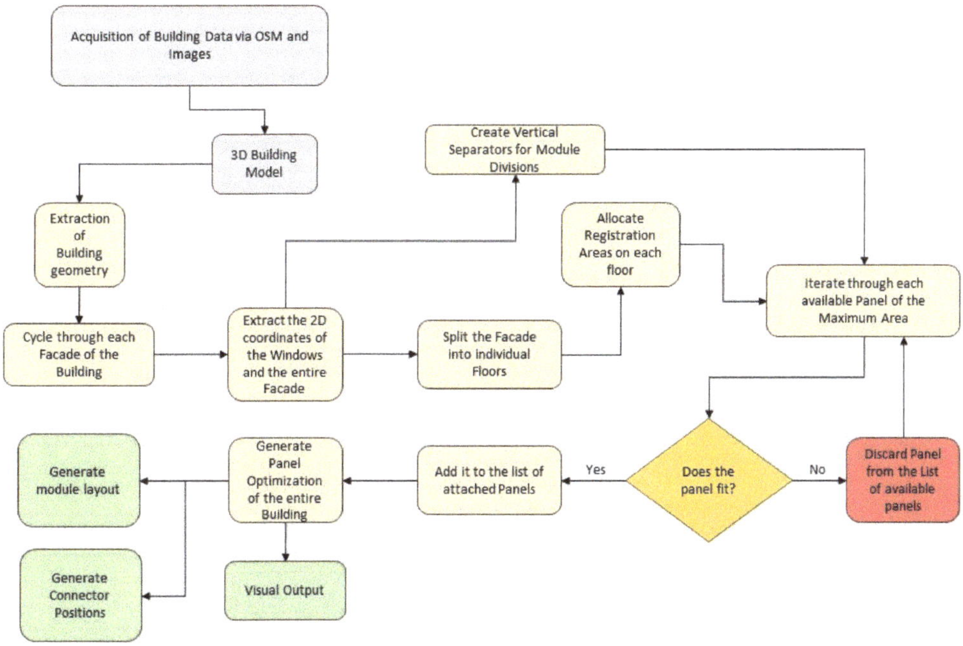

Figure 2. Flow-chart of the solution RG1.2. The inputs are shown in gray, the outputs in green, and the steps required in yellow.

Before applying RG1.2, the ENSNARE project has defined a technique to draft the building model online. In Figure 3, one of the four pictures of a demo building is shown. This demo-building is part of the ENSNARE research project and it is located in Milan.

Figure 3. Actual photo of a demo-building in Milan. With images like this and the OSM floor-plan, the building model can be generated as shown in Figure 4.

The building model is generated with the images of the building and the OSM information. The building model type that is achieved is shown in Figure 4.

Figure 4. Building model generated with building images and OSM, in this case, the demo-building in Milan.

This building model is the primary input for RG1.2. In the next subsections, each of the steps and the code of RG1.2 are explained.

2.1. Tools Associated with the Plugin

The FreeCAD plugin ENSNARE_MODULE_PLACEMENT comprises several tools that facilitate the generation of the layout of prefabricated solar panel modules with the output of the 3D building model. Each of these tools serves primarily two sets of functions. Namely, a set of tools to extract the relevant information about the 3D building geometries and another set of tools to place solar panels based on the geometric information as explained in Figure 5.

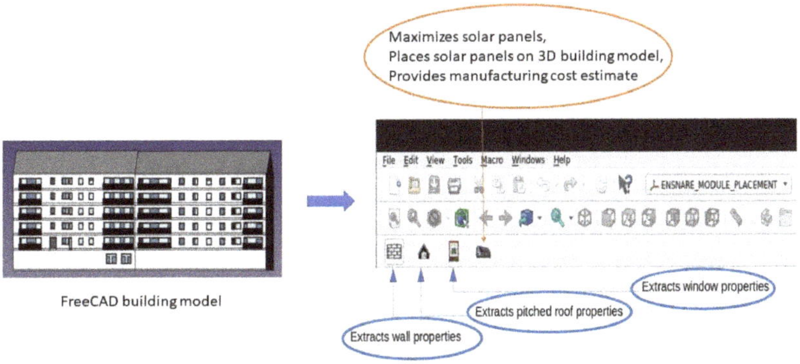

Figure 5. Several tools and functions of the plugin.

2.1.1. Extraction of Relevant Information

This research has devised an algorithm for maximizing the number of solar panels of diverse dimensions that can be accommodated on a two-dimensional facade. To implement this algorithm, it is necessary to extract pertinent information related to the building's facade from the 3D building model generated by the plugin team in ENSNARE.

ENSNARE_MODULE_PLACEMENT is equipped with several tools that enable the extraction of essential information from the online 3D building model, as depicted in Figure 5. The subsequent subsection provides a comprehensive description of these tools:

1. *Tool: exportWallProperties*

This tool facilitates the extraction of essential data related to a building facade, such as its height, width, and four corner points. The user can access this tool by selecting the desired facade and clicking on the corresponding icon. Upon execution, the tool generates a JSON file containing the aforementioned information. This tool is implemented in Listing 1 as shown below:

Listing 1. A Python function which generates a JSON file containing building geometries.

```

def exportWallProperties():

    clicked_face = freecad_utils.get_clicked_face()
    facade_height,
    facade_width,
    wall_vertices_list = face_dimensions(clicked_face)
    wall_dict = { "wall_height": facade_height,
                "wall_width": facade_width,
              "vertexes_anticlockwise":wall_vertices_list}

    with open(f"{constants.path}/wall_properties.json","w")
             as write_file:
        json.dump(wall_dict, write_file, indent=4)

    FreeCADGui.Selection.clearSelection()
```

This function uses another function internally: *face_dimensions*, which shows how exactly the corner vertices and the height and width are determined from the FreeCAD data. This function returns those values to the user. This procedure is described in Listing 2:

Listing 2. A Python function integral to the previous one.

```python

def face_dimensions(face: 'Part.Face') ->
                    Tuple[float, float]:
    vertexes = [v.Point for v in face.Vertexes]
    corner_points_list_bottom = []
    corner_points_list_top = []
    max_z = 0
    min_z = 0
    for point in vertexes:

        if point.z == min_z:
            corner_points_list_bottom.append(point)
        if point.z > max_z:
            max_z = point.z

    for point in vertexes:

        if point.z == max_z:
            corner_points_list_top.append(point)

    if corner_points_list_bottom[0].x >
            corner_points_list_bottom[1].x:
        right_bottom_vertex = corner_points_list_bottom[0]
        left_bottom_vertex = corner_points_list_bottom[1]
    else:
        right_bottom_vertex = corner_points_list_bottom[1]
        left_bottom_vertex = corner_points_list_bottom[0]

    if corner_points_list_top[0].x >
            corner_points_list_top[1].x:
        right_top_vertex = corner_points_list_top[0]
        left_top_vertex = corner_points_list_top[1]
    else:
        right_top_vertex = corner_points_list_top[1]
        left_top_vertex = corner_points_list_top[0]

    wall_vertices = [left_bottom_vertex,
    right_bottom_vertex, right_top_vertex, left_top_vertex]

    ref_point = wall_vertices[0]
    for point in wall_vertices[1:]:
        if point.z == ref_point.z:
            fac_width = abs(point.sub(ref_point).Length)
        else:
            fac_height = abs(FreeCAD.Vector(ref_point.x,
            ref_point.y, point.z).sub(ref_point).Length)

    assert fac_height and fac_width

    wall_index_list = []
    for index, point in enumerate(vertexes):
        if point == wall_vertices[0]
        or point == wall_vertices[1]
        or point == wall_vertices[2]
        or point == wall_vertices[3]:
            wall_index_list.append(index)

    wall_vertices_list = []
    for item in wall_vertices:
        item_list = math_utils_.vector2list(
                                item,scale=0.001)
        wall_vertices_list.append(item_list)

    return fac_height, fac_width, wall_vertices_list
```

2. *Tool: exportTriangularRoofProperties*

This tool serves the purpose of extracting information related to pitched roofs. In some cases, buildings consist of facades with pitched roofs on top. To obtain relevant information such as the height and width of the facade, the topmost point of the roof, and the vertices of the roof, the user must select the appropriate roof and click on the corresponding icon.

The code for generating the required information from a pitched roof is presented in Listing 3:

Listing 3. A Python tool which extracts relevant geometric information associated with pitched roofs.

```

def face_dimensions_roof_triangle(face: 'Part.Face') ->
                                Tuple[float, float]:
    vertexes = [v.Point for v in face.Vertexes]
    corner_points_list_bottom = []
    corner_points_list_top = []

    z_list_vertexes = []

    for point in  vertexes:
        point.z = math_utils_.truncate(point.z, 4)
        z_list_vertexes.append(point.z)

    max_z = max(z_list_vertexes)
    min_z = min(z_list_vertexes)

    for point in vertexes:

        if point.z == min_z:
            corner_points_list_bottom.append(point)
        if point.z > max_z:
            max_z = point.z

    for point in vertexes:

        if point.z == max_z:
            corner_points_list_top.append(point)

    # roof_start
    print(corner_points_list_top)
    print(corner_points_list_bottom)
    # roof_end
    wall_vertices = [corner_points_list_bottom[0],
                     corner_points_list_bottom[1],
                     corner_points_list_top[0]]
    roof_topmost_point = corner_points_list_top[0][2]*1000
    fac_width = abs(corner_points_list_bottom[1].sub(
                corner_points_list_bottom[0]).Length)
    side1 = abs(corner_points_list_top[0].sub(
                corner_points_list_bottom[0]).Length)
    side2 = abs(corner_points_list_top[0].sub(
                corner_points_list_bottom[1]).Length)

    area = math_utils_.area_triangle(side1, side2,
                                fac_width)
    fac_height = 2 * area / fac_width

    assert fac_height and fac_width

    wall_vertices_list = []
    for item in wall_vertices:
        item_list = math_utils_.vector2list(item,
                                scale=0.001)
        wall_vertices_list.append(item_list)

    return fac_height, fac_width, wall_vertices_list,
        roof_topmost_point
```

3. *Tool: exportLBCProperties*

This tool is intended to determine the left bottom corner points of a building facade. The need for this tool arises due to the manner in which FreeCAD software version 0.19.3 stores information. In later stages, a 2D projection of the facade is required, and it is necessary to know which portion of the 3D facade corresponds to its left bottom corner. Since it is challenging to select and extract the left bottom corner point from FreeCAD, the sketch file of the left bottom-most window must be chosen, and the button needs to be clicked to obtain the left bottom points of the facade.

4. *Tool: exportRBCProperties*

This tool has a similar purpose as the previous one, with the difference being that it tracks the right bottom points of the building facade. The need for this tool arises because some building facades have no windows on their left side, only on the right. Since the previous tool uses the corner window positions to determine the corners of the facade for efficient 2D projection, it cannot be used in such cases.

5. *Tool: exportWindowProperties*

This tool is utilized to collect and store the position, length, and width of all the windows on a given facade. The user selects the sketch files of all the windows present on the corresponding facade and clicks on this tool. The information regarding the position, length, and width of each window is then collectively stored in a JSON file for further use.

2.1.2. Maximizing the Solar Panels

This particular section describes the workbench's functionality to maximize the total number of possible solar panel modules that can be placed on the facade. Additionally, it also provides a total manufacturing cost estimate for this entire project. This entire functionality is achieved by pressing a single button on the workbench, as described in detail below.

6. *Tool: maximize_solar_panel*

This tool's main function is to maximize the number of solar panels that can be fit in a selected facade and also to provide a manufacturing cost estimate. It serves a variety of functions in order to achieve that. Firstly, **it projects the entire 3D facade into 2D**. For that, it calculates the slope and reference point of the facade by taking into consideration the four corner vertices and the left bottom point of the facade (obtained from previous tools). This procedure is implemented in Listing 4 as shown below:

Listing 4. A Python function which performs a geometric projection.

```

def get_slope_reference_point():

    height_wall = wall_dict["vertexes_anticlockwise"][3][2]
    list_x = [wall_dict["vertexes_anticlockwise"][0][0],
              wall_dict["vertexes_anticlockwise"][1][0],
              wall_dict["vertexes_anticlockwise"][2][0],
              wall_dict["vertexes_anticlockwise"][3][0]]

    list_y = [wall_dict["vertexes_anticlockwise"][0][1],
              wall_dict["vertexes_anticlockwise"][1][1],
              wall_dict["vertexes_anticlockwise"][2][1],
              wall_dict["vertexes_anticlockwise"][3][1]]


    arr_x = np.asarray(list_x)
    arr_y = np.asarray(list_y)

    for value in left_bottom_dict:
        if sketch in value["label"]:
            point = value["placement"]["origin"]
            i_x = (np.abs(arr_x - point[0])).argmin()
            i_y = (np.abs(arr_y - point[1])).argmin()

```

```
    if i_x == 0:
        if list_x[0]!= list_x[3]:
            wall_dict["vertexes_anticlockwise"][3] =
                    [list_x[i_x], list_y[i_x], height_wall]
    elif i_x == 1:
        if list_x[1]!=list_x[2]:
            wall_dict["vertexes_anticlockwise"][2] =
                    [list_x[i_x], list_y[i_x], height_wall]
    elif i_x == 2:
        if list_x[2]!=list_x[1]:
            wall_dict["vertexes_anticlockwise"][1] =
                    [list_x[i_x], list_y[i_x], 0.0]
    else:
        if list_x[3]!=list_x[0]:
            wall_dict["vertexes_anticlockwise"][0] =
                        [list_x[i_x], list_y[i_x], 0.0]

    p0 = np.array(wall_dict["vertexes_anticlockwise"][0])
    p1 = np.array(wall_dict["vertexes_anticlockwise"][1])

    dist = np.linalg.norm(p0 - p1)
    fac_width = dist * 1000
    wall_vertices_list=wall_dict["vertexes_anticlockwise"]
    first_wall_vertex_xy =
          [wall_dict["vertexes_anticlockwise"][0][0],
          wall_dict["vertexes_anticlockwise"][0][1]]
    reference_point_ = [list_x[i_x], list_y[i_x], 0]
    reference_point_xy = [reference_point_[0],
                          reference_point_[1]]

    point1 = wall_dict["vertexes_anticlockwise"][0]
    point2 = wall_dict["vertexes_anticlockwise"][1]
    point_x = [point1[0], point2[0]]
    point_y = [point1[1], point2[1]]

    slope, intercept = np.polyfit(point_x,point_y,1)
    origin = [0,0]
    change_point_value = [origin[0] - reference_point_[0],
                         origin[1] - reference_point_[1]]
    reference_point =
             [reference_point_[0]+change_point_value[0],
              reference_point_[1]+change_point_value[1]]

    return slope, change_point_value, reference_point,
           fac_width, wall_vertices_list
```

This tool is designed to extract essential information required for generating the 2D facade and converting it back into 3D after the placement of solar panel modules. It collects the reference point and slope information required for generating the 2D facade, and the 3D orientation of the facade is extracted from its left bottom point. This information is crucial for the placement of the optimized number of solar panel modules in 3D space. Additionally, the tool calculates the 2D positions of all the windows based on the provided data.

Subsequently, the floors within the facade are calculated using a proprietary algorithm as shown in Listing 5. This algorithm considers the list of windows along with their 2D positions and the height of the facade. The algorithm proceeds as follows:

1. Firstly, it calculates all the top and bottom 2D positions of each window.
2. Then, it separates all the bottom and top positions into two distinct lists.
3. Subsequently, the topmost corner positions of a window on a specific floor are computed, and the same applies to the lowermost corner positions.
4. The corresponding topmost corner points of each floor are compared against the bottom corner point. The floor's height is then determined as the average of those two values.

Listing 5. A Python function which returns a list of floors.

```python

def get_floors(window_list, height):

    window_corner_list =
    get_top_bottom_points_for_each_window(window_list_all)

    top_corner_list=top_corner_as_list(window_corner_list)
    bottom_corner_list=
    bottom_corner_as_list(window_corner_list)

    max_top_corner_new =
    get_max_corner_points(top_corner_list)

    max_top_corner_new.sort()

    min_bottom_corner_new =
    get_min_corner_points(bottom_corner_list)

    min_bottom_corner_new.sort()

    if max_top_corner_new[0] > min_bottom_corner_new[0]:
        min_bottom_corner_new.pop(0)

    index_list_min = []
    index_list_max = []

    if len(max_top_corner_new)<len(min_bottom_corner_new):
        for index, value in enumerate(max_top_corner_new):
            if value > min_bottom_corner_new[index]:
                index_ =
                min_bottom_corner_new.
                index(min_bottom_corner_new[index])

                index_list_min.append(index_)
        min_bottom_corner_new =
        [i for j, i in enumerate(min_bottom_corner_new)
        if j not in index_list_min]

    elif len(max_top_corner_new)>len(min_bottom_corner_new):
        for index, value in enumerate
                        (min_bottom_corner_new):
            if value > max_top_corner_new[index]:
                index_ =  max_top_corner_new.
                index(max_top_corner_new[index])

                index_list_max.append(index_)

        max_top_corner_new =
        [i for j, i in enumerate(max_top_corner_new)
        if j not in index_list_max]

    else:
         min_bottom_corner_new = min_bottom_corner_new
         max_top_corner_new = max_top_corner_new

    if len(min_bottom_corner_new)>len(max_top_corner_new):
        min_bottom_corner_new.pop(-1)

    floor_list = []
    for (item1, item2) in zip(max_top_corner_new,
                          min_bottom_corner_new):
        floor_list.append((item1+item2) / 2)

    floor_list = np.array(floor_list)
    difflist_floor = np.diff(floor_list)
    difflistbool_floor =
```

```
        difflist_floor[difflist_floor<1600]

    index_list_ = []
    for item in difflistbool_floor:

        index= np.where(difflist_floor == item)
        for item2 in index:
            for item3 in item2:
                index_list_.append(int(item3))

    floor_list =
    [i for j, i in enumerate(floor_list)
    if j not in index_list_]
    floor_index_list = []

    for value in floor_list:
        if value < 1600:
            index = floor_list.index(value)
            floor_index_list.append(index)

    floor_list =
    [i for j, i in enumerate(floor_list)
    if j not in floor_index_list]
    floor_list.insert(0, 0.0) # hardcoded
    floor_list.insert(-1,height)
    floor_list.sort()


    return floor_list
```

Then, a text file is generated to provide comprehensive information about the permissible panel and module dimensions, window list within a facade, floors in the facade, as well as the height and width of the facade in 2D, all of which are generated from the workbench. This information is then utilized to implement the solar panel maximization algorithm, which is described in the next section.

2.2. Solar Panel Optimization Algorithm

The goal of this section is to explain the Solar Panel Optimization algorithm, implemented using Python. Let us first look at the nature of the problem at hand. Essentially, we are given the following data:

- A finite collection of solar panels that are both photovoltaic and collector.
- A building facade (this information also includes the positions of the various windows and balconies present on the facade.)
- A set of engineering requirements that a panel, if attached, must satisfy.

The goal (it should be immediately clear that we are dealing with 2-dimensional objects), then, is to place the panels on the facade such that the total area covered is maximized. It should be immediately clear, that this is a **Constrained Optimization** problem. The easiest way to generate this maximized placement is by brute-forcing: go through all possible combinations of panel placements and then pick the one that covers the maximum possible area. However, as the number of panels increases and the facades become more complicated, this brute-force approach becomes highly undesirable as it is inefficient and slow. The next best thing is a greedy approach.

2.3. A Rough Overview of the Algorithm

We will first employ a "divide and conquer" approach. Given a facade, we will break the optimization problem into smaller chunks.

- **STEP 1:** divide the entire facade into floors. A typical floor is shown in Figure 6.
- **STEP 2:** For each floor, we will use a "dynamic programming" approach: given a floor, we will begin an iteration that will start from the leftmost point of the floor and

end at the right end of the floor. At each stage of the iteration, we will try placing a panel of the largest area from the list of available panels, and progressively work towards the smaller ones.

- **STEP 3:** Once we have placed all the panels, we will create the vertical module divisions, that is, the separation that defines the length of each prefabricated module. The length limitation in ENSNARE project is 6 meters, due to the prefabrication processes of the industrial partners. Each module might contain several windows/panels. These module separations need to adhere to certain engineering requirements—there is an upper limit and a lower limit to the module separations.

Note that we also need to attach registration areas for the panels. These are the yellow rectangles in Figure 7. The goal of the registration areas is to host the pipes and cables of the solar panels. Also, if needed, to have the possibility to open and access from the exterior of the facade these pipes and cables, for instance, in case of emergency.

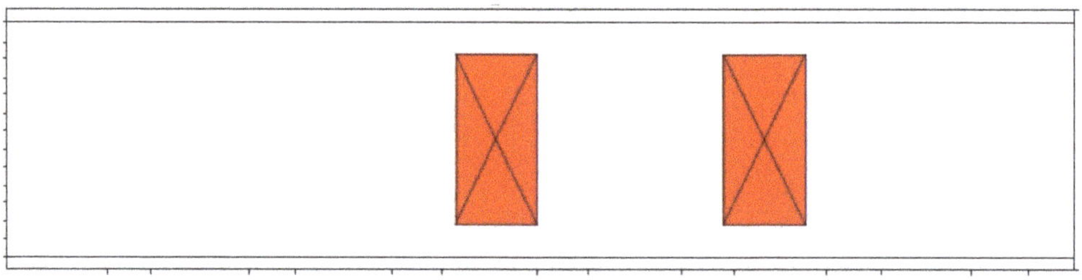

Figure 6. An example of a typical facade section from floor to floor. In this case, the floor has two windows (in red).

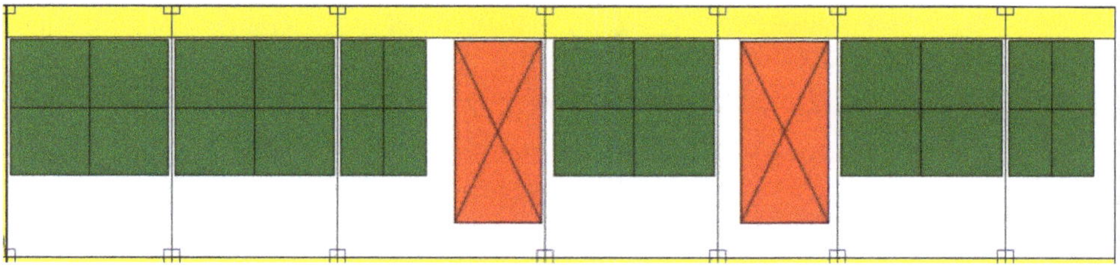

Figure 7. A typical facade section from floor to floor, in this case, with windows and the added prefabricated modules with the registration area in yellow, and the solar panels in green.

2.4. Pythonic Implementation of Basic Objects

For the purposes of this algorithm, we will be mostly dealing with objects such as windows, panels, registration areas, etc. All of these objects have a certain distinct physical property in common: they are all rectangular. That is, we need precisely four two-tuples to describe every such rectangular object. It is important to note that this rectangular nature is essential for the algorithm. Since the optimization problem is highly geometric in nature, we will be required to implement a wide array of useful geometric properties. We have implemented **RectangularObject** using dataclasses as described in Listing 6.

The various class properties will be used extensively throughout the code. We have omitted a few class methods in Listing 6 for the sake of brevity. One such class method is **intersects (self,cls)**, described in Listing 7, returns a boolean object to decide if two rectangular objects are intersecting or not:

Listing 6. The base class for all the rectangular objects that we will encounter in this project.

```
@dataclass(kw_only= True)
class RectangularObject:

    object_type: str = field(init = False)
    object_coordinates: list

    @property
    def area(self) -> float:
        return (self.length)*(self.height)

    @property
    def right_end_point(self) -> float:
        return self.object_coordinates[3][0]

    @property
    def left_end_point(self) -> float:
        return self.object_coordinates[0][0]

    @property
    def top_end_point(self) -> float:
        return self.object_coordinates[1][1]

    @property
    def bottom_end_point(self) -> float:
        return self.object_coordinates[0][1]

    @property
    def length(self) -> float:
        return abs(self.object_coordinates[3][0] - self.object_coordinates[0][0])

    @property
    def height(self) -> float:
        return abs(self.object_coordinates[1][1] - self.object_coordinates[0][1])
```

Listing 7. Class method to decide if two rectangular objects are intersecting.

```
#to check if two rectangular objects intersect
def intersects(self,cls) -> bool:
    A = self.left_end_point > cls.right_end_point or cls.left_end_point > self.right_end_point
    B = self.bottom_end_point > cls.top_end_point or cls.bottom_end_point > self.top_end_point
    if A or B:
        return False
    else:
        return True
```

Further, the use of **dataclasses** increases the performance of the code. Other rectangular objects, such as floors, are first of all child classes of the rectangular object class. They might have additional properties, as Listing 8 demonstrates:

Listing 8. Python implementation of a facade. A facade can consist of floors, panels, windows, etc.

```
    @dataclass(kw_only= True)
    class Facade(RectangularObject):

        object_type = ENSNARE.FACADE
        has_pitched_roof:bool = False
        has_panels:bool = False
        has_module_divisions:bool = True
        floor_list = []
        pitched_roof:PitchedRoof = field(init = False)
```

As we will see in later sections, the algorithm will act on a facade object.

2.5. Specific Requirements of the ENSNARE Project

The various modules and panels have to adhere to certain engineering requirements which is essential to the manufacturing process, which is given by the manufacturing partners of the project:

- Sizes of the registration areas;
- The length of a given module must be at least the MIN_MODULE_LENGTH and at most MAX_MODULE_LENGTH;
- The sizes of the inner and outer profile areas;
- The panels cannot be of arbitrary size: there is a finite amount of panels of specific lengths and widths.

All this information is contained in the **requirements.py** file. A typical example is demonstrated in Listing 9 (note that the REQUIREMENTS class also inherits from float, since these are floating point numbers).

Listing 9. requirements.py, which contains the various manufcaturing requirements.

```
class REQUIREMENTS(float,Enum):
"engineering requirments for the ENSNARE project"

INNER_PROFILE = 100
OUTER_PROFILE = 40
STEP_SIZE = 10
MAX_MODULE_LENGTH = 3300
MIN_MODULE_LENGTH = 1500


AVAILABLE_PANEL_DATA = [
        (1066, 1756),
        (1756, 1086),
        (1551, 1756),
        (1756, 1551),
        (2036, 1756),
        (1756, 2056),
    ]

class REGISTRATION_AREA_SIZE(float,Enum):
    "sizes of the available registration areas"

    VERTICAL = 400
    HORIZONTAL = 400
```

The facade information, which might comprise floors, windows, and balconies, is provided in the form of a text file. This text file consists of a dictionary, which contains the coordinates of the above-mentioned objects. The general structure of the text file is shown in Listing 10:

Listing 10. A typical facade text file. We have provided a simplified version for the sake of brevity.

```
{
    "facade": {
        "length": 22_000,
        "height": 2600,
        "windows": [
            {
                "xy": (2_200, 400),
                "length": 2_000,
                "height": 1400
            },
            {
```

```
                    "xy": (5_750, 400),
                    "length": 1_950,
                    "height": 1400
                } }
```

Before the main algorithm can do anything, we first need to translate this facade information into Pythonic objects. A helper function that converts this into a JSON object (which Python can understand) is implemented in Listing 11:

Listing 11. A helper function to convert the text file into a Python dictionary.

```

def read_from_textfile(file_name:str) -> dict:

    file = open(file_name, "r")
    text_data = file.read()
    dictionary = ast.literal_eval(text_data)
    return dictionary
```

Once this is performed, extracting the Pythonic objects from this is fairly straightforward. This is achieved via the function **return_facade_from_textfile**, which accepts the text file location as an argument and returns a **Facade** object. This is implemented in Listing 12:

Listing 12. Function which reads the facade text file and returns a Python Facade object.

```

def return_facade_from_textfile(text_file_location:str) -> Facade:

#storing the textfile as a python dictionary
facade_data_dict = read_from_textfile(file_name = text_file_location)

#creating the facade
facade_length = facade_data_dict["facade"]["length"]
facade_height = facade_data_dict["facade"]["height"]
facade_coordinates = np.array([(0,0),
                               (0,facade_height),
                               (facade_length,facade_height),
                               (facade_length,0)])

new_facade = Facade(object_coordinates=facade_coordinates)
```

2.6. The Main Algorithm I: Placing Panels

The function responsible for placing panels on a given floor is the **place_panel** function, which accepts a **Floor** object as an argument and returns an object of the same type. The **Floor** class, as described in Listing 13, is a child class of the **RectangularObject** class, with a few further attributes to store the list of windows and panels present on the floor.

Listing 13. Python implementation of a Floor class. This is a simplified version.

```

@dataclass(kw_only=True)
class Floor(RectangularObject):

    object_type = ENSNARE.FLOOR
    window_list:list[RectangularObject]
    floor_module_separation:list
    panel_list:list[RectangularObject]
    registration_area_side:RectangularObject
    registration_area_top:RectangularObject
```

The **place_panel** function places the panel starting from the left endpoint of the floor. The rate at which it moves along the right is given by the step-size parameter. For the sake of clarity, we have broken this up into several snippets as described in Listings 14–17.

Listing 14. A snippet of the panel placing function.

```
def place_panel(floor:Floor) -> Floor:
    #placing the panels with a step size given by the inner profile size
    current_distance = starting_point

    while current_distance < floor.right_end_point:

        #creating a temporary copy for each iteration
        temp_panel_list = [panel for panel in available_panels]
```

At each stage of the iteration, we try to place the panel of the maximum area—this is the greedy approach that we referred to earlier.

Listing 15. A snippet of the panel placing function.

```
#trying to fit all the panels until we run out
while len(temp_panel_list) != 0:
    #starting with the panel of maximum area
    panel = max(temp_panel_list, key=attrgetter('panel_area'))
```

For each panel, we try to see if it fits at the given position of the iteration: we check if it intersects any windows and panels. We also see if it sticks out of the floor.

Listing 16. A snippet of the panel placing function.

```
#instead of checking if the panel touches the windows
#we will check if the panel touches the window shell
modified_window_list = [window.inner_profile_shell for window in floor.window_list]

#now check if this panel fits
panel_exceeds_floor = test_panel.bottom_end_point < (floor.bottom_end_point+REQUIREMENTS.OUTER_PROFILE)
panel_touches_shell = any(shell.intersects(test_panel) for shell in modified_window_list)
panel_exceeds_length = test_panel.right_end_point > (floor.right_end_point-REQUIREMENTS.OUTER_PROFILE)
```

If it fits, we will add this panel to the list of panels associated with this floor. Otherwise, we discard this panel and see if there are any smaller panels available. If there are no smaller panels available, we move on to the next stage of the iteration.

Listing 17. A snippet of the panel placing function.

```
    #i.e panel does not fit
    if panel_exceeds_floor or panel_touches_shell or panel_exceeds_length:
        #if this is the smallest panel and it is hitting a window
        if len(temp_panel_list) == 1 and panel_touches_shell:
            #finding the problematic window
            problematic_windows = find_intersecting_objects(test_panel, modified_window_list)
            new_distance = max(shell.right_end_point for shell in problematic_windows)
            current_distance = new_distance
            break

        temp_panel_list.remove(panel)
        continue

    #if it fits
    else:
```

```
                floor.panel_list.append(test_panel)
                current_distance = current_distance + REQUIREMENTS.OUTER_PROFILE+
        test_panel.length
                break

        current_distance = current_distance + step_size
```

2.7. The Main Algorithm II: Placing Modules

Once the panels have been placed, a function called the **create_module_seprations** tries to create the module separations: it creates the separations using an object called the **VerticalSeparator** which is implemented in Listing 18. This is simply a vertical line given by the top point, bottom point, and its distance from the y-axis.

Listing 18. A Vertical Separator.

```

    @dataclass
    class VerticalSeparator:

        top:float = 0
        bottom:float = 0
        distance:float = 0
```

The **create_module_seprations**, described in Listing 19, first collects the left/right endpoints of the windows and panels present on the floor.

Listing 19. A snippet of the create_vertical_separations function.

```

def create_module_separation(floor:Floor) -> Floor:
    #First, we collect all the rectangular objects on this floor
    #Then, we collect their right/left end points
    end_point_list = []
    for window in floor.window_list:
        end_point_list.append(window.right_end_point+REQUIREMENTS.
    OUTER_PROFILE)
        end_point_list.append(window.left_end_point-REQUIREMENTS.
    OUTER_PROFILE)
    for panel in floor.panel_list:
        end_point_list.append(panel.right_end_point+REQUIREMENTS.
    OUTER_PROFILE)
        end_point_list.append(panel.left_end_point-REQUIREMENTS.OUTER_PROFILE
    )
```

From this list, the function attempts to create pairings x, y (where x and y are members of this list) in such a way that the following three conditions are satisfied:

1. $x < y$
2. $|x - y| \leq$ MAX_MODULE_LENGTH
3. $|x - y| \geq$ MIN_MODULE_LENGTH

This is implemented in Listing 20 shown below:

Listing 20. A snippet of the create_vertical_separations function.

```

    #Main module separation algorithm
    start_point = end_point_list[0]
    module_separation.append(start_point)
    for current_point in end_point_list:
        current_length = current_point-start_point
        if current_length <= max_mod_length and current_length >=
    min_mod_length:
            module_separation.append(current_point)
            start_point = current_point
```

```
    floor.floor_module_separation = module_separation
    return floor
```

So far, the algorithm acts on each floor: it places the panels then it places the modules (i.e., the vertical separations). Once this is performed for each floor, we can combine them to obtain a panel placement on the entire facade.

2.8. Generating Outputs

The output of the solar panel maximization algorithm generates a visual plot of the entire facade. This is achieved using Matplotlib (see Figure 8).

It is also possible to generate a JSON file providing detailed information regarding the 2D coordinates of each module and panel in the facade. Currently, four different designs of prefabricated modules are being used in the project as shown in Figure 9.

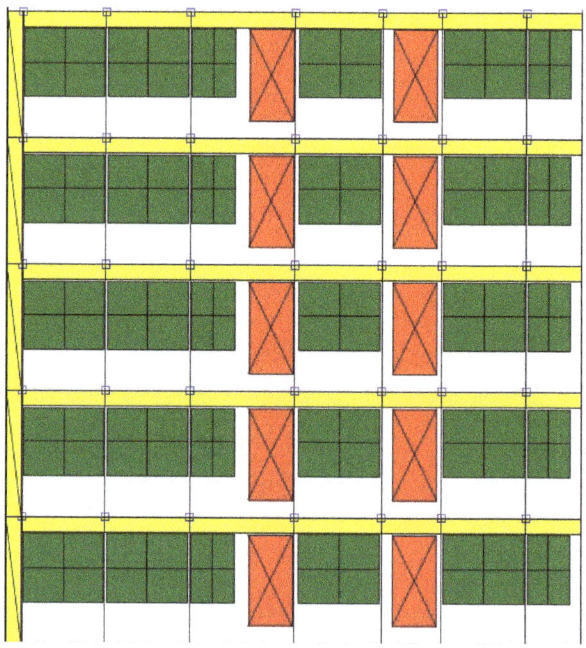

Figure 8. A typical facade with windows in red, and the layout of the added prefabricated modules with solar panels in green and the registration area in yellow. The tiny squares determine the location of the connectors or anchors.

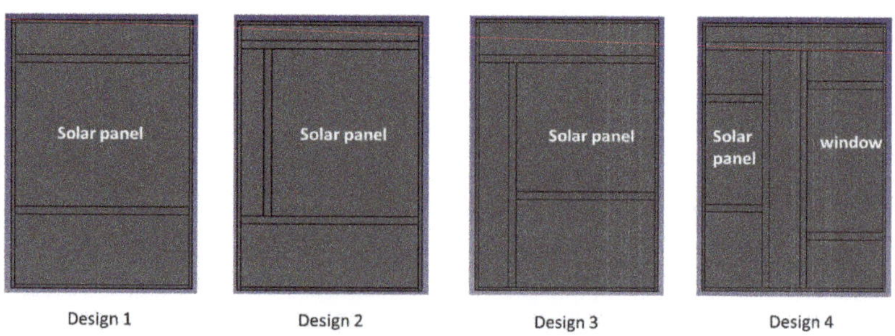

Figure 9. Four prefabricated modules were designed in FreeCAD.

Comparing the module and panel dimensions from the data generated by the algorithm, the most appropriate design (out of the four available designs) is chosen. These four designs are hard-coded using simply Python and FreeCAD tools. Now, the 3D projection algorithm is applied and the already existing 3D model in FreeCAD now gets updated by the solar panel modules as shown in Figure 10.

Furthermore, two Excel files are generated as well. One of them provides the connector positions corresponding to each module and window in the facade. The other file provides intricate details regarding the dimensions of each component of each module in the facade, which are essential for robotic assembly. These data can also provide a total manufacturing cost estimate for the project.

Figure 10. Output of the code described in this section RG1.2, that includes prefabricated layout definition including solar panels and registration areas in the building model of demo-building in Milan, Italy.

3. Addressing RG1.4 by Re-Configuring the Layout of the Prefabricated Modules

In the next step of the building renovation project, the engineering team visits the site and obtains accurate measurements of the building. The technique used in the ENSNARE project is based on using targets on the critical parts of the building, such as the location of the connectors, windows, and other building edges. Once the engineering team has measured the location of the targets (see Figure 11), the building model measurements need to be readjusted, and accordingly, the layout of the prefabricated modules too.

Adjusting the layout of prefabricated modules manually is time-consuming work. The Flowchart described in Figure 12 describes the automated steps to reduce this time.

Data pre-processing may be necessary depending on the structure and orientation of the data provided. The pre-processing steps taken may include generating a transformation matrix to multiply the provided data and transform it into a form compatible with the algorithm. Additionally, it is important to note that the connector positions are provided, rather than the actual window positions on the facade. As a result, it is necessary to determine the correct window dimensions and positions before applying the algorithm. This is achieved by subtracting or adding the connector offset length to the provided data in order to derive the required positions and dimensions of windows.

Figure 11. Capturing the real building images with AprilTags, which are located in the critical points of the facade. Example of the demo building in Milan.

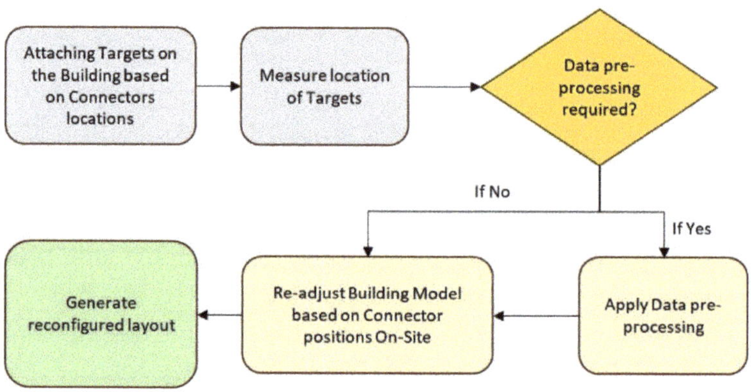

Figure 12. Flow-chart of the solution RG1.4. The inputs are shown in gray, the outputs in green, and the steps required in yellow.

In order to generate the text file with correct information, the following functions were implemented:

1. First, the data provided needed to be converted into 2D because otherwise the algorithm cannot be applied. Hence, the slope of the facade is determined, along with its width, height, and floors from the provided data. This is achieved through Listing 21 shown below:

Listing 21. A Python function which determines the facade slope.

```

def get_slope_width(panel_list):
    reference_point_3d = [0.0, 0.0, 0.0]
    x_value_ref = 0.0
    y_value_ref = 0.0
    z_value_ref = 0.0
    y_value_list_floors = []
    y_value_start = 0.0
    for item in panel_list:
        y_value = item['y/m']
        if y_value > 0.0:
```

```
            y_value = 0.0
            item['y/m'] = 0.0
        if abs(y_value - 0.0) < 1.0:
            x_value_ref =  item['x/m']
            y_value_ref =  item['y/m']
            z_value_ref =  item['z/m']

        if y_value < y_value_start:
            y_value_start = y_value
            x_value_start = 0.0
            if abs(x_value_start - item['x/m']) < 1.0:
                top_most_point_x = item['x/m']
                top_most_point_y = item['y/m']
                top_most_point_z = item['z/m']
        y_value_list_floors.append(abs(item['y/m']*1000))

    point_1 = np.array(reference_point_3d)
    point_2 = np.array([x_value_ref,
                        y_value_ref,
                        z_value_ref])
    point_3 = np.array([top_most_point_x,
                        top_most_point_y,
                        top_most_point_z])

    dist_hor = np.linalg.norm(point_1 - point_2)
    fac_width = dist_hor * 1000

    dist_vert = np.linalg.norm(point_1 - point_3)
    fac_height = dist_vert * 1000

    point_x = [point_1[0], point_2[0]]
    point_y = [point_1[2], point_2[2]]
    slope, intercept = np.polyfit(point_x,point_y,1)

    groups = []
    for item in y_value_list_floors:
        if len(groups) == 0:
            groups.append([item])
            continue
        curr_group = groups[-1]
        if all(abs(item - x) < 10 for x in curr_group):
            curr_group.append(item)
        else:
            groups.append([item])

    floor_list = []
    for floor_level in groups:
        floor_list.append(mean(floor_level))

    floor_list[0] = 0.0
    floor_list[-1] = fac_height

    return slope, fac_width, fac_height, floor_list
```

2. Then, the window list is generated, as shown in Listing 22, from the provided data through the code shown below. Once this window and other important information pertaining to the facade width, height, and floors have been obtained, the text file is generated in order to apply the algorithm.

Listing 22. A Python function which generates a text file containing the list of windows.

```

def generate_window_list(window_info):
    windows_list = []
    for item in window_info:
        individual_window_dict = {}
        if item['y/m'] > 0.0:
            item['y/m'] = 0.0
```

```
        perpendicular = abs(item['y/m'])
        point = [item['x/m'], item['z/m']]
        ref_point = [0.0, 0.0]
        x_coord = math.dist(point, ref_point)
        individual_window_dict["xy"] =
        (x_coord*1000, perpendicular*1000)
        windows_list.append(individual_window_dict)


    window_final_list = []
    for chunk in chunker_longest(windows_list, 4):
        chunk_list = list(chunk)

        # sort out the x and y values of the
        4 corner points of the window
        x_coords = []
        y_coords = []
        for item in chunk_list:
            x_coords.append(item['xy'][0])
            y_coords.append(item['xy'][1])
        x_coords.sort()
        y_coords.sort()

        # Take the window as the largest approximation
        of all the corner points
        if x_coords[0] >= x_coords[1]:
            LBP_x = x_coords[0]
            LTP_x = x_coords[0]
        else:
            LBP_x = x_coords[1]
            LTP_x = x_coords[1]

        if x_coords[2] >= x_coords[3]:
            RBP_x = x_coords[2]
            RTP_x = x_coords[2]
        else:
            RBP_x = x_coords[3]
            RTP_x = x_coords[3]

        if y_coords[0] >= y_coords[1]:
            LBP_y = y_coords[0]
            RBP_y = y_coords[0]
        else:
            LBP_y = y_coords[1]
            RBP_y = y_coords[1]

        if y_coords[2] >= y_coords[3]:
            LTP_y = y_coords[2]
            RTP_y = y_coords[2]
        else:
            LTP_y = y_coords[3]
            RTP_y = y_coords[3]

        # Calculating the corner points of the window
        Left_Bottom_Point = [LBP_x, LBP_y]
        Left_Top_Point = [LTP_x, LTP_y]
        Right_Bottom_Point = [RBP_x, RBP_y]
        Right_Top_Point = [RTP_x, RTP_y]

        # Creating dictionary for each window
        window_dict_single = {}
        window_dict_single['xy'] = Left_Bottom_Point
        window_dict_single['length'] =
          abs(Left_Bottom_Point[0] - Right_Bottom_Point[0])
        window_dict_single['height'] =
          abs(Left_Bottom_Point[1] - Left_Top_Point[1])

        window_final_list.append(window_dict_single)
```

```
    return window_final_list
```

It must be said that the prefabricated module layout is as detailed as for being used for manufacturing purposes, including profiles, enclosure boards, insulation and, of course, solar panels (see Figure 13).

Figure 13. Building model and detailed prefabricated module layout of the demo-building in Milan.

4. Tests and Results of the Solutions

The main objective of the solutions was to reduce time. In order to test the efficiency of the solutions, up to 10 residential buildings have been used to test the aforementioned tools, as explained in Figure 14. Applying the algorithm takes a few minutes, depending on the complexity of the building.

As an example, a complex residential building like the demo building in Milan, which has approximately 3000 m^2 of building envelope was monitored. Defining the prefabricated layout manually can take up to 1020 h (0.34 $\times$ 3000) according to previous manual tests [23]. With the solutions described in this paper, it can take about 25 min in each of the steps (RG 1.2 and RG 1.4).

With this technique, the arduous work of defining the layout of prefabricated modules for facade renovation can be reduced to a minimum. By utilizing these tools, residential building owners, promoters, and engineers can make informed decisions regarding the installation of prefabricated modules with solar panels, thereby ensuring that the renovation project is both economically viable and energy-efficient.

Figure 14. Residential buildings where the tools were applied. In the columns in the left and center, the initial inputs are shown, that is images of the building and the building models. In the column in the right, the building models with the prefabricated module layouts are shown, that is, the output of RG1.2 and RG1.4.

5. Conclusions and Further Developments

The results described in this paper have successfully addressed the research gaps **RG1.2** and **RG1.4**. The time consumed for generating the prefabricated facade module layout is minimal with a commercial and updated computer/processor.

The various tools and techniques introduced will help semi-automate the process of making buildings more energy efficient. Future research will include:

1. Improving the optimization algorithm. Currently, there are certain facades where the algorithm performs poorly. This can be circumvented by trying a different approach, instead of the greedy approach as discussed in this paper. It is possible to re-formulate this problem as an optimal packing problem [26,27].
2. As discussed before, the methods discussed in this paper have not been fully extended to include more diverse building types, such as buildings with pitched roofs. This is one possible avenue for further advancements.

As a final conclusion, it must be commented that the techniques described in this paper will be integrated with an online tool that will include not only the techniques in RG1.1, RG1.2, and RG1.3, but also energy calculations. All this will contribute to reducing the energy consumption of the residential building stock. However, it must be said that these automated procedures developed in RG1.2 and RG1.4 cannot exchange the experience and knowledge of human designers. Aesthetic criteria, for instance, is not considered yet. Therefore, the designs and layouts must always be supervised and, if needed, amended.

Author Contributions: Conceptualization, K.I.; methodology, K.I.; software, S.D., A.S. and K.I.; validation, K.I., S.D. and A.S.; resources, K.I. and T.B.; writing—original draft preparation, S.D. and K.I.; writing—review and editing, K.I. and A.S.; visualization, K.I.; supervision, C.H.; project administration, K.I. and T.B.; funding acquisition, K.I. and T.B. All authors have read and agreed to the published version of the manuscript.

Funding: The research presented in this paper was originally carried out in the context of the H2020 project ENSNARE. This project has received funding from the European Union's Horizon 2020 research and innovation program under grant agreement No. 958445.

Data Availability Statement: Data are contained within the article.

Acknowledgments: The authors would like to thank Agi Hidri for the assistance with the figures.

Conflicts of Interest: The authors declare no conflict of interest.

Abbreviations

The following abbreviations are used in this manuscript:

RES	Renewable Energy Sources
CAD	Computer-Aided Design
OSM	Open Street Maps

References

1. European Commission. *Focus: Energy Efficiency in Buildings*; European Commission: Brussels, Belgium, 2020.
2. Garay Martinez, R.; Astudillo Larraz, J. Performance assessment of façade integrated glazed air solar thermal collectors. *Energy Procedia* **2017**, *115*, 353–360. [CrossRef]
3. Urbikain, M.K. Energy efficient solutions for retrofitting a residential multi-storey building with vacuum insulation panels and low-E windows in two European climates. *J. Clean. Prod.* **2020**, *269*, 121459. [CrossRef]
4. Hillebrand, G.; Arends, G.; Streblow, R.; Madlener, R.; Mueller, D. Development and design of a retrofit matrix for office buildings. *Energy Build.* **2014**, *70*, 516–522. [CrossRef]
5. Mateus, D.; Duarte, J. A grammar-based system for building envelope design to maximize PV performance. In Proceedings of the 10th Conference on Advanced Building Skins, Bern, Switzerland, 10–11 October 2016.
6. Mateus, D.; Duarte, J.P.; Romão, L. Energy-Based Design: A Digital Design System for the Design of Energy-Harvesting Building Envelopes. In Proceedings of the XXV International Conference of the Ibero-American Society of Digital Graphics (SIGraDi 2021), Online, 8–12 November 2021. [CrossRef]
7. Stiny, G. Introduction to shape and shape grammars. *Environ. Plan. Plan. Des.* **1980**, *7*, 343–351. [CrossRef]

8. Barco-Santa, A.; Vareilles, É.; Gaborit, P.; Aldanondo, M. Building renovation adopts mass customization: Configuring insulating envelopes. *J. Intell. Inf. Syst.* **2017**, *49*, 119–146. [CrossRef]
9. Tsai, I.; Kim, Y.; Seike, T. Decision-making consideration in energy-conservation retrofitting strategy for the opening of existing building in Taiwan. *AIJ J. Technol. Des.* **2017**, *55*, 963. [CrossRef]
10. Du, H.; Huang, P.; Jones, P. Modular facade retrofit with renewable energy technologies: The definition and current status in Europe, Energy and Buildings. *Energy Build.* **2019**, *205*, 109543. [CrossRef]
11. D'Oca, S.; Ferrante, A.; Ferrer, C.; Pernetti, R.; Gralka, A.; Sebastian, R.; Op't Veld, P. Technical, financial, and social barriers and challenges in deep building renovation: Integration of lessons learned from the H2020 cluster projects. *Buildings* **2018**, *8*, 174. [CrossRef]
12. ENSNARE Consortium. 2021. Available online: https://www.ensnare.eu (accessed on 29 September 2023).
13. *DIN 18202*; Toleranzen im Hochbau–Bauwerke. Deutsches Institut für Normung (DIN): Berlin, Germany, 2013.
14. Skibniewski, M.; Hendrickson, C. Analysis of Robotic Surface Finishing Work on Construction Site. *J. Constr. Eng. Manag.* **1988**, *114*, 53–68. [CrossRef]
15. Balaguer, C.; Abderrahim, M.; Carlos, B.; Mohamed, A. Trends in robotics and automation in construction. In *Robotics and Automation in Construction*; InTech: Rijeka, Croatia, 2008. [CrossRef]
16. Warszawski, A. Economic implications of robotics in building. *Build. Environ.* **1985**, *20*, 73–81. [CrossRef]
17. Hu, R.; Iturralde, K.; Linner, T.; Zhao, C.; Pan, W.; Pracucci, A.; Bock, T. A Simple Framework for the Cost–Benefit Analysis of Single-Task Construction Robots Based on a Case Study of a Cable-Driven Facade Installation Robo. *Buildings* **2021**, *11*, 8. [CrossRef]
18. Pan, M.; Linner, T.; Pan, W.; Cheng, H.; Bock, T. Structuring the context for construction robot development through integrated scenario approach. *Autom. Constr.* **2020**, *114*, 103174. [CrossRef]
19. Pan, M.; Linner, T.; Pan, W.; Cheng, H.; Bock, T. A framework of indicators for assessing construction automation and robotics in the sustainability context. *J. Clean. Prod.* **2018**, *182*, 82–95. [CrossRef]
20. Pan, M.; Linner, T.; Pan, W.; Cheng, H.; Bock, T. Influencing factors of the future utilisation of construction robots for buildings: A Hong Kong perspective. *J. Build. Eng.* **2020**, *330*, 101220. [CrossRef]
21. BERTIM, D2.5. Efficient Mass Manufacturing and Installation of Prefabricated Modules. 2016.
22. Hephaestus Consortium, Hephaestus–EU H2020 Project. 2017. Available online: https://www.hephaestusproject.eu (accessed on 29 September 2023).
23. Iturralde, K.; Gambao, E.; Bock, T. Compilation and assessment of automated façade renovation. In Proceedings of the 38th International Symposium on Automation and Robotics in Construction (ISARC), Dubai, United Arab Emirates, 2–4 November 2021. [CrossRef]
24. Iturralde, K.; Shen, J.; Bock, T. AprilTag detection for building measurement. In Proceedings of the 40th International Symposium on Automation and Robotics in Construction, Chennai, India, 3–9 July 2023; pp. 589–592. [CrossRef]
25. The FreeCAD Team, FreeCAD Your Own 3D Parametric Modeler, (n.d.). Available online: https://www.freecadweb.org (accessed on 29 September 2023).
26. Huang, E.; Korf, R.E. Optimal Rectangle Packing: An Absolute Placement Approach. *J. Artif. Intell. Res.* **2013**, *46*, 47–87. [CrossRef]
27. Barco Santa, A.; Fages, J.G.; Vareilles, E.; Aldanondo, M.; Gaborit, P. Open Packing for Facade-Layout Synthesis Under a General Purpose Solver. In Proceedings of the CP 2015—21st Interna- tional Conference on the Principles and Practice of Constraint Programming, Cork, Ireland, 31 August–4 September 2015; Volume 9255, pp. 508–523. [CrossRef]

buildings

Article

Soil–Structure Interaction Consideration for Base Isolated Structures under Earthquake Excitation

Arcan Yanik * and Yalcincan Ulus

Department of Civil Engineering, Istanbul Technical University, 34469 Istanbul, Turkey
* Correspondence: yanikar@itu.edu.tr; Tel.: +90-212-2856790

Abstract: This study aims to analytically implement base isolation with soil–structure interaction (SSI) on a sample structure and to develop a very simple solution to add these combined effects into the mass, damping and stiffness matrices of the structure. A spectrum analysis is also carried out considering the base-isolated structures and SSI. Dynamic simulations are performed throughout the study. In these simulations, three shear frame structures with different properties are considered. The strong ground motions selected for these analyses are eighteen different events with far-fault and near-fault components. In addition, four different base and soil structure combination cases are taken into account. These four analytical cases are a conventional structure with a fixed base and with SSI and a seismically isolated structure with or without the SSI. The numerical results showed that when SSI is considered, the effectiveness of the base isolation system may decrease, and the effect is prominent in softer soil conditions.

Keywords: base isolation; soil–structure interaction; earthquake; seismic; passive control

check for updates

Citation: Yanik, A.; Ulus, Y. Soil–Structure Interaction Consideration for Base Isolated Structures under Earthquake Excitation. *Buildings* **2023**, *13*, 915. https://doi.org/10.3390/buildings13040915

Academic Editor: Nerio Tullini

Received: 27 February 2023
Revised: 28 March 2023
Accepted: 29 March 2023
Published: 30 March 2023

1. Introduction

Base isolation (BI), also known as seismic isolation, is a widely used technique for protecting structures from earthquake effects. It is mostly used in critical facilities that need continuous operation even after severe earthquakes. Examples of these structures are hospitals, airports, nuclear power plants and government buildings. BI is also used in highway engineering structures such as bridges and viaducts. During earthquakes, the superstructure and the foundation are effectively decoupled from each other with the help of BI systems, resulting in much lower base shear forces and inter-story drifts [1].

Soil–structure interaction (SSI), also known as soil–foundation–structure interaction, is a design concept that takes the responses of structure, foundation and soil media into account. SSI is a concept that is often neglected in regard to design structures. Conventionally, structures are considered fixed to the ground, and the response of the underlying soil is neglected. This is because the effects of soil are usually considered beneficial to the structure, so neglecting soil can be considered conservative and safe. In FEMA 356 [2], soil–structure interaction is classified as a rare case that may modify the spectral response of the structure, particularly in soft soil conditions. There are two main approaches for evaluating SSI, one of which is the direct approach, where the soil is modeled as a continuum of the structure. The other is called the substructure approach, where the properties of soil are calculated via simple impedance functions or springs and dashpots. Soil properties that are calculated include the stiffness and damping characteristics of the soil–structure interaction. The substructure approach to impedance functions requires the foundation to be rigid and allows for the formulation of soil–structure interaction. In this paper, the substructure approach is used.

One of the first studies that investigated the effect of SSI in base-isolated structures was by Constantinou and Kneifati [3], who conducted a parametric study on a linear 1-degree-of-freedom structure. Bycroft [4] introduced approaching solutions and analysis for the SSI when high frequencies are involved. Novak and Henderson [5] made use of the equations of motion for a structure by evaluating the SSI effect. Pappin et al. [6] highlighted the importance of the site effects on performance-based design. Mylonakis and Gazetas [7] investigated the use of SSI. They found that the effects of SSI are not necessarily beneficial and can be detrimental instead. Tongaonkar and Jangid [8] studied the effects of SSI on bridges. They concluded that considering SSI effects on isolated bridges is beneficial to the bridge and may result in cost reduction in design. When analyzing the damaging earthquake of 1999 in Athens, Mylonakis et al. [9] reached a conclusion that the effects of a surface layer made of soft rock and stiff soil were a detrimental factor. Syngros [10] analyzed the history of two different cases related to the seismic response of piles, supporting bridges and pile groups; the Ohba Ohashi bridge; and the Fukae Section of the Hanshin Expressway. He concluded that the type of soil changed the seismic waves, and the ground surface motion encountered by the specific bridge was greater; moreover, a higher frequency was measured in that portion. During the study of a 3D nonlinear model of BI, Deb [11] took SSI into consideration by using a near-fault earthquake. A detailed study related to an isolated structure and SSI with an unbounded section was conducted by Tsai et al. [12]. Dicleli et al. [13] studied the effects of SSI on two different types of bridges, one with heavy superstructure and light substructure and the other with opposite properties. They concluded that SSI effects should be investigated in soft soil conditions, regardless of the bridge type. Liu et al. [14] conducted a study in the time domain that investigates the effects of wind in high-rise structures with tuned mass dampers, considering SSI effects. They found that SSI affects the effectiveness of the dampers in soft soil conditions. Spyrakos et al. [15] studied the effects of SSI in buildings in the frequency domain. They found that SSI effects on the damping of the isolated structures are negligible, and the damping depends on BI damping characteristics.

Jeremic et al. [16] investigated the effects of SSI with varying soil properties on a four-span bridge. Karabörk et al. [17] conducted a study that investigates the effects of SSI on an isolated structure. Kausel [18] investigated the SSI history and phenomena. Genes et al. [19] conducted a study in order to analyze SSI effects in two different reinforced concrete buildings. Matinmanesh and Asheghabadi [20] took into consideration a plane finite element and used three different earthquake data with different frequency contents. They concluded that all types of soil have an effect on bedrock movement, but with different rates. Manolis and Markou [21] presented a combination of BI and SSI. They concluded that the BI performance depends on the type of soil. Giarlelis and Mylonakis [22] conducted a study related to the role of SSI in the elastic performance of multistory buildings. Li et al. [23] studied SSI in base-isolated structures in the frequency domain using the substructure approach. Luco [24] studied the effects of SSI on a structure that has a nonlinear BI system and concluded that SSI amplifies the response of the isolated structures. Tsai et al. [25] investigated the effects of SSI and damping in isolated structures. Yanik [26] proposed an instantaneous optimal control performance index for active control of structures under seismic excitation. Ashiquzzaman and Hong [27] studied the SSI effects on isolated nuclear power plant containment buildings. Zhou and Wei [28] conducted a similar study to Ashiquzzaman and Hong [27], where they studied SSI effects on nuclear power plants. They concluded that SSI directly affects the performance of isolation systems [28]. Xuefei et al. [29] investigated the optimal placement of passive control devices by considering SSI. Forcellini [30] studied the SSI effects on base-isolated buildings using OpenSees [31] framework and found that SSI effects are more pronounced in softer soil conditions. Dai et al. [32] studied the effects of SSI in a specific base-isolated bridge. They also found that in isolated structures, SSI effects are more pronounced. El-Sinawi [33] evaluated the effectiveness of magneto-rheological dampers and BI under seismic excitation. In a recent study, Yanik [34] mentioned that taking SSI into consideration reduces the effectiveness of magneto-rheological dampers. Zhenxia and Haiping [35] studied SSI effects on base-isolated structures using the finite element method. They found that the performance of BI systems is correlated with soil conditions. Some interesting recent numerical research about

SSI effects on tall buildings is presented by Forcellini [36]. As stated in [36], SSI effects have been extensively studied for low-rise buildings; however, SSI effects in tall buildings are a concept that needs to be investigated more. Another recent study on semi-active control by considering SSI is given by Jalali et al. [37]. Maleska and Beben [38] investigated the behavior of soil–steel composite bridges under earthquake excitation in their interesting research. Jishuai et al. [39] predicted the influence of SSI on reinforced concrete buildings by using neural networks. Liguo et al. [40] proposed a new framework for tuned mass damper systems with SSI effects. Yulin et al. [41] investigated the earthquake response of multi-span bridges by taking into account abutment–soil–foundation-structure interactions.

At present, there are a limited number of studies that investigate SSI and BI together, and the results of these studies are not always consistent with each other. This study aims to analyze SSI and BI effects on two-dimensional shear frame structures with the assumption of stories being rigid diaphragms. In addition, one of the most important aims of this paper is to analyze the SSI effect on high-rise buildings with base isolation. Furthermore, another aim is to analytically propose a new simple matrix formulation to study combined SSI effects and base isolation together and test the computational efficiency of this formulation. Analyses are conducted on three different shear frame structures, which are five, ten and forty stories high. The Newmark average acceleration method is used for dynamic analysis. In the next section, the formulation of the problem is presented.

2. Formulation of the Problem

In this section, the formulation of the problem is given in a detailed way. The Newmark average acceleration method is used for dynamic analysis. The soil model is defined below.

2.1. Representing Soil Structure

To take into account SSI in the analysis, the "spring and dashpot method" is used. Spring and dashpot analysis replaces the soil with springs and dashpots with horizontal, vertical, rotational and torsional degrees of freedom. In this study, the horizontal and rotational degrees of freedom are considered.

Calculations of the stiffness and damping coefficients of springs and dashpots are achieved by using the frequency-independent impedance functions [42]. As seen from the equations below, both structure and soil properties contribute to the calculations.

$$k_s = \frac{8Gr}{2-\nu} \tag{1}$$

$$k_r = \frac{8Gr^3}{3(1-\nu)} \tag{2}$$

$$c_s = \frac{4.6}{2-\nu}\rho V_S r^2 \tag{3}$$

$$c_r = \frac{0.4}{1-\nu}\rho V_S r^4 \tag{4}$$

In the equations above, the "s" subscript denotes "swaying", which is the horizontal component, and the "r" subscript denotes "rocking", which is the rotational component. k_s and k_r are swaying and rocking stiffness of the soil, respectively. Similarly, c_s and c_r are swaying and rocking and damping the soil, respectively. V_s, ρ and ν are the shear wave velocity, mass density and Poisson's ratio of the soil. G is the shear modulus of the soil, which can be calculated with the following equation:

$$G = \rho V_S^2 \tag{5}$$

r is the radius of the circular foundation. For a rectangular foundation with the dimensions of a and b, an equivalent r value can be calculated by equalizing areas of the rectangle and the circle, such that:

$$\pi r^2 = ab \tag{6}$$

$$r = \sqrt{\frac{ab}{\pi}} \tag{7}$$

The matrix formulation of the structure is defined below.

2.2. Matrix Formulation of the Structure

In this section, the formulation of the mass, stiffness and damping matrices of the structure is presented. In order to represent soil properties and BI in mass, stiffness and damping matrices, the corresponding values must be embedded into the matrices. An n-story fixed base shear building is shown in Figure 1. Conventional mass (**M**), damping (**C**) and stiffness (**K**) matrices for a fixed-based shear building can be written as:

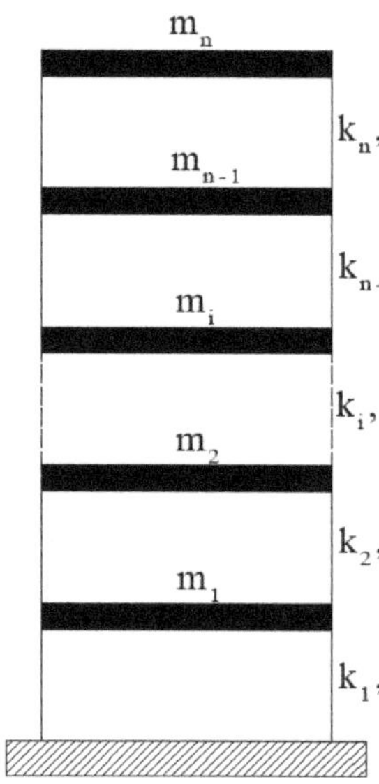

Figure 1. Fixed-based shear building.

$$\mathbf{M} = \begin{bmatrix} m_1 & 0 & 0 & \cdots & 0 \\ 0 & m_2 & 0 & \cdots & 0 \\ 0 & 0 & \ddots & \cdots & 0 \\ \vdots & \vdots & \vdots & m_{n-1} & 0 \\ 0 & 0 & 0 & 0 & m_n \end{bmatrix}_{n\times n} \tag{8}$$

$$\mathbf{C} = \begin{bmatrix} c_1 + c_2 & -c_2 & 0 & \cdots & 0 \\ -c_2 & c_2 + c_3 & -c_3 & \cdots & 0 \\ 0 & -c_3 & \ddots & \cdots & 0 \\ \vdots & \vdots & \vdots & c_{n-1} + c_n & -c_n \\ 0 & 0 & 0 & -c_n & c_n \end{bmatrix}_{n\times n} \tag{9}$$

$$\mathbf{K} = \begin{bmatrix} k_1+k_2 & -k_2 & 0 & \cdots & 0 \\ -k_2 & k_2+k_3 & -k_3 & \cdots & 0 \\ 0 & -k_3 & \ddots & \cdots & 0 \\ \vdots & \vdots & \vdots & k_{n-1}+k_n & -k_n \\ 0 & 0 & 0 & -k_n & k_n \end{bmatrix}_{n\times n} \tag{10}$$

where k_1 to k_n are the stiffnesses of the stories, c_1 to c_n are the damping of the stories, and m_1 to m_n are the mass of each story of an n-story structure. In the next section, the formulation for the base-isolated structure is presented.

2.3. Base Isolated Structure

The base-isolated shear building is shown in Figure 2. Properties of the BI can be implemented with the formulation given above by simply taking the BI layer as the first story of the structure.

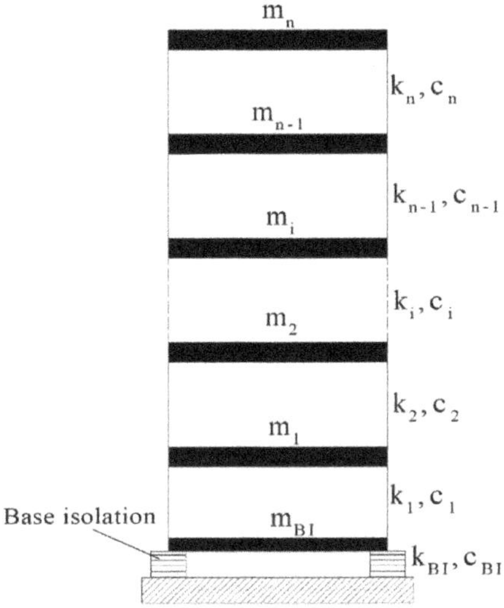

Figure 2. Base isolated shear building.

Mass ($\mathbf{M_{BI}}$), damping ($\mathbf{C_{BI}}$) and stiffness ($\mathbf{K_{BI}}$) matrices of a base-isolated n-story shear building can be written as:

$$\mathbf{M_{BI}} = \begin{bmatrix} m_{BI} & 0 & 0 & 0 & \cdots & 0 \\ 0 & m_1 & 0 & 0 & \cdots & 0 \\ 0 & 0 & m_2 & 0 & \cdots & 0 \\ 0 & 0 & 0 & \ddots & \cdots & 0 \\ \vdots & \vdots & \vdots & \vdots & m_{n-1} & 0 \\ 0 & 0 & 0 & 0 & 0 & m_n \end{bmatrix}_{(n+1)\times(n+1)} \tag{11}$$

$$\mathbf{C_{BI}} = \begin{bmatrix} c_{BI}+c_1 & -c_1 & 0 & 0 & \cdots & 0 \\ -c_1 & c_1+c_2 & -c_2 & 0 & \cdots & 0 \\ 0 & -c_2 & c_2+c_3 & -c_3 & \cdots & 0 \\ 0 & 0 & -c_3 & \ddots & \cdots & 0 \\ \vdots & \vdots & \vdots & \vdots & c_{n-1}+c_n & -c_n \\ 0 & 0 & 0 & 0 & -c_n & c_n \end{bmatrix}_{(n+1)\times(n+1)} \tag{12}$$

$$\mathbf{K_{BI}} = \begin{bmatrix} k_{BI}+k_1 & -k_1 & 0 & 0 & \cdots & 0 \\ -k_1 & k_1+k_2 & -k_2 & 0 & \cdots & 0 \\ 0 & -k_2 & k_2+k_3 & -k_3 & \cdots & 0 \\ 0 & 0 & -k_3 & \ddots & \cdots & 0 \\ \vdots & \vdots & \vdots & \vdots & k_{n-1}+k_n & -k_n \\ 0 & 0 & 0 & 0 & -k_n & k_n \end{bmatrix}_{(n+1)\times(n+1)} \tag{13}$$

where k_{BI} is the stiffness, c_{BI} is the damping and m_{BI} is the mass of the BI system. BI adds one degree of freedom to the fixed base structure; therefore, the dimensions of all matrices are increased by one. SSI implementation to the expressions given above is defined in the next section.

2.4. SSI Implementation

The spring–dashpot model of an n-story shear building considering SSI is shown in Figure 3. m_b and I_b are the mass and the mass moment of inertia of the foundation, respectively. I_1 to I_n are the mass moment of inertia of each floor. k_s is the swaying stiffness, k_r is the rocking stiffness, c_s is the swaying damping and c_r is the rocking damping of the foundation-soil medium. h_1 to h_n are story heights with respect to the ground. Swaying values are the horizontal components of the foundation–soil medium, whereas rocking values are rotational components. Mass, damping and stiffness matrices with SSI implementation can be written as follows [43]:

$$\mathbf{M_{SSI}} = \begin{bmatrix} \mathbf{M} & \mathbf{M_v} & \mathbf{M_h} \\ \mathbf{M_v^T} & \left(m_b + \sum\limits_{i=1}^{n} m_i\right) & \left(\sum\limits_{i=1}^{n} m_i h_i\right) \\ \mathbf{M_h^T} & \left(\sum\limits_{i=1}^{n} m_i h_i\right) & \left(\sum\limits_{i=1}^{n} m_i h_i^2 + I_b + \sum\limits_{i=1}^{n} I_i\right) \end{bmatrix}_{(n+2)\times(n+2)} \tag{14}$$

$$\mathbf{C_{SSI}} = \begin{bmatrix} \mathbf{C} & 0 & 0 \\ 0 & c_s & 0 \\ 0 & 0 & c_r \end{bmatrix}_{(n+2)\times(n+2)} \tag{15}$$

$$\mathbf{K_{SSI}} = \begin{bmatrix} \mathbf{K} & 0 & 0 \\ 0 & k_s & 0 \\ 0 & 0 & k_r \end{bmatrix}_{(n+2)\times(n+2)} \tag{16}$$

where $\mathbf{M_{SSI}}$ is the mass matrix, $\mathbf{C_{SSI}}$ is the damping matrix and $\mathbf{K_{SSI}}$ is the stiffness matrix of the shear building considering SSI. In addition, $\mathbf{M_v}$ and $\mathbf{M_h}$ are defined as follows:

$$\mathbf{M_v} = \begin{bmatrix} m_1 \\ m_2 \\ \vdots \\ m_{n-1} \\ m_n \end{bmatrix}_{n\times 1} \tag{17}$$

$$\mathbf{M_h} = \begin{bmatrix} m_1 h_1 \\ m_2 h_2 \\ \vdots \\ m_{n-1} h_{n-1} \\ m_n h_n \end{bmatrix}_{n \times 1} \tag{18}$$

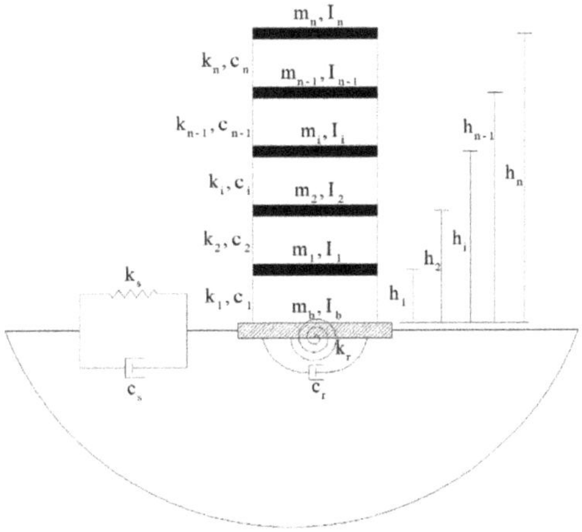

Figure 3. Shear building considering SSI.

SSI consideration implements two additional degrees of freedom to the fixed base structure; therefore, the dimensions of all matrices are increased by two. In the next section, the proposed simple formulation is presented.

2.5. Proposed Simple Formulation for SSI with Base Isolation

SSI in base-isolated structures can be represented by adjusting the SSI matrix formulation. This case is depicted in Figure 4. Fixed-based **M**, **C** and **K** matrices that are used in SSI matrices can be substituted with base-isolated $\mathbf{M_{BI}}$, $\mathbf{C_{BI}}$ and $\mathbf{K_{BI}}$ matrices. The height of the BI system, h_{BI}, must be included in the mass matrix considering SSI and BI together.

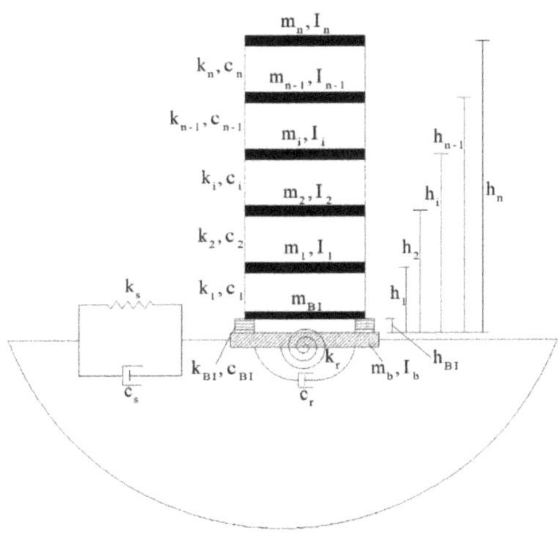

Figure 4. Shear building considering SSI and base isolation together.

The simple contribution of our research is the consideration of SSI and BI together in the formulation. With the help of an extensive number of numerical simulations that are performed by using many different earthquakes and soil types, it has been shown in the next sections that this type of formulation is computationally efficient. This simple formulation was not studied before. The details of the simple formulation are given below.

Mass, damping and stiffness matrices considering SSI and BI together can be written as follows:

$$\mathbf{M_{SSI,BI}} = \begin{bmatrix} \mathbf{M_{BI}} & \mathbf{M_{v,BI}} & \mathbf{M_{h,BI}} \\ \mathbf{M^T_{v,BI}} & \left(m_b + \sum\limits_{i=1}^{n} m_i\right) & \left(\sum\limits_{i=1}^{n} m_i h_i\right) \\ \mathbf{M^T_{h,BI}} & \left(\sum\limits_{i=1}^{n} m_i h_i\right) & \left(\sum\limits_{i=1}^{n} m_i h_i^2 + I_b + \sum\limits_{i=1}^{n} I_i\right) \end{bmatrix}_{(n+3)\times(n+3)} \tag{19}$$

$$\mathbf{C_{SSI,BI}} = \begin{bmatrix} \mathbf{C_{BI}} & 0 & 0 \\ 0 & c_s & 0 \\ 0 & 0 & c_r \end{bmatrix}_{(n+3)\times(n+3)} \tag{20}$$

$$\mathbf{K_{SSI,BI}} = \begin{bmatrix} \mathbf{K_{BI}} & 0 & 0 \\ 0 & k_s & 0 \\ 0 & 0 & k_r \end{bmatrix}_{(n+3)\times(n+3)} \tag{21}$$

where $\mathbf{M_{SSI,BI}}$ is the mass matrix, $\mathbf{C_{SSI,BI}}$ is the damping matrix and $\mathbf{K_{SSI,BI}}$ is the stiffness matrix of the shear building considering SSI and BI together. $\mathbf{M_{v,BI}}$ and $\mathbf{M_{h,BI}}$ can be presented as:

$$\mathbf{M_{v,BI}} = \begin{bmatrix} m_{BI} \\ m_1 \\ m_2 \\ \vdots \\ m_{n-1} \\ m_n \end{bmatrix}_{(n+1)\times 1} \tag{22}$$

$$\mathbf{M_{h,BI}} = \begin{bmatrix} m_{BI} h_{BI} \\ m_1 h_1 \\ m_2 h_2 \\ \vdots \\ m_{n-1} h_{n-1} \\ m_n h_n \end{bmatrix}_{(n+1)\times 1} \tag{23}$$

While SSI introduces two additional degrees of freedom to the fixed base structure, BI adds one; therefore, the dimensions of all matrices are increased by three in total. Equations of the motion of the structure under seismic excitation are given below.

2.6. Equations of the Motion of the Structure under Earthquake Excitation

Four different equations of motion are presented in this section to cover all four analysis cases; these cases are fixed base, base-isolated, fixed base considering SSI and structure with BI considering SSI. The equation of motion for the fixed base analysis case can be written as follows:

$$\mathbf{M}(t) + \mathbf{C}\dot{u}(t) + \mathbf{K}u(t) = -\mathbf{m}^{*}{}_{g}(t) \tag{24}$$

where $\mathbf{m}^*$ is a vector containing the diagonal of $\mathbf{M}$, which happens to be the mass value

$$\mathbf{m}^* = \begin{bmatrix} m_1 \\ m_2 \\ \vdots \\ m_{n-1} \\ m_n \end{bmatrix}_{n \times 1} \tag{25}$$

Equations of motion for the base-isolated analysis case can be written by substituting the conventional fixed base matrices with a base-isolated matrix:

$$\mathbf{M_{BI}}(t) + \mathbf{C_{BI}}\dot{u}(t) + \mathbf{K_{BI}}u(t) = -\mathbf{m}^*_{\mathbf{BI}}g(t) \tag{26}$$

where $\mathbf{m}^*_{\mathbf{BI}}$ is a vector containing the diagonal elements of $\mathbf{M_{BI}}$:

$$\mathbf{m}^*_{\mathbf{BI}} = \begin{bmatrix} m_{BI} \\ m_1 \\ m_2 \\ \vdots \\ m_{n-1} \\ m_n \end{bmatrix}_{(n+1) \times 1} \tag{27}$$

$$\mathbf{M_{SSI}}(t) + \mathbf{C_{SSI}}\dot{u}(t) + \mathbf{K_{SSI}}u(t) = -\mathbf{m}^*_{\mathbf{SSI}}g(t) \tag{28}$$

where $\mathbf{m}^*_{\mathbf{SSI}}$ is a vector containing the diagonal elements of $\mathbf{M}$ with $(m_b + \sum_{i=1}^{n} m_i)$ and $(\sum_{i=1}^{n} m_i h_i)$ from $\mathbf{M_{SSI}}$ [43]:

$$\mathbf{m}^*_{\mathbf{SSI}} = \begin{bmatrix} m_1 \\ m_2 \\ \vdots \\ m_{n-1} \\ m_n \\ \left(m_b + \sum\limits_{i=1}^{n} m_i\right) \\ \left(\sum\limits_{i=1}^{n} m_i h_i\right) \end{bmatrix}_{(n+2) \times 1} \tag{29}$$

For the joint case of SSI and BI, the equation of motion is presented as given below:

$$\mathbf{M_{SSI,BI}}(t) + \mathbf{C_{SSI,BI}}\dot{u}(t) + \mathbf{K_{SSI,BI}}u(t) = -\mathbf{m}^*_{\mathbf{SSI,BI}}g(t) \tag{30}$$

where $\mathbf{m}^*_{\mathbf{SSI,BI}}$ is a vector containing the diagonal of $\mathbf{M_{BI}}$ with the additions of $(m_b + \sum_{i=1}^{n} m_i)$ and $(\sum_{i=1}^{n} m_i h_i)$ from $\mathbf{M_{SSI,BI}}$ [43]:

$$\mathbf{m}^*_{\mathbf{SSI,BI}} = \begin{bmatrix} m_{BI} \\ m_1 \\ m_2 \\ \vdots \\ m_{n-1} \\ m_n \\ \left(m_b + \sum\limits_{i=1}^{n} m_i\right) \\ \left(\sum\limits_{i=1}^{n} m_i h_i\right) \end{bmatrix}_{(n+3) \times 1} \tag{31}$$

In the equations defined above, ${}_g(t)$ is the horizontal ground motion acceleration. $u(t)$, (t) and (t) are the displacement, velocity and acceleration vectors of the structure.

In order to perform the dynamic analysis under seismic excitation with or without consideration of SSI and by taking into account all base conditions, the following approach was used. The mathematical equations defined in this section are implemented in a code that was created in MATLAB software. The accuracy of the MATLAB code was checked by comparing the results with SAP2000 software by considering a simple shear building example. Moreover, MATLAB code did not create any instability problems during the performance of an extensive number of simulations. It was fast and efficient by using the proposed simple formula defined in Section 2.5. A numerical example is defined in the next section.

3. Numerical Example

Analyses were carried out on three different shear frame structures, which have five, ten and forty stories. For each building, four different base conditions were considered. These conditions consist of a fixed base, an isolated base, an SSI-included base and an isolated base considering SSI. The models of these cases are shown in Figures 1–4, respectively. It should be noted here that Figures 1–4 represent general cases and are created for buildings with n number of stories. For SSI cases, four different soil conditions, very soft, soft, medium and dense, are considered. All cases are tested with a suite of earthquakes with different rupture distances. Soil properties of each building can be calculated with the equations listed in Section 2.1. Soil properties are taken and adapted from [14], which have been used in different studies previously [43–45]. Soil properties of five-and ten-story buildings are the same because the foundation dimensions are the same. Soil properties are given in Table 1. Soil properties of five- and ten-story buildings are given in Table 2. Soil properties of the forty-story building are given in Table 3. Three different shear frame buildings are used in the analysis. Building properties of the forty-story building, five-story building and ten-story building are given in Tables 4–6, respectively. Building properties are chosen by us to be in line with real-life building properties. Base isolation properties such as mass m_b, stiffness k_b and damping c_b are calculated with the following expressions. In the equations below, m is the mass, and k is the stiffness of the first story above the isolation level. m, k and c values are given in Tables 4–6. The following equation is taken from [1]. The critical damping ratio for base isolation is taken as 10%.

$$\begin{aligned} m_b &= m;\ k_b = 0.05k; \\ c_b &= \xi_b 2\sqrt{m_{total}\,k_b} \end{aligned} \tag{32}$$

Table 1. Soil properties.

Soil	ν	ρ ton/m^3	V_S m/s^2	G kN/m^2
Very Soft	0.49	1.60	50	4000
Soft	0.49	1.80	100	18,000
Medium	0.48	1.90	300	171,000
Dense	0.33	2.40	500	600,000

Table 2. Soil properties of the five- and ten-story buildings.

Soil	k_s kN/m	k_r kN/m	c_s kNs/m	c_r kNs/m
Very Soft	2.39×10^5	3.00×10^7	3.49×10^5	1.01×10^6
Soft	1.07×10^6	1.35×10^8	7.87×10^5	2.28×10^6
Medium	1.02×10^7	1.26×10^9	2.47×10^6	7.09×10^6
Dense	3.24×10^7	3.43×10^9	4.74×10^6	1.16×10^7

Table 3. Soil properties of the forty-story buildings.

Soil	k_s kN/m	k_r kN/m	c_s kNs/m	c_r kNs/m
Very Soft	4.78×10^5	2.40×10^8	2.79×10^6	1.62×10^7
Soft	2.15×10^6	1.08×10^9	6.30×10^6	3.66×10^7
Medium	2.03×10^7	1.01×10^{10}	1.98×10^7	1.14×10^8
Dense	6.48×10^7	2.74×10^{10}	3.79×10^7	1.86×10^8

Table 4. Properties of the forty-story building.

story height, h_i (m)	4 to 160
story mass, m_i (ton)	980
story stiffness, k_i (kN/m)	2.1×10^6 to 0.99×10^6
story damping, c_i (kNs/m)	42.6×10^3 to 20×10^3
story inertia, I_i (ton·m^2)	1.31×10^5
foundation mass, m_0 (ton)	1960
foundation inertia, I_0 (ton·m^2)	1.96×10^5

Table 5. Properties of the five-story building.

story height, h_i (m)	4 to 20
story mass, m_i (ton)	300
story stiffness, k_i (kN/m)	3.5×10^5 to 1.50×10^5
story inertia, I_i (ton·m^2)	7.5×10^3
foundation mass, m_0 (ton)	300
foundation inertia, I_0 (ton·m^2)	7.5×10^5

Table 6. Properties of the ten-story building.

story height, h_i (m)	4 to 40
story mass, m_i (ton)	300
story stiffness, k_i (kN/m)	7.0×10^5 to 3.00×10^5
story inertia, I_i (ton·m^2)	7.5×10^3
foundation mass, m_0 (ton)	300
foundation inertia, I_0 (ton·m^2)	7.5×10^5

A large number of earthquakes are used in the analysis. All the earthquake data were downloaded from the PEER Ground Motion Database [46]. Corresponding earthquakes include near-field records with rupture distances of less than 20 km and far-field records with rupture distances of more than 20 km. Earthquake events used in this study are listed in Table 7. The specific earthquake recording names and recording station information are presented in the first two columns of Table 8. The number of records can also be obtained from Table 8. Table 8 is given in the next section. More information about consistent approaches in selecting earthquake records for seismic hazard mitigation studies can be obtained from [47–49]. Moreover, the most recent approaches concerning base isolation systems can be found in [50].

We wanted to compare different soil cases with uniform building and uniform base isolation parameters. Therefore, we used the same type of base isolation properties in every building and soil type in this study.

Dynamic simulations are performed by using Newmark's average acceleration time stepping method. In addition to time history analysis, spectrum analysis was also conducted. In the following section, numerical results that were obtained by performing dynamic simulations using the sample building defined in this section are presented. The earthquakes defined in Table 7 are also used in the following section.

Table 7. Earthquake events considered in the analysis.

Event	Year	Magnitude (M_w)	Mechanism
Imperial Valley, US	1940	6.95	Strike Slip
Imperial Valley, US	1979	6.53	Strike Slip
Loma Prieta, US	1989	6.93	Reverse Oblique
Manjil, IR	1990	7.37	Strike Slip
Cape Mendocino, US	1992	7.01	Reverse
Erzincan, TR	1992	6.69	Strike Slip
Northridge, US	1994	6.69	Reverse
Dinar, TR	1995	6.40	Normal
Kobe, JP	1995	6.90	Strike Slip
Chi-Chi, TW	1999	7.62	Reverse Oblique
Düzce, TR	1999	7.14	Strike Slip
Hector Mine, US	1999	7.13	Strike Slip
Kocaeli, TR	1999	7.51	Strike Slip
Iwate, JP	2008	6.90	Reverse
Darfield, NZ	2010	7.00	Strike Slip
El Mayor Cucapah, MX	2010	7.20	Strike Slip
Christchurch, NZ	2011	6.20	Reverse Oblique

Table 8. Maximum roof displacement values of the ten-story building.

Earthquake	Station	Soil	Fixed (m)	BI (m)	BI Reduction (%)	SSI (m)	SSI_BI (m)	BI Reduction w/SSI (%)
Imperial Valley, 1940	El Centro Array #9	Very Soft	0.185	0.078	57.86	0.075	0.053	29.29
Imperial Valley, 1940	El Centro Array #9	Soft	0.185	0.078	57.86	0.101	0.078	23.07
Imperial Valley, 1940	El Centro Array #9	Medium	0.185	0.078	57.86	0.163	0.078	52.06
Imperial Valley, 1940	El Centro Array #9	Dense	0.185	0.078	57.86	0.179	0.078	56.26
Imperial Valley, 1979	Calipatria Fire Station	Very Soft	0.070	0.022	68.48	0.022	0.016	26.70
Imperial Valley, 1979	Calipatria Fire Station	Soft	0.070	0.022	68.48	0.028	0.019	32.25
Imperial Valley, 1979	Calipatria Fire Station	Medium	0.070	0.022	68.48	0.057	0.022	61.18
Imperial Valley, 1979	Calipatria Fire Station	Dense	0.070	0.022	68.48	0.069	0.022	68.11
Imperial Valley, 1979	Delta	Very Soft	0.103	0.051	50.32	0.053	0.046	12.30
Imperial Valley, 1979	Delta	Soft	0.103	0.051	50.32	0.063	0.044	29.62
Imperial Valley, 1979	Delta	Medium	0.103	0.051	50.32	0.084	0.051	39.14
Imperial Valley, 1979	Delta	Dense	0.103	0.051	50.32	0.093	0.051	44.87
Loma Prieta, 1989	Alameda Naval Air Stn Hanger	Very Soft	0.195	0.036	81.59	0.026	0.030	−14.31
Loma Prieta, 1989	Alameda Naval Air Stn Hanger	Soft	0.195	0.036	81.59	0.067	0.038	43.62
Loma Prieta, 1989	Alameda Naval Air Stn Hanger	Medium	0.195	0.036	81.59	0.162	0.036	77.66
Loma Prieta, 1989	Alameda Naval Air Stn Hanger	Dense	0.195	0.036	81.59	0.185	0.036	80.52
Loma Prieta, 1989	Capitola	Very Soft	0.261	0.057	78.07	0.102	0.043	57.31
Loma Prieta, 1989	Capitola	Soft	0.261	0.057	78.07	0.235	0.060	74.32
Loma Prieta, 1989	Capitola	Medium	0.261	0.057	78.07	0.238	0.059	75.32
Loma Prieta, 1989	Capitola	Dense	0.261	0.057	78.07	0.256	0.058	77.38
Erzincan, 1992 (EW)	Erzincan	Very Soft	0.516	0.250	51.55	0.232	0.197	15.14
Erzincan, 1992 (EW)	Erzincan	Soft	0.516	0.250	51.55	0.374	0.244	34.67
Erzincan, 1992 (EW)	Erzincan	Medium	0.516	0.250	51.55	0.481	0.250	47.90
Erzincan, 1992 (EW)	Erzincan	Dense	0.516	0.250	51.55	0.494	0.250	49.31
Erzincan, 1992 (NS)	Erzincan	Very Soft	0.571	0.249	56.40	0.231	0.196	15.26
Erzincan, 1992 (NS)	Erzincan	Soft	0.571	0.249	56.40	0.372	0.243	34.74
Erzincan, 1992 (NS)	Erzincan	Medium	0.571	0.249	56.40	0.479	0.249	47.95
Erzincan, 1992 (NS)	Erzincan	Dense	0.571	0.249	56.40	0.533	0.249	53.29
Cape Mendocino, 1992	Cape Mendocino	Very Soft	0.285	0.106	62.85	0.126	0.111	12.50
Cape Mendocino, 1992	Cape Mendocino	Soft	0.285	0.106	62.85	0.194	0.108	44.37
Cape Mendocino, 1992	Cape Mendocino	Medium	0.285	0.106	62.85	0.256	0.106	58.53
Cape Mendocino, 1992	Cape Mendocino	Dense	0.285	0.106	62.85	0.276	0.106	61.57
Cape Mendocino, 1992	Shelter Cove Airport	Very Soft	0.009	0.004	52.19	0.006	0.003	50.17

Table 8. *Cont.*

Earthquake	Station	Soil	Fixed (m)	BI (m)	BI Reduction (%)	SSI (m)	SSI_BI (m)	BI Reduction w/SSI (%)
Cape Mendocino, 1992	Shelter Cove Airport	Soft	0.009	0.004	52.19	0.006	0.003	41.20
Cape Mendocino, 1992	Shelter Cove Airport	Medium	0.009	0.004	52.19	0.008	0.004	44.65
Cape Mendocino, 1992	Shelter Cove Airport	Dense	0.009	0.004	52.19	0.009	0.004	50.99
Northridge, 1994	Anacapa Island	Very Soft	0.017	0.002	85.23	0.005	0.002	59.54
Northridge, 1994	Anacapa Island	Soft	0.017	0.002	85.23	0.007	0.002	71.38
Northridge, 1994	Anacapa Island	Medium	0.017	0.002	85.23	0.014	0.002	83.04
Northridge, 1994	Anacapa Island	Dense	0.017	0.002	85.23	0.016	0.002	84.52
Northridge, 1994	Canoga Park	Very Soft	0.637	0.320	49.77	0.297	0.252	15.26
Northridge, 1994	Canoga Park	Soft	0.637	0.320	49.77	0.478	0.312	34.73
Northridge, 1994	Canoga Park	Medium	0.637	0.320	49.77	0.614	0.320	47.87
Northridge, 1994	Canoga Park	Dense	0.637	0.320	49.77	0.631	0.320	49.24
Kobe, 1995	Abeno	Very Soft	0.107	0.025	76.62	0.034	0.016	53.33
Kobe, 1995	Abeno	Soft	0.107	0.025	76.62	0.058	0.024	58.10
Kobe, 1995	Abeno	Medium	0.107	0.025	76.62	0.080	0.025	68.67
Kobe, 1995	Abeno	Dense	0.107	0.025	76.62	0.097	0.025	74.17
Kobe, 1995	HIK	Very Soft	0.103	0.029	71.54	0.032	0.016	48.03
Kobe, 1995	HIK	Soft	0.103	0.029	71.54	0.076	0.022	71.56
Kobe, 1995	HIK	Medium	0.103	0.029	71.54	0.092	0.029	68.64
Kobe, 1995	HIK	Dense	0.103	0.029	71.54	0.099	0.029	70.56
Dinar, 1995	Balikesir	Very Soft	0.005	0.002	60.24	0.003	0.001	52.65
Dinar, 1995	Balikesir	Soft	0.005	0.002	60.24	0.007	0.002	74.38
Dinar, 1995	Balikesir	Medium	0.005	0.002	60.24	0.005	0.002	65.64
Dinar, 1995	Balikesir	Dense	0.005	0.002	60.24	0.005	0.002	62.85
Dinar, 1995	Dinar	Very Soft	0.309	0.110	64.31	0.156	0.061	61.02
Dinar, 1995	Dinar	Soft	0.309	0.110	64.31	0.225	0.103	54.45
Dinar, 1995	Dinar	Medium	0.309	0.110	64.31	0.269	0.110	59.31
Dinar, 1995	Dinar	Dense	0.309	0.110	64.31	0.306	0.110	64.02
Kocaeli, 1999	Arcelik	Very Soft	0.053	0.016	70.20	0.017	0.014	16.73
Kocaeli, 1999	Arcelik	Soft	0.053	0.016	70.20	0.034	0.016	53.71
Kocaeli, 1999	Arcelik	Medium	0.053	0.016	70.20	0.052	0.016	69.52
Kocaeli, 1999	Arcelik	Dense	0.053	0.016	70.20	0.053	0.016	70.43
Kocaeli, 1999	Istanbul	Very Soft	0.014	0.010	31.95	0.009	0.009	1.23
Kocaeli, 1999	Istanbul	Soft	0.014	0.010	31.95	0.016	0.010	40.94
Kocaeli, 1999	Istanbul	Medium	0.014	0.010	31.95	0.014	0.010	28.99
Kocaeli, 1999	Istanbul	Dense	0.014	0.010	31.95	0.013	0.010	25.07
Chi-Chi, 1999	CHY006	Very Soft	0.362	0.073	79.85	0.087	0.060	31.67
Chi-Chi, 1999	CHY006	Soft	0.362	0.073	79.85	0.183	0.071	61.13
Chi-Chi, 1999	CHY006	Medium	0.362	0.073	79.85	0.344	0.073	78.76
Chi-Chi, 1999	CHY006	Dense	0.362	0.073	79.85	0.357	0.073	79.57
Chi-Chi, 1999	CHY101	Very Soft	0.216	0.080	62.68	0.092	0.103	−11.95
Chi-Chi, 1999	CHY101	Soft	0.216	0.080	62.68	0.103	0.094	8.96
Chi-Chi, 1999	CHY101	Medium	0.216	0.080	62.68	0.173	0.082	52.78
Chi-Chi, 1999	CHY101	Dense	0.216	0.080	62.68	0.197	0.081	58.97
Duzce, 1999	Bursa	Very Soft	0.848	0.433	48.95	0.401	0.341	15.03
Duzce, 1999	Bursa	Soft	0.848	0.433	48.95	0.643	0.422	34.35
Duzce, 1999	Bursa	Medium	0.848	0.433	48.95	0.820	0.433	47.22
Duzce, 1999	Bursa	Dense	0.848	0.433	48.95	0.840	0.433	48.45
Duzce, 1999	Duzce	Very Soft	0.632	0.316	49.95	0.295	0.249	15.42
Duzce, 1999	Duzce	Soft	0.632	0.316	49.95	0.474	0.309	34.88
Duzce, 1999	Duzce	Medium	0.632	0.316	49.95	0.610	0.317	48.03
Duzce, 1999	Duzce	Dense	0.632	0.316	49.95	0.626	0.317	49.42
Manjil, 1990	Abbar	Very Soft	0.110	0.079	28.09	0.061	0.060	1.50
Manjil, 1990	Abbar	Soft	0.110	0.079	28.09	0.078	0.087	−10.94
Manjil, 1990	Abbar	Medium	0.110	0.079	28.09	0.092	0.082	10.35
Manjil, 1990	Abbar	Dense	0.110	0.079	28.09	0.099	0.081	18.94
Manjil, 1990	Rudsar	Very Soft	0.057	0.015	73.12	0.026	0.012	55.12
Manjil, 1990	Rudsar	Soft	0.057	0.015	73.12	0.053	0.015	71.25
Manjil, 1990	Rudsar	Medium	0.057	0.015	73.12	0.061	0.015	75.50
Manjil, 1990	Rudsar	Dense	0.057	0.015	73.12	0.058	0.015	73.63
Hector Mine, 1999	Banning	Very Soft	0.011	0.006	40.51	0.006	0.004	34.64
Hector Mine, 1999	Banning	Soft	0.011	0.006	40.51	0.006	0.006	8.82
Hector Mine, 1999	Banning	Medium	0.011	0.006	40.51	0.010	0.006	35.18
Hector Mine, 1999	Banning	Dense	0.011	0.006	40.51	0.010	0.006	38.88
Hector Mine, 1999	Indio	Very Soft	0.073	0.044	39.99	0.046	0.035	24.68

Table 8. *Cont.*

Earthquake	Station	Soil	Fixed (m)	BI (m)	BI Reduction (%)	SSI (m)	SSI_BI (m)	BI Reduction w/SSI (%)
Hector Mine, 1999	Indio	Soft	0.073	0.044	39.99	0.082	0.046	43.52
Hector Mine, 1999	Indio	Medium	0.073	0.044	39.99	0.067	0.044	34.91
Hector Mine, 1999	Indio	Dense	0.073	0.044	39.99	0.071	0.044	38.34
Iwate, 2008	AKT023	Very Soft	0.186	0.036	80.50	0.043	0.030	31.37
Iwate, 2008	AKT023	Soft	0.186	0.036	80.50	0.085	0.035	58.73
Iwate, 2008	AKT023	Medium	0.186	0.036	80.50	0.161	0.036	77.71
Iwate, 2008	AKT023	Dense	0.186	0.036	80.50	0.179	0.036	79.83
Iwate, 2008	IWT010	Very Soft	0.139	0.067	51.61	0.062	0.043	30.69
Iwate, 2008	IWT010	Soft	0.139	0.067	51.61	0.080	0.065	18.65
Iwate, 2008	IWT010	Medium	0.139	0.067	51.61	0.129	0.068	46.80
Iwate, 2008	IWT010	Dense	0.139	0.067	51.61	0.137	0.068	50.50
El Mayor-Cucapah, 2010	Chihuahua	Very Soft	0.133	0.040	69.61	0.038	0.053	−38.85
El Mayor-Cucapah, 2010	Chihuahua	Soft	0.133	0.040	69.61	0.134	0.046	65.74
El Mayor-Cucapah, 2010	Chihuahua	Medium	0.133	0.040	69.61	0.138	0.041	70.22
El Mayor-Cucapah, 2010	Chihuahua	Dense	0.133	0.040	69.61	0.142	0.041	71.50
El Mayor-Cucapah, 2010	Michoacan de Ocampo	Very Soft	0.332	0.113	65.99	0.107	0.095	11.68
El Mayor-Cucapah, 2010	Michoacan de Ocampo	Soft	0.332	0.113	65.99	0.184	0.117	36.35
El Mayor-Cucapah, 2010	Michoacan de Ocampo	Medium	0.332	0.113	65.99	0.276	0.114	58.63
El Mayor-Cucapah, 2010	Michoacan de Ocampo	Dense	0.332	0.113	65.99	0.315	0.113	63.95
Darfield, 2010	Canterbury Aero Club	Very Soft	0.056	0.037	35.09	0.045	0.029	34.56
Darfield, 2010	Canterbury Aero Club	Soft	0.056	0.037	35.09	0.052	0.039	24.88
Darfield, 2010	Canterbury Aero Club	Medium	0.056	0.037	35.09	0.050	0.037	25.14
Darfield, 2010	Canterbury Aero Club	Dense	0.056	0.037	35.09	0.053	0.037	30.06
Darfield, 2010	DSLC	Very Soft	0.218	0.064	70.61	0.053	0.046	12.41
Darfield, 2010	DSLC	Soft	0.218	0.064	70.61	0.071	0.071	0.30
Darfield, 2010	DSLC	Medium	0.218	0.064	70.61	0.196	0.066	66.20
Darfield, 2010	DSLC	Dense	0.218	0.064	70.61	0.215	0.065	69.82
Christchurch, 2011	ADCS	Very Soft	0.017	0.003	81.89	0.006	0.003	46.34
Christchurch, 2011	ADCS	Soft	0.017	0.003	81.89	0.006	0.003	49.92
Christchurch, 2011	ADCS	Medium	0.017	0.003	81.89	0.014	0.003	78.63
Christchurch, 2011	ADCS	Dense	0.017	0.003	81.89	0.016	0.003	80.98
Christchurch, 2011	CECS	Very Soft	0.009	0.006	38.67	0.009	0.003	64.57
Christchurch, 2011	CECS	Soft	0.009	0.006	38.67	0.007	0.004	31.37
Christchurch, 2011	CECS	Medium	0.009	0.006	38.67	0.008	0.006	30.16
Christchurch, 2011	CECS	Dense	0.009	0.006	38.67	0.009	0.006	36.30

4. Numerical Results

The numerical results section is concentrated on the acceleration, velocity and displacement time histories of the structures. Dynamic simulations are performed for every soil condition given above. In addition, the results are presented separately for each soil condition. The comparisons are carried out between the conventional structures and the structure with BI. Figure 5 shows the top story acceleration time history responses of the five-story building for the 1989 Loma Prieta Earthquake, considering different soil conditions. We should mention here that the accelerations are relative accelerations and not absolute acceleration responses. The maximum velocities are also investigated, but they are not shown here, and no comparative velocity analysis is given in this section because of space constraints. However, for all mentioned earthquakes, building heights and soil cases mentioned in this paper earlier, the velocity investigation is carried out, and the outcomes are considered in the conclusion section. The displacement time histories for different soil conditions for the five-story building under the effect of the Loma Prieta earthquake and the maximum displacement of each story are not shown for this building, as the maximum roof displacements for every case defined are shown in Table 8. The maximum displacement curves are only shown for the forty-story building in this section. Figure 6 shows the maximum relative acceleration of the ten-story building under the effect of the Kocaeli 1999 earthquake. The top story displacement time history for the ten-story building under the effect of the Kocaeli earthquake is given in Figure 7. For the forty-story building under the effect of the 1979 Imperial Valley Earthquake, the top story acceleration time histories and maximum acceleration of each story are shown in Figures 8 and 9, respectively. Moreover, in Figure 10, the top story displacement time histories are shown, while in Figure 11, the maximum story displacements are presented for the 1979 Imperial Valley earthquake.

Considering the acceleration results (the majority of the results are not shown in this paper) for five-, ten- and forty-story buildings, BI reduces the responses significantly in all cases. This is more of an expected outcome. In addition, SSI reduces the acceleration values, but these effects are only pronounced in softer soil conditions. Considering the velocity curves for five-, ten- and forty-story buildings (these curves are not shown here), the results are similar to the acceleration results; BI reduces the responses considerably. It can also be claimed that SSI reduces the velocity values, but again, these effects are only pronounced in softer soil conditions. In very soft soil conditions, SSI effects can be observed clearly, but when moving towards denser soil, the effects disappear, and the structures behave like they are fixed based.

Considering the displacement results for five-, ten- and forty-story buildings, although the base isolated displacements might be higher than in fixed base cases, large displacements occur at the BI layer, providing low inter-story displacements, and the structures behave linearly. This may also mean that the BI absorbs most of the seismic energy by itself. However, an energy-based analysis was not carried out here. This is also an anticipated characteristic of BI. It can also be mentioned that SSI reduces the displacement values; like the acceleration and velocity investigations, these effects are only pronounced in softer soil conditions. This is the expected result, as the dense soil case has the closest behavior to a fixed base case among all soil cases studied in this paper.

Maximum roof displacement values were calculated for each earthquake, and results for the ten-story building are presented in Table 8. More specific outcomes with numerical percentages are given in this part. For the base isolated case, the displacement of the isolation system was subtracted from the story displacements. The sixth column of Table 8 shows the fixed base response reduction percentage of the base isolated case. For this comparison, SSI was not included. Without considering SSI in the base isolated cases, the response reduction percentage ranges from 28% to 85%, which is a significant maximum reduction percentage. If we consider the ninth column of Table 8, which is the base response reduction percentage of the base isolated case considering SSI, this case compares the structure without BI with SSI and the structure with BI considering SSI. For this case, the response reduction ranges from −38% to almost 85%. Moreover, −38% corresponds to the El Mayor-Cucapah 2010 earthquake and is a very soft soil condition case. Additionally, this percentage shows that BI increased the uncontrolled structure displacement. However, this is one of the only four extreme cases out of 132 analyses. We only wanted to interpret the negative percentage meaning here. However, if the ninth column is considered specifically, it tells us that BI usage still significantly decreases the responses of upper stories, although these reductions are not as high as in the cases in which SSI is not considered. Another important finding from this table, which can be seen clearly, is that for dense soil cases, the response reductions are almost identical to those for the fixed base cases. If we compare the sixth and ninth columns of Table 8, the difference in the response reduction percentage for the fixed base case and dense soil case ranges from 9% to −2%. Furthermore, −2% means that in one of the dense soil cases, the response reduction is higher than in the case without SSI. This may also be considered as one of the four extreme cases and not an expected result.

It can be understood from the results that the effectiveness of BI is generally reduced when SSI is considered. Only 12 out of 132 analyses showed that the BI reduction is clearly increased considering SSI. In the next section, the BI spectrum is given.

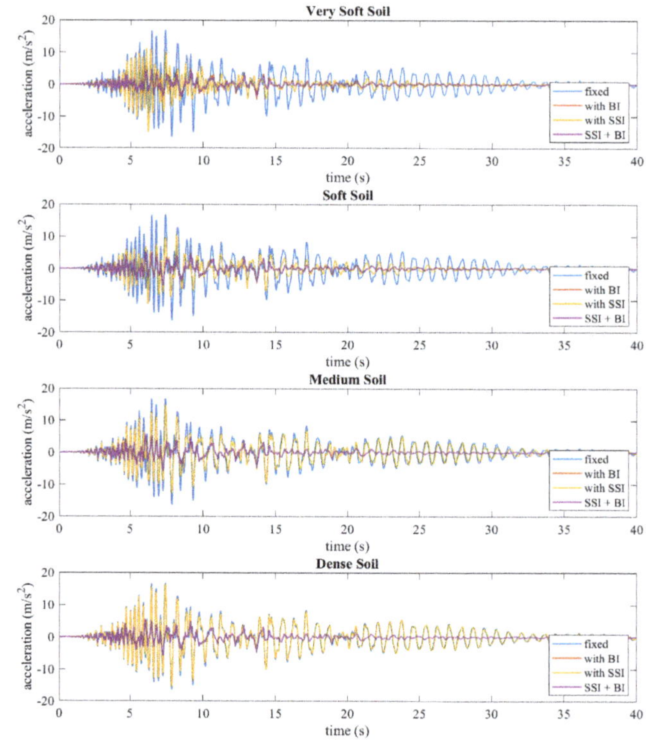

Figure 5. Top story acceleration–time history of five-story buildings (Loma Prieta, 1989).

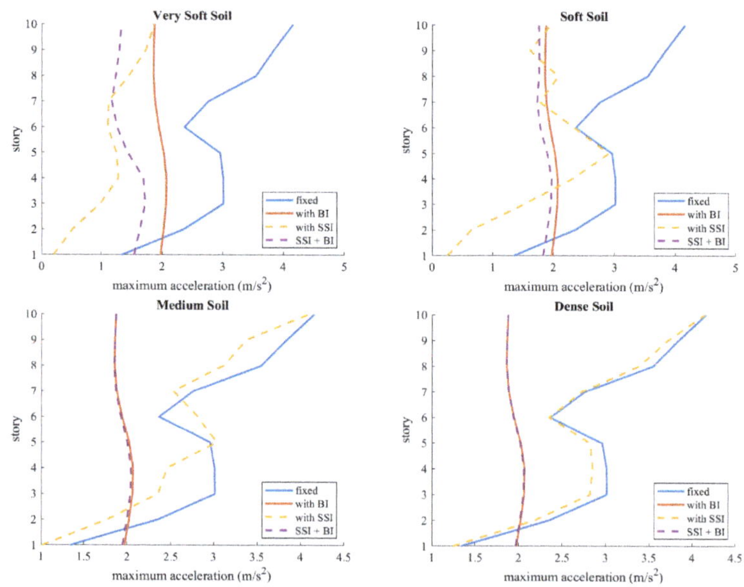

Figure 6. Maximum acceleration of each story for the ten-story building (Kocaeli, 1999).

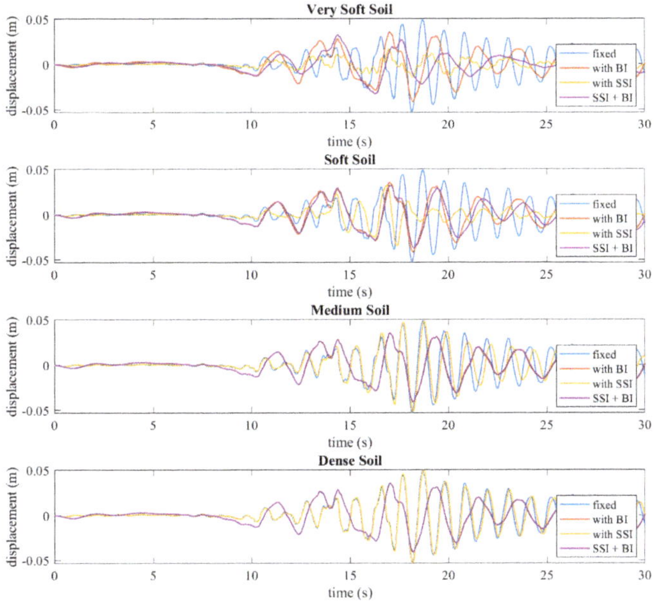

Figure 7. Top story displacement–time history of the ten-story building (Kocaeli, 1999).

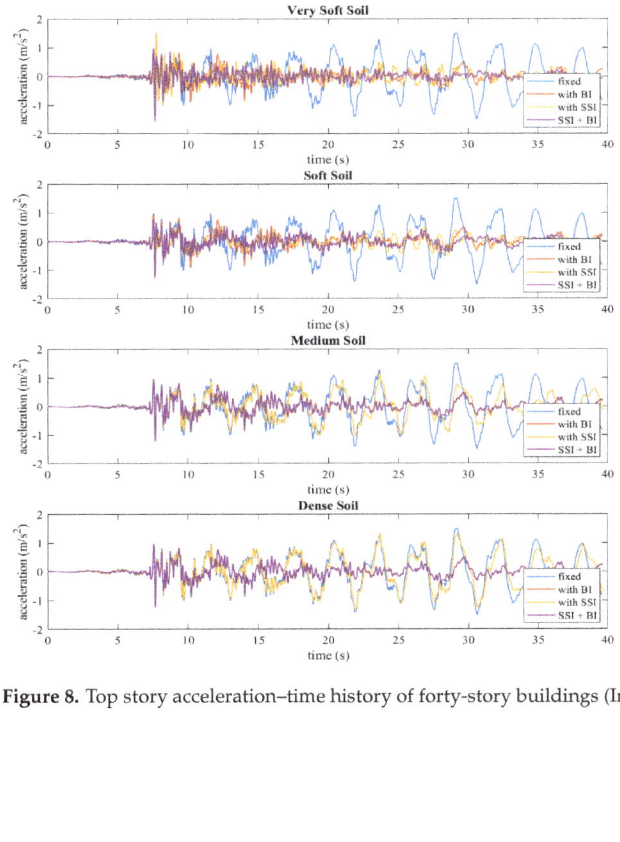

Figure 8. Top story acceleration–time history of forty-story buildings (Imperial Valley, 1979).

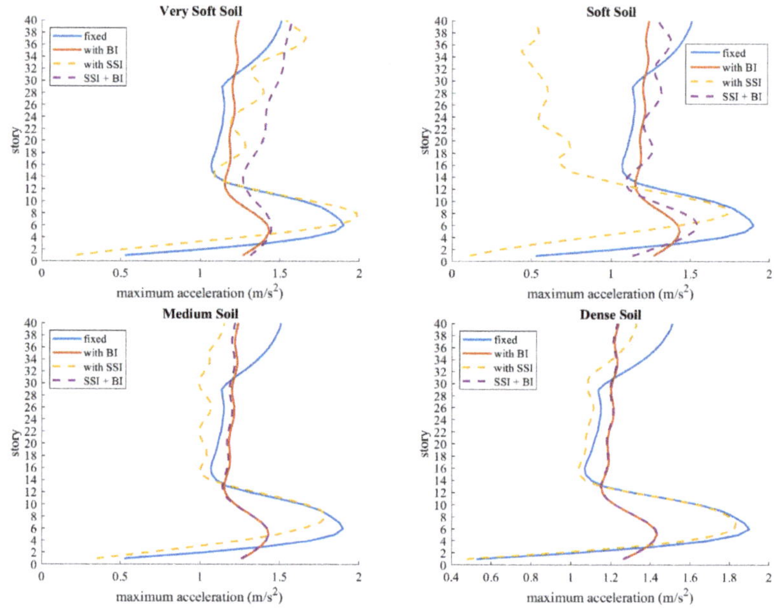

Figure 9. Maximum acceleration of each story for the forty-story building (Imperial Valley, 1979).

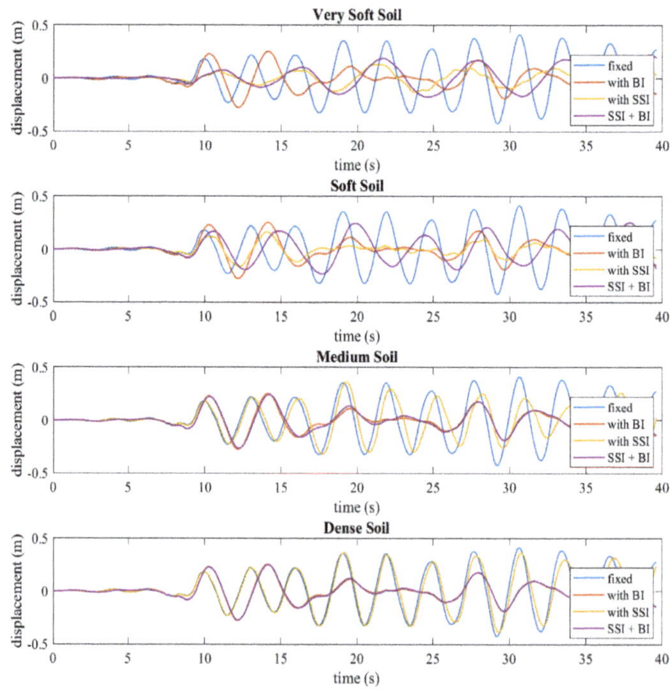

Figure 10. Top story displacement–time history of forty-story buildings (Imperial Valley, 1979).

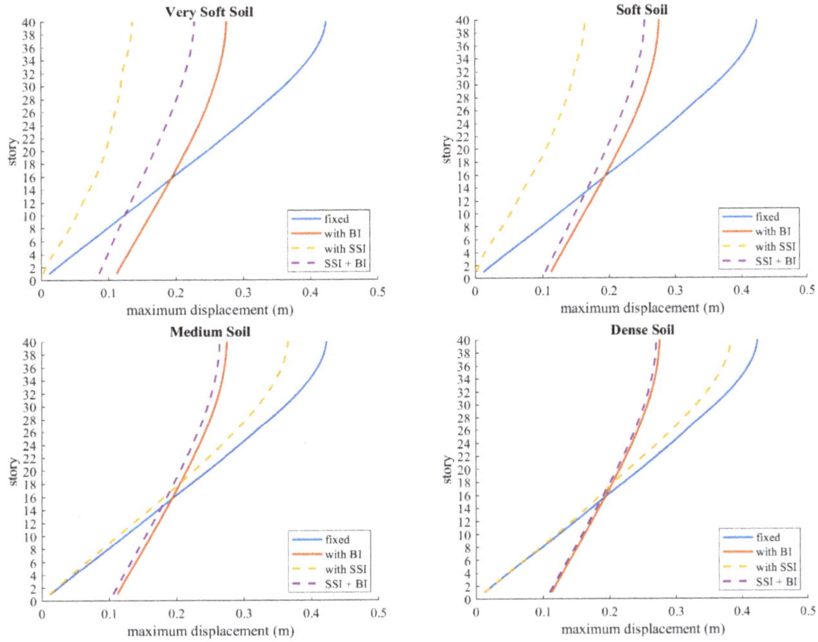

Figure 11. Maximum displacement of each story for the forty-story building (Imperial Valley, 1979).

Spectral Analysis

Spectral analyses are conducted for the five-story building considering the 1940 Imperial Valley earthquake. Spectral acceleration and displacement graphics can be seen in Figure 12. The results show that the period elongation feature of BI and SSI lowers the spectral acceleration response of the structure. BI displacements are higher than fixed conditions, which is the same as the time history results. We can also see that SSI effects are more noticeable in softer soil conditions, in a similar manner to time history results and that the effects diminish when we move towards denser soil. Table 9 shows the detailed results of the spectral analyses. In Table 9, T is the period, and SA, SV and SD are the spectral acceleration, spectral velocity and spectral displacement, respectively. If we consider Table 9 and Figure 12 for very soft soil conditions, the values of SA range from 1.84 to 5.34 with respect to the base conditions; for soft soil conditions, values of SA range from 1.92 to 5.34 with respect to the base conditions, for medium soil values of SA range from 1.92 to 5.45, which are the maximum obtained SA values among all the different soil conditions. Lastly, for dense soil, SA values range from 1.92 to 5.39. It can be observed from the table that when the period of the structure increases, SA decreases. This is applicable to all different soil conditions. BI cases have higher SA in all soil conditions. However, for spectral displacements, we can observe that when the period increases, the spectral displacement also increases. BI implementation increases SD, and this is the expected result. For all different soil types, this increase can be observed. The conclusions obtained from this study are presented in the next section.

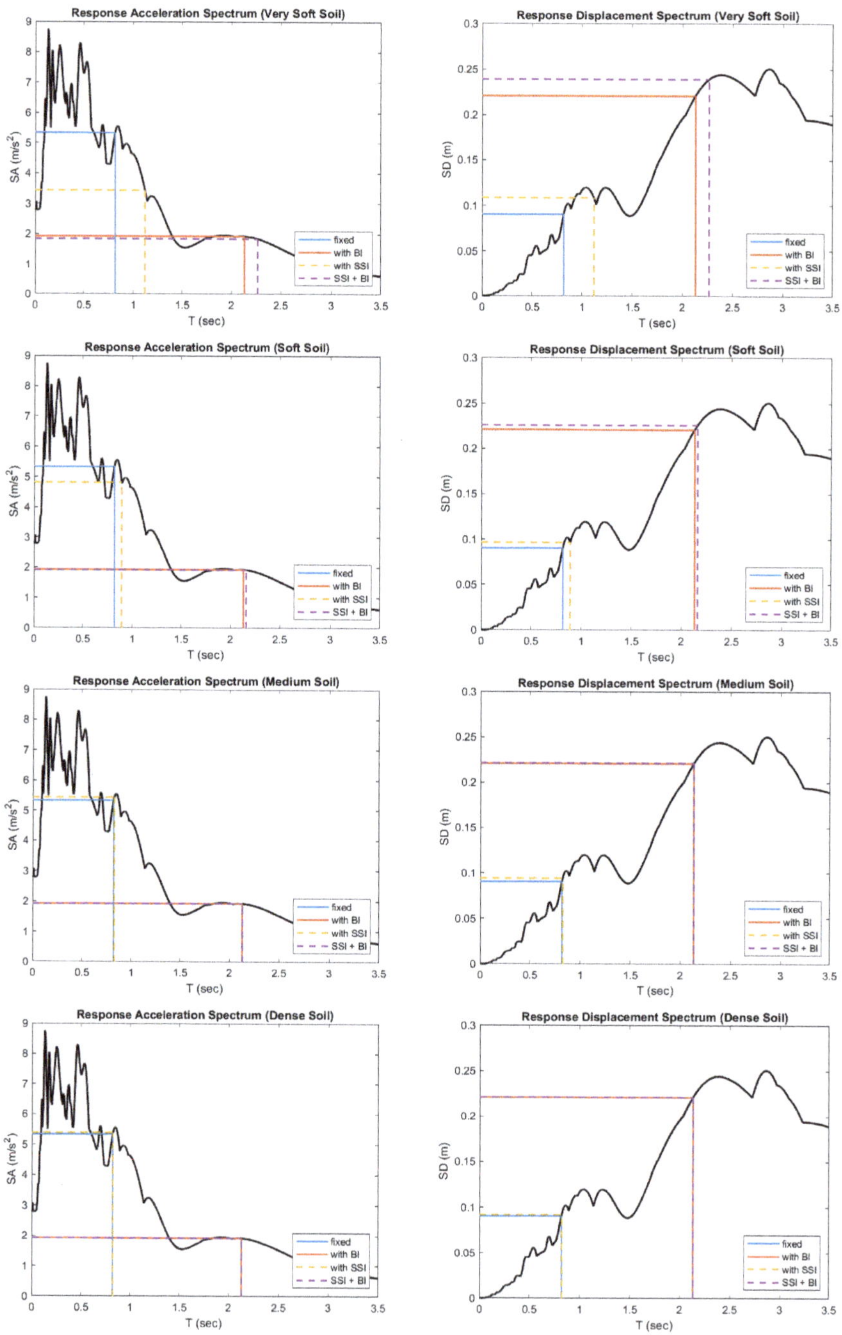

Figure 12. Spectral acceleration and displacement graphics for every soil condition.

Table 9. Spectral analysis results of the five-story building (Imperial Valley, 1940).

Soil	Base	T (s)	SA (m/s^2)	SV (m/s)	SD (m)
Very Soft Soil	fixed	0.82	5.34	0.70	0.09
Very Soft Soil	BI	2.13	1.93	0.65	0.22
Very Soft Soil	SSI	1.12	3.44	0.61	0.11
Very Soft Soil	SSI+BI	2.26	1.84	0.66	0.24
Soft Soil	fixed	0.82	5.34	0.70	0.09
Soft Soil	BI	2.13	1.93	0.65	0.22
Soft Soil	SSI	0.89	4.82	0.68	0.10
Soft Soil	SSI+BI	2.16	1.92	0.66	0.23
Medium Soil	fixed	0.82	5.34	0.70	0.09
Medium Soil	BI	2.13	1.93	0.65	0.22
Medium Soil	SSI	0.83	5.45	0.72	0.09
Medium Soil	SSI+BI	2.13	1.92	0.65	0.22
Dense Soil	fixed	0.82	5.34	0.70	0.09
Dense Soil	BI	2.13	1.93	0.65	0.22
Dense Soil	SSI	0.82	5.39	0.70	0.09
Dense Soil	SSI+BI	2.13	1.92	0.65	0.22

5. Conclusions

A large number of numerical analyses are conducted for three different buildings. With respect to the numerical results that are given in this paper, the following conclusions are obtained:

- BI may greatly reduce the acceleration, velocity and displacements of structures that are induced by earthquakes.
- Although the total displacement might be higher than in a fixed structure, most of it occurs on the isolation system, and the superstructure moves as a whole, resulting in much lower inter-story drift compared to the fixed base structure.
- SSI may modify the acceleration, velocity and displacement responses of structures.
- The results show that SSI mostly reduces the effects of earthquakes. In order to stay on the safer side, design codes do not specify SSI analysis procedures; they often briefly state that SSI can sometimes modify earthquake responses, and for those rare cases, it should be investigated.

This study's results are in line with these assumptions. For a small number of cases, SSI may slightly increase the response of structures. The results also show that SSI effects are much more pronounced in soft soil conditions and hardly ever present in dense soil conditions. In dense soil conditions, the response of structures eminently approaches that of fixed base conditions. Based on this study, SSI affects the performance of BI systems. The effectiveness of a BI system is reduced when SSI is considered. According to this study, both far and near-fault earthquakes lead to responses that have similar characteristics. The heights of the structures also do not have any significant effects when SSI and BI are considered together. Spectral analysis results are in line with the time history results.

The results obtained from this study are also compared with the existing literature. In this research, it is seen that SSI affects the performance of the isolation system. The same outcome was also obtained in [27,28,35]. Additionally, SSI effects were found to be more pronounced in soft soil conditions, which is similar to the findings presented in [13,14]. Moreover, our study includes many results about SSI effects on tall buildings. This is an important aspect of SSI research, as it was stated in [36] that SSI effects in tall buildings is a concept that needs to be investigated more.

Lastly, it should be mentioned that, by performing an extensive number of dynamic simulations that were coded based on the simple proposed formulation defined in Section 2.5, it was seen that the solution is computationally efficient, and it does not create any stability problems for the numerical model. For future studies, three-dimensional models considering SSI and BI can be studied. The formulation given in this study may also be generalized for three-dimensional structures considering SSI as a future study.

Author Contributions: Conceptualization, A.Y.; methodology, A.Y.; software, Y.U.; validation, Y.U. and A.Y.; formal analysis, Y.U.; investigation, A.Y. and Y.U.; resources, Y.U.; data curation, Y.U.; writing—original draft preparation, A.Y. and Y.U.; writing—review and editing, A.Y.; visualization, A.Y. and Y.U.; supervision, A.Y. All authors have read and agreed to the published version of the manuscript.

Funding: This research received no external funding.

Data Availability Statement: The data presented in this study are available on request from the corresponding author.

Conflicts of Interest: The authors declare no conflict of interest.

References

1. Chopra, A.K. *Dynamics of Structures, Theory and Applications to Earthquake Engineering*, 4th ed.; Pearson Education Limited: Upper Saddle River, NJ, USA, 2013.
2. American Society of Civil Engineers (ASCE). *Fema 356 Prestandard and Commentary for the Seismic Rehabilitation of Buildings*; Federal Emergency Management Agency: Washington, DC, USA, 2000.
3. Constantinou, M.C.; Kneifati, M.C. Dynamics of Soil-Base-Isolated-Structure Systems. *J. Struct. Eng.* **1988**, *114*, 211–221. [CrossRef]
4. Bycroft, G.N. Soil-structure interaction at higher frequency factors. *Earthq. Eng. Struct. Dyn.* **1977**, *5*, 235–248. [CrossRef]
5. Novak, M.; Henderson, P. Base-isolated buildings with soil-structure interaction. *Earthq. Eng. Struct. Dyn.* **1989**, *18*, 751–765. [CrossRef]
6. Pappin, J.W.; Lubkowski, Z.A.; King, R.A. The Significance of Site Response Effects on Performance Based Design. In Proceedings of the 12th World Conference on Earthquake Engineering, Auckland, New Zealand, 30 January–4 February 2000.
7. Mylonakis, G.; Gazetas, G. Seismic Soil-Structure Interaction: Beneficial or Detrimental? *J. Earthq. Eng.* **2000**, *4*, 277–301. [CrossRef]
8. Tongaonkar, N.; Jangid, R. Seismic response of isolated bridges with soil–structure interaction. *Soil Dyn. Earthq. Eng.* **2003**, *23*, 287–302. [CrossRef]
9. Mylonakis, G.; Voyagaki, E.; Price, T. Damage Potential of the 1999 Athens, Greece, Accelerograms. *Bull. Earthq. Eng.* **2003**, *1*, 205–240. [CrossRef]
10. Syngros, K. Seismic Response of Piles and Pile-Supported Bridge Piers Evaluated through Case Histories. Ph.D. Thesis, The City University of New York, New York, NY, USA, 2004.
11. Deb, S. Seismic base isolation—An overview. *Curr. Sci.* **2004**, *87*, 1426–1430. Available online: http://www.jstor.org/stable/24109483 (accessed on 5 July 2019).
12. Tsai, C.S.; Chen, C.-S.; Chen, B.-J. Effects of unbounded media on seismic responses of FPS-isolated structures. *Struct. Control Health Monit.* **2004**, *11*, 1–20. [CrossRef]
13. Dicleli, M.; Albhaisi, S.; Mansour, M.Y. Static Soil–Structure Interaction Effects in Seismic-Isolated Bridges. *Pract. Period. Struct. Des. Constr.* **2005**, *10*, 22–33. [CrossRef]
14. Liu, M.-Y.; Chiang, W.-L.; Hwang, J.-H.; Chu, C.-R. Wind-induced vibration of high-rise building with tuned mass damper including soil–structure interaction. *J. Wind. Eng. Ind. Aerodyn.* **2008**, *96*, 1092–1102. [CrossRef]
15. Spyrakos, C.; Koutromanos, I.; Maniatakis, C. Seismic response of base-isolated buildings including soil–structure interaction. *Soil Dyn. Earthq. Eng.* **2009**, *29*, 658–668. [CrossRef]
16. Jeremić, B.; Jie, G.; Preisig, M.; Tafazzoli, N. Time domain simulation of soil-foundation-structure interaction in non-uniform soils. *Earthq. Eng. Struct. Dyn.* **2009**, *38*, 699–718. [CrossRef]
17. Karabörk, T.; Deneme, İ.Ö.; Bilgehan, R.P. Dynamic soil structure interaction analysis for base isolated structures. *Erciyes Univ. J. Inst. Sci. Technol.* **2010**, *26*, 77–87.
18. Kausel, E. Early history of soil–structure interaction. *Soil Dyn. Earthq. Eng.* **2010**, *30*, 822–832. [CrossRef]
19. Genes, M.; Doğanay, E.; Bikçe, M.; Kaçın, S. Soil-Structure Interaction in RC Frame Buildings from Strong-Motion Recordings. *KSU J. Eng. Sci.* **2011**, *14*, 1–7. Available online: http://jes.ksu.edu.tr/en/download/article-file/180955 (accessed on 9 August 2019).
20. Matinmanesh, H.; Asheghabadi, M.S. Seismic Analysis on Soil-Structure Interaction of Buildings over Sandy Soil. *Procedia Eng.* **2011**, *14*, 1737–1743. [CrossRef]
21. Manolis, G.D.; Markou, A. A distributed mass structural system for soil-structure-interaction and base isolation studies. *Arch. Appl. Mech.* **2012**, *82*, 1513–1529. [CrossRef]
22. Giarlelis, C.; Mylonakis, G. The role of soil-structure interaction in the inelastic performance of multi-story buildings. In Proceedings of the 9th International Conference on Structural Dynamics, EURODYN 2014, Porto, Portugal, 30 June–2 July 2014.
23. Li, C.P.; Liu, W.Q.; Wang, S.G.; Du, D.S. Analysis of Soil-Structure Interaction (SSI) Effects on Seismic Response of Base-Isolated Structures. *Adv. Mater. Res.* **2010**, *163–167*, 4199–4207. [CrossRef]
24. Luco, J.E. Effects of soil–structure interaction on seismic base isolation. *Soil Dyn. Earthq. Eng.* **2014**, *66*, 167–177. [CrossRef]
25. Tsai, C.S.; Hsueh, C.I.; Su, H.C. Roles of soil-structure interaction and damping in base-isolated structures built on numerous soil layers overlying a half-space. *Earthq. Eng. Eng. Vib.* **2016**, *15*, 387–400. [CrossRef]

26. Yanik, A. Absolute Instantaneous Optimal Control Performance Index for Active Vibration Control of Structures under Seismic Excitation. *Shock Vib.* **2019**, *2019*, 4207427. [CrossRef]
27. Ashiquzzaman; Hong, K.-J. Simplified Model of Soil-Structure Interaction for Seismically Isolated Containment Buildings in Nuclear Power Plant. *Structures* **2017**, *10*, 209–218. [CrossRef]
28. Zhou, Z.; Wei, X. Seismic Soil-Structure Interaction Analysis of Isolated Nuclear Power Plants in Frequency Domain. *Shock Vib.* **2016**, *2016*, 6127895. [CrossRef]
29. Zhao, X.; Wang, S.; Du, D.; Liu, W. Optimal Design of Viscoelastic Dampers in Frame Structures considering Soil-Structure Interaction Effect. *Shock Vib.* **2017**, *2017*, 9629083. [CrossRef]
30. Forcellini, D. Seismic assessment of a benchmark based isolated ordinary building with soil structure interaction. *Bull. Earthq. Eng.* **2018**, *16*, 2021–2042. [CrossRef]
31. McKenna, F. OpenSees: A Framework for Earthquake Engineering Simulation. *Comput. Sci. Eng.* **2011**, *13*, 58–66. [CrossRef]
32. Dai, W.; Rojas, F.; Shi, C.; Tan, Y. Effect of soil structure interaction on the dynamic responses of base isolated bridges and comparison to experimental results. *Soil Dyn. Earthq. Eng.* **2018**, *114*, 242–252. [CrossRef]
33. El-Sinawi, A.H.; AlHamaydeh, M.H.; Jhemi, A.A. Optimal Control of Magnetorheological Fluid Dampers for Seismic Isolation of Structures. *Math. Probl. Eng.* **2013**, *2013*, 251935. [CrossRef]
34. Yanik, A. Seismic control performance indices for magneto-rheological dampers considering simple soil-structure interaction. *Soil Dyn. Earthq. Eng.* **2020**, *129*, 105964. [CrossRef]
35. Zhenxia, S.; Haiping, D. The Analysis of Seismic Response for Base-isolated Structure by LS-DYNA. Indian Institute of Technology Kanpur. Available online: http://www.iitk.ac.in/nicee/wcee/artical/14_14-0204.pdf (accessed on 7 August 2019).
36. Forcellini, D. Seismic fragility of tall buildings considering soil structure interaction (SSI) effects. *Structures* **2022**, *45*, 999–1011. [CrossRef]
37. Jalali, H.H.; Farzam, M.F.; Gavgani, S.A.M.; Bekdaş, G. Semi-active control of buildings using different control algorithms considering SSI. *J. Build. Eng.* **2023**, *67*, 105956. [CrossRef]
38. Maleska, T.; Beben, D. Behaviour of Soil–Steel Composite Bridges under Strong Seismic Excitation with Various Boundary Conditions. *Materials* **2023**, *16*, 650. [CrossRef]
39. Wang, J.; Xie, Y.; Guo, T.; Du, Z. Predicting the Influence of Soil–Structure Interaction on Seismic Responses of Reinforced Concrete Frame Buildings Using Convolutional Neural Network. *Buildings* **2023**, *13*, 564. [CrossRef]
40. Jin, L.; Li, B.; Lin, S.; Li, G. Optimal Design Formula for Tuned Mass Damper Based on an Analytical Solution of Interaction between Soil and Structure with Rigid Foundation Subjected to Plane SH-Waves. *Buildings* **2022**, *13*, 17. [CrossRef]
41. Deng, Y.; Ge, S.; Lei, F. Effects of Pounding and Abutment Behavior on Seismic Response of Multi-Span Bridge Considering Abutment-Soil-Foundation-Structure Interactions. *Buildings* **2023**, *13*, 260. [CrossRef]
42. Datta, T.K. *Seismic Analysis of Structures*; John Wiley & Sons (Asia) Pte Ltd.: Singapore, 2010.
43. Pioldi, F.; Salvi, J.; Rizzi, E. Refined FDD modal dynamic identification from earthquake responses with Soil-Structure Interaction. *Int. J. Mech. Sci.* **2017**, *127*, 47–61. [CrossRef]
44. Bekdaş, G.; Nigdeli, S.M. Metaheuristic based optimization of tuned mass dampers under earthquake excitation by considering soil-structure interaction. *Soil Dyn. Earthq. Eng.* **2017**, *92*, 443–461. [CrossRef]
45. Salvi, J.; Pioldi, F.; Rizzi, E. Optimum Tuned Mass Dampers under seismic Soil-Structure Interaction. *Soil Dyn. Earthq. Eng.* **2018**, *114*, 576–597. [CrossRef]
46. Pacific Earthquake Engineering Research Center. PEER NGA-West2 Database. Available online: https://ngawest2.berkeley.edu/ (accessed on 2 July 2019).
47. Cardone, D.; Viggiani, L.; Perrone, G.; Telesca, A.; Di Cesare, A.; Ponzo, F.; Ragni, L.; Micozzi, F.; Dall'Asta, A.; Furinghetti, M.; et al. Modelling and Seismic Response Analysis of Existing Italian Residential RC Buildings Retrofitted by Seismic Isolation. *J. Earthq. Eng.* **2023**, *27*, 1069–1093. [CrossRef]
48. Cardone, D.; Conte, N.; Dall'Asta, A.; Di Cesare, A.; Flora, A.; Lamarucciola, N.; Micozzi, F.; Ponzo, F.C.; Ragni, L. RINTC-E project: The seismic risk of existing Italian RC buildings retrofitted with seismic isolation. In Proceedings of the 7th ECCOMAS Thematic Conference on Computational Methods in Structural Dynamics and Earthquake Engineering, COMPDYN 2019, Crete, Greece, 24–26 June 2019.
49. Iervolino, I.; Spillatura, A.; Bazzurro, P. Seismic Reliability of Code-Conforming Italian Buildings. *J. Earthq. Eng.* **2018**, *22*, 5–27. [CrossRef]
50. Calvi, P.M.; Calvi, G.M. Historical development of friction-based seismic isolation systems. *Soil Dyn. Earthq. Eng.* **2018**, *106*, 14–30. [CrossRef]

buildings

MDPI

Article

Weaving Octopus: An Assembly–Disassembly-Adaptable Customized Textile Hybrid Prototype

Ziqi Cui †, Siman Zhang †, Salvatore Viscuso * and Alessandra Zanelli

TextilesHUB Research Laboratory, Department of Architecture, Building Environment and Construction Engineering (DABC), Politecnico di Milano, 20133 Milano, Italy; ziqi.cui@mail.polimi.it (Z.C.); siman.zhang@mail.polimi.it (S.Z.); alessandra.zanelli@polimi.it (A.Z.)

* Correspondence: salvatore.viscuso@polimi.it

† These authors have contributed equally to this work.

Abstract: As global challenges evolve rapidly, lightweight architecture emerges as an effective and efficient solution to meet rapidly changing needs. Textiles offer flexibility and sustainability, addressing spatial requirements in urban and residential designs, particularly in underutilized areas. This study developed a user-friendly and customizable textile hybrid structure prototype by exploring different weaving methods to find more flexible and adaptable solutions. The research adopts a three-stage process: concept design, parametric simulation prototype, and physical scale-up testing. Methodologies include Finite Element Analysis (FEA) for assessing structural bending and tensile behavior, evolutionary computation for multi-objective optimization, Arduino for enabling interactive dynamic and lighting systems, and a website interface for bespoke decisions. Results revealed a groundbreaking textile hybrid prototype, applicable individually or collectively, with flexible assembly and disassembly in various scenarios. The prototype also offers an eco-friendly, cost-efficient facade renovation solution, enhancing aesthetics and providing shading benefits. The research encompasses interactive lightweight construction design, bending-active textile hybrids, form-finding, circular economy, and mass customization, contributing to advances in lightweight construction design while promoting sustainable practices in textile architecture.

Keywords: lightweight structure; assemble and disassemble; textile hybrids; weaving method; parametric design; form-finding; adaptive design; interactive system

Citation: Cui, Z.; Zhang, S.; Viscuso, S.; Zanelli, A. Weaving Octopus: An Assembly–Disassembly-Adaptable Customized Textile Hybrid Prototype. *Buildings* **2023**, *13*, 2413. https://doi.org/10.3390/buildings13102413

Academic Editors: Carol Monticelli, Anna Stefańska, Daniel Mateus, Carlos C. Duarte and Nuno D. Cortiços

Received: 7 August 2023
Revised: 18 September 2023
Accepted: 20 September 2023
Published: 22 September 2023

1. Introduction

Lightweight temporary constructions offer potential solutions through innovative recyclable materials, bioengineering principles, and advancements in manufacturing technology, promising more sustainable construction methods [1–5]. The integration of digital modeling and simulation tools in the architectural and engineering domains has significantly enhanced design and analysis capabilities, leading to cost reduction, shorter construction times, and improved building performance and sustainability [5–10].

In temporary constructions, textiles offer several advantages, including lightweight properties, flexibility, and cost-effectiveness [9–11]. Moreover, textiles hold immense potential for optimizing lightweight, structural, and non-structural applications, including canopies, pavilions, and facade elements [8–14]. Notably, projects like CITA's Hybrid Tower, Isoropia, have further expanded computational methods, scale, and structural innovation, bringing prototypes into the realm of large-scale manufacturing and real-world certification [15–17]. These structures are characterized by various benefits that are easy to install, making them ideal for rapid installation and disassembly, thereby addressing the demand for quick, adaptable building solutions [12–14,17–19]. Furthermore, the inherent flexibility of textiles enables their adaptive use in dynamic interactive installations [20–23].

The existing research on textile architecture predominantly revolves around machine-knitted fabrics, with limited attention given to handcrafted woven textiles. Additionally, comprehensive studies on optimizing design and manufacturing processes throughout the lifecycle for mass customization in different spatial types are lacking. Therefore, our research emphasizes the exploration of lightweight textile architectural prototypes based on weaving, aiming to achieve multifunctional adaptability with an efficient sustainable workflow.

Therefore, our research emphasizes the exploration of lightweight textile architectural prototypes based on weaving, aiming to achieve multifunctional adaptability. The design research considers the following five points: Design for Assembly (DfA), Design for Disassembly (DfDisa), Design for Change (DfCh) [24], Material-based Computational Design (MCD) [25], and Circular Economy Design Principles [26–28].

For experimentation, we use the recyclable, replaceable polymer materials provided by PolRe@ Company. This resilient textile material comprises an inner core for structure and an external jacket for cover, protection, and decoration. The tubular textile with an extruded polymer structural core exhibits high mechanical resistance and recyclability. Bending elements were applied with Glass Fiber Reinforced Polymer (GFRP), adhering to common standard sizes while considering the ratio of flexural strength to stiffness on a logarithmic scale. This approach was guided by the principles of "Ashby diagrams" on common building materials with a ratio of strength to stiffness [29]. Furthermore, PolRe@ and GFRP offer a range of sizes and profiles that can be customized and conveniently chosen to align with the unique requirements of building or furniture applications.

The use of elastic deformation in 20th-century architecture, particularly in the context of double-curved shell structures, had several significant implications and applications [30]. Elastic deformation was employed as an economic construction method primarily for double-curved membranes and shell structures.

In more recent times, advancements in simulation techniques have opened up new possibilities in architectural design and construction. These developments have expanded the use of elastic deformation beyond double-curved shells to various other applications in bending-active structures. Generally, examples of bending-active structures include the following [31]:

- Catenoids and grid shells: These are architectural elements with curved or grid-like forms that can adapt to different loads and environmental conditions through elastic deformation.
- Bent structural components with membranes: The combination of bent structural components and tensioned membranes can create innovative architectural forms that respond dynamically to changing conditions.

Today, economic reasons such as advantages in transportation and the assembling process, as well as the performance and adaptability of structures, support the use of active bending. The advantages of bending-active structures lie, however, not only in the possibility of generating complex curved geometries for static structures but also in the shape adaptation possibilities, based on reversible elastic deformation [32]. Summing up, the adaptivity of active bending systems can result from the combination of two main design requirements:

- Adaptivity in construction: Providing the right material properties and a reversible deformation process, active bending may also be used for adaptive structures that can be installed with different sizes and geometries, thus allowing a large tolerance during the construction stage.
- Adaptivity in use: These structures can change shape or move during their service life, in response to external stimuli, providing opportunities for dynamic architectural design.

In addition, the availability of industrial manufacturing processes for semi-finished products like Fiber-Reinforced Polymer (FRP) through methods like pultrusion has made these materials more economically viable for use in bending-active structures. FRP offers lightweight, durable, and flexible properties that are well-suited for such applications.

In order to generate a comprehensive understanding of these potentials, the research follows an iterative workflow of designing, prototyping, testing, and scaling up, employing a systematic approach to continuously improve and refine designs to align with desired objectives and perform optimally in real-world scenarios.

Concept Generation

The Weaving Octopus (WO) concept is inspired by soft-bodied creatures, enabling the structure to adapt flexibly to complex scenarios and meet diverse user needs. The WO prototype composed of skeletons and skin and created a flexibility that can respond in real-time to human behavior and the surrounding environment, adding convenience and vitality to urban corners and vacant lots. This multifunctional prototype can be used individually as decorative furniture, lighting devices, or art installations. Additionally, the WO units can be modularly assembled to adapt flexibly to larger spatial environments (Figure 1).

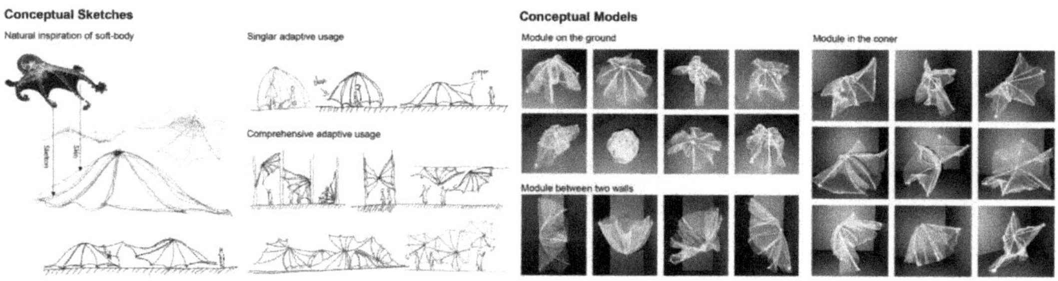

Figure 1. Left: WO conceptual sketches for adaptive use in urban scale. **Right**: WO conceptual physical models applied in different scenarios (on the ground, between two walls, and in the corner).

In the initial experiment, the model utilized only PolRe@ for the skin and skeleton for recycling efficiency. The weaving-based skeleton did not exhibit the desired mechanical behavior. Thus, we introduced another elastic material for the skeleton while retaining PolRe@ for the skin, resulting in a textile hybrid structure. This "textile hybrid" integrates bending and form-active systems based on textile material behavior, leading to a lightweight and self-stable structure [15,33–35].

The concept model (Figure 1) shows the WO prototype's adaptability for different spatial typology, such as squares, walls, and corners, addressing functional deficiencies in urban spaces caused by changing demands and activating overlooked areas.

2. Materials and Methods

This section focuses on the material and methodology employed during the WO prototype design phase. Figure 2 illustrates the workflow of this stage, which involved both physical model testing and digital model simulations to conduct experiments and explorations.

Due to the limitations of the operating area and experimental equipment, a 1:10 scale demonstrator was constructed to test the stability, flexibility, and material usage of different weaving methods for generating WO prototype structures. Additionally, a single-board microcontroller system was integrated into the 1:10 physical model to verify the dynamic interactivity potential of different skin weaving methods.

The form-finding process for construction structures can be accomplished through both physical prototype models and digital simulations. The primary objective is to determine the equilibrium shape of the prestressed mesh, which serves as input for assessing the structural stability under external load conditions [36–40]. The parametric digital model used various computational design tools and platforms for form-finding, weaving algorithm development, and mechanical performance simulations.

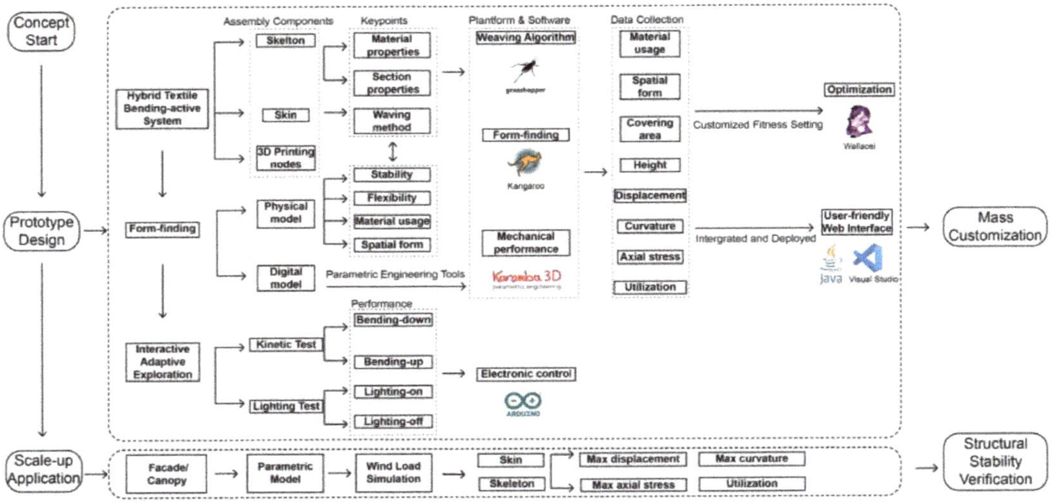

Figure 2. The working flowchart.

Due to limited material mechanical property data for PolRe, we adopted parameters from polyethylene terephthalate (PET) in the actual computational process for form finding and mechanical performance analysis [41]. This decision is based on the similarity between PolRe and PET, as PolRe's core is also made of PET material, albeit with enhanced strength of skin. Additionally, properties for GFRP were obtained by referencing existing research materials, including tensile strength and Young's modulus and shear modulus [42,43].

Data on material usage, covering area, displacement, and other relevant parameters were collected through these simulations. The results were the foundation for subsequent multi-objective optimization, where customized fitness settings were employed to aid in design customization and optimization decisions. Moreover, using Java and Visual Studio allowed the simulation process to be implemented into an interactive web interface, simplifying, and visualizing the mass customization process.

In the scale-up application stage, we conducted simulations using Karamba to analyze the mechanical behavior of a parameterized textile facade composed of WO prototypes under horizontal wind loads. In the specific context of architectural applications, we adapted the dimensions and cross-sections of PolRe and GFRP while incorporating material parameters to perform more precise mechanical analysis calculations. This approach allowed us to theoretically validate the stability of this hybrid textile-bending active structure in the construction context. The detailed computational process and results will be presented in Section 4.3.

2.1. Bending-Active Skeleton

To satisfy high elasticity and strength ratios for skeletons in the WO prototype, the Ashby diagram was utilized to select the appropriate material [29]. GFRP was chosen for its bending-active properties after considering the material's performance and availability. The research on skeleton bending initially began with single-bent rod elements and progressed to multi-bent rod elements (Figure 3). Kangaroo was employed for quick form-finding, while Karamba 3D facilitated accurate Finite Element Analysis (FEA). Ultimately, a planar hexagonal configuration formed by six bent rods was selected as the skeletons' footprints for further investigations.

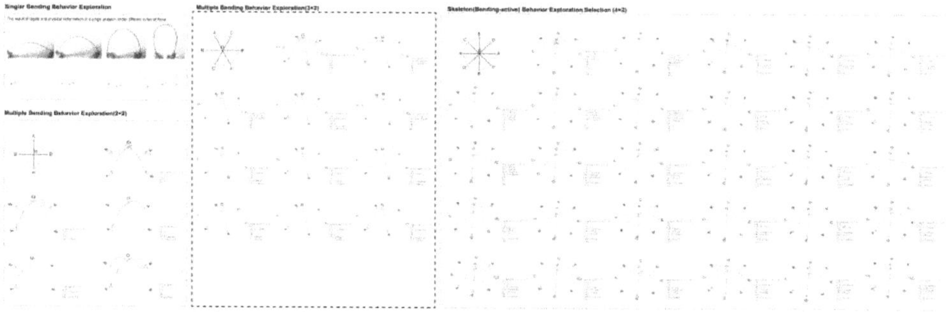

Figure 3. Left up: The physical and digital form-finding experiments for singular skeletons. **Left down**: The digital form-finding experiments of 2 * 2 skeleton. **Middle**: The digital form-finding experiments of 2 × 3 skeleton (selected). **Right**: The digital form-finding experiments of 2 × 4 skeleton.

2.2. Form-Active Weaving Surface

The investigation into the weaving methods for both the skeleton and skin components of the WO prototype was carried out using both physical and digital models. The use of weaving-based surfaces offers advantages in terms of easy assembly and disassembly, as well as reducing the need for additional connection components.

The weaving system on the bending-active skeleton utilized two PolRe@ fibers intertwined with each other. The logic of weaving on the skeleton surface in the physical model was translated into a weaving algorithm in Grasshopper (GH) for further exploration and analysis (Figure 4).

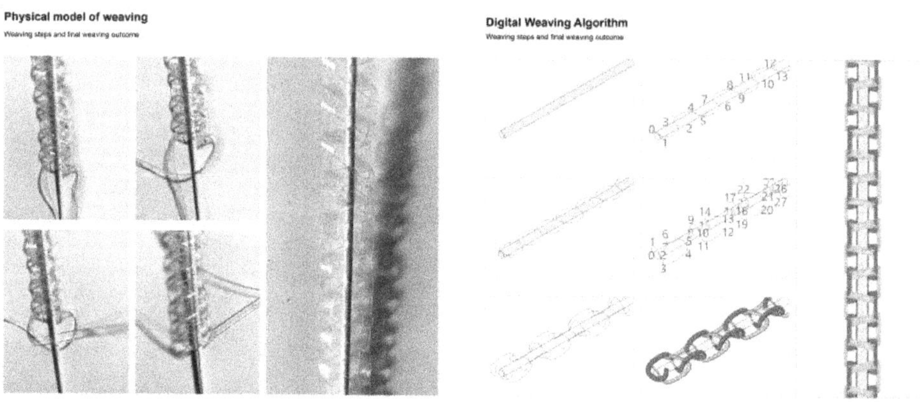

Figure 4. Weaving method on the skin (**Left**: physical model, **Right**: digital model), number in the right figure shows the sequent of the list of the weaving algorithm.

The experiment of form-active weaving skin expanded from 2D diagrams to 3D models, exploring three different weaving methods (A, B, and C). Manual bending experiments were performed on weaving models using methods A, B, and C to test flexibility and stability. The physical and digital models provided data on material usage and form configuration with different weaving methods while maintaining the same bending behavior of the skeleton (Figure 5). After evaluating parameters such as flexibility, stability, material usage, and spatial form, methods B and C were selected for subsequent overall application analysis and optimization experiments.

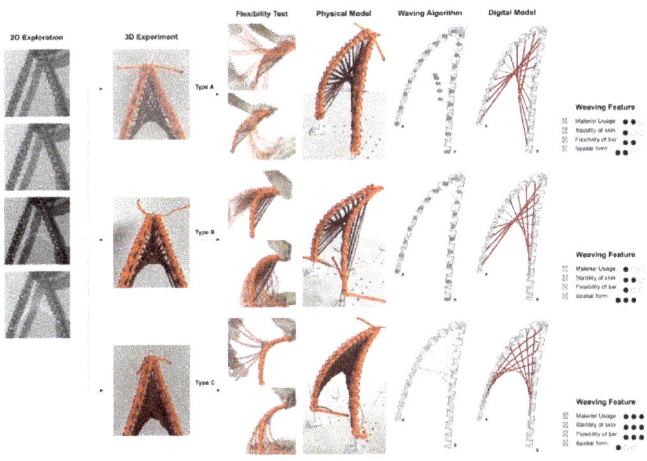

Figure 5. Weaving method on skin with 3 different weaving method A, B and C.

2.3. Computational-Aided Design

2.3.1. Form-Finding and Parametric Process

Parametric control played a crucial role in exploring the morphological variations of the WO prototype. After finalizing the skin's weaving method and the skeleton's footprints, the overall form-finding process was conducted. Figure 6 illustrates the parametric design workflow based on weaving method B.

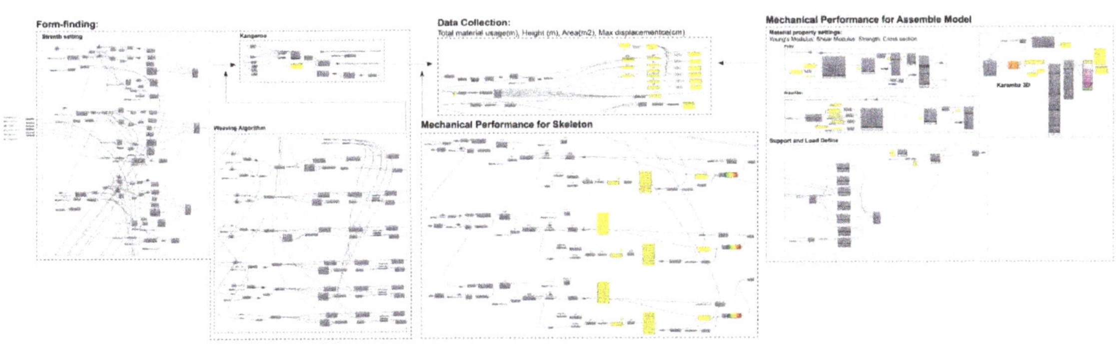

Figure 6. Parametric control flow exampled with weaving method B.

The process started with using Kangaroo for fast form-finding, considering both the skin strain and skeleton elasticity. The next step involved using Karamba 3D to obtain rapid Finite Element Analysis (FEA) engineering performance results and to visualize the assembled WO components. Finally, GH tools were used to collect data on material usage, covering area, height, max displacement, max curvature, and other relevant parameters, which were then utilized in the subsequent optimization and customization process.

The parametric design process enabled the efficient generation of a wide variety of WO units by simply adjusting the input related to the skeleton's load conditions. However, manually adjusting and selecting the WO form consumed significant computational time each time the input was changed. To streamline and expedite the customization process, there was a need for an optimized and user-friendly interface.

The input end of the GH battery pack allows for the adjustment and control of the length of each woven skin. Using the Kangaroo solver, we obtain preliminary form-finding results. This initial form serves as input for Karamba's mechanical analysis, where we collect data such as axial stress displacement and utilization for various WO configurations. Additionally, information regarding the coverage area, height, and material usage can be derived from the form-finding process using the GH base calculator. In the subsequent multiple objective optimizations, we focus on data relevant to design objectives, particularly for temporary pavilions. This includes considerations such as available space, material usage, and maximum displacement. The weights of these three fitnesses are equally emphasized in the objective optimization process.

2.3.2. Multiple Objective Optimization

The optimization process utilized the evolutionary multi-objective optimization engine, Wallacei, within Grasshopper. This plugin allows users to set multiple different fitness criteria to cater to diverse requirements. The optimization process was demonstrated by optimizing the WO prototype as an assembled pavilion. Thus, the main objectives for this optimization were to ensure less material consumption, better mechanical performance, and sufficient available space for people to stay inside. To address these objectives, three constraints and fitness components were set (Figure 7):

- Minimizing total material usage: a sum length of materials for weaving skin used in every 2 skeletons.

$$\text{Total Material usage (m)} = \text{Material usage } (AOC + A'OC' + A'OB + AOB' + BOC + B'OC')$$

- Better mechanical performance: minimal–maximum displacement assessed in Karamba 3D. The maximum displacement in Karamba 3d can be calculated using the following equation:

$$\delta\_max = 5qL4/384EI$$

(δ_max: maximum displacement at the center of the beam, q: uniform load, L: length of the beam, E: Young's modulus of the beam material, I: moment of inertia of the beam's cross-section.)

- Sufficient available space: endure the ratio of the final form's area to its height within the range of 0.058 to 2.444. The available space can be calculated using the following equation:

$$As = |H/A(c) - 0.058| + |H/A(c) - 2.44| - (2.44 - 0.058)$$

$$H/A(c) = H/Area$$

$$H/A(min) = 0.058$$

$$H/A(max) = 2.44$$

(As: available space for fitness 03, A(c): value of current state area, H: height, A: area)

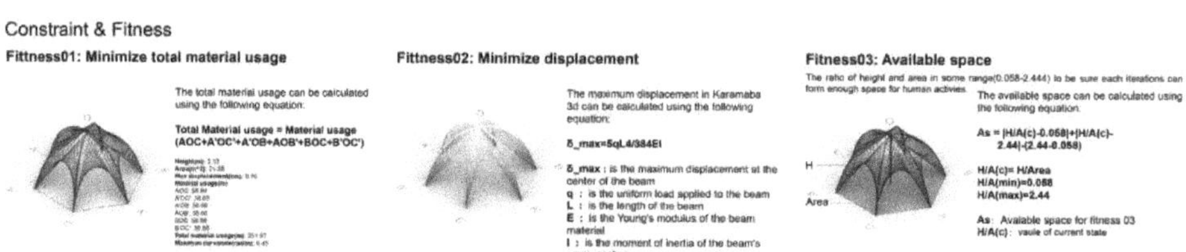

Figure 7. Constraints and fitness setting of the optimization process.

The optimization results will be presented in Section 3, showcasing the iteration of optimization of the WO prototype that meets the specified criteria for an assembled pavilion.

2.4. User-Friendly Website Interface

To enhance user experience and facilitate customization while obtaining essential parameters, we developed a user-friendly web interface. The process involves packaging complex algorithms using the Hops component in Grasshopper for Rhino7.0, developing web page functionality using the Java programming language in Visual Studio Code 2022, adding web components, and utilizing the Rhino Compute API to construct the web interface (Figure 8). Rhino Compute, acting as a Geometry Server, enables seamless integration and communication between Rhino and Visual Studio, ensuring efficient cross-platform compatibility. Through the web interface, users gain the convenience of directly adjusting input parameters. As a result, they receive real-time responses concerning the model and material usage data, which proves instrumental in informed design decision making. The user-friendly web interface thus plays a pivotal role in enhancing the accessibility and efficiency of the design process.

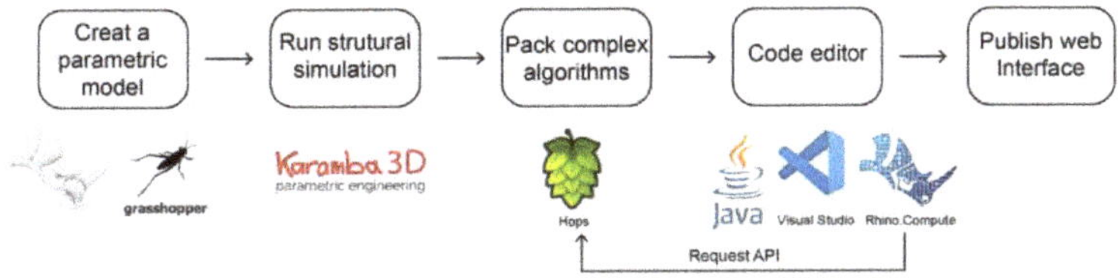

Figure 8. Workflow to build a user-friendly web interface.

2.5. Interactive Electronic Control

The WO prototype further explored the potential of textile hybrid applications through single-board microcontroller kits for electronic interactive control. Utilizing Arduino board and Arduino IDE2.2.1 software, the study investigated dynamic interactivity and illuminated interactions (Figure 9). The experimentation involved various electronic components such as Ultrasonic Sensor, MG996R servo, and LED light sources. By programming the Arduino IDE in C++, the control of lighting switches or servo speed and direction was achieved by reading distance data from the Ultrasonic Sensor via the serial monitor, allowing interactive electronic control of the WO prototype.

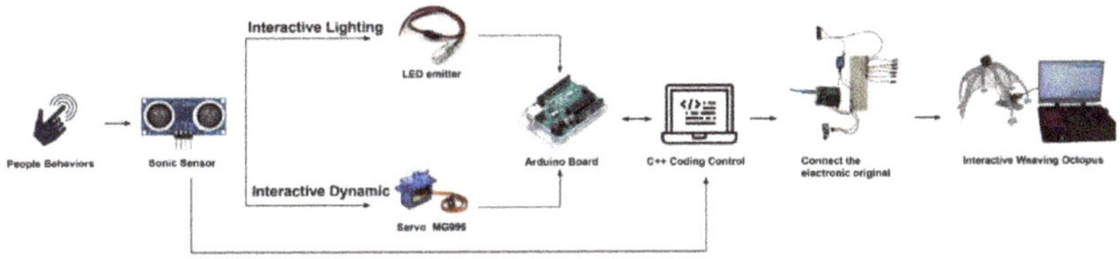

Figure 9. Workflow to integrate interactive electronic control system into the Weaving Octopus prototype.

3. Results

The study developed two prototypes, type B and type C, which applied the different weaving methods mentioned above (Figure 10, Left). We first compare the parameters of the results when the base area formed by the type B and type C is the same, that is, when the fulcrum positions of the skeletons are fixed (Figure 10, Right). From Figure 10, it can be observed that the max displacement formed by type C is about 8.11 cm, which is much larger than type B, which is 0.76 cm. The max utilization of B and C's skeletons are 0.3% and 1.4%. Therefore, B is more stable than C. In addition, the heights of the internal space formed by B and C are 3.13 and 1.9 m, and the total weaving consumables are 351.97 m and 291.98 m. The binding force of C weaving method to the skeleton is not as strong as that of B. We also designed different experiments to further explore the characteristics of these two. According to their characteristics and potential, the specific application of these two prototypes is discussed.

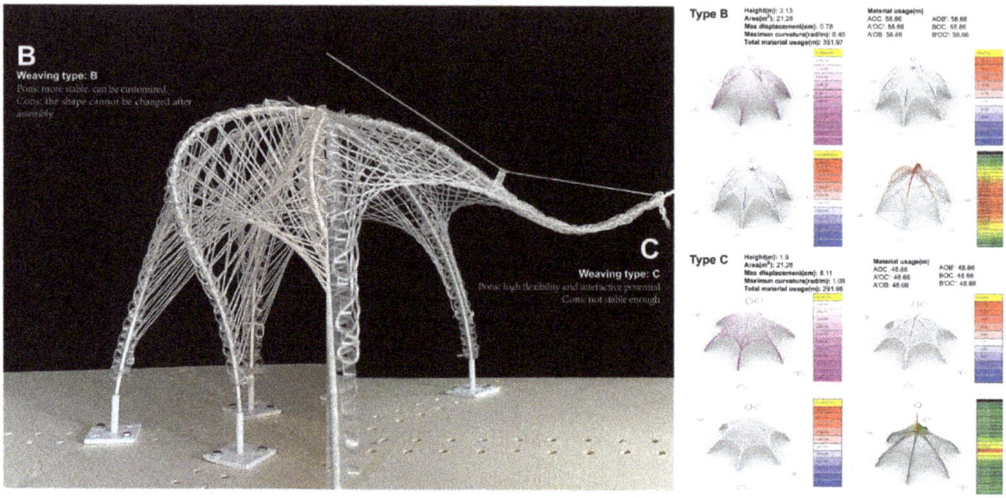

Figure 10. Left: The physical model comparison between prototype B and prototype C. **Right**: The structure simulation comparison of prototype B and prototype C in terms of displacement, utilization, stress, and curvature.

3.1. *Type B Prototype*

3.1.1. Form-Finding and Results

In our physical model testing, we confirmed the prototype's stability relying solely on skeleton elasticity and weaving constraints. To evaluate the overall stability of type B and type C structures, we conducted simulations using Karamba3D. The results, presented in Figure 10 (right) show that the type B weaving method exhibited superior stability compared to type C in maximum displacement and maximum curvature. Type B adopts weaving method B, which can effectively control the relative distance between the skeletons and provide higher stability for the whole structure. However, once this method is used, the result will be a fixed shape, and the skeleton will not have mobility.

In the first experiment of type B, the length of the weaving material between every two skeletons is uniformly shrunk at the same time, and we can obtain the simulation results in GH as follows (Figure 11): Controlling the weaving material consumption can effectively change the shape of the prototype. We control the initial length to one and gradually adjust the ratio of "length" in Kangaroo to shrink the weaving length. Its prototype gradually changes from a relatively flat plane to resemble a cocoon. These weaving structures bind the entire skeleton. With shrinkage, the max curvature in the structure gradually increases (from 0.155 to 0.727), and the max displacement formed by the entire structure is gradually smaller (from 1.84 to 0.488 cm).

To further analyze the prototype, the second experiment is to change the weaving material between the two skeletons unevenly, to observe how the control of weaving will affect the final shape of the prototype. The results are impressive; stochastic control over the weave allows us to obtain more diverse variant states of the prototype. Figure 12 (Figure 12, Left) is only a part of the many variation results, but the actual situation is more abundant. This means that we can exploit this property for adaptive applications of this prototype. Each result gives real-time feedback of the following parameters: the actual weaving consumables between each group of skeletons, the total weaving consumables, the enclosed area, and the height of the internal space. We selected some variation results to further analyze their mechanical properties in Karamba 3D. Taking Figure 12

(Figure 12, Right) as an example, it can form a good internal space for human activities. The total weaving material is 358.46 m; the max displacement is 1.35 cm, which is within the acceptable range. The area and height of the inner space are 18.9 sqm and 3.1 m. That is to meet the scale of human activities.

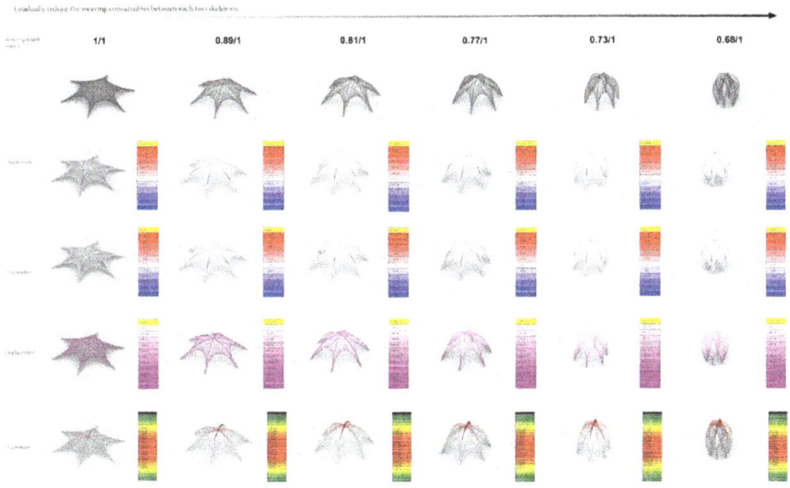

Figure 11. Gradually reduce the weaving consumables between the two skeletons.

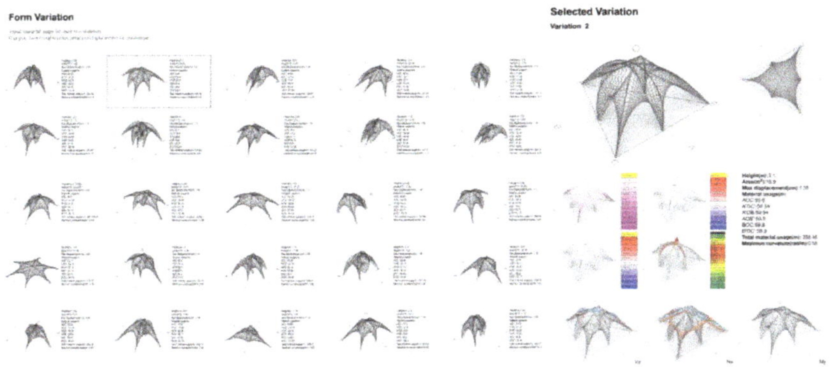

Figure 12. Left: Form variation. Input: material usage for each two skeletons. Output: form/height/area/curvature/displacement of the prototype. **Right**: Taking one of the variations as an example for further analysis.

3.1.2. Optimization and Results

In the above form-finding process, structural stability, weaving consumables, and whether the variation result can form a reasonable space to be used fare the parameters that need to be compared in each variation process. Simple comparisons cannot capture the changing rules of these parameters, so this research assumes whether there are certain specific shapes that can form an available space, making the variation results less consumable and more stable. By using Wallacei, a multi-objective optimization plug-in in Grasshopper, the research obtained the following results.

During the optimization process using Wallacei, we performed a total of 24,000 iterations at the beginning, which took 20 h. After around 4000 iterations, the stated outcome of multiple optimizations comes to the convergence criteria. As the number of iterations increases, the results of the last few iterations are closer to the ideal results in the three

Fitness items (Figure 13: the red line represents the initial iteration, and the blue line represents the last few iterations). The optimization direction is to minimize the weaving consumables, minimize the structural displacement, and ensure that the available space is within a reasonable range.

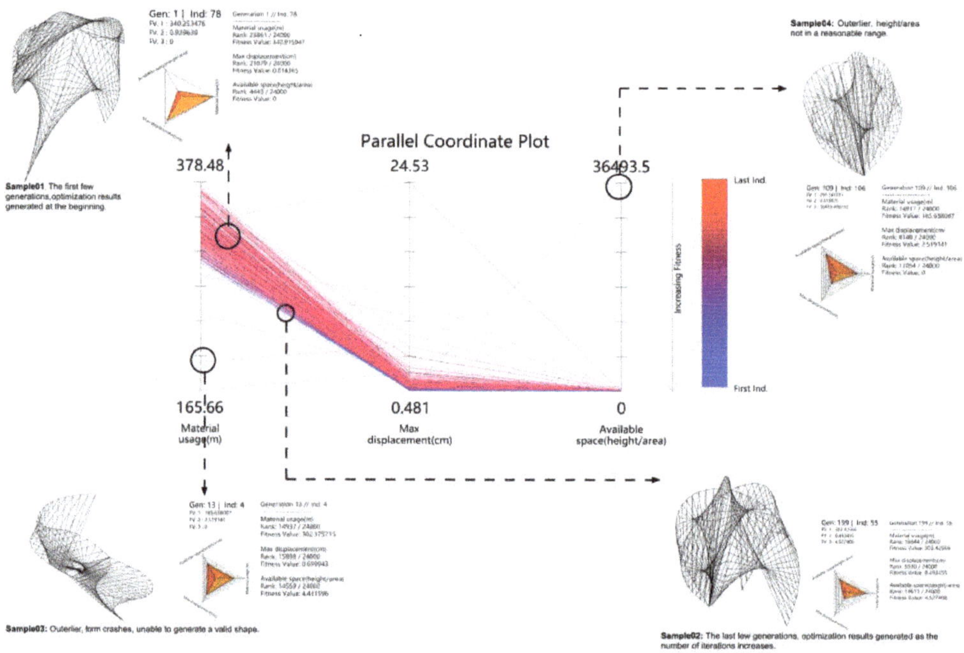

Figure 13. Parallel Coordinate Plot (PCP) graph of optimization results.

By observing all the iteration results, the optimization process controls the available space (Fitness03) so that most of the results are at or close to the XY plane (Figure 14, Left). This means that the value of fitness03 is close to or equal to 0, from above in Section 2.3.2 (Figure 7), that is, the ratio of length to area of each result is between 0.058 and 2.444, which can be regarded as a space that can provide human activities. For weaving consumables, the final optimized results are concentrated between 29 and 305 m (Figure 14, Middle: the blue peak in graph). For the max displacement in the structure, the final optimized results are concentrated between 0.4 and 0.6 cm (Figure 14, Right: the blue peak in graph).

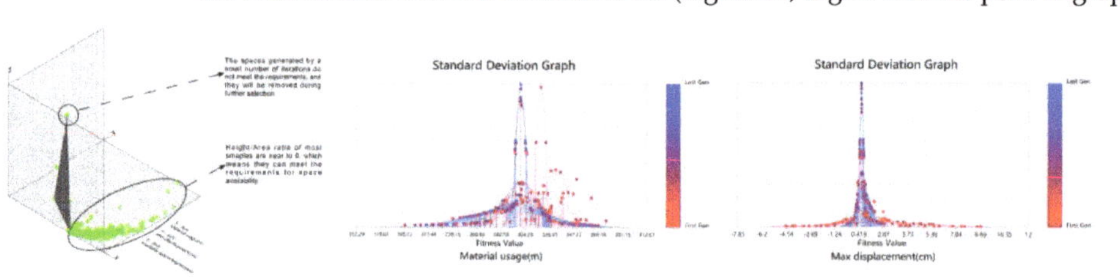

Figure 14. Left: 3D plot graph. X-Fitness01: material usage; Y-Fitness02: max displacement in the structure; Z-Fittness03: available space. **Middle**: standard Deviation Graph of fitness01. **Right**: standard Deviation Graph of fittness02.

In the Parallel Coordinate Plot (PCP) graph, each iteration is a polyline, reflecting the size of its corresponding fitness value. The Parallel Coordinate Plot (PCP) obtained in this study will generally appear in the following situations (Figure 13): sample01 is from the first iterations' result; sample02 is from the last iterations' results. Sample03 shows an unexpected situation, and checking this result found that all the skeletons were flattened on a plane and could not form a valid shape. This may be due to the crash of the Kangaroo solver, so this study will not consider this case. Sample04 exemplifies a situation where the available area is seriously unsatisfied. The foot points of the generated shape all overlap at one point, making the shape unable to form an internal cavity and not enough to provide internal activity space.

Finally, we compare the result of the initial iteration with the last optimization. Taking Generation01 as an example (Figure 15, black line in the left), a total of 96 different results were obtained after deduplication (Figure S1). The shape of the prototype is closer to an exaggerated form. In these cases, the value of the material consumables and the structural displacement are relatively large. Observing the last iteration Generation199 (Figure 15, black line in the right), a total of 17 different results were obtained after deduplication (Figure S2). As the evolutionary direction moves closer to the fitness goal, the shape of each prototype is closer together, forming a similar cocoon shape. At the same time, the material consumables and structural stability are optimized compared to the original generation results.

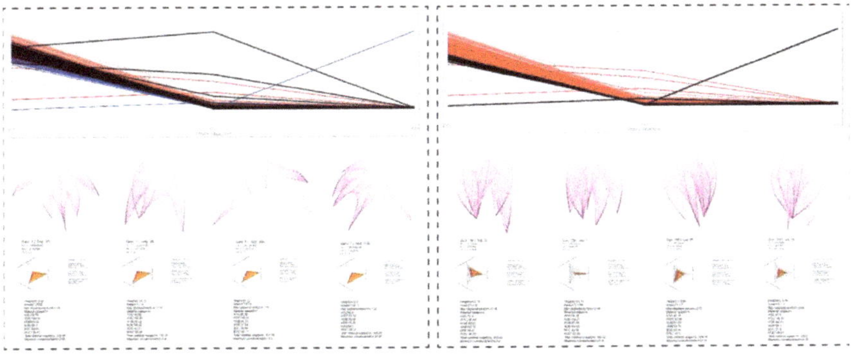

Figure 15. Left: Result from generation 01. **Right**: Result from generation 199.

All the optimization results are recorded in the form of an Excel table in Table S1. In this experiment, some more efficient prototypes were obtained through optimization, which will be an intuitive result in the selection of prototypes. However, considering the practical application, our ideal prototype is not some specific form but a variety of possibilities to meet the different needs in practical applications.

3.2. *Type C Prototype*

Dynamic Behavior Experiment and Results

Type C adopts weaving method C, which provides less stability than method B, but the advantage is that the woven prototype has the possibility of flexible movement with a single skeleton while maintaining its basic shape. Inspired by the structure of human fingers [44], we first restore the movement principle through a physical model (Figure 16, Left): Add a component that can constrain the control line in the middle of the skeleton, and change the lifting and gathering state of the skeleton by shrinking or relaxing the control line at the vertex. The movement of a single skeleton can be easily and flexibly controlled (Figure 16, Middle). Applying it to the entire structure, the simulation results are as follows (Figure 16, Right). We concluded that due to the difference in the weaving method, C provides limited constraints on its skeleton, allowing the skeleton to move freely to take advantage of the bending-active structure. Therefore, it has the potential to become a flexible structure in interactive and robot design [45].

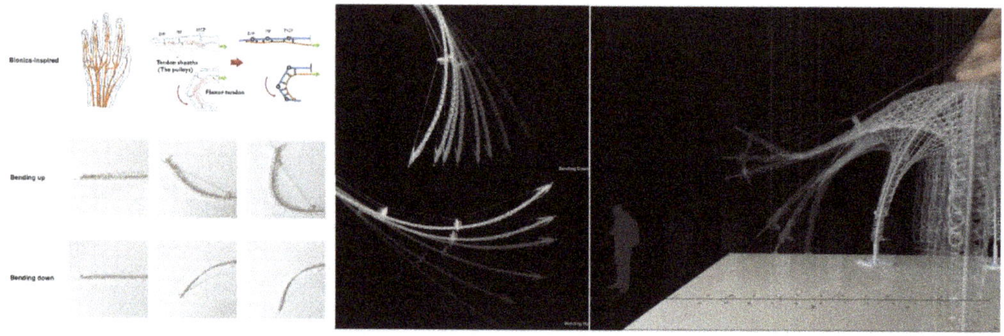

Figure 16. Left: Skeleton movement principle and experiment. **Middle**: Single skeleton movement effect display. **Right**: Physically simulate the motion state of the skeleton on the entire structure.

3.3. Self-Combination Exploration

As an independent unit, WO has good adaptability, flexibility, a changeable shape, and shape controllability. In the modular design, it has the ability of various splicing with other units, which makes it have a very wide application range. The flexible self-combination capability was not discussed too much in this study, but it is worth noting that the prototype, as an independent module, has many possibilities to connect with other modules. The connection methods are mainly divided into the following (Figure 17, Left): two modules share one foot point; two modules can be connected through two different foot points; two modules are connected through up to three different foot points. The connection method and the changeable modules determine that the combination results will have many different changes, reflecting the flexibility and adaptability of the combination structure. In addition, the layout of the plan formed by the aggregation of many modules is also diverse in order to meet the venue construction of different temporary activities (Figure 17, Right). Considering the modular architectural form, this prototype is also meaningful and valuable.

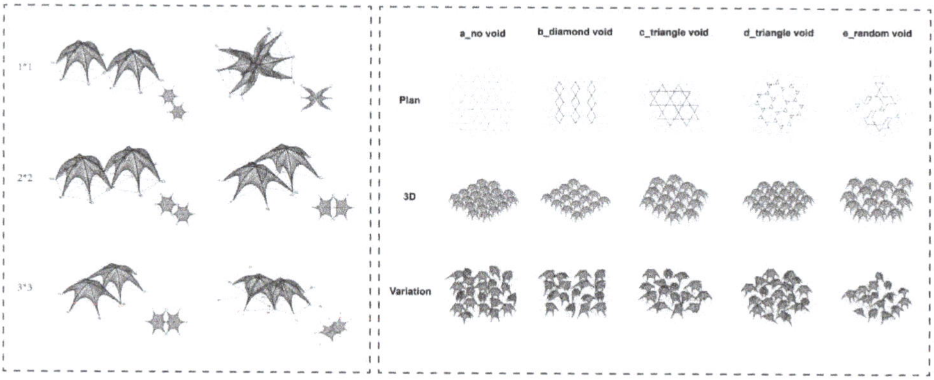

Figure 17. Variations of self-combination and variations of layout configuration.

4. Applications

The WO prototypes under the B and C weaving methods show different mechanical properties and development potential. This chapter will explore the application of WO prototypes based on this research from small-scale furniture decorations to architectural facades to large-scale urban installations.

4.1. Web Interface Assist Customization

With its stable structure and controllable shape, type B can be widely used in light devices that require customization [46]. We have developed a set of user-customized processes in this application scenario (Figure 18): Users adjust the slider on this interactive web to explore their desired shape, and the web page will provide real-time feedback on material usage and mechanical properties to assist in product construction. After the user's order is generated, the manufacturer will send materials and assembly manuals to help the user complete the construction of the product. These materials will also be recycled at the end of the product's life cycle. Variations of the prototype will have different sizes—large, medium, and small,—and they can be applied in various fields such as urban furniture, pavilions, interior decoration, and even clothing (Figure 18).

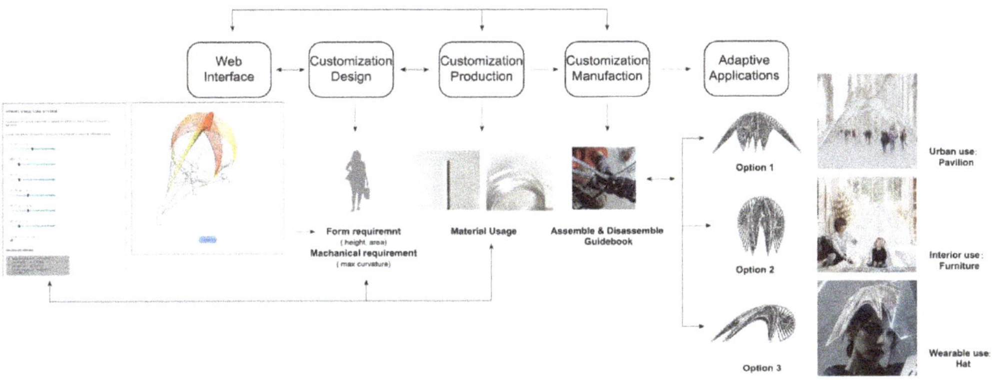

Figure 18. Customization workflow through web interface.

4.2. Interactive Demo

The type C weaving method is more flexible. The application based on this weaving method focuses on exploring the WO prototype for interacting with human behavior and the environment.

The dynamic results presented as photographs (Figure 19, Left, and Right up) demonstrate a demo-like experiment. Due to the size limitation, this model does not use real PolRe$^{@}$ material but uses PVC wire instead. This experimental demo verified that the WO prototype's interactive possibilities could be further improved into a mature product. In addition, the research also introduces the lighting system in terms of interaction (Figure 19, Middle and Right down). The weaving material was changed into an optical fiber tape in this demo, and the light source was at both ends of the weaving structure. In practical applications, materials can be selected based on specific needs, providing greater versatility.

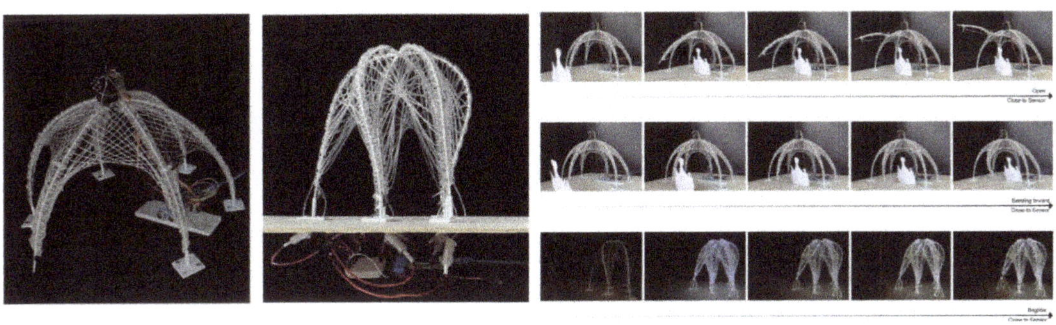

Figure 19. Dynamic and lighting interactive system display.

4.3. Kinetic Façade

This application explores the potential of implementing a dynamic textile facade [47–49] using the WO prototype. As a practical example, we selected the Pedagogy in the Secondary Education School Universidad Católica and integrated the textile hybrid facade comprised of WO units onto the building's elevation (Figure 20, Left). The WO modules can be opened or closed manually or automatically, offering aesthetic enhancements and practical sunshade functionalities. This dynamic feature adds an interactive dimension to the building's facade, accommodating changing environmental conditions and user preferences.

Figure 20. The application of the interactive system on the building façade.

This chapter further verified the feasibility of the prototype in specific scenarios through mechanical simulation of the facade design. In order to simplify the calculation, we cut the simulation range into a building facade of 6 × 6 m. These 12 connected facade elements are considered as a structural whole for calculation. The focus of the simulation is to compare the stress in the skeleton and weaving system of the WO facade under normal state and wind load, as well as the deformation displacement of the whole structure. In the parametric structural engineering tool, Karamba 3D, we preset that the structure consists of three parts: control rope, GFRP skeleton, and weaving system. Detailed material selection and input parameters in Karamba 3D are as shown in Table 1.

Table 1. Material selection and input parameters in Karamba 3D.

Element (Material)	Diameter (d) or Width (w) & Thickness (h) [cm]	Specific Weight Gamma (KN/m^3)	In-Plane Shear Modulus G12 (KN/cm^2)	Transverse Shear Modulus G31, G32 (KN/cm^2)	Yield Strength fy1, fy2 (KN/cm^2)	Tensile Strength ft1, ft2 (KN/cm^2)	Compressive Strength fc1, fc2 (KN/cm^2)
Skeleton (GFRP, orthotropic)	d = 3	E1 = 3656.9 E2 = 1092.4	1026	1026	45	74	−3112.3
Weaving system/Control rope (PET, isotropic)	w = 0.4/w = 0.8 h = 0.1	E1 = E2 = 109.5	52.5	52.5	5.52	5.52	−9.8

During the simulation process, the stress conditions of each component and anchor points are shown in Figure 21, Left. Wind load is calculated with reference to such formula:

$$F = 0.00256\ Cd\ V^2 A$$

(F = wind force (lb), Cd = drag coefficient, V = wind velocity (mi/h), A = projected area (ft^2)) [50].

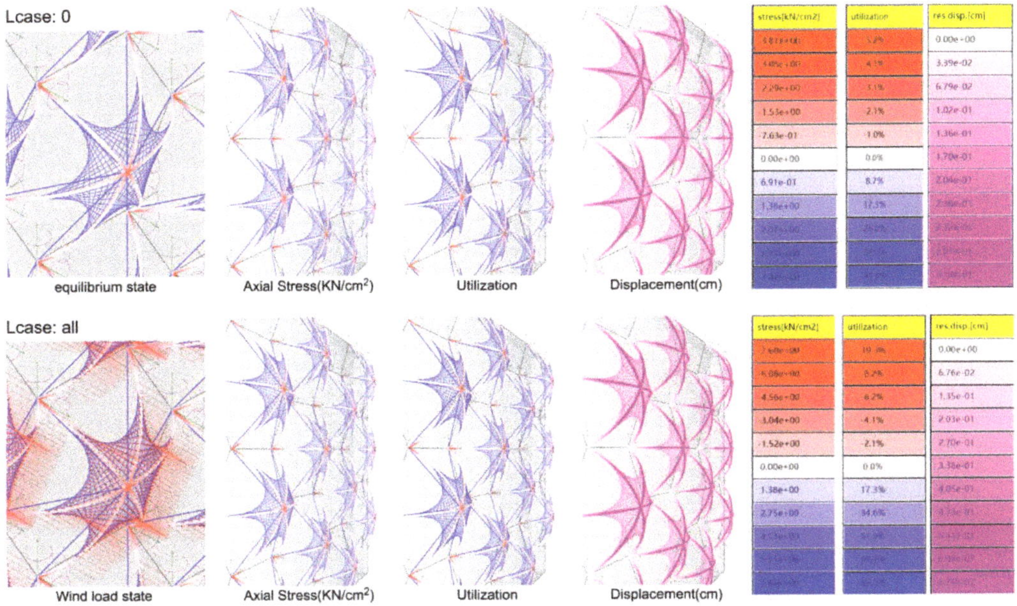

Figure 21. Comparison of the stress on the facade system under normal state (Lcase: 0) and wind load (Lcase: all).

The wind speed of 24 m/s is used to test the designed adaptive structure. This value is obtained from Eurocode EN 1991-1-4:2005 [51], with action on the structures. Wind loads in the simulation will be applied in the direction perpendicular to the building facade. After calculation, the comparison of the value of stresses and the displacement in this structure between equilibrium and wind load can be seen in Figure 21. The maximum utilization in both states does not exceed 100%. Under the action of wind load, the maximum displacement of the structure is approximately 0.676 cm, which is far less than the maximum threshold allowed for structural deformation.

The maximum value of tensile stress in this weaving system is 3.02 kN/cm^2, in control rope is 2.85 kN/cm^2 (Table 2). The yield strength of PET is 5.52 kN/cm^2, which means this weaving system can withstand relatively high wind speeds. The maximum value of the axial stress in the skeleton is 3.47 kN/cm^2 (Table 2). The yield strength of GFRP is 45 kN/cm^2. Through the above simulation and calculation, it can be seen that the structure as a whole can distribute high loads without crushing. It is feasible to apply this structure to building facades.

Table 2. Comparison of results of different structural elements under normal state and wind loads.

Element (Material)	Max Axial Stress (KN/cm^2) Lcase0/Lcase1	Max Bending Moment (KNm) Lcase0/Lcase1	Max Utilization (%) Lcase0/Lcase1	Max Displacement (cm) Lcase0/Lcase1
Skeleton (GFRP, orthotropic)	3.446532/3.477179	±0.095258/±0.9623	0.084179/0.083935	0.3394/0.3363
Weaving system (PET, isotropic)	3.029832/3.023727	0/0	0.765444/0.767842	0.3394/0.3363
Control rope (PET, isotropic)	2.858931/2.856777	0/0	0.408419/0.408111	0.2729/0.2723

4.4. Adaptive Canopy

In the last application exploration, a 1:1 physical model of the WO prototype was fabricated using PolRe@ to explore its adaptive use in different indoor space types. The 1:1 WO model was installed in the open space and stairwell of Polimi Textiles Lab (Figure 22), forming furniture or an interior decoration ceiling for resting or viewing according to different space types. This successful implementation confirms the WO prototype's adaptability for various indoor and outdoor spatial environments.

Figure 22. The 1:1 scaled-up model in Polimi Textiles HUB Lab.

5. Discussion

Developing and designing the "Weaving Octopus" typology presents innovative approaches for sustainable and environmentally friendly textile temporal construction and decoration. However, there are limitations in the research that require further investigation and consideration.

5.1. Contribution

This research thoroughly explores and discusses the textile hybrid structure of bending-active and weaving. Unlike traditional weaving methods, this research creates a new weaving algorithm based on our irregular surface. Different weaving methods employed for the skin result in distinct dynamic behavioral characteristics, allowing users to choose suitable methods based on desired states.

This study develops two weaving methodologies with two distinct characteristics. "Weaving Octopus" (WO) demonstrates strong adaptability, excellent flexibility, ease of self-connection, and a wide range of controllable forms. Notably, WO fulfills the diverse needs of urban blank spaces and finds applications in installation design, interior design, and building facades.

The study establishes a comprehensive research-application workflow that begins with material research and progresses through physical experiments and computer simulations. Different application potentials are then explored based on the prototype's characteristics. The research also culminates in developing a small interactive demo, a user-friendly web interface, and a 1:1 scale physical model to verify the prototype's practical application potential. These experimental results are well implemented in practical applications.

From a life cycle perspective [27], prototypes in this study are sustainable. The woven skin and structural elements minimize connection joints and materials, enabling easy material recovery and recycling. The weaving technique allows convenient assembly and disassembly [26,52], with a parametric design flow ensuring precise control over the final effect and materials [46], significantly reducing waste and time costs. These solutions address the circular economy and environmental challenges of temporary lightweight architecture.

5.2. Limitation

Due to the manual hand weaving involved in this prototype, achieving precise control over the weaving accuracy is challenging. Additionally, the weaving methodology requires users to invest time in learning the techniques, limiting its universal applicability. Possible solutions could involve combining robotic arms for precise weaving manufacturing [53] or integrating virtual reality (VR), augmented reality (AR), and mixed reality (XR) technologies to assist users in achieving precise manufacturing [54].

The practical application of the prototype requires further consideration of specific details. For example, the detailed design of anchor points for outdoor installations and the ability to withstand snow and rain loads require examination. These factors need to be addressed to ensure the structural integrity and performance of the "Weaving Octopus" in outdoor environments.

Parametric tools and optimization software greatly aided the project's research. However, to simplify the problem and facilitate the calculation, the optimization of the type B prototype does not traverse all the possibilities of variations. We think that the current iteration is basically enough for us to observe its optimization direction and draw rough conclusions. The prototype also did not take wind loads and other live loads into account during the Karamba 3D simulations. The mechanical analysis should be recalculated when considering different cases of application.

5.3. Future Work

This paper does not further explore the modular application of this prototype, but this paper presents some inspiring possibilities through simple simulations. In future in-depth research, it is necessary to consider the detailed and feasible connection methods between modules, the overall mechanical conditions of the aggregation results, the effective combination methods, and application scenarios between groups.

While this typology exhibits dynamic and variable potential, currently it lacks responsiveness to climatic conditions. Future research could focus on developing responses to solar forces and rain based on the dynamic behavior of the WO prototype. This would enable the realization of dynamic shading components, solar energy collection, and rainwater harvesting.

6. Conclusions

The research conducted on the "Weaving Octopus" typology confirms the feasibility and adaptivity of a textile hybrid system based on a bending-active structural system, achieved through innovative weaving methodologies. At the construction level, through specific facade applications, we have theoretically demonstrated the structural stability of the WO bending-active textile-hybrid structure. This offers a promising solution to address the diverse challenges faced by the world in terms of environmental concerns and human settlements. Moreover, it contributes significantly to advancing lightweight structural design and promoting sustainable practices in textile architecture.

Supplementary Materials: The following supporting information can be downloaded at: https://www.mdpi.com/article/10.3390/buildings13102413/s1, Figure S1: All the results from generation01; Figure S2: All the results from generation199; Table S1: Wallaice_24000results.

Author Contributions: Conceptualization, Z.C., S.Z., S.V. and A.Z.; Methodology, Z.C., S.Z. and S.V.; Software, Z.C. and S.Z.; Validation, Z.C. and S.Z.; Formal analysis, Z.C. and S.Z.; Investigation, Z.C. and S.Z.; Resources, Z.C. and S.Z.; Data curation, Z.C. and S.Z.; Writing—original draft, Z.C. and S.Z.; Writing—review & editing, Z.C., S.Z. and S.V.; Visualization, Z.C. and S.Z.; Supervision, S.V.

and A.Z.; Project administration, A.Z.; Funding acquisition, A.Z. All authors have read and agreed to the published version of the manuscript.

Funding: This work was supported by DABC R&D founding and TextilesHUB laboratory of Politecnico di Milano.

Data Availability Statement: No new data were created.

Acknowledgments: We would like to thank the PolRe® company and Textiles HUB Lab for their generous material and equipment support, which enabled us to carry out our experiments and prototype development.

Conflicts of Interest: The authors declare no conflict of interest.

References

1. Popescu, M.; Rippmann, M.; Liew, A.; Reiter, L.; Flatt, R.J.; Van Mele, T.; Block, P. Structural design, digital fabrication and construction of the cable-net and knitted formwork of the KnitCandela concrete shell. *Structures* **2021**, *31*, 1287–1299. [CrossRef]
2. De Vita, M.; Beccarelli, P.; Laurini, E.; De Berardinis, P. Performance Analyses of Temporary Membrane Structures: Energy Saving and CO2 Reduction through Dynamic Simulations of Textile Envelopes. *Sustainability* **2018**, *10*, 2548. [CrossRef]
3. Shu, Q.; Middleton, W.; Dörstelmann, M.; Santucci, D.; Ludwig, F. Urban Microclimate Canopy: Design, Manufacture, Installation, and Growth Simulation of a Living Architecture Prototype. *Sustainability* **2020**, *12*, 6004. [CrossRef]
4. Lienhard, J.; Schleicher, S.; Poppinga, S.; Masselter, T.; Milwich, M.; Speck, T.; Knippers, J. Flectofin: A hingeless flapping mechanism inspired by nature. *Bioinspir. Biomim.* **2011**, *6*, 045001. [CrossRef]
5. Meiklejohn, E.; Hagan, B.; Ko, J. Rapid Sketching of Woven Textile Behavior: The Experimental Use of Parametric Modeling and Interactive Simulation in the Weaving Process. *Comput. Aided Des.* **2022**, *149*, 103263. [CrossRef]
6. Sippach, T.; Dahy, H.; Uhlig, K.; Grisin, B.; Carosella, S.; Middendorf, P.J.P. Structural optimization through biomimetic-inspired material-specific application of plant-based natural fiber-reinforced polymer composites (NFRP) for future sustainable lightweight architecture. *Polymers* **2020**, *12*, 3048. [CrossRef]
7. Zanelli, A.; Monticelli, C.; Viscuso, S. Closing the Loops in Textile Architecture: Innovative Strategies and Limits of Introducing Biopolymers in Membrane Structures. In *Regeneration of the Built Environment from a Circular Economy Perspective*; Della Torre, S., Cattaneo, S., Lenzi, C., Zanelli, A., Eds.; Springer International Publishing: Cham, Switzerland, 2020; pp. 263–276.
8. Lüling, C.; Rucker-Gramm, P.; Weilandt, A.; Beuscher, J.; Nagel, D.; Schneider, J.; Maier, A.; Bauder, H.-J.; Weimer, T. Advanced 3D Textile Applications for the Building Envelope. *Appl. Compos. Mater.* **2021**, *29*, 343–356. [CrossRef]
9. Cherif, C. *Textile Materials for Lightweight Constructions*; Springer: Berlin/Heidelberg, Germany, 2016.
10. Monticelli, C.; Zanelli, A. Material saving and building component efficiency as main eco-design principles for membrane architecture: Case—Studies of ETFE enclosures. *Archit. Eng. Des. Manag.* **2021**, *17*, 264–280. [CrossRef]
11. Gasparini, K. Digital Hybridisation in Adaptive Textiles for Public Space. *Textiles* **2022**, *2*, 436–446. [CrossRef]
12. Mazzola, C.; Stimpfle, B.; Zanelli, A.; Canobbio, R. TemporActive Pavillion: First Loop of Design and Prototyping of an Ultra-lightweight Temporary Architecture. In *Proceedings of the TensiNet Symposium 2019 "Softening the Habitats. Sustainable Innovation in Minimal Mass Structures and Lightweight Architectures"*; Maggioli: Rimini, Italy, 2019; pp. 390–401.
13. Ahlquist, S.; Lienhard, J.; Knippers, J.; Menges, A. Exploring Material Reciprocities for Textile-Hybrid Systems as Spatial Structures. In *Proceedings of the Prototyping Architecture: The Conference Papers*; The Building Centre Trust: London, UK, 2013; pp. 187–210.
14. Boulic, L.; D'Acunto, P.; Bertagna, F.; Castellón, J.J. Form-driven design of a bending-active tensile façade system. *Int. J. Space Struct.* **2020**, *35*, 174–190. [CrossRef]
15. Ahlquist, S.; Lienhard, J.; Knippers, J.; Menges, A. Physical and Numerical Prototyping for Integrated Bending and Form-active Textile Hybrid Structures. In Proceedings of the Rethinking Prototyping: Proceeding of the Design Modelling Symposium, Berlin, Germany, 30 September 2013; University of Stuttgart: Stuttgart, Germany; pp. 1–14.
16. Lienhard, J.; Knippers, J.J.J.o.t.I.A.f.S.; Structures, S. Bending-active textile hybrids. *J. Int. Assoc. Shell Spat. Struct.* **2015**, *56*, 37–48.
17. La Magna, R.; Fragkia, V.; Längst, P.; Lienhard, J.; Noël, R.; Šinke Baranovskaya, Y.; Tamke, M.; Ramsgaard Thomsen, M. Isoropia: An Encompassing Approach for the Design, Analysis and Form-Finding of Bending-Active Textile Hybrids. In *Proceedings of IASS Annual Symposia*; International Association for Shell and Spatial Structures (IASS): Madrid, Spain, 2018; pp. 1–8.
18. Kamble, Z.; Behera, B.K. Sustainable hybrid composites reinforced with textile waste for construction and building applications. *Constr. Build. Mater.* **2021**, *284*, 122800. [CrossRef]
19. Thomsen, M.R.; Tamke, M.; Deleuran, A.H.; Tinning, I.K.F.; Evers, H.L.; Gengnagel, C.; Schmeck, M. Hybrid tower, designing soft structures. In *Modelling Behaviour: Design Modelling Symposium 2015*; Springer: Berlin/Heidelberg, Germany, 2015; pp. 87–99.
20. Millentrup, V.; Thomsen, M.R.; Nicholas, P. Actuated Textile Hybrids: Textile smocking for designing dynamic force equilibria in membrane structures. In Proceedings of the eCAADe: Architecture in the Age of the 4th Industrial Revolution, Porto, Portugal, 11–13 September 2019; pp. 521–530. Available online: https://ecaadesigradi2019.arq.up.pt/ (accessed on 6 August 2023).
21. Mankus, M. Tactics for temporary uses of buildings and spaces in architecture. Typology and sociocultural value. *Moksl. Liet. Ateitis Sci. Future Lith.* **2015**, *7*, 132–140. [CrossRef]

22. Ahlquist, S.; Ketcheson, L.; Colombi, C. Multisensory Architecture: The Dynamic Interplay of Environment, Movement and Social Function. *Archit. Des.* **2017**, *87*, 90–99. [CrossRef]
23. Kim, M.-J.; Kim, B.-G.; Koh, J.-S.; Yi, H. Flexural biomimetic responsive building façade using a hybrid soft robot actuator and fabric membrane. *Autom. Constr.* **2023**, *145*, 104660. [CrossRef]
24. Roxas, C.L.C.; Bautista, C.R.; Dela Cruz, O.G.; Dela Cruz, R.L.C.; De Pedro, J.P.Q.; Dungca, J.R.; Lejano, B.A.; Ongpeng, J.M.C. Design for Manufacturing and Assembly (DfMA) and Design for Deconstruction (DfD) in the Construction Industry: Challenges, Trends and Developments. *Buildings* **2023**, *13*, 1164. [CrossRef]
25. Yazici, S.; Tanacan, L. Material-based computational design (MCD) in sustainable architecture. *J. Build. Eng.* **2020**, *32*, 101543. [CrossRef]
26. *ISO 20887:2020*; Sustainability in Buildings and Civil Engineering Works—Design for Disassembly and Adaptability—Principles, Requirements and Guidance. ISO: Geneva, Switzerland, 2020.
27. Roberts, M.; Allen, S.; Clarke, J.; Searle, J.; Coley, D. Understanding the global warming potential of circular design strategies: Life cycle assessment of a design-for-disassembly building. *Sustain. Prod. Consum.* **2023**, *37*, 331–343. [CrossRef]
28. Dams, B.; Maskell, D.; Shea, A.; Allen, S.; Driesser, M.; Kretschmann, T.; Walker, P.; Emmitt, S. A circular construction evaluation framework to promote designing for disassembly and adaptability. *J. Clean. Prod.* **2021**, *316*, 128122. [CrossRef]
29. Lienhard, J. Bending-Active Structures: Form-Finding Strategies Using Elastic Deformation in Static and Kinetic Systems and the Structural Potentials Therein. Ph.D. Thesis, University of Stuttgart, Stuttgart, Germany, 2014.
30. Menges, A. *Emergence: Morphogenetic Design Strategies*; Wiley & Sons: Hoboken, NJ, USA, 2004; AD—Architectural Design; pp. 62–63.
31. Alpermann, H.; Lafuente Hernández, E.; Gengnagel, C. Case-Studies of Arched Structures Using Actively-Bent Elements. In Proceedings of the IASS-APCS Symposium: From Spatial Structures to Space Structures, Seoul, Korea, 21 May 2012.
32. Ahlquist, S.; Menges, A. Frameworks for Computational Design of Textile Micro-Architectures and Material Behavior in Forming Complex Force-Active Structures. In Proceedings of the ACADIA 2013 Conference, Waterloo, Belgium, 24 October 2013; pp. 281–292.
33. Lienhard, J.; Eversmann, P. New hybrids—From textile logics towards tailored material behaviour. *Archit. Eng. Des. Manag.* **2021**, *17*, 169–174. [CrossRef]
34. Lienhard, J.; Ahlquist, S.; Knippers, J.; Menges, A. Extending the Functional and Formal Vocabulary of Tensile Membrane Structures through the Interaction with Bending-Active Elements. In Proceedings of the Tensinet Symposium, Istanbul, Turkey, 2 December 2013.
35. Ahlquist, S. Sensory material architectures: Concepts and methodologies for spatial tectonics and tactile responsivity in knitted textile hybrid structures. *Int. J. Archit. Comput.* **2016**, *14*, 63–82. [CrossRef]
36. Linkwitz, K. Formfinding by the "direct approach" and pertinent strategies for the conceptual design of prestressed and hanging structures. *Int. J. Space Struct.* **1999**, *14*, 73–87. [CrossRef]
37. Linhard, J.; Bletzinger, K.-U. "Tracing" the equilibrium—Recent advances in numerical form finding. *Int. J. Space Struct.* **2010**, *25*, 107–116. [CrossRef]
38. Veenendaal, D.; Block, P. An overview and comparison of structural form finding methods for general networks. *Int. J. Solids Struct.* **2012**, *49*, 3741–3753. [CrossRef]
39. Argyris, J.H.; Angelopoulos, T.; Bichat, B. A general method for the shape finding of lightweight tension structures. *Comput. Methods Appl. Mech. Eng.* **1974**, *3*, 135–149. [CrossRef]
40. Maurin, B.; Motro, R. The surface stress density method as a form-finding tool for tensile membranes. *Eng. Struct.* **1998**, *20*, 712–719. [CrossRef]
41. Mikail, O. Mechanical and Thermal Properties of HDPE/PET Microplastics, Applications, and Impact on Environment and Life. In *Advances and Challenges in Microplastics*; El-Sayed, S., Ed.; IntechOpen: Rijeka, Croatia, 2023; Chapter 5.
42. Gao, H.; Sun, Y.; Jian, J.; Dong, Y.; Liu, H. Study on mechanical properties and application in communication pole line engineering of glass fiber reinforced polyurethane composites (GFRP). *Case Stud. Constr. Mater.* **2023**, *18*, e01942. [CrossRef]
43. Smits, J. Fiber-Reinforced Polymer Bridge Design in the Netherlands: Architectural Challenges toward Innovative, Sustainable, and Durable Bridges. *Engineering* **2016**, *2*, 518–527. [CrossRef]
44. Lurie-Luke, E. Product and technology innovation: What can biomimicry inspire? *Biotechnol. Adv.* **2014**, *32*, 1494–1505. [CrossRef]
45. Schleicher, S. *Bio-Inspired Compliant Mechanisms for Architectural Design: Transferring Bending and Folding Principles of Plant Leaves to Flexible Kinetic Structures*; University of Stuttgart: Stuttgart, Germany, 2015.
46. Bonev, M.; Hvam, L.; Clarkson, J.; Maier, A. Formal computer-aided product family architecture design for mass customization. *Comput. Ind.* **2015**, *74*, 58–70. [CrossRef]
47. Hosseini, S.M.; Mohammadi, M.; Rosemann, A.; Schröder, T.; Lichtenberg, J. A morphological approach for kinetic façade design process to improve visual and thermal comfort: Review. *Build. Environ.* **2019**, *153*, 186–204. [CrossRef]
48. Barozzi, M.; Lienhard, J.; Zanelli, A.; Monticelli, C. The Sustainability of Adaptive Envelopes: Developments of Kinetic Architecture. *Procedia Eng.* **2016**, *155*, 275–284. [CrossRef]
49. Hosseini, S.M.; Fadli, F.; Mohammadi, M. Biomimetic Kinetic Shading Facade Inspired by Tree Morphology for Improving Occupant's Daylight Performance. *J. Daylighting* **2021**, *8*, 65–85. [CrossRef]

50. Kabošová, L.; Kormaníková, E.; Kmet', S.; Katunský, D. Shape-Changing Tensegrity-Membrane Building Skin. In Proceedings of the MATEC Web of Conferences; EDP Sciences: Les Ulis, France, 2020; p. 00046.
51. *EN 1991-1-4*; 2005: Eurocode 1: Actions on Structures—Part 1–4: General Actions—Wind Actions. European Standard Committee: Brussel, Italy, 2005.
52. Cheshire, D. *The Handbook to Building a Circular Economy*; Routledge: New York, NY, USA, 2021.
53. Brugnaro, G.; Baharlou, E.; Vasey, L.; Menges, A. Robotic Softness: An Adaptive Robotic Fabrication Process for Woven Structures. In *Acadia 2016 Posthuman Frontiers: Data, Designers, and Cognitive Machines, Proceedings of the 36th Annual Conference of the Association for Computer Aided Design in Architecture Paperback, Ann Arbor, Michigan, 22 October 2016*; Acadia: Ann Arbor, MI, USA, 2016.
54. Stals, A.; Caldas, L. State of XR research in architecture with focus on professional practice—A systematic literature review. *Archit. Sci. Rev.* **2020**, *65*, 138–146. [CrossRef]

buildings

Review

Carbon Fibre-Reinforced Polymer (CFRP) Composites in Civil Engineering Application—A Comprehensive Review

Dhanasingh Sivalinga Vijayan [1], Arvindan Sivasuriyan [1], Parthiban Devarajan [1], Anna Stefańska [2,*], Łukasz Wodzyński [2] and Eugeniusz Koda [2]

1 Department of Civil Engineering, Aarupadai Veedu Institute of Technology-Vinayaka Mission Research Foundation, Paiyanoor 603104, India; vijayan@avit.ac.in (D.S.V.); sivarvind@gmail.com (A.S.); parthi92bhde@gmail.com (P.D.)

2 Institute of Civil Engineering, Warsaw University of Life Sciences, 02-787 Warsaw, Poland; lukasz_wodzynski@sggw.edu.pl (Ł.W.); eugeniusz_koda@sggw.edu.pl (E.K.)

* Correspondence: anna_stefanska@sggw.edu.pl

Abstract: In civil engineering, carbon fibre-reinforced polymer (CFRP) composites have emerged as a promising alternative to conventional materials. The article provides a comprehensive overview of the application of CFRP composites in various building structural elements and their characteristics and properties, such as their fatigue and corrosion resistance, stiffness and high strength, and incorporation of temperature factors. The advantages and disadvantages of CFRP composites and the current trends and prospects for CFRP composites in the construction sector are discussed. In addition, the article compares various studies on CFRP composites to shed light on their performance and potential limitations. This paper aims to provide useful information to researchers and practitioners interested in using CFRP composites in civil engineering applications. In addition, the article discusses emerging materials in CFRP, such as nanostructured carbon fibres, hybrid fibre reinforcement, and self-sensing CFRP. Additionally, the paper outlines how CFRP composites promote sustainability by increasing structural durability and longevity.

Keywords: civil engineering; CFRP composites; strength; stiffness; corrosion resistance; fatigue resistance; temperature factors

check for updates

Citation: Vijayan, D.S.; Sivasuriyan, A.; Devarajan, P.; Stefańska, A.; Wodzyński, Ł.; Koda, E. Carbon Fibre-Reinforced Polymer (CFRP) Composites in Civil Engineering Application—A Comprehensive Review. *Buildings* **2023**, *13*, 1509. https://doi.org/10.3390/buildings13061509

Academic Editor: Chenggao Li

Received: 14 May 2023
Revised: 2 June 2023
Accepted: 7 June 2023
Published: 12 June 2023

1. Introduction

Carbon fibre-reinforced polymer (CFRP) composites have recently gained popularity in different engineering applications, particularly civil engineering. CFRP composites are ideally suited for civil engineering structures due to their exceptional mechanical properties, high durability, and light weight. There has been a significant increase in the use of CFRP composites in the construction of bridges, buildings, and other infrastructure projects over the past few decades [1]. CFRP composites consist of carbon fibres woven together and then impregnated with a resin to create a strong and durable material. The carbon fibres provide high tensile strength, stiffness, and fatigue resistance, while the resin matrix protects against environmental factors such as moisture and ultraviolet radiation. The resultant composite material is light and has a high strength-to-weight ratio, making it ideal for structural applications [2,3].

An increased strength-to-weight ratio is one of the most significant benefits of CFRP composites. CFRP composites have a much higher strength-to-weight ratio than conventional construction materials such as concrete and steel. This means that CFRP composites can provide the same strength and durability as steel and concrete with less material, resulting in lighter and more efficient structures. This is especially advantageous in applications where weight is a determining factor, such as bridges and tall buildings. Durability is another advantage of CFRP composites [4]. The resistance of CFRP composites to corrosion and environmental degradation makes them ideal for use in harsh environments. In addition, they have a high fatigue resistance, which allows them to withstand repeated loading cycles without degrading [5]. This is especially important for bridges subject to constant movement and heavy loads.

Additionally, CFRP composites are highly adaptable, making them ideal for complex structures [6]. Carbon fibres can be interwoven in various patterns and orientations to produce a material tailored to specific structural requirements. This enables the creation of structures that are both durable and lightweight, as well as aesthetically pleasing [7].

One of the most important factors in CFRP composites' use in civil engineering is their usability in large-scale construction such as bridges [8] or buildings foundation [9]. Bridges are subjected to various loads and environmental factors, making them an ideal application for CFRP composites [10], strengthening existing structures [11]. CFRP composites can reinforce concrete and steel structures, increasing their load-carrying capacity and extending their service life. CFRP composites can also create lightweight, durable, and aesthetically pleasing bridge decks [12]. The beams reinforced with carbon fibre-reinforced polymer (CFRP) in geopolymer concrete (GC) subjected to load-deflection with and without stirrups were investigated by Mehdi et al. (2023) and adopted a bending test of the three-point method under similar conditions. The retrofitted GC beams have a greater capacity for load-carrying, deflection, and ductility than the reference beams without CFRP. The study's experimental and analysed data indicated that the impact of CFRP was more significant in terms of shear strength [13]. Figure 1 illustrates three distinct wrapping methods: side wrapping, U-wrapping, and complete wrapping in CFRP.

Figure 1. Three wrapping techniques for CFRP in concrete (from left): side, U-wrapping, and complete wrapping.

Pan et al. (2018) researched the morphology of magnesium (Mg) alloys and CFRP/Mg laminate alloys to assess their ability to resist galvanic corrosion, as well as their failure mode and interlaminate failure load. The study found that when the silicate is removed from the electrolyte, a pitted oxide film develops from the ceramic oxide film. This pitted film, present in CFRP laminates, significantly improved the peel strength by approximately 0.5 times compared to the oxide film and served as an admirable shield in terms of galvanic corrosion in laminates of CFRP/Mg [14]. Shen et al. (2022) analysed the CFRP grid by considering the cyclic behaviour in aspect ratio, reinforcement ratio, and reinforcement configuration. For this analysis, two reinforced concrete specimens, 13 concrete shear walls, and 11 grided samples with varying proportions of 1.01 to 2.20 are examined beneath the inverted loading. From the conclusion, it has been evident that the reduced aspect ratio will encourage ductility efficiently by 13.8% to 36% and show the finest stress distribution. Samples with the CFRP grid resemble reduced lasting distortion and increased displacement and load-carrying capacity in the concrete shear wall [15]. Sung Won et al. (2022) determined the reinforcement of compression in the bars of CFRP usage in RC columns by experimenting with 24 short columns subjected to elevated temperatures (0, 150, 300, and 450 °C) and concentric loading after the cooling of the column. The experimental result shows that the CFRP bars used in the column subjected to the elevated temperature show increased compressive resistance of 3 to 15% at a lower temperature of 150 °C,

7 to 13.6% at an increased temperature of 300 °C, and 50% at 450 °C [16]. Guo et al. (2023) conducted a study in which full-scale hollow RC box girders with different degrees of damage strengthened using CFRP with prestress of various levels were investigated regarding flexural behaviour. The study involved experiments and finite element analysis (FEA) [17]. Numerical simulations evaluated the flexural behaviour of the girders, including modes of failure, yield and maximum capacities, and deflections. Four box girders were tested. The hollow box girder, which had only minor damage and no reinforcement, served as a control specimen. The other three damaged box girders were reinforced with CFRP, with prestressing levels of 30%, 40%, and 60%, respectively. The findings indicated that the implementation of prestressed CFRP effectively enhanced the yield and ultimate capacities of the box girders. The study also revealed the remarkable strengthening effect of slightly damaged box girders reinforced with prestressed CFRP. CFRP is preferable to other FRP composites due to its superior tensile strength, elasticity modulus, fatigue strength, tensile strength, fire resistance, and chemical resistance. It is ideally suited for structural applications that require stiffness, rigidity, cyclic loading, tensile strength, fire resistance, and chemical resistance. With an improved strength-to-weight ratio, corrosion resistance, fatigue resistance, and customization options, CFRP has undergone significant development. It has found use in the construction of bridges, where it strengthens buildings, makes decks that are lightweight, and increases capacity for carrying loads.

The additional study must address several significant research gaps in the CFRP composites field. First, the long-term resistance to galvanic corrosion and performance of CFRP/Mg laminate alloys must be thoroughly evaluated. This research should consider various environmental conditions and optimise alloy compositions to improve resistance. Second, optimal design parameters and configurations of CFRP grids for various structural applications and loading conditions must be investigated. Examining a broader range of aspect ratios, reinforcement ratios, and reinforcement configurations can enhance the ductility, stress distribution, and overall performance of grid-reinforced CFRP structures. Finally, additional research is necessary to fully comprehend the behaviour of CFRP-reinforced concrete columns at elevated temperatures. This research encompasses various temperature ranges and loading conditions, allowing for a greater comprehension of mechanical properties, fire resistance, and durability. Eliminating these research gaps will significantly advance the knowledge and application of CFRP composites in civil engineering, developing more durable and dependable structures.

1.1. Material Properties of CFRP Composites

The exceptional material properties of CFRP composites make them suitable for various engineering applications. They have high tensile strength, rigidity, and fatigue resistance, are lightweight, corrosion-resistant, can be customised, and have excellent fatigue resistance. These characteristics contribute to the expanding use of CFRP composites in civil engineering applications such as bridge construction, structural reinforcement, and the creation of lightweight and resilient components.

Meizhong Wu et al. (2022) conducted a study to examine the static and fatigue shear behaviours of concrete beams with CFRP strips in place of steel stirrups. The results demonstrated that concrete beams with CFRP strip stirrups exhibit similar static shear behaviour to RC beams. Still, CFRP strip stirrups have superior fatigue life compared to RC beams, which significantly retards the onset of deflection, concrete cracks, and stirrup strain. Using CFRP strip stirrups can enhance the shear resistance of concrete beams under static and fatigue loading [18]. Honghan Dong et al. (2020) studied the corrosion damage caused by a hygrothermal environment to pile foundations. The CFRP-CFST pile is a composite structure comprised of external CFRP sheets and internal concrete-filled steel tubes. A system simulating a high-temperature, humid environment was designed to conduct corrosion experiments with these specimens. The mechanical properties and corrosion resistance increased when the concrete-filled steel tube was externally bonded to CFRP sheets, as determined by the test results. The experimental findings demonstrate that the CFRP-CFST pile is an effective method for protecting piles from corrosion and can be widely used for high-pile wharves in hostile environments [19]. Ananthkumar et al. (2020) investigated the effectiveness of composite materials, such as CFRP and GFRP, in preventing rebar corrosion. They used 300-mm-tall, 100-mm-diameter CFRP cylinders as a wrap and GFRP powder as an additive in concrete. The cylinders were subjected to 60 days of accelerated corrosion using a solution of 0.5 M HCl and 3% NaCl. The corrosion rate was computed, and the results indicated that CFRP and GFRP exhibited excellent corrosion resistance. The appropriate corrosion protection material is proposed [20].

1.2. CFRP Engineering Applications and Future Development in the Construction Industry

CFRP engineering applications and future development in construction and concrete will advance the industry. CFRP composites are used to reinforce, retrofit, and rehabilitate concrete structures, increasing their load-carrying capacity and durability. CFRP's lightweight construction potential makes high-rise buildings and long-span bridges more efficient and sustainable. Future development uses CFRP for unique configurations and innovative structural systems to create resilient and environmentally friendly infrastructure.

1. Construction firms increasingly use CFRP composites for structural reinforcement. They are widely used to strengthen and repair concrete bridges, columns, and beams. CFRP composites increase the durability of concrete structures by increasing load-carrying capacity, flexural and tensile strength, and durability;
2. Advanced retrofitting and rehabilitation techniques will advance CFRP construction. CFRP composites improve infrastructure performance and durability. Damaged structures can meet modern design requirements and withstand higher loads by externally bonding CFRP laminates or wraps to concrete elements;
3. CFRP composites make lightweight construction possible. Their high strength-to-weight ratio makes lightweight structures structurally sound. CFRP composites reduce dead loads on foundations and supporting systems, improving efficiency and sustainability. High-rise buildings and long-span bridges benefit from weight reduction;
4. Future advancements in CFRP in construction will involve investigating novel structural systems. Using CFRP cables, grids, and fabrics, novel structural configurations are created that optimise load distribution, increase structural stiffness, and improve overall performance. These innovative systems offer design flexibility, allowing for the construction of distinctive, visually striking structures with enhanced strength, durability, and sustainability;
5. CFRP construction's future is sustainable. CFRP composites reduce material and energy consumption and prolong the structure's lifespan, promoting sustainability. CFRP composites are lightweight, reducing transportation costs and carbon emissions. CFRP technology will help achieve sustainable construction goals and build resilient infrastructure as it advances.

1.3. Overview of CFRP Composites in Civil Engineering

This review article explores various uses of carbon fibre-reinforced polymer (CFRP) composites in civil engineering. It focuses on their usage in building structural components such as slabs, beams, shear walls, columns, etc. The article begins with an overview of the advantages and limitations of CFRP composites, including their increased stiffness and strength, resistance to fatigue and corrosion, and the influence of temperature factors on their performance. Additionally, the article examines the characteristics and properties of CFRP composites and compares the findings of various studies to provide a comprehensive understanding of their effectiveness in civil engineering applications. The article discusses current trends and future outlooks for CFRP composites in civil engineering. It provides insights for researchers and practitioners interested in utilising these materials in their projects.

1.4. Advantages and Limitations of CFRP Composites in Civil Engineering

Composites of CFRP are progressively used in the branches of civil engineering and structural uses. These materials offer a range of advantages and limitations that engineers and designers need to consider when selecting them for their projects [21,22]. Here are five key points to consider:

Lightweight: CFRP composites are incredibly lightweight and can offer up to five times the strength-to-weight ratio of traditional building materials such as steel or concrete. This makes them ideal for use in various structures and high-rises where weight is a concern.

Corrosion-resistant: CFRP composites have increased corrosion resistance, making them perfect for environmental use when subjected to corrosion and moisture elements. This makes them a good choice for bridges, marine structures, and other applications.

High initial cost: One of the significant restrictions of composites made of CFRP is their high initial cost compared to traditional building materials. This can make them cost-prohibitive for some projects, particularly those with tight budgets.

Brittle behaviour: CFRP composites can exhibit brittle behaviour under certain conditions, limiting their use in applications requiring high-impact resistance. Engineers must carefully consider the application and design for the specific use of the composite.

Limited fire resistance: CFRP composites can have limited fire resistance, limiting their use in specific applications. However, using specialised coatings and other treatments can help improve their fire resistance and make them suitable for more applications.

2. Types of FRP Composites Used in Civil Engineering

Fibre-reinforced polymer (FRP) composites are utilised in various fields of civil engineering because of their increased durability, greater corrosion, and strength-to-weight ratio [23,24]. Here are some common types of FRP composites adopted in civil engineering:

- Glass fibre-reinforced polymer (GFRP);
- Basalt fibre-reinforced polymer (BFRP);
- Aramid fibre-reinforced polymer (AFRP);
- Carbon fibre-reinforced polymer (CFRP).

CFRP composites have been chosen for review due to their widespread use in various civil engineering applications and their potential for future growth [25]. They are increasingly used in applications such as bridge repair and retrofitting, seismic strengthening, and reinforcement of concrete structures [26,27]. Their greater corrosion resistance, improved strength-to-weight ratio, and higher durability make them suitable for these applications [28]. However, their high initial cost and limited fire resistance must also be considered [29]. By reviewing the remarks on using CFRP composites in civil engineering, engineers and designers can better understand their potential benefits and limitations and make informed decisions about their use in their projects. Figure 2 details the various stages of connecting CFRP strips to concrete to enhance strength. Rui Guo et al. (2023) conducted a 403-day experiment on the accelerated ageing of a carbon/glass fibre-reinforced hybrid rod in deionized water at 40 °C, 60 °C, and 80 °C. The results demonstrated that hybrid rods' water absorption and diffusion behaviour conform to a two-stage model, with resin relaxation and interfacial debonding resulting in a decrease in interfacial strength and glass transition temperature. The plasticization effect is reversible with the removal of bonding water after drying, whereas the interfacial debonding is irreversible. Long-term life evaluation revealed that the interface shear strength of hybrid rod shells has a rapid degradation rate and reaches a stable level of 62%, making it the most important design parameter for bridges [30].

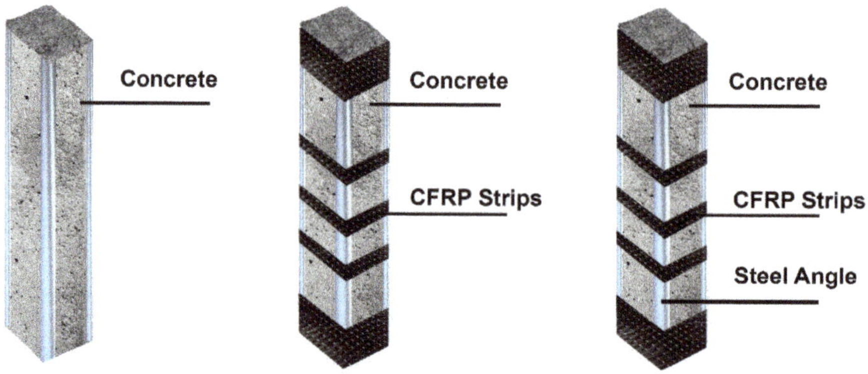

Figure 2. Stages of connecting CFRP strips to concrete for strength enhancement.

2.1. Carbon Fibre-Reinforced Polymer (CFRP) Composites

2.1.1. Application of CFRP to Buildings' Structural Elements

CFRP increasingly strengthens and rehabilitates structural elements such as slabs, columns, beams, and shear walls [31,32]. By adding CFRP to these elements, engineers can improve their capacity for load-carrying, durability, and seismic resistance while reducing their weight and thickness [33,34]. In this way, CFRP is helping to extend the lifespan of existing buildings and increase their safety and resilience in the face of natural disasters and other hazards [35,36].

Beams

In concrete beams, CFRP can be applied as external reinforcement or lamination to improve flexural strength and ductility; reduce the amount of conventional reinforcement required; and delay or prevent premature failure caused by corrosion or other forms of degradation. Ying et al. (2022) experimented with a new signal processing method to improve the damage diagnosis technique of detecting damage using a prestressed NSM CFRP beam transducer. It depends on the greater decomposition of the variational modal of the wavelet-tunable Q-factor transform [37]. An experiment shows that NSM CFRP beams and an analogue signal serve as studies on numerical cases to evaluate the proposed technique. The findings demonstrate that the ambient noise can be eliminated and the damage feature of the NSM CFRP beam can be separated using the variational modal of the wavelet-tunable Q-factor transform, which perfectly leads to damage assessment in the beam. Lu et al. (2022) investigated the peak deflection, failure pattern, and impact force peak value by using a drop hammer in a three-point bending test to determine the behaviour of impact in CFRP-strengthened RC beams (CFRP-RC beams) and reinforced concrete (RC) beams [38]. The developed strain crack was observed using a high-speed camera and the Digital Image Correlation (DIC) method. Varying the hammer height of impact and the bonding of CFRP in the beam leads to different outcomes regarding dynamic response and failure pattern. Thus, the crack will be decreased efficiently by adding CFRP to the beam, and impact velocity will be improved with the reduced inhibition effect. In Figure 3, the application of CFRP reinforcement to a concrete beam is demonstrated using adhesive bonding, which is intended to increase the strength of the concrete.

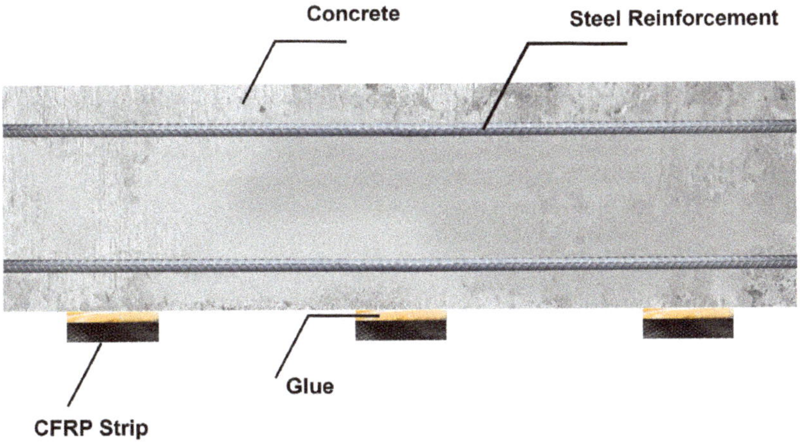

Figure 3. Adhesive bonding of CFRP reinforcement to concrete beam for improved strength.

Zhu et al. (2023) studied the CFRP-strengthened concrete beam using electromechanical impedance (EMI)-based monitoring of the interfacial performance by sustained loading and a wet-dry cycle test. The signal generated by EMI leads to changes in the interfacial performance under monotonic bending. Thus, the EMI method is effective even without EMI data [39]. Qiang et al. (2023) experimented with the three composite beams with two CFRP plates of 3 mm or 2 mm with a prestress of 15% or 10% using the test of 4 bending to determine flexure behaviour in steel composite beams. Before the analysis, the CFRP plate was under supervision for 100 h, leading to 3% strength usage from CFRP [40]. The analysed findings indicate that the CFRP of prestressed beams will significantly develop stiffness, ultimate load, and yielding load, and the ratio of beams strengthened is 80% at BS-1 and BS-3 under proper anchorage. Yu et al. (2022) suggested the bending failure mode using the data of 122 strengthened sheets of CFRP in non-corroded RC beams and 96 strengthened sheets of CFRP in corroded RC beams to evaluate the capacity of bending in the CFRP RC beam [41]. The corrosion degree-based and strengthening ratio-based approaches are suggested to determine the failures of four balanced and five failure modes. The result indicates the proposed failure mode calculation to help decide the bending failure and bending capacity of the RC beams with sheets of CFRP.

Wang et al. (2023) analysed the strengthening behaviour of CFRP-strengthened steel beams by adopting a three-dimensional finite model to encourage the mechanical response under a static four-point bending load [42]. A coded trapezoidal mixed-mode cohesive zone model (CZM) in the user subroutine ABAQUS of UMAT was used to design the failure of ductile adhesive in the strengthened beam using CFRP, and this method is found to be more efficient because of its increased stress distribution, fracture energy, and reduced stress. Guo et al. (2023) discussed the behaviour of load deflection, ultimate debonding, and load capacity by absorbing the 14 beams that contain 1 notched beam, 1 intact beam, and 12 CFRP retrofitted beams under varying temperatures of about 20 °C to 80 °C [43]. Bond slip behaviour was observed for different CFRP strain measurements, and the debonding load varies based on temperature change; that is, it decreased at 80 °C and increased for the temperature from 20 °C to 60 °C, and the energy of interfacial fracture reduced with the improvement of temperature. Liu et al. (2023) proposed the determination of the long-run behaviour of recycled aggregate concrete (RAC) beams with CFRP by adopting the finite element analysis method and evaluating the tendon relaxation, shrinkage, concrete creep, and tension stiffening [44]. The numerical evaluation was carried out depending on the superposition principle to assess the difference in strain and stress. The result shows that the long-term deflection in the RAC-CFRP beam can be reduced by improving the prestress load, and the axial shortening is the primary cause of deflection.

Jin et al. (2022) developed a 3D mesoscale simulation method to identify the failure mechanism of strengthened CFRP RC beams [45]. The stirrup and CFRP fibre ratios were introduced to determine the size effect and shear strength characteristics. The findings suggest that the increased ratio of the stirrup and CFRP fibre will increase strength gain, and the nominal shear strength will be reduced with the improvement of the section size. Huang et al. (2022) developed a ductility controllable device to predict the static and flexural behaviour by using drop weight impact and four-point bending in RC beams of end anchors of type H RC beams prestressed strengthened CFRP [46]. CFRP bars are used in place of reinforcement bars. The device indicates the ultimate resistance, ductility improvement, and overload indication. Multiple impacts were analysed based on the 3D non-linear finite model method. This method uses high-strength CFRP material effectively. Lam et al. (2023) studied the reinforcing methods against carbon fibre-reinforced polymer (CFRP) plate web buckling for steel beams [47]. The researchers used these four single-coped steel beams in the presence and absence of reinforced CFRP for experimental analysis. A CFRP plate can increase the load-bearing capacity, and the reinforcement effect is greatly effective with CFRP layers. It is analysed using a FEM model. Zhan et al. (2023) demonstrated the strengthening effect in conventional RC and CFRP-strengthened beams by adopting the quasi-static loading and drop hammer impact tests [48]. They evaluated the residual load-carrying mechanism based on a high-resolution explicit tool. The findings show that the CFRP sheet will decrease damage in the beam mid-span, residual displacement, and the removal of impact damage. The presence of CFRP will reduce the impact of post-impact energy absorption, residual stiffness, and resistance. The beam can be affected by the greater impact energy.

Hasan et al. (2023) investigated the hybrid method of shear-deficient RC beams to determine the consequences of curing. The shear capacity was reduced by strengthening the beams with CFRP [49]. Thus, the complete coupling of the beam using CFRP will improve the beam's shear performance effectively. Depending on the shear and bending tests, a design strategy was developed to adjust the RC bridge corrosion. Alabdulhady et al. (2022) analysed eight RC-supported beams with one layer of CFRP based on flexure load to determine the failure mode mechanism, stiffness, load carrying capacity, and flexural behaviour with varying strength of compression (fc) of 21.10, 36.10, 48.20, and 68.50 MPa to describe the reduced, normal, and increased strength [50]. The result indicates that the compression strength in concrete was inversely proportional to the CFRP behaviour. The comparison of experimental and ACI 440.2R-17 was within $\pm$16%, respectively.

Using CFRP composites, Lokman Gemi et al. (2019) strengthened prefabricated purlins against shear damage caused by vertical loading. The failure mode of the CFRP-reinforced purlins was dominated by bending damage, and, depending on the CFRP wrapping, the vertical loading capacity was increased by up to 59%. Damage analysis was also performed on the CFRP composite [51]. Yasin Onuralp et al. (2021) investigated the behaviour of prefabricated concrete purlins (PCPs) through numerical modelling using the finite element programme ABAQUS. The parameters longitudinal steel reinforcement ratio, shear friction reinforcement ratio, bending reinforcement ratio, suspension reinforcement ratio, concrete and steel mechanical properties, pre-stressing level, CFRP ply orientation, the number of CFRP plies, and CFRP composite material properties were chosen. Compared to the effects of the parameters related to CFRP, the results of the parameters related to reinforced concrete were found to be minimal. The general FRP layout is proposed to delay or prevent shear cracks in the beams, and numerical analyses validate the proposed layout [52]. Ceyhun Aksoylu et al. (2020) utilised two CFRP applications to reinforce circular-holed shear-deficient beams. The results demonstrated that a D/H ratio of 0.30 decreased load-carrying capacity while increasing ductility. Various configurations of CFRP enhance load-bearing capacity and ductility. At a D/H ratio of 0.64, no CFRP-based strengthening option was effective [53]. Pultruded GFRP composite beams infilled with hybrid fibre-reinforced concrete under four-point loading are analysed experimentally, analytically, and numerically in this work. Experimental variables included pultruded GFRP box profiles, conventional steel bars, hybrid bars, and externally wrapped GFRP. Always use hybrid reinforcements [54,55]. Yasin et al. (2022) examined reinforced concrete beams with circular holes and CFRP-strengthened failures [56,57]. Gemi et al. (2022) examined CFRP-reinforced, shear-deficient, under-balanced reinforced concrete beams with rectangular cross-sections. The ideal strip for wf/sf was defined by rules, but the beam needs to reach 0.82 to attain sufficient shear reserve value [58,59]. Emrah Madenci et al. (2023) examined the elastic properties of textile-based composites with carbon nanotube (CNT) additions. CNT enhanced axial tensile force and bending capacity in experiments, whereas MWCNT raised tensile modulus by 9% [60,61]. The study examines how wrapping the composite beam affects the reinforced concrete beam's shear strength and load deflection. Three-point and four-point loading evaluated nine hybrid beams with varied shear span-to-depth ratios. The composite beam's ductility and strength were greatly improved by the GFRP wraps, which reached beam depth-related levels [62,63].

Columns

The load-bearing and ductility capacities of any structure improve by using CFRP to wrap around the concrete column [64]. This technique enhances the composite material's ability to resist failure and cracking, resulting in a more robust structure [65]. Tang et al. (2022) accompanied the theoretical and analysed determination of axial compression of a concrete-filled double-skin tube (CFDST) stub column curbed with carbon fibre-reinforced polymer (CFRP) with the stainless-steel outer tube [66]. The result depends on comparing the evaluation conducted for the CFRP-curbed CFDST and the square CFDST. In conclusion, it has been proven that the increase in CFRP layers will result in an improvement of 17% in the ultimate bearing capacity of the column. Additionally, a new method based on the twin shear unified strength theory and method of limit equilibrium was suggested to forecast the bearing capacity. Li et al. (2022) described the characteristics of bending and compression by different load eccentricity and slenderness ratios in nine specimens of CFRP-strengthened square concrete-filled steel tubular (SCFST) composite columns [67]. They compared them with SCFST columns without CFRP. The investigation was conducted regarding various aspects such as strength reduction factor, moment-curvature response, longitudinal strain response, peak load, and failure mode. The result is then verified with various existing design approaches, such as the empirical equation, GB50936, and the AISC-LRFD. From the investigational analysis, it has been concluded that the design approach based on GB50936 gives a more accurate value than the other two methods. However, it also shows that this method does not apply to columns with a higher slenderness ratio. The result indicates that improved energy dissipation will lead to an increased ultimate bearing capacity of SCFST-CFRP as compared to that of the SCFST column.

Tang et al. (2023) determined the compression behaviour of a CFRP-curbed concrete-filled double-skin stainless-steel tube (CFDSST) stub column by evaluating the strain response and axial load shortening curves [68]. Typical failure modes were identified by developing the FE model along with the experiment, and the comparison was made with CFDSST without CFRP. The study found that CFRP-imposed CFDSST will efficiently enhance compression behaviour and local buckling restraint compared to CFDSST without CFRP. A proposed model was also suggested based on the principle of superposition to examine the ultimate bearing capacity. This model has been proven to accurately predict the ultimate load. Zhou et al. (2023) experimented with the impact resistance behaviour by comparing the cantilever columns made of RC and CFRP grid-reinforced engineered cementitious composites (ECC) columns [69]. Dynamic behaviour was developed in ECC by using the continuous surface cap model (CSCM). The parametric examination was conducted based on horizontal impact loading and showed that the CFRP grid-reinforced ECC was prone to lower shear failure than the ECC curbed column and had increased shear capacity compared with the RC column. Figure 4 shows how CFRP reinforcement can enhance the strength and durability of concrete columns, increasing their load-carrying capacity and resistance to external forces.

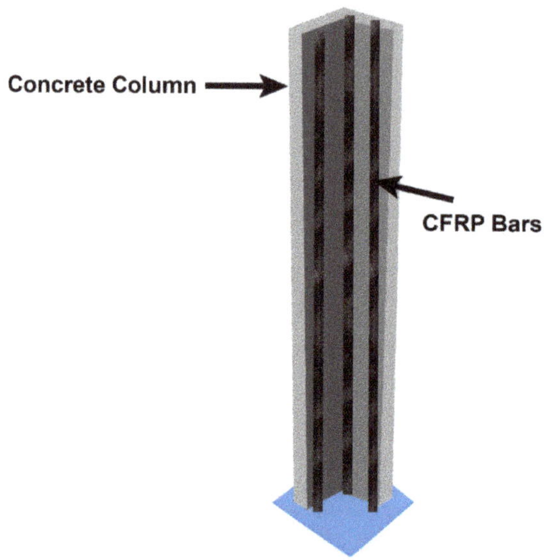

Figure 4. Enhancing strength and durability of concrete columns through CFRP reinforcement for increased load-carrying capacity and resistance to external forces.

Samy et al. (2022) experimented with the load-carrying capacity of a rectangular and square column by adopting a new strengthening technique that enriches the CFRP jacket efficiency covering the column [70]. This 13-column specimen was analysed for varying profile and shape of cross section based on the varying radius of 10 mm, 20 mm, and 30 mm, increased load carrying capacity of 35%, 54%, and 80% with a varying ellipsoidal profile of 10 mm to 30 mm, and increased CFRP jacket by 12%, 29%, and 50%. In conclusion, it has been found that the traditional method of covering the column will increase efficiency by 60%. In contrast, the new technique of CFRP jackets will increase efficiency by 80%. Cao et al. (2023) studied the buckling properties of CFRP strengthened. High-strength steel (HSS) welded T-section column [71]. A sample of an 800 MPA column of T-section HSS welded from six specimens was analysed for axial compression by considering the three crucial factors: width to the thickness of the plate, count of CFRP layers, and slenderness of column. The bearing capacity FE model was described using the ANSYS software for analysis. From the results, it has been concluded that the CFRP layer will successfully decrease the flexural deformation by 56.5%, the torsional deformation by 90.4%, and local buckling of the specimen. The capacity of the bearing of the sample will increase by 21%, 27.6%, and 31.5% for one, two, and three layers of CFRP, respectively. Tan et al. (2023) inspected the flexural strength of the RC column by attaching the prefabricated plates of CFRP vertically and a concrete jacket around the column [72]. Cyclic loading was applied to five different samples to determine the strain distribution, and the cyclic performance of the column was strengthened using CFRP vertical plates. The result shows that the specimen without CFRP will show a two-times lower cracking load and fail in the compression area by developing the plastic hinges compared to those with CFRP. It has been declared that load-carrying capacity will increase by 17% and 15% based on 2-layer and 1-layer CFRP plates attached to the column compared to the standard column, respectively.

Chen et al. (2022) experimented with the axial compression behaviour test of the CFRP steel tube column with 18 specimens, including the nine that contain the CFRP internally, based on the factors of number and position of CFRP [73]. Samples of 139 axial compressions were gathered and analysed to determine the equation, and a FEM model was created to compare the results of the equation. The analysis proved that 7.3%, 12.55%, and 10.6% increased the ultimate load capacity of the specimens with the growing count of CFRP layers. Mohammed and Abebe (2022) determined the finite element analysis (FEA) for determining the blast resistance for RC columns confined with CFRP and the standard RC column [74]. Studies have been carried out in terms of considering the reinforcement detail schemes, the height of bursts, concrete compressive strengths, 0/90 CFRP-strengthened RC columns, and standoff distances. From the result, it has been identified that the smaller the scaled distance, the larger the lateral displacement and failure of shear, and the increase of layers of 0/90 CFRP will result in a smaller blast.

Rodriguez et al. (2021) experimented with the beam-column connection using two different beam-column joint systems. Initially, it was tested without damage [75]. Later, two rehabilitation techniques were conducted using ultra-high-performance mortar with steel fibres and carbon fibre-reinforced polymer connected externally. Various loadings were used to detect load-deformation capacity in terms of equivalent viscous damping, drift capacity, energy dissipation, cracking, and visual damage. The findings show that in columns C1 and C2, the capacity of load carrying developed after rehabilitation by around 15% and 20%, followed by decreased ductility ratios in 21 and 30% of the specimens. For the equal drift ratio, the damage indices were found to be low; for the 3% ratio, the damage indices were found to be 0.68 for rehabilitated specimens and 0.94 for real specimens. Xiaong et al. (2023) studied the ultimate and yield capacity of the concrete-filled steel tubes (CFST) stub columns of preloaded circular columns strengthened with CFRP [76]. Forty-eight models were assessed for parametric study to obtain the accuracy and reliability of the sample by comparing the FE model analysis and theoretical equation with the experimental results, and it was identified that the effectiveness of strengthening is greater for ultimate strength and lower for yield strength when the CFRP is used to strengthen the precast CFST stub column.

Slab

CFRP can enhance the flexural strength and structural integrity of concrete slabs by bonding them to the underside, thus reducing deflection [77]. Türer et al. (2023) analysed the performance of CFRP strips in strengthening flat slabs by considering the strain behaviour, energy dissipation capacity, maximum bearing capacity, and initial stiffness as the parameters [78]. They developed a model using ABAQUS; for this, nineteen slab samples were created to determine the position and size, placed adjacent to the column, and strengthened using CFRP strips with fan-type anchors. After the analysis, the load-carrying capacity was improved by 50% in the presence of anchored CFRP strips. Zhou et al. (2023) determined the usage of CFRP sheets and a self-locking device to strengthen the RC slab at the end anchorage to arrest the deboning [79]. A four-point bending test was performed for six one-way RC slabs with CFRP bonding, which leads to increased ultimate load and the bonding being improved efficiently by 46% by adopting the hybrid anchored (HA) CFRP method of strengthening. With the increased length of the bond, the usage of the CFRP rate will fall to 28%, and it requires more examination in cases of strengthening with CFRP sheets.

Yazdani et al. (2021) experimented with the enactment of prestressed self-consolidating concrete (SCC) slabs that are reinforced with carbon fibre-reinforced polymer (CFRP) sheets [80]. Factors such as energy absorption, force-deflection curves, and cracking behaviour were considered for this study. The analysis shows that the presence of a single CFRP layer sheet will improve energy absorption and flexural strength by 71% and 30%, respectively, and decrease the width of the crack by 23%. In contrast, using two-layer CFRP will not effectively perform in the case of flexural capacity. When there is an improvement in the eccentricity ratio, the load-bearing capacity also improves by 80%. Thus, the findings suggest combining a single CFRP layer improves ductility, load-bearing capacity, and deferred deboning because of less tensile strain and cracking. Azevedo et al. (2022) determined the fire resistance of reinforced concrete (RC) slabs analysed with carbon fibre reinforced polymer (CFRP) strips by using three various methods: continuous reinforcement embedded at the ends (CREatE), externally bonded reinforcement (EBR), and near-surface mounting (NSM) and compared the results obtained [81]. The CREatE method shows greater fire resistance for about 24 min than EBR with 2 min and NSM with 16 min without protection. The critical temperature of glass transition varies based on the modulus curves for CREatE, NSM, and EBR as 3.0 Tg, 1.0 Tg, and 2.5 Tg, respectively. The depiction in Figure 5 shows how CFRP reinforcement can be used in slabs to improve the capacity of load-carrying, prolong the service life of existing structures, and enhance the performance of new ones.

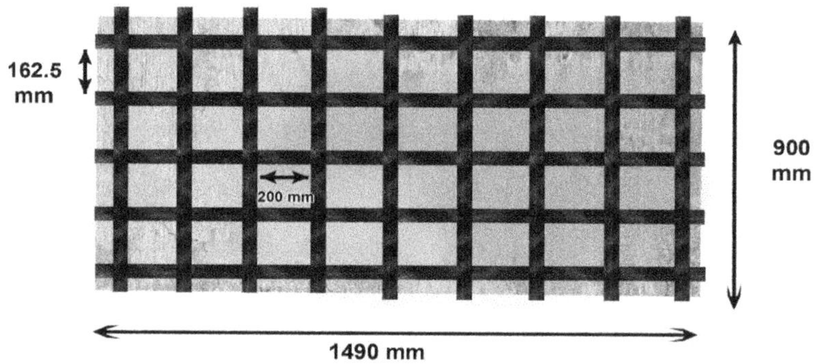

Figure 5. Application of CFRP reinforcement in slabs.

Yang et al. (2023) experimented with the resistance to blast in the presence of CFRP-strengthened RC structures depending on the coupled Lagrange–Euler (CLE) method [82]. Parameters such as charge weight, retrofitted scheme, retrofitted material, and retrofitted were considered to determine blast resistance ability in the air-baked slabs retrofitted with CFRP. The analysis suggests that the present CFRP will reduce the flying debris and deformation in the underwater explosion, thereby showing better blast resistance. Assad et al. (2022) experimented with the numerical method that arouses the fire by using two various methods, namely externally bonded (EB) and near-surface mounted (NSM), and compared the experimental data derived by using a three-dimensional non-linear finite element model (FE) to verify the structural and thermal behaviour of the RC slabs with CFRP [83]. From the analysis, it has been identified that the NSM method indicates the best result in terms of fire resistance and static loading, with an improvement of 15% of the ultimate load compared with the EBR method.

Bielak (2023) presented a test setup to determine if the dowel behaviour of CFRP grids matches the real method, and the impact of pre-strain, pre-crack width, and shear crack in the tension region based on the effective area was analysed [84]. Based on the test setup, the conclusion has been drawn that the major load will depend on the formed crack pattern. Ghayeb et al. (2023) studied the influence of using CFRP sheets in the tension region of two-way RC slabs [85]. Punching shear was developed using the static load in the strengthened and unstrengthened slabs. The test result shows that the sample with a CFRP-strengthened slab offers better shear resistance, increased ductility and load displacement, and increased punching capacity. The presence of CFRP will improve the ultimate strength by 30% and the ductility by 50 to 73% compared to the sample without CFRP. Breveglieri et al. (2021) analysed the RC structures in the laboratory and the environmental exposure method [86]. In the outdoor method, prestressed slabs with CFRP strips and non-prestressed slabs were kept in the environment for four years. The same has been tested in the laboratory for a certain period to determine the ultimate slab strength. From the analysis of the two observations, it has been observed that long-term exposure does not impact the capacity of the strengthened slab to carry loads.

Alrousan and Alnemrawi (2022) analysed the reinforcement performance of the bridge's deck under the consequences of reinforcement with CFRP and ASR damage [87]. The NLFEA technique is used to predict the experimental data with reinforcement ratios

(0.38, 0.46, and 0.57%) and the stages of first, second, third, and without ASR damage. From the technique, it has been found that the ratios of 0.38% and 0.46% will result in increased capacity of load carrying, durability, and serviceability, a reduced rate of 2 and 3 levels of ASR damage, and a linearly reduced relation between the ASR level and the energy of the slab.

Shear Wall

Bonding CFRP to the surface of shear walls can enhance their resistance to seismic forces, shear capacity, and stiffness, increasing safety and durability [88]. Shen et al. (2022) inquired about the distortion action of concrete shear walls by adopting a method depending on the uniaxial shear-flexure model (USFM) [89]. By using this method, it has been found that there is increased lateral load due to the variation between cyclic and monotonic loading, with a varying peak load from 4.9% to 10.8%. The new suggested method accurately identifies the peak load point, strength, initial stiffness, and displacement response. The result indicates that the flexural component shows 65% failure of total deformation compared to the slip and shear components, and further analysis is required for the other components. Sahebjam and Showkati (2016) examined the effect of polymer fibre by applying quasi-static cyclic loading to perforated carbon fibre-reinforced polymer–steel composite shear walls [90]. Four single-bay and single-story perforated shear walls and three composite shear walls with an aspect ratio of 1.33 and varied fibre direction were analysed by considering the ductility, load-carrying capacity, and stiffness by a hysteresis curve. The curve shows that the fibre in the tension area will affect the factors of stiffness and ductility but neglect the fibre orientation.

El-Kashif et al. (2019) determined the shear behaviour of RC shear walls undergoing lateral load by adopting the FEM in ANSYS [91]. A model of seven walls from two specimens was developed and investigated using reverse cyclic and monotonic loading to set the concrete confinement, ductility, strength of flexure, and shear capacity. The result shows that using fibre-reinforced polymer (FRP) material will efficiently remove the shear failure caused by its brittle nature. Comparing experimental and numerical results will lead to a similar conclusion.

Altin et al. (2013) discussed the shear deficiency in RC walls on hysteresis behaviour by adopting one specimen without retrofitting and four samples with retrofitting along with CFRP strips of aspect ratio 1.5 and testing based on the parallel strip, horizontal, and X-shape to increase the ductility and strength [92]. From the analysis, the horizontal stripes contribute to the development of plastic hinges because of flexural hysteresis, and the X-shape indicates an unsuccessful premature shear wall. Using CFRP strips increases displacement capacity and limits shear crack development. Hatami et al. (2012) experimented with the non-linear shear behaviour of composite steel shear walls reinforced with carbon fibres (CSSW) and steel shear walls (SSW) [93]. The properties of fibre content and panel width were investigated both experimentally and numerically. The findings suggest that the increased fibre will lead to higher stiffness, carbon absorption, capacity, and strength with reduced ductility values and a greater influence on SSW.

In contrast, a wider panel width will improve CSSW and SSW behaviour. Yu et al. (2023) experimented with bolted connection laying in CFRP steel composite shear wall by adopting the cyclic load on samples of two 1/3-scale steel plate shear wall (SPSW) and related them to the pure SPSW by finite element analysis to determine energy dissipation capacity, stiffness, and the bearing capacity of ultimate and found them to be higher by 4.2%, 7.7%, and 23% [94]. The aspect ratio, CFRP layer, and angle orientation were identified and calculated. The suggestion indicates that the improved angle orientation increased the capacity of energy dissipation and ductility. The aspect ratio to be used should lie between 1 and 1.5 for the CFRP composite shear wall.

Meftah et al. (2006) determined the seismic analysis of 20-story RC coupled shear walls subjected to 3 earthquakes with varying fibre [95]. Adhesives and adherents are used as shear wall members and are analysed using the mixed finite element method. The lateral deflection due to fibre arrangement leads to dynamic behaviour change, which is also analysed in the RC coupled shear wall. Furthermore, an investigation is required to strengthen the RC couple shear wall. Huang et al. (2020) analysed about six shear walls in which one of the wall sample is a wall of RC reinforced with steel and the rest is a CFRP-reinforced grid in the horizontal and vertical configuration under various factors such as reinforcement configuration, aspect ratio, and horizontal reinforcement ratio with various cyclic loadings [96]. This specimen undergoes compression diagonally, and the aspect ratio ranges from 1 to 1.4. The result shows that the horizontal CFRP grid arranged horizontally indicates increased concrete confinement, decreased deformation, and increased shear resistance. A model named Truss Arch was created to determine the shear capacity of the CFRP gridded shear wall.

Yu and Zhu (2023) developed a composite sandwich plate specimen made of polyethene terephthalate (PET), steel plate shear wall (SPSW), and carbon fibre reinforced polymer (CFRP) and compared it with non-stiffened traditional SPSW for hysteresis behaviour and the mechanism of failure by performing the quasi-static test [97]. The comparison suggests that the sandwiched plate's stiffness and energy dissipation are improved by 34% and 38.44% for the drift load of 2%, respectively, and the plane deformation decreased by 93%. Thus, the sandwiched CFRP corrugated plates provide a better anti-buckling position.

3. Characteristics and Properties of CFRP Composites

3.1. Durabilityh of CFRP

3.1.1. Corrosion Resistance

CFRP is a corrosion-resistant material due to its inert properties, high strength, and resistance to harsh environments [98]. Wrapping CFRP around corroded structures can enhance their durability and lifespan, reducing the need for costly repairs and replacements [99]. Karim et al. (2020) conducted a 13-week salt spray corrosion test to identify the effectiveness of various rivet Zn-Ni and Almac coating types in corrosion behaviour and the degradation of joint strength in CFRP/Aluminium joints [100]. The experimental result shows that self-piercing rivet (SPR) joints with Zn–Ni-coated joints will deliver three times lower strength losses when compared with Almac-coated joints, increasing the corrosion resistance more effectively than that of Almac. Huang et al. (2023) experimented with the bonding behaviour of concrete and corroded steel with CFRP cathodic protection [101]. Tests have been conducted considering the bond performance and three ICCP current densities, including linear polarisation, pull-out, induced current cathodic protection (ICCP), and accelerated corrosion tests. The experimental result indicates a gradual increase in bond strength by 20% with a decrease in the ratio of pre-corrosion by 50.7% and ICCP current density. Shao et al. (2023) studied the characteristics of horizontal bearing and the life and durability of the CFRP composite piles [102]. For this, the durability stage is separated into the starting stage of reinforcement corrosion and the propagation of the crack. It is analysed using the Monte Carlo simulation method and chloride diffusion model, whereas thick-walled cylinder theory is used to identify the concrete cracking layer. With a different ratio of replacement mode, a pile bending test is carried out. Findings show that for CFRP composite piles, the lower the reduction ratio, the higher the replacement ratio of flexural stiffness. Meanwhile, when there is an increment in the replacement ratio of CFRP reinforcement, there will be a decrement in the shear force and the moment of pile bending.

Ren et al. (2022) researched the fatigue properties and mechanical effects of aggressive corrosion in CFRP strand sheet/steel double strap joints with different bond lengths [103]. By adopting a quasi-static tensile protocol, the specimen is exposed to the environment for 24, 48, and 72 h, and the bond strength, stiffness, and ultimate tensile strength are calculated. The analysis showed that there is an increment in fatigue performance and a reduced rate of corrosion with the presence of an increased length of the bond. The specimen exposed to 72 h in the corrosive environment with improved bond length leads to improved strength of 11% and stiffness of 14%, which are higher when compared with 24 and 48 h. observation.

Wu et al. (2021) discussed the properties of glass fibre sheets (GFS) on the bond between CFRP and steel and their corrosion behaviour [104]. Experimental tests such as fatigue, a static test to determine the bond behaviour, and an accelerated corrosion test to assess the efficiency of GFS were conducted in three different types depending on the presence and absence of the GFS. The result suggested that the GFS declined the fatigue bond performance and increased the behaviour of the static bond, which led to the presence of GFS as an adhesive layer that will reduce galvanic corrosion in the CFRP steel bonding.

Hu et al. (2022) used carbon fibre-reinforced polymer (CFRP) for its better electrochemical and mechanical properties as an anodic component in impressed current cathodic protection (ICCP) [105]. Measurement of electrochemical properties was conducted for 19 RC cylinders in terms of ICCP densities of current (5, 20, and 80 mA/m^2) degrees of pre-corrosion (3%, 6%, and 12%) by using different methods of protection such as CFRP wrapping, epoxy coating, and non-protection. It has resulted in ICCP and CFRP wrapping being much more efficient than epoxy coating for the suitable current density. For the pre-corroded sample of 3%, the density of current is 5 mA/m^2 and for 6 and 12% of the pre-corroded sample, 20 mA/m^2 is optimal. Thus, CFRP wrapping could decrease the transmission path of steel and increase corrosion resistance by thickening the concrete core.

Chen et al. (2022) discussed the mechanical and corrosive behaviour of riveted joints and CFRP/aluminum stacks using the cyclic salt spray test [106]. Joints are subjected to an environment conducive to corrosion for four weeks. It was found that when there is a development of interfacial corrosion in CFRP/Al stacks, there is a gradual decrease in failure displacement of around 14 mm with a 3% reduction in maximum load, which shows the behaviour of durability and corrosion in the rivet joints of CRP/Al stacks. Wang et al. (2022) experimented with three samples that included a chloride attack of 2 years in combined piles of RC (AC piles), CFRP-bonded RC piles (CP piles), ordinary piles of RC (UC piles) for detecting the free ion chloride concentration (Cf) profile [107]. For this, the indoor test was conducted by creating a marine simulation system, and the piles were tested using a numerical model depending on Fick's II law. The result shows that the Cf of ordinary RC piles will be higher than that of CFRP-bonded piles, which has a positive effect on decreasing the chloride content and sheltering the pile from external chloride attack. Sou et al. (2021) studied mechanical degradation in CFRP/Al bolted joints with epoxy and PVC film corrosion protection [108]. The specimen is exposed at 30 degrees Celsius to a 3.5% sodium chloride solution for about 0 to 8 weeks. Double-lap bearings and material were analysed using a step profiler and micro-CT. From the analysis, it has been concluded that the epoxy coating joints show more favourable protection as that PVC film shows more failure at joints and thus results in the epoxy coating being a better protective covering with less deterioration to damage.

3.1.2. Fatigue Resistance

Mohabeddine et al. (2022) suggested a new approach based on fatigue cyclic degradation to determine the life of fatigue in CFRP-bonded retrofitted metallic detail patches [109]. Degradation of the fatigue cycle was considered for the new fatigue damage accumulation model approach. As per the new approach, it has been identified that the increased stress level led to a decreased extension ratio because of the effect of fatigue loading. These approaches can be used effectively using the finite element method. Doroudi et al. (2021) experimented with numerical and practical strategies for determining the behaviour of CFRP-cracked steel plates under crack loading [110]. Four CFRP-strengthened specimens and one cracked steel specimen without CFRP were analysed for the conduct of fatigue in terms of CFRP-to-steel bonded joints, fatigue-life extension, and failure modes under the loading of fatigue behaviour. The CFRP-strengthened plate shows higher fatigue life depending on the developed bond slip model than the specimen without CFRP.

Gadomski and Pyrzanowski (2016) discussed the difficulty in finding the CFRP structure deformation at the fatigue destruction timing, leading to the decrement of structure stiffness [111]. The moment has been analysed for this analysis, which gives an inappropriate result. A new approach based on the electrical resistance has been suggested by measuring it under varied static and periodic load conditions. The bending moment, eight voltages, and centre deflection of the probe were determined. As per the experimental and numerical conclusions, the change in electrical resistance gives the best result in CFRP fatigue destruction of structure compared to the bending moment. He et al. (2022) studied the failure mechanism and fatigue behaviour in CFRP/Al single-lap joints depending on the digital image correlation (DIC) system to record the fatigue process [112]. Under varied reliabilities, fatigue behaviour was discussed based on the well bull distribution theory. It was found that the bonding portion stiffness in the lap joint is four times greater than that of other parts. A crack was initially developed at the Al lap end with 70% and 50% stress levels, and the stress level was reduced with the increment of adhesive failure proportion.

Li et al. (2022) inspected the fatigue properties of CFRP plates [113]. The experiment utilised four kinds of stress fatigue and strengthening configurations at five different levels of corrosion damage in terms of crack propagation, fatigue fractography, and fatigue life. The result shows that the presence of CFRP will increase fatigue life and decrease crack growth in the corroded steel plates. The development also leads to an extension of fatigue greater than 85.3 times in the corroded steel of the unpatched plate and two times in the uncorded steel plate. Li et al. (2022) presented the experimental and theoretical analysis of flexural behaviour, loss on prestress, and interfacial stress of obtaining prestressed carbon fibre reinforced polymer (CFRP) used in steel structures by applying 25% prestress with a solution of 3.5% NaCl in the sample [114]. It has been proven that prestressing CFRP will suspend the bonding at the interface by decreasing the interfacial stress, resulting in a reduction of loss due to prestress of 2.7% in the sample subjected to wet/dry cycles (WDCs).

Lesiuk et al. (2017) experimented with fatigue cracks by comparing CFRP patches in the beam of mild-rimmed steel and puddled iron used 100 years before [115]. A hybrid approach was suggested for determining the fatigue crack based on energy dissipation. It has been evident that the growth of fatigue increases in the steel used earlier compared with the modern steel. The suggested approach results in the use of CFRP for strengthening the fatigue crack, which is appropriate for old structures, and further investigation is required to evaluate the optimal conditions of the CFRP patches.

Vavouliotis et al. (2011) determined the fatigue loading effect in quasi-isotropic carbon fibre-reinforced laminates (CFRs) by absorbing the electromechanical response [116]. The epoxy matrix with multi-wall carbon nanotube (MWCNT) was examined and related to epoxy CRP at three stress levels. These are then associated with acoustic emission and stiffness degradation to predict the occurrence of damage. The presence of a characteristic damage state (CDS) associated with electrical resistance leads to a reduction in stiffness, initially indicating the rest of life independently from the applied stress level and showing an increased confidence coefficient (R2). Kotrotsos et al. (2023) determined the behaviour of delamination on fatigue resistance in the CFRPs by adopting two modes (Mode I and Mode II) under different loading conditions and adopting the fatigue onset life test based on Paris Law with varied displacement and constant amplitude [117]. From the investigation, it has been shown that the BMI resin-modified CFRS offers better resistance against the delamination that occurs.

3.1.3. The Incorporation of CFRP in Temperature Factors

CFRP is known for its excellent thermal conductivity and low thermal expansion, making it a valuable material in temperature-sensitive applications [118]. Adding CFRP to structures exposed to extreme temperatures can reduce thermal stress, enhance stability, and improve overall performance [119]. Li et al. (2023) studied the moisture–heat coupling effect in nano-SiO_2 adhesive specimens and the CFRP-steel lap joint bonding using scanning electron microscopy (SEM) [120]. Due to the increased ageing in a 25 °C water bath, the adhesive glass transition temperatures (Tg,s, Tg,t) became reduced because of the variation in the temperature. The reduced effect of bonding occurs due to the reduction of shear s73 strength. Once the water bath is carried out, the interface toughness is reduced, reducing shear stress transfer. Al-Abdwais and Al-Mahaidi (2022) experimented with the effectiveness of modified compendious material to withstand the load at increased temperatures depending on the presence of CFRP reinforcement [121]. The beams are tested under constant service loads and elevated temperatures until failure occurs. These are then compared with the numerical value. The analysis shows that the epoxy adhesive will perform less well than cementitious adhesive at an increased temperature.

Yoo et al. (2022) determined the bond strength in ultra-high-strength concrete (UHPC) with CFRP and compared it with residual steel bars [122]. A 150 mm × 150 mm × 150 mm specimen was tested at elevated temperatures of 150 °C and 250 °C and at ambient temperature. The findings show that increased thermal temperature leads to decreased bond stress. Bond strength improved by 9% when subjected to a 250 °C temperature that was elevated. After analysing the experimental results with the CMR model and the BPE model, the bond-slip is increased in CMR than in BPE, indicating that the bond stress of residue in CFRP bars in UHPC meets the requirements mentioned in ACI 440.6 M after heating. Wang et al. (2023) determined the analytical solution for debonding CFRP to the steel/concrete interface to detect the thermal effects based on thermal and mechanical loads [123]. The proposed analytical solution is examined for CFRP thickness, CFRP elastic modulus, and thermal effects with four different forms of experimental data. The conclusion shows that the suggested solution effectively determines the response of debonding. The bond length is strongly linked to thermal effects when using stiff and thick CFRP materials to strengthen the concrete. Kaiser et al. (2022) discussed the failure mechanism and mechanical performance at different temperatures, such as elevated temperature, cold temperature, and room temperature, of titanium adhesive tubular lap joints (TLJs) and thin-walled CFRP [124]. Experimental design, a static tensile test, and finite element analysis were conducted to detect the damage mode in detail. The findings show that at elevated temperatures, a failure and the delamination of CFRP that occurred at the inserted end are formed because of shear stress when titanium deformation occurs. Thus, the adhesive bond damage mechanism's behaviour is affected majorly at high cold and hot temperatures.

Peng et al. (2023) evaluated the specimens of three concrete-filled square steel tubulars and the specimen of eight concrete-filled square CFRP-steel tubulars (CF–S-CFRP-ST) regarding compressive strength and a layer of CFRP [125]. Depending on the displacement curve of shear force, elastic stiffness and load-carrying capacity can be improved by increasing compression strength. Improving CFRP layers leads to a change in elastic stiffness and shear capacity. The experimental value is then evaluated using ABAQUS, and the formula for assessing the capacity of shear of CF-S-CFRP-ST is well correlated with the experimental value. Jahani et al. (2022) analysed and compared 23 RC beams of near-surface mounted (NSM) carbon FRP (CFRP) for the dependent based on time behaviour [126]. Temperature (20 and 50 °C), steel reinforcement ratio, and CFRP strengthening area are the factors considered for evaluation. Based on the deflections that show the temperature increase will not have any influence.

In contrast, the strengthening area has less effect on the deflections based on the dependence of time, which is identified experimentally depending on the age-adjusted effective modulus method (AEMM). Alkhawaldeh and Al-Rousan (2022) developed 12 Rc beams grouped under three forms of four joints at ambient temperature and temperature ranges of 400 °C and 600 °C [127]. Adding the CFRP layers one by one as one, two, and three are subjected to a quasi-static loading test. Structural performance depends on parameters such as stiffness degradation, energy dissipation, ductility of displacement, load of horizontal lateral displacement, and displacement ductility. After the experimental analysis, it has been proven that the presence of CFRP layers plays efficiently with the increase in heat damage at the RC beam-column joint, which leads to improved load capacity, increased lateral displacement, improved dissipation of energy, and delayed secant degradation of stiffness.

4. Emerging Materials in CFRP Composites

4.1. Nanostructured Carbon Fibres

Nanostructured CFRP carbon fibres are being studied for concrete reinforcement. Nanometer-diameter fibres have high tensile strength and modulus. CFRP-concrete composites may benefit from nanostructured carbon fibres. Yanming Li et al. (2019) investigated how nano-SiO_2/carbon fibre reinforcement affects oil well cement performance to improve its adaptability to oil well pressure. Compressive, tensile, modulus of elasticity, deflection, pull-out, and bridging effects improved [128]. Pitcha Jongvivatsakul et al. (2022) examined whether CNTs can improve concrete-CFRP bonding. Epoxy with 0.5% SWCNTs and 1.0% MWCNTs improved bonding strength, ultimate slip, effective bond length, bond stress-slip relationship, interfacial fracture energy, and crack formation [129]. Yuhang Du et al. (2022) studied the preparation, dispersion, change laws, and effect mechanisms of the dynamic compressive strength of modified carbon nanotube-fiber reinforcements (MCNF). Carbon nanotubes are easier to deposit on the negative electrode, and MCNF dispersion in an alkaline environment increases with polycarboxylate superplasticizer content. MCNF concrete had 14.0–35.5% higher dynamic compressive strength than untreated concrete, peaking at 0.3% MCNF content [130].

4.2. Hybrid Fiber Reinforcement

Hybridization creates high-performance composites by combining fibres. Hybrid fibre reinforcements such as carbon-glass or carbon-aramid are being investigated for CFRP-concrete applications. This approach uses complementary fibre properties to optimise composite performance. Rajai Al-Rousan et al. (2021) used the CFRP sheet to reinforce RC beams internally. The study assessed the CFRP sheet's flexural performance and efficiency as primary or supplemental flexural longitudinal steel reinforcement. Internal strengthening significantly improved most parameters [131]. Milad Abolfazli et al. (2023) assessed the bond strength of fibre-reinforced polymer (FRP) tubes and seawater sea sand concrete (SWSSC) following exposure to varying temperatures. The testing of 27 samples included glass FRP, carbon FRP, and hybrid glass-carbon FRP. The tubes with the strongest and weakest bond strengths were GFRP and CFRP, respectively, when exposed to elevated temperatures [132].

4.3. Self-Sensing CFRP

Self-sensing CFRP composites monitor strain and damage. Conductive fillers such as carbon nanotubes or graphene can give CFRP composites electrical conductivity. Measurements of electrical properties can detect structural damage or deformation in real-time. Using electrical insulation techniques, Sang-Hak Lee et al. (2020) propose and fabricate a CFRP-based shape memory alloy hybrid composite (SMAHC) beam with embedded shape memory alloy (SMA) actuators. Self-sensing-based deflection control was well controlled, but there was a slight delay in response time. Future research will concentrate on eliminating this delay [133]. Akira Todoroki et al. (2014) proposed a self-sensing time domain reflectometry (TDR) method for CFRP plates that uses a narrow-strip line to determine the transverse location of the damage. The findings apply to damage monitoring [134]. Pyeong-Su Shin et al. (2023) compared the self-sensing of CFRP and dual fibre composite (DFC) in a three-point bending test. The change in CFRP's electrical resistance (CER) trend was similar to that of dual fibre composite (DFC), but the interface between two fractured CFs had a greater impact. The interface between CFs in composite materials impacted the self-sensing of CFRP [135].

4.4. High-Modulus Carbon Fibres

Traditional CFRP materials use standard-modulus carbon fibres with 200–250 GPa tensile modulus. However, 300-GPa-plus high-modulus carbon fibres are being developed. High-modulus fibres stiffen concrete structures, improving load transfer. Isamu Yoshitake et al. (2020) investigated near-surface-mounted (NSM) strengthening for wheel-loaded cantilevered-reinforced concrete (RC) bridge deck slabs. Ultra-high-modulus (455 GPa) CFRP rods were used to strengthen bonds. Monotonic and cyclic loadings tested 15 RC beam (160 mm) specimens. The flexural fatigue test showed that the strengthened beams survived 2 million cycle loadings [136]. Nitin Lamba et al. (2023) investigated the mechanical properties of high-strength concrete with recycled CFRP fibres. The hardened properties of concrete test samples were determined using the Ultrasonic Pulse Velocity Test (UPVT) and compressive strength tests. Crystalline materials were characterised using X-ray diffraction analysis, while mechanical properties were evaluated using linear regression analysis [137].

5. Promoting Sustainability through Enhanced Structural Durability and Longevity

Incorporating CFRP composites into concrete promotes sustainability by increasing structural durability, decreasing energy demands, reinforcing existing structures, and decreasing cement consumption. These characteristics contribute to sustainable building practises and are consistent with efforts to fight climate change. Zhang et al. (2019) evaluated the mechanical performance and efficiency of epoxy-coated CFRP-reinforcement laminates in various configurations of recycled concrete beams. The outcomes demonstrated enhanced ductility, load-bearing capacity, fracture toughness, and fracture energy. Utilising digital image correlation (DIC), strain evolution and fracture propagation were captured. There is a need for pilot tests to increase confidence. High-quality recycling programmes should be implemented [138]. Chen Xiong et al. (2021) examine the viability of collaboratively using recycled CFRP fibre-reinforced rubberized concrete (RFRRC). The results of the experiments indicate an increase in compressive strength, ductility, flexural toughness, impact resistance, and energy absorption capacity. The ecological evaluation reveals a decrease in CO_2 emissions and an increase in mechanical properties [139]. Sebastian George Maxineasa et al. (2015) evaluate and compare the environmental performances of an unreinforced reinforced concrete (RC) beam with those of various CFRP flexural strengthening techniques using the Life Cycle Assessment (LCA) methodology. The results indicate that the environmental impact of all evaluated CFRP strengthening solutions is significantly less than that of the RC beam. The cement and steel reinforcement manufacturing phases have the greatest environmental impact. The paper concludes that the use of composite materials can contribute significantly to the sustainable development of the building industry [140].

According to Prathamesh Khorgade et al. (2022), CFRP is an excellent substitute for conventionally used steel in the construction industry. This study examined the environmental impact of two bridge systems, Rosensteinsteg II and a flyover over German highway A-20 and found that cradle-to-gate CO_2 emissions were reduced by 28% and 18% for conventional building materials and CFRP, respectively. This suggests that prestressed CFRP bridges are the most environmentally viable option, particularly for bridges that do not experience heavy vehicular loads [141]. Zhuo Tang et al. (2020) studied the compressive behaviour of geopolymeric recycled aggregate concrete (RAC) encased in CFRP jackets. Increasing the thickness of the CFRP jacket resulted in a significant increase in the compressive strength and ultimate strain of geopolymer concrete. Empirical stress and strain models have been recommended to predict the final condition of CFRP-confined geopolymeric concrete [142].

6. Disadvantages of CFRP Composites

Numerous benefits are associated with the use of CFRP composites in construction, but it is essential to consider the material's disadvantages. Poor transverse shear and weak fire resistance are two significant disadvantages of CFRP. Understanding these limitations is crucial to making informed decisions and designing CFRP structures effectively in the construction industry.

1. Their relatively low transverse shear resistance is one of the disadvantages of CFRP composites. It is well known that CFRP materials have a lower shear strength in

the transverse direction than their high tensile strength in the longitudinal direction. This restriction must be considered during structural design to ensure adequate reinforcement and load distribution;
2. CFRP composites are fire-prone. CFRP is strong at ambient temperatures but degrades at high temperatures. CFRP's organic resin matrix degrades easily in fires. Fireproofing CFRP structures requires coatings or encapsulation;
3. CFRP composites cannot redistribute loads such as steel or concrete. CFRP fails suddenly without warning or plastic deformation when it reaches its load capacity. This behaviour may require structural redundancy or progressive collapse prevention strategies;
4. CFRP costs more than steel and concrete. Carbon fibre production and resin impregnation increase costs. In cost-effective construction projects, this cost factor can affect CFRP adoption;
5. UV radiation can degrade CFRP composite resin matrices, reducing their mechanical properties. Over time, sunlight and outdoor conditions can discolour, delaminate, and reduce performance. For CFRP structures to last, UV protection measures such as coatings or UV-resistant additives must be taken.

7. Current Trends and Future Outlook for CFRP Composites in Civil Engineering

CFRP composites are rapidly gaining acceptance in the construction industry for strengthening and repairing concrete structural elements. Here are five current trends and future outlooks for CFRP composites in concrete structural elements:

1. Increasingly, CFRP composites are used in seismic retrofitting initiatives for existing buildings to improve earthquake resistance. This trend is anticipated to continue as more buildings are identified as vulnerable to seismic hazards;
2. Developing new manufacturing techniques for CFRP composites is anticipated to increase production efficiency and reduce costs. This could increase the use of CFRP composites in the construction industry;
3. Integration of CFRP composites with other building systems, such as Building Information Modelling (BIM) and other digital technologies, is on the rise. This can assist in optimising the design and construction process, decreasing waste, and enhancing project outcomes;
4. There is potential for using CFRP composites in new applications, including soil reinforcement, bridge construction, and other infrastructure projects. CFRP composites are already used in a variety of concrete structural elements;
5. The construction industry is increasingly concerned with sustainability, and CFRP composites are viewed as a more sustainable alternative to conventional building materials. As more building owners and developers prioritise sustainable design and construction, the demand for CFRP composites will increase;
6. CFRP is a unique material that demonstrates minimal changes in its mechanical properties under varying temperature conditions, making it suitable for structures exposed to high temperatures or rapid temperature fluctuations. This characteristic enables the use of CFRP in applications including aerospace components, industrial buildings, and high-temperature storage facilities;
7. Highly resistant to corrosion, CFRP composites are ideal for marine environments. They do not corrode or deteriorate in saltwater, ensuring long-lasting durability and dependable performance in mooring systems. In addition, they have high fatigue resistance, allowing them to withstand cyclic loading and stress cycles, thereby reducing the risk of material fatigue and extending the service life of mooring systems. This makes CFRP a dependable material for maintaining the stability and safety of marine structures and vessels.

8. Conclusions

CFRP has emerged as a high-performance composite material with numerous advantages in the construction industry. Its exceptional properties, which include a high modulus of elasticity, superior strength-to-weight ratio, fatigue strength, and tensile strength compared to other FRP composites, make it highly desirable for various structural applications.

The ductility and flexural strength of concrete columns, beams, and slabs can be substantially improved by employing CFRP as external reinforcement and lamination. In addition to reducing reliance on conventional reinforcement, this method mitigates premature failure due to corrosion or degradation. Additionally, CFRP's exceptional fire and chemical resistance increases its suitability for harsh environments.

Despite obstacles such as the relatively high cost and the need for specialised installation techniques, the advantages of CFRP in construction make it an attractive material for the industry's future. Its use can potentially improve structures' safety, durability, and sustainability, thereby promoting resilience in the face of climate change and other environmental factors.

As manufacturing processes continue to advance and CFRP becomes more accessible, the construction industry is anticipated to emphasise the development and adoption of CFRP for structural applications. This trend will permit the development of resilient and sustainable structures to withstand future challenges. With continued research and innovation, CFRP has the potential to revolutionise the construction industry and make it more resilient.

Author Contributions: Conceptualization, D.S.V., P.D. and A.S. (Arvindan Sivasuriyan); methodology, E.K. and A.S. (Anna Stefańska); validation, A.S. (Arvindan Sivasuriyan) and E.K.; investigation, D.S.V., P.D. and Ł.W.; resources, D.S.V., P.D., Ł.W. and A.S. (Anna Stefańska); data curation, A.S. (Arvindan Sivasuriyan) and A.S. (Anna Stefańska); writing—original draft preparation, D.S.V., P.D., A.S. (Arvindan Sivasuriyan), A.S. (Anna Stefańska), E.K. and Ł.W.; writing—review and editing, A.S. (Arvindan Sivasuriyan) and A.S. (Anna Stefańska); visualization, D.S.V., P.D. and A.S. (Arvindan Sivasuriyan); supervision, A.S. (Arvindan Sivasuriyan) and E.K.; project administration, A.S. (Arvindan Sivasuriyan), A.S. (Anna Stefańska) and E.K. All authors have read and agreed to the published version of the manuscript.

Funding: This research received no external funding.

Data Availability Statement: The data presented in this study are available upon request from the corresponding authors and shared after his consideration.

Conflicts of Interest: The authors declare no conflict of interest.

References

1. Kiersnowska, A.; Fabianowski, W.; Koda, E. The Influence of the Accelerated Aging Conditions on the Properties of Polyolefin Geogrids Used for Landfill Slope Reinforcement. *Polymers* **2020**, *12*, 1874. [CrossRef] [PubMed]
2. Chalot, A.; Michel, L.; Ferrier, E. Experimental Study of External Bonded CFRP-Concrete Interface under Low Cycle Fatigue Loading. *Compos. Part B Eng.* **2019**, *177*, 107255. [CrossRef]
3. Lee, T.; Jeong, S.; Woo, U.; Choi, H.; Jung, D. Experimental Evaluation of Shape Memory Alloy Retrofitting Effect for Circular Concrete Column Using Ultrasonic Pulse Velocity. *Int. J. Concr. Struct. Mater.* **2023**, *17*, 13. [CrossRef]
4. Cruz, R.; Correia, L.; Cabral-Fonseca, S.; Sena-Cruz, J. Durability of Bond of EBR CFRP Laminates to Concrete under Real-Time Field Exposure and Laboratory Accelerated Ageing. *Constr. Build. Mater.* **2023**, *377*, 131047. [CrossRef]
5. Al-Mawed, L.K.; Hamad, B.S. Experimental and Numerical Assessments of Slab-Column Connections Strengthened Using Bonded Hemp Fiber Fabric Sheets. *Int. J. Concr. Struct. Mater.* **2023**, *17*, 8. [CrossRef]
6. Obaidat, Y.T.; Barham, W.; Obaidat, A.T.; Abuzakham, H. Improving the Shear Capacity of Recycled Aggregate Concrete Beams with NSM-CFRP Strip. *Pract. Period. Struct. Des. Constr.* **2023**, *28*, 04023016. [CrossRef]
7. Yang, J.; Lu, S.; Zeng, J.J.; Wang, J.; Wang, Z. Durability of CFRP-Confined Seawater Sea-Sand Concrete (SSC) Columns under Wet-Dry Cycles in Seawater Environment. *Eng. Struct.* **2023**, *282*, 115774. [CrossRef]
8. Li, G.; Li, X.; Fang, C.; Wang, J.; Liu, R. Dynamic Behavior of Concrete-Filled Steel Tube Cantilever Columns Stiffened with Encased Carbon Fiber Reinforced Plastic Profile Subjected to Lateral Impact Load. *Int. J. Impact Eng.* **2023**, *177*, 104561. [CrossRef]
9. Godlewski, T.; Mazur, Ł.; Szlachetka, O.; Witowski, M.; Łukasik, S.; Koda, E. Design of Passive Building Foundations in the Polish Climatic Conditions. *Energies* **2021**, *14*, 7855. [CrossRef]

10. Hadigheh, S.A.; Ke, F.; Fatemi, H. Durability Design Criteria for the Hybrid Carbon Fibre Reinforced Polymer (CFRP)-Reinforced Geopolymer Concrete Bridges. *Structures* **2022**, *35*, 325–339. [CrossRef]
11. Abas Golham, M.; Al-Ahmed, A.H.A. Behavior of GFRP Reinforced Concrete Slabs with Openings Strengthened by CFRP Strips. *Results Eng.* **2023**, *18*, 101033. [CrossRef]
12. Zhou, S.C.; Demartino, C.; Xu, J.J.; Xiao, Y. Effectiveness of CFRP Seismic-Retrofit of Circular RC Bridge Piers under Vehicular Lateral Impact Loading. *Eng. Struct.* **2021**, *243*, 112602. [CrossRef]
13. Ozturk, M.; Sengun, K.; Arslan, G. CFRP Contribution to Load-Carrying Capacity of Retrofitted Geopolymer Concrete Beams. *Structures* **2023**, *48*, 1391–1402. [CrossRef]
14. Pan, Y.; Wu, X.; Huang, Z.; Wu, G.; Sun, S.; Ye, H.; Zhang, Z. A New Approach to Enhancing Interlaminar Strength and Galvanic Corrosion Resistance of CFRP/Mg Laminates. *Compos. Part A Appl. Sci. Manuf.* **2018**, *105*, 78–86. [CrossRef]
15. Shen, J.; Huang, Z.; Song, X.; Lin, H. Cyclic Behavior of Concrete Shear Wall with CFRP Grid-Steel Reinforcement. *Compos. Struct.* **2022**, *297*, 115938. [CrossRef]
16. Yoo, S.W.; Choo, J.F. Behavior of CFRP-Reinforced Concrete Columns at Elevated Temperatures. *Constr. Build. Mater.* **2022**, *358*, 129425. [CrossRef]
17. Guo, X.; Zeng, L.; Zheng, X.; Li, B.; Deng, Z. Flexural Behavior of Damaged Hollow RC Box Girders Repaired with Prestressed CFRP. *Materials* **2023**, *16*, 3338. [CrossRef]
18. Wu, M.; Yuan, F.; Guo, S.; Li, W.; Chen, G.; Zhou, Y.; Huang, Z.; Yang, X. Experimental Investigation of the Shear Behaviour of Concrete Beams with CFRP Strip Stirrups under Static and Fatigue Loading. *Structures* **2022**, *41*, 1602–1615. [CrossRef]
19. Dong, H.; Zhou, Y.; Zhuang, N. Study on Corrosion Characteristics of Concrete-Filled CFRP-Steel Tube Piles under Hygrothermal Environment. *Adv. Mater. Sci. Eng.* **2020**, *2020*, 4849038. [CrossRef]
20. Ananthkumar, M.; Mini, K.M.; Prakash, C.; Sharma, S.V.; Krishnaa, A.C.B. Study on the Efficiency of CFRP and GFRP in Corrosion Resistance of Rebar Embedded in Concrete. *IOP Conf. Ser. Mater. Sci. Eng.* **2020**, *872*, 012137. [CrossRef]
21. Seyhan, E.C.; Goksu, C.; Saribas, I.; Ilki, A. Hybrid Use of Externally Embedded FRP Reinforcement for Seismic Retrofitting of Substandard RC Columns. *J. Compos. Constr.* **2023**, *27*, 04023022. [CrossRef]
22. Shakir, Q.M.; Alsaheb, S.D.; Farsangi, E.N. Rehabilitation of Deteriorated Reinforced Self-Consolidating Concrete Brackets and Corbels Using CFRP Composites: Diagnosis and Treatment. *J. Build. Pathol. Rehabil.* **2023**, *8*, 16. [CrossRef]
23. Hanif, M.U.; Seo, S.-Y.; Van Tran, H.; Senghong, K. Monitoring and Characterizing the Debonding in CFRP Retrofitted RC Beams Using Acoustic Emission Technology. *Dev. Built Environ.* **2023**, *14*, 100141. [CrossRef]
24. Moosa, M.K.; Ali, A.Y. Experimental Investigation on The Transform The Simply Supported Girders to Continuous Girder by Using The UHPC Cast in Place Joint. *KSCE J. Civ. Eng.* **2023**, *27*, 1697–1707. [CrossRef]
25. Samy, K.; Fawzy, A.; Founda, M.A. Strengthening of Historic Reinforced Concrete Columns Using Concrete and FRP Jacketing Techniques. *Asian J. Civ. Eng.* **2022**, *24*, 885–896. [CrossRef]
26. Al-Saawani, M.A.; El-Sayed, A.K.; Al-Negheimish, A.I. Inclined FRP U-Wrap Anchorage for Preventing Concrete Cover Separation in FRP Strengthened RC Beams. *Arab. J. Sci. Eng.* **2022**, *48*, 4879–4892. [CrossRef]
27. Tudjono, S.; Prastiwi, Y.A.; Selalatu, D.T. Experimental Study of the Effect of CFRP String on Shear Reinforcement of Reinforced Concrete Beams. In Proceedings of the 6th International Conference On Science And Technology (ICST21): Challenges and Opportunities for Innovation Research on Science Materials, and Technology in the COVID-19 Era, Mataram, Indonesia, 2 November 2021.
28. Kim, Y.J.; Hassani, A. Stepped Reinforced Concrete Beams Retrofitted with Carbon Fiber-Reinforced Polymer Sheets and Ultra-High-Performance Concrete. *Struct. J.* **2023**, *120*, 91–104. [CrossRef]
29. Murad, Y.Z.; Al-Mahmood, H.; Tarawneh, A.; Aljaafreh, A.J.; AlMashaqbeh, A.; Abdel Hadi, R.; Shabbar, R. Shear Strengthening of RC Beams Using Fabric-Reinforced Cementitious Matrix, Carbon Plates, and 3D-Printed Strips. *Sustainability* **2023**, *15*, 4293. [CrossRef]
30. Guo, R.; Li, C.; Xian, G. Water Absorption and Long-Term Thermal and Mechanical Properties of Carbon/Glass Hybrid Rod for Bridge Cable. *Eng. Struct.* **2023**, *274*, 115176. [CrossRef]
31. Xue, J.; Lavorato, D.; Tarantino, A.M.; Briseghella, B.; Nuti, C. Rebar Replacement in Severely Damaged RC Bridge Column Plastic Hinges: Design Criteria and Experimental Investigation. *J. Struct. Eng.* **2023**, *149*, 04023007. [CrossRef]
32. Saharan, S.; Kaur, G.; Bansal, P.P. Confined Ultrahigh-Performance Fibre-Reinforced Concrete in Retrofitted Beam–Column Joint: Experimental Study. *Mag. Concr. Res.* **2023**, *75*, 217–233. [CrossRef]
33. Saljoughian, A.; Mostofinejad, D.; Raji, A. Retrofit of Concrete Columns with Fibre Strips Using Grooves and Corner Battens. *Proc. Inst. Civ. Eng.-Struct. Build.* **2023**, *176*, 203–217. [CrossRef]
34. Kim, K.-M.; Park, S.-W. Tensile Behavior of Textile Reinforced Concrete Members Reinforced with a Carbon Grid by Different Manufacturing Methods. *J. Korea Concr. Inst.* **2023**, *35*, 111–121. [CrossRef]
35. Xue, X.; Makota, C.; Khalaf, O.I.; Jayabalan, J.; Samui, P.; Abdulsahib, G.M. Machine Learning Approach for Prediction of Lateral Confinement Coefficient of CFRP-Wrapped RC Columns. *Symmetry* **2023**, *15*, 545. [CrossRef]
36. Khan, A.-R.; Nasir, R.; Fareed, S. Simulation of Reinforced Concrete Columns Strengthened with CFRP Wraps. *Int. J. Civ. Eng.* **2023**, *21*, 299–313. [CrossRef]
37. Yin, X.; Huang, Z.; Liu, Y. Damage Features Extraction of Prestressed Near-Surface Mounted CFRP Beams Based on Tunable Q-Factor Wavelet Transform and Improved Variational Modal Decomposition. *Structures* **2022**, *45*, 1949–1961. [CrossRef]

38. Lu, J.; Zhang, Y.; Duan, L.; Huo, Y.; Liu, H. Dynamic Behavior of CFRP Strengthen RC Beams Based on Digital Image Correlation Technology. *Eng. Fract. Mech.* **2022**, *271*, 108597. [CrossRef]
39. Zhu, M.; Li, X.; Deng, J.; Peng, S. Electrochemical Impedance Based Interfacial Monitoring for Concrete Beams Strengthened with CFRP Subjected to Wetting–Drying Cycling and Sustained Loading. *Constr. Build. Mater.* **2023**, *366*, 130238. [CrossRef]
40. Qiang, X.; Chen, L.; Jiang, X. Experimental and Theoretical Study on Flexural Behavior of Steel–Concrete Composite Beams Strengthened by CFRP Plates with Unbonded Retrofit Systems. *Compos. Struct.* **2023**, *309*, 116763. [CrossRef]
41. Yu, X.Y.; Jiang, C.; Zhang, W.P. Failure Mode-Based Calculation Method for Bending Bearing Capacities of Corroded RC Beams Strengthened with CFRP Sheets. *Eng. Struct.* **2022**, *271*, 114946. [CrossRef]
42. Wang, Z.; Xian, G.; Yue, Q. Finite Element Modeling of Debonding Failure in CFRP-Strengthened Steel Beam Using a Ductile Adhesive. *Compos. Struct.* **2023**, *311*, 116818. [CrossRef]
43. Guo, D.; Gao, W.Y.; Liu, Y.L.; Dai, J.G. Intermediate Crack-Induced Debonding in CFRP-Retrofitted Notched Steel Beams at Different Service Temperatures: Experimental Test and Finite Element Modeling. *Compos. Struct.* **2023**, *304*, 116388. [CrossRef]
44. Liu, X.; Yu, W.; Huang, Y.; Yang, G.; You, W.; Gao, L.; Song, J. Long-Term Behaviour of Recycled Aggregate Concrete Beams Prestressed with Carbon Fibre-Reinforced Polymer (CFRP) Tendons. *Case Stud. Constr. Mater.* **2023**, *18*, e01785. [CrossRef]
45. Jin, L.; Zhang, J.; Li, D.; Du, X. Meso-Scale Analysis of Shear Performance and Size Effect of CFRP Sheets Strengthened RC Beams. *Structures* **2022**, *45*, 1630–1645. [CrossRef]
46. Huang, Z.; Deng, W.; Li, R.; Chen, J.; Sui, L.; Zhou, Y.; Zhao, D.; Yang, L.; Ye, J. Multi-Impact Performance of Prestressed CFRP-Strengthened RC Beams Using H-Typed End Anchors. *Mar. Struct.* **2022**, *85*, 103264. [CrossRef]
47. Lam, C.C.; Zhang, Y.; Gu, J.; Cai, J. Reinforcing Strategies of CFRP Plate for Coped Steel Beams against Local Buckling. *Eng. Struct.* **2023**, *275*, 115227. [CrossRef]
48. Zhang, J.; Wu, J.; Du, W.; Tong, C.; Zhu, Z.; Jing, Y. Residual Load-Carrying Performance of CFRP Strengthened RC Beam after Drop Hammer Impact. *Int. J. Impact Eng.* **2023**, *175*, 104547. [CrossRef]
49. Hasan, M.A.; Akiyama, M.; Kojima, K.; Izumi, N. Shear Behaviour of Reinforced Concrete Beams Repaired Using a Hybrid Scheme with Stainless Steel Rebars and CFRP Sheets. *Constr. Build. Mater.* **2023**, *363*, 129817. [CrossRef]
50. Alabdulhady, M.Y.; Ojaimi, M.F.; Chkheiwer, A.H. The Efficiency of CFRP Strengthening and Repair System on the Flexural Behavior of RC Beams Constructed with Different Concrete Compressive Strength. *Results Eng.* **2022**, *16*, 100763. [CrossRef]
51. Gemi, L.; Aksoylu, C.; Yazman, Ş.; Özkılıç, Y.O.; Arslan, M.H. Experimental Investigation of Shear Capacity and Damage Analysis of Thinned End Prefabricated Concrete Purlins Strengthened by CFRP Composite. *Compos. Struct.* **2019**, *229*, 111399. [CrossRef]
52. Özkılıç, Y.O.; Yazman, Ş.; Aksoylu, C.; Arslan, M.H.; Gemi, L. Numerical Investigation of the Parameters Influencing the Behavior of Dapped End Prefabricated Concrete Purlins with and without CFRP Strengthening. *Constr. Build. Mater.* **2021**, *275*, 122173. [CrossRef]
53. Aksoylu, C.; Yazman, Ş.; Özkılıç, Y.O.; Gemi, L.; Arslan, M.H. Experimental Analysis of Reinforced Concrete Shear Deficient Beams with Circular Web Openings Strengthened by CFRP Composite. *Compos. Struct.* **2020**, *249*, 112561. [CrossRef]
54. Gemi, L.; Madenci, E.; Özkılıç, Y.O. Experimental, Analytical and Numerical Investigation of Pultruded GFRP Composite Beams Infilled with Hybrid FRP Reinforced Concrete. *Eng. Struct.* **2021**, *244*, 112790. [CrossRef]
55. Arslan, M.H.; Yazman, Ş.; Hamad, A.A.; Aksoylu, C.; Özkılıç, Y.O.; Gemi, L. Shear Strengthening of Reinforced Concrete T-Beams with Anchored and Non-Anchored CFRP Fabrics. *Structures* **2022**, *39*, 527–542. [CrossRef]
56. Özkılıç, Y.O.; Aksoylu, C.; Yazman, Ş.; Gemi, L.; Arslan, M.H. Behavior of CFRP-Strengthened RC Beams with Circular Web Openings in Shear Zones: Numerical Study. *Structures* **2022**, *41*, 1369–1389. [CrossRef]
57. Gemi, L.; Madenci, E.; Özkılıç, Y.O.; Yazman, Ş.; Safonov, A. Effect of Fiber Wrapping on Bending Behavior of Reinforced Concrete Filled Pultruded GFRP Composite Hybrid Beams. *Polymers* **2022**, *14*, 3740. [CrossRef]
58. Gemi, L.; Alsdudi, M.; Aksoylu, C.; Yazman, Ş.; Özkılıç, Y.O. Optimum Amount of CFRP for Strengthening Shear Deficient Reinforced Concrete Beams. *Steel Compos. Struct.* **2022**, *43*, 735–757. [CrossRef]
59. Madenci, E.; Özkılıç, Y.O.; Aksoylu, C.; Safonov, A. The Effects of Eccentric Web Openings on the Compressive Performance of Pultruded GFRP Boxes Wrapped with GFRP and CFRP Sheets. *Polymers* **2022**, *14*, 4567. [CrossRef] [PubMed]
60. Madenci, E.; Özkılıç, Y.O.; Aksoylu, C.; Asyraf, M.R.M.; Syamsir, A.; Supian, A.B.M.; Mamaev, N. Buckling Analysis of CNT-Reinforced Polymer Composite Beam Using Experimental and Analytical Methods. *Materials* **2023**, *16*, 614. [CrossRef] [PubMed]
61. Aksoylu, C.; Özkılıç, Y.O.; Madenci, E.; Safonov, A. Compressive Behavior of Pultruded GFRP Boxes with Concentric Openings Strengthened by Different Composite Wrappings. *Polymers* **2022**, *14*, 4095. [CrossRef]
62. Özkılıç, Y.O.; Gemi, L.; Madenci, E.; Aksoylu, C.; Kalkan, İ. Effect of the GFRP Wrapping on the Shear and Bending Behavior of RC Beams with GFRP Encasement. *Steel Compos. Struct.* **2022**, *45*, 193–204. [CrossRef]
63. Madenci, E.; Özkılıç, Y.O.; Aksoylu, C.; Asyraf, M.R.M.; Syamsir, A.; Supian, A.B.M.; Elizaveta, B. Experimental and Analytical Investigation of Flexural Behavior of Carbon Nanotube Reinforced Textile Based Composites. *Materials* **2023**, *16*, 2222. [CrossRef]
64. Patel, T.K.; Salla, S.R.; Vasanwala, A.; Patel, D. Experimental Study on the Stress–Strain Behavior of Partially and Fully Wrapped Axially Loaded Square RC Columns Strengthened with BFRP. In *Sustainable Building Materials and Construction*; Lecture Notes in Civil Engineering; Kondraivendhan, B., Modhera, C.D., Matsagar, V., Eds.; Springer: Singapore, 2022; pp. 179–186.

65. Xu, J.; Tan, C.; Aboutaha, R.S. Experimental Investigations of Fire-Damaged RC Columns Retrofitted with CFRP or Steel Jackets. In *EASEC16*; Lecture Notes in Civil Engineering; Wang, C.M., Dao, V., Kitipornchai, S., Eds.; Springer: Singapore, 2021; pp. 1607–1614.
66. Tang, H.; Wang, H.; Liu, R.; Zou, X.; Jia, Y. Axial Compression Behavior of CFRP-Confined Square Concrete-Filled Double Skin Tube Stub Columns with Stainless Steel Outer Tube. *Ocean Eng.* **2022**, *266*, 112871. [CrossRef]
67. Li, G.; Sun, X.; Yang, Z.; Fang, C.; Qiu, Z. Buckling Behavior of Slender Square Concrete Filled Steel Tubular Columns Strengthened with CFRP Profile under Combined Compression and Bending. *J. Build. Eng.* **2022**, *53*, 104563. [CrossRef]
68. Tang, H.; Zou, X.; Yue, Z.; Liu, Y. Compressive Behavior of CFRP-Confined Concrete-Filled Double-Skin Stainless-Steel Tube Stub Columns. *Ocean Eng.* **2023**, *271*, 113735. [CrossRef]
69. Zhou, C.; Wang, W.; Zheng, Y.; Liu, X.; Cao, H.; Hui, Y. Dynamic Behavior of RC Columns Confined with CFRP Grid-Reinforced ECC Subjected to Lateral Low-Velocity Impact. *Int. J. Impact Eng.* **2023**, *172*, 104402. [CrossRef]
70. Samy, K.; Fouda, M.A.; Fawzy, A.; Elsayed, T. Enhancing the Effectiveness of Strengthening RC Columns with CFRP Sheets. *Case Stud. Constr. Mater.* **2022**, *17*, e01588. [CrossRef]
71. Cao, X.; Chen, Y.; Wang, H.; Cheng, C.; Zhou, X.; Zhang, H.; Kim, S.E.; Kong, Z. Experimental and Numerical Investigation of 800 MPa HSS Welded T-Section Column Strengthened with CFRP. *Thin-Walled Struct.* **2023**, *184*, 110510. [CrossRef]
72. Tan, C.; Jiang, X.; Qiang, X.; Xu, G. Flexural Strengthening of Full-Scale RC Columns with Adhesively-Bonded Longitudinal CFRP Plates: An Experimental Investigation. *J. Build. Eng.* **2023**, *67*, 105969. [CrossRef]
73. Chen, Z.; Pang, Y.; Xu, R.; Zhou, J.; Xu, W. Mechanical Performance of Ocean Concrete-Filled Circular CFRP-Steel Tube Columns under Axial Compression. *J. Constr. Steel Res.* **2022**, *198*, 107514. [CrossRef]
74. Mohammed, T.A.; Abebe, S. Numerical Investigation on CFRP Strengthening and Reinforcement Bar Detailing of RC Columns to Resist Blast Load. *Heliyon* **2022**, *8*, e10059. [CrossRef]
75. Rodríguez, V.; Guerrero, H.; Alcocer, S.M.; Tapia-Hernández, E. Rehabilitation of Heavily Damaged Beam-Column Connections with CFRP Wrapping and SFRM Casing. Soil Dyn. *Earthq. Eng.* **2021**, *145*, 106721. [CrossRef]
76. Xiong, C.N.; Shao, Y.B.; Tong, L.W.; Dai, K.S.; Luo, Y.X. Static Strength of CFRP-Strengthened Preloaded Circular Concrete-Filled Steel Tube Stub Column Columns—Part II: Theoretical and Numerical Analysis. *Thin-Walled Struct.* **2023**, *184*, 110547. [CrossRef]
77. Afefy, H.M.; Kassem, N.M.; Taher, S.E.-D.F. Retrofitting of Defected Closure Strips for Full-Depth Precast Concrete Deck Slabs Using EB-CFRP Sheets. *Pract. Period. Struct. Des. Constr.* **2019**, *24*, 1–12. [CrossRef]
78. Türer, A.; Mercimek, Ö.; Anıl, Ö.; Erbaş, Y. Experimental and Numerical Investigation of Punching Behavior of Two-Way RC Slab with Different Opening Locations and Sizes Strengthened with CFRP Strip. *Structures* **2023**, *49*, 918–942. [CrossRef]
79. Zhou, C.; Wang, L.; Wang, Y.; Fang, Z. Experimental Study on the Flexural Strengthening of One-Way RC Slabs with End-Buckled and/or Externally Bonded CFRP Sheets. *Eng. Struct.* **2023**, *282*, 115832. [CrossRef]
80. Yazdani, S.; Asadollahi, S.; Shoaei, P.; Dehestani, M. Failure Stages in Post-Tensioned Reinforced Self-Consolidating Concrete Slab Strengthened with CFRP Layers. *Eng. Fail. Anal.* **2021**, *122*, 105219. [CrossRef]
81. Azevedo, A.S.; Firmo, J.P.; Correia, J.R.; Chastre, C.; Biscaia, H.; Franco, N. Fire Behaviour of CFRP-Strengthened RC Slabs Using Different Techniques—EBR, NSM and CREatE. Compos. *Part B Eng.* **2022**, *230*, 109471. [CrossRef]
82. Yang, G.; Fan, Y.; Wang, G.; Cui, X.; Li, Q.; Leng, Z.; Deng, K. Mitigation Effects of Air-Backed RC Slabs Retrofitted with CFRP Subjected to Underwater Contact Explosions. *Ocean Eng.* **2023**, *267*, 113261. [CrossRef]
83. Assad, M.; Hawileh, R.A.; Abdalla, J.A. Modeling the Behavior of CFRP-Strengthened RC Slabs under Fire Exposure. *Procedia Struct. Integr.* **2022**, *42*, 1668–1675. [CrossRef]
84. Bielak, J. On the Role of Dowel Action in Shear Transfer of CFRP Textile-Reinforced Concrete Slabs. *Compos. Struct.* **2023**, *311*, 116812. [CrossRef]
85. Ghayeb, H.H.; Atea, R.S.; Al-Kannoon, M.A.A.; Lee, F.W.; Wong, L.S.; Mo, K.H. Performance of Reinforced Concrete Flat Slab Strengthened with CFRP for Punching Shear. Case Stud. *Constr. Mater.* **2023**, *18*, e01801. [CrossRef]
86. Breveglieri, M.; Czaderski, C. RC Slabs Strengthened with Externally Bonded CFRP Strips under Long-Term Environmental Exposure and Sustained Loading. Part 2: Laboratory Experiments. *Compos. Part C Open Access* **2021**, *6*, 100210. [CrossRef]
87. Alrousan, R.Z.; Alnemrawi, B.R. The Behavior of Alkali-Silica Reaction-Damaged Full-Scale Concrete Bridge Deck Slabs Reinforced with CFRP Bars. *Results Eng.* **2022**, *16*, 100651. [CrossRef]
88. Ebadi-Jamkhaneh, M.; Kontoni, D.-P.N. Numerical Finite Element Investigation of Thin Steel Shear Walls Retrofitted with CFRP Layers under Reversed Cyclic Loading. *J. Build. Pathol. Rehabil.* **2022**, *7*, 62. [CrossRef]
89. Shen, J.; Huang, Z.; Song, X.; Yao, Y. Deformation Performance Analysis of Concrete Shear Wall with CFRP Grids Based on the Modified Uniaxial Shear-Flexural Model. *J. Build. Eng.* **2022**, *54*, 104621. [CrossRef]
90. Sahebjam, A.; Showkati, H. Experimental Study on the Cyclic Behavior of Perforated CFRP Strengthened Steel Shear Walls. *Arch. Civ. Mech. Eng.* **2016**, *16*, 365–379. [CrossRef]
91. El-Kashif, K.F.O.; Adly, A.K.; Abdalla, H.A. Finite Element Modeling of RC Shear Walls Strengthened with CFRP Subjected to Cyclic Loading. *Alex. Eng. J.* **2019**, *58*, 189–205. [CrossRef]
92. Altin, S.; Anil, Ö.; Kopraman, Y.; Kara, M.E. Hysteretic Behavior of RC Shear Walls Strengthened with CFRP Strips. *Compos. Part B Eng.* **2013**, *44*, 321–329. [CrossRef]
93. Hatami, F.; Ghamari, A.; Rahai, A. Investigating the Properties of Steel Shear Walls Reinforced with Carbon Fiber Polymers (CFRP). *J. Constr. Steel Res.* **2012**, *70*, 36–42. [CrossRef]

94. Yu, J.G.; Zhu, S.Q.; Feng, X.T. Seismic Behavior of CFRP-Steel Composite Plate Shear Wall with Edge Reinforcement. *J. Constr. Steel Res.* **2023**, *203*, 107816. [CrossRef]
95. Meftah, S.A.; Yeghnem, R.; Tounsi, A.; Adda bedia, E.A. Seismic Behavior of RC Coupled Shear Walls Repaired with CFRP Laminates Having Variable Fibers Spacing. *Constr. Build. Mater.* **2007**, *21*, 1661–1671. [CrossRef]
96. Huang, Z.; Shen, J.; Lin, H.; Song, X.; Yao, Y. Shear Behavior of Concrete Shear Walls with CFRP Grids under Lateral Cyclic Loading. *Eng. Struct.* **2020**, *211*, 110422. [CrossRef]
97. Yu, J.-G.; Zhu, S.-Q. Study on Seismic Behavior of Oblique CFRP Corrugated Plate-Steel Plate Light Weight Sandwich Composite Shear Wall. *Structures* **2023**, *48*, 2062–2081. [CrossRef]
98. Thomas, C.A.; Baskar, K. Testing and Evaluation of Bond Surface Profile Influencing the CFRP Strengthening of Steel Members. *J. Test. Eval.* **2018**, *46*, 20170195. [CrossRef]
99. Zhang, E.Q.; Tang, L.; Bernin, D.; Jansson, H. Effect of the Paste-Anode Interface under Impressed Current Cathodic Protection in Concrete Structures. *Mater. Corros.* **2018**, *69*, 1104–1116. [CrossRef]
100. Karim, M.A.; Bae, J.H.; Kam, D.H.; Kim, C.; Choi, W.H.; Park, Y. Do Assessment of Rivet Coating Corrosion Effect on Strength Degradation of CFRP/Aluminum Self-Piercing Riveted Joints. *Surf. Coat. Technol.* **2020**, *393*, 125726. [CrossRef]
101. Huang, X.; Zhou, Y.; Zheng, X.; Xing, F.; Sui, L.; Hu, B. Bond Performance between Corroded Steel Bars and Concrete in Cathodic Protection System with CFRP as Anode. *Compos. Struct.* **2023**, *309*, 116739. [CrossRef]
102. Shao, W.; Sun, Q.; Xu, X.; Yue, W.; Shi, D. Durability Life Prediction and Horizontal Bearing Characteristics of CFRP Composite Piles in Marine Environments. *Constr. Build. Mater.* **2023**, *367*, 130116. [CrossRef]
103. Ren, X.; Sherif, M.M.; Wei, Y.; Lyu, Y.; Sun, Y.; Ozbulut, O.E. Effect of Corrosion on the Tensile and Fatigue Performance of CFRP Strand Sheet/Steel Double Strap Joints. *Eng. Struct.* **2022**, *260*, 114240. [CrossRef]
104. Wu, C.; Yu, Y.Z.; ho Tam, L.; Orr, J.; He, L. Effect of Glass Fiber Sheet in Adhesive on the Bond and Galvanic Corrosion Behaviours of CFRP-Steel Bonded System. *Compos. Struct.* **2021**, *259*, 113218. [CrossRef]
105. Hu, J.; Wang, S.; Lu, Y.; Li, S. Investigation on Efficiency of Cathodic Protection Applied on Steel in Concrete Cylinder with CFRP Wrap Serving as Anode. *Case Stud. Constr. Mater.* **2022**, *17*, e01389. [CrossRef]
106. Chen, Y.; Li, M.; Su, T.; Yang, X. Mechanical Degradation and Corrosion Characterization of Riveted Joints for CFRP/Al Stacks in Simulated Marine Environments. *Eng. Fail. Anal.* **2022**, *137*, 106382. [CrossRef]
107. Wang, Y.; Chen, H.; Li, Y.; Chen, J.; Zhuang, N. Numerical and Experimental Investigation on the Chloride Ion Resistance of Reinforced Concrete Piles Externally Bonded with CFRP Sheets under Dry-Wet Cycles. *Constr. Build. Mater.* **2022**, *359*, 129521. [CrossRef]
108. Suo, H.; Cheng, H.; Liang, B.; Deng, K.; Luo, B.; Zhang, K.; Chen, H. The Mechanical Degradation Mechanism of CFRP/Al Double-Lap Bolted Joints (with and without Corrosion Protections) after Seawater Ageing. *Compos. Struct.* **2021**, *276*, 114561. [CrossRef]
109. Mohabeddine, A.; Correia, J.; Montenegro, P.A.; De Jesus, A.; Castro, J.M.; Calçada, R.; Berto, F. An Approach for Predicting Fatigue Life of CFRP Retrofitted Metallic Structural Details. *Int. J. Fatigue* **2022**, *154*, 106557. [CrossRef]
110. Doroudi, Y.; Fernando, D.; Hosseini, A.; Ghafoori, E. Behavior of Cracked Steel Plates Strengthened with Adhesively Bonded CFRP Laminates under Fatigue Loading: Experimental and Analytical Study. *Compos. Struct.* **2021**, *266*, 113816. [CrossRef]
111. Gadomski, J.; Pyrzanowski, P. Experimental Investigation of Fatigue Destruction of CFRP Using the Electrical Resistance Change Method. *Meas. J. Int. Meas. Confed.* **2016**, *87*, 236–245. [CrossRef]
112. He, Z.; Luo, Q.; Li, Q.; Zheng, G.; Sun, G. Fatigue Behavior of CFRP/Al Adhesive Joints—Failure Mechanisms Study Using Digital Image Correlation (DIC) Technique. *Thin-Walled Struct.* **2022**, *174*, 109075. [CrossRef]
113. Li, A.; Xu, S.; Wang, Y.; Wu, C.; Nie, B. Fatigue Behavior of Corroded Steel Plates Strengthened with CFRP Plates. *Constr. Build. Mater.* **2022**, *314*, 125707. [CrossRef]
114. Li, J.; Zhu, M.; Deng, J. Flexural Behaviour of Notched Steel Beams Strengthened with a Prestressed CFRP Plate Subjected to Fatigue Damage and Wetting/Drying Cycles. *Eng. Struct.* **2022**, *250*, 113430. [CrossRef]
115. Lesiuk, G.; Katkowski, M.; Duda, M.; Królicka, A.; Correia, J.A.F.O.; De Jesus, A.M.P.; Rabiega, J. Improvement of the Fatigue Crack Growth Resistance in Long Term Operated Steel Strengthened with CFRP Patches. *Procedia Struct. Integr.* **2017**, *5*, 912–919. [CrossRef]
116. Vavouliotis, A.; Paipetis, A.; Kostopoulos, V. On the Fatigue Life Prediction of CFRP Laminates Using the Electrical Resistance Change Method. *Compos. Sci. Technol.* **2011**, *71*, 630–642. [CrossRef]
117. Kotrotsos, A.; Geitona, A.; Kostopoulos, V. On the Mode I and Mode II Fatigue Delamination Growth of CFRPs Modified by Electrospun Bis-Maleimide Resin. Compos. *Sci. Technol.* **2023**, *237*, 110000. [CrossRef]
118. Barile, C.; Casavola, C.; Vimalathithan, P.K.; Pugliese, M.; Maiorano, V. Thermomechanical and Morphological Studies of CFRP Tested in Different Environmental Conditions. *Materials* **2018**, *12*, 63. [CrossRef]
119. Al-Tamimi, A.K.; Hawileh, R.A.; Abdalla, J.A.; Rasheed, H.A.; Al-Mahaidi, R. Durability of the Bond between CFRP Plates and Concrete Exposed to Harsh Environments. *J. Mater. Civ. Eng.* **2015**, *27*, 04014252. [CrossRef]
120. Li, Y.; Ma, X.; Li, H.; Zheng, H.; Li, C.; Gao, Y.; Long, J.; Li, Z. Effect of Moisture-Heat Coupling on Mechanical Behavior of Nano-SiO_2 Adhesives and CFRP-Steel Lap Joints. *Thin-Walled Struct.* **2023**, *183*, 110391. [CrossRef]
121. Al-Abdwais, A.H.; Al-Mahaidi, R.S. Evaluation of High Temperature Endurance of RC Beams Retrofitted with NSM Technique Using CFRP Composites and Modified Cement-Based Adhesive. *Eng. Struct.* **2022**, *264*, 114445. [CrossRef]

122. Yoo, S.J.; Kim, Y.H.; Yuan, T.F.; Yoon, Y.S. Evaluation of Residual Bond Behavior of CFRP and Steel Bars Embedded in UHPC after Exposure to Elevated Temperature. *J. Build. Eng.* **2022**, *56*, 104768. [CrossRef]
123. Wang, Y.J.; Wu, Z.M.; Liu, H.B.; Zhang, Q.M.; Yang, S.T.; Li, Y.C. Influence of Thermal Effects on Debonding Response of CFRP-to-Concrete/Steel Interfaces under Thermal and Mechanical Loads: An Analytical Solution. *Compos. Struct.* **2023**, *303*, 116333. [CrossRef]
124. Kaiser, I.; Zhang, C.; Tan, K.T. Mechanical Behavior and Failure Mechanisms of CFRP and Titanium Tubular Adhesive Lap Joints at Extreme Temperatures. *Compos. Struct.* **2022**, *290*, 115528. [CrossRef]
125. Peng, K.; Wang, Q.; Shao, Y. Test on Shearing Performance of Concrete Filled Square CFRP-Steel Tube. *Ocean Eng.* **2023**, *274*, 114065. [CrossRef]
126. Jahani, Y.; Baena, M.; Codina, A.; Barris, C.; Torres, L. Time-Dependent Behavior of NSM CFRP-Strengthened RC Beams under Different Service Temperatures. *Compos. Struct.* **2022**, *300*, 116106. [CrossRef]
127. Alkhawaldeh, A.A.; Al-Rousan, R.Z. Upgrading Cyclic Response of Heat-Damaged RC Beam-Column Joints Using CFRP Sheets. *Case Stud. Constr. Mater.* **2022**, *17*, e01699. [CrossRef]
128. Li, Y.; Guo, X.; Yang, J.; Li, M. Preparation of Nano-SiO_2/Carbon Fiber-Reinforced Concrete and Its Influence on the Performance of Oil Well Cement. *Int. J. Polym. Sci.* **2019**, *2019*, 2783018. [CrossRef]
129. Jongvivatsakul, P.; Thongchom, C.; Mathuros, A.; Prasertsri, T.; Adamu, M.; Orasutthikul, S.; Lenwari, A.; Charainpanitkul, T. Enhancing Bonding Behavior between Carbon Fiber-Reinforced Polymer Plates and Concrete Using Carbon Nanotube Reinforced Epoxy Composites. *Case Stud. Constr. Mater.* **2022**, *17*, e01407. [CrossRef]
130. Du, Y.; Lu, S.; Xu, J.; Xia, W.; Wang, T.; Wang, Z. Experimental Study of Impact Mechanical and Microstructural Properties of Modified Carbon Fiber Reinforced Concrete. *Sci. Rep.* **2022**, *12*, 12928. [CrossRef]
131. Al-Rousan, R.; Ababneh, A.; Alhassan, M. Hybrid CFRP-Steel for Enhancing the Flexural Behavior of Reinforced Concrete Beams. *J. King Saud Univ.-Eng. Sci.* **2021**, *33*, 459–470. [CrossRef]
132. Abolfazli, M.; Ivan John Reyes, R.; Choong, D.; Bazli, M.; Rajabipour, A.; Pourasiabi, H.; Arashpour, M. Bond Behaviour between CFRP, GFRP, and Hybrid C-GFRP Tubes and Seawater Sea Sand Concrete after Exposure to Elevated Temperatures. *Constr. Build. Mater.* **2023**, *392*, 131884. [CrossRef]
133. Lee, S.H.; Kim, S.W. Self-Sensing-Based Deflection Control of Carbon Fibre-Reinforced Polymer (CFRP)-Based Shape Memory Alloy Hybrid Composite Beams. *Compos. Struct.* **2020**, *251*, 112544. [CrossRef]
134. Todoroki, A.; Kurokawa, H.; Mizutani, Y.; Matsuzaki, R.; Yasuoka, T. Self-Sensing Time Domain Reflectometry Method for Damage Monitoring of a CFRP Plate Using a Narrow-Strip Transmission Line. *Compos. Part B Eng.* **2014**, *58*, 59–65. [CrossRef]
135. Shin, P.S.; Baek, Y.M.; Kim, J.H.; Kwon, D.J. The Factor Influencing Self-Sensing Property of Carbon Fiber. *Compos. Sci. Technol.* **2023**, *238*, 110017. [CrossRef]
136. Yoshitake, I.; Hasegawa, H.; Shimose, K. Monotonic and Cyclic Loading Tests of Reinforced Concrete Beam Strengthened with Bond-Improved Carbon Fiber Reinforced Polymer (CFRP) Rods of Ultra-High Modulus. *Eng. Struct.* **2020**, *206*, 110175. [CrossRef]
137. Lamba, N.; Raj, R.; Singh, P. Mechanical Response of Recycled Carbon Fiber Reinforced Polymer Fibers in High-Strength Concrete. *Mater. Today Proc.* **2023**, *78*, 603–607. [CrossRef]
138. Zhang, L.W.; Sojobi, A.O.; Liew, K.M. Sustainable CFRP-Reinforced Recycled Concrete for Cleaner Eco-Friendly Construction. *J. Clean. Prod.* **2019**, *233*, 56–75. [CrossRef]
139. Xiong, C.; Li, Q.; Lan, T.; Li, H.; Long, W.; Xing, F. Sustainable Use of Recycled Carbon Fiber Reinforced Polymer and Crumb Rubber in Concrete: Mechanical Properties and Ecological Evaluation. *J. Clean. Prod.* **2021**, *279*, 123624. [CrossRef]
140. Maxineasa, S.G.; Taranu, N.; Bejan, L.; Isopescu, D.; Banu, O.M. Environmental Impact of Carbon Fibre-Reinforced Polymer Flexural Strengthening Solutions of Reinforced Concrete Beams. *Int. J. Life Cycle Assess.* **2015**, *20*, 1343–1358. [CrossRef]
141. Khorgade, P.; Rettinger, M.; Burghartz, A.; Schlaich, M. A Comparative Cradle-to-Gate Life Cycle Assessment of Carbon Fiber-Reinforced Polymer and Steel-Reinforced Bridges. *Struct. Concr.* **2022**, *24*, 1737–1750. [CrossRef]
142. Tang, Z.; Li, W.; Tam, V.W.Y.; Yan, L. Mechanical Performance of CFRP-Confined Sustainable Geopolymeric Recycled Concrete under Axial Compression. *Eng. Struct.* **2020**, *224*, 111246. [CrossRef]

MDPI
St. Alban-Anlage 66
4052 Basel
Switzerland
www.mdpi.com

Buildings Editorial Office
E-mail: buildings@mdpi.com
www.mdpi.com/journal/buildings

www.ingramcontent.com/pod-product-compliance
Lightning Source LLC
Chambersburg PA
CBHW061601220926
51850CB00005B/389